Advanced Mineralogy

Already Published in this Series:

Volume 1 Composition, Structure, and Properties of Mineral Matter: Concepts, Results, and Problems

Volume 2 Methods and Instrumentations: Results and Recent Developments

Volume 3 Mineral Matter in Space, Mantle, Ocean Floor, Biosphere, Environmental Management, and Jewelry

To be Published in this Series:

Volume 4 Processes of Mineral Formation Frontiers in Experimental Research and Evolution in Geological History

Volume 5 Minerals as a Source of Metals, Energy, and Materials

Springer

Berlin
Heidelberg
New York
Barcelona
Budapest
Hong Kong
London
Milan
Paris
Santa Clara
Singapore
Tokyo

Arnold S. Marfunin (Ed.)

Advanced Mineralogy

Volume 3
Mineral Matter in Space, Mantle, Ocean Floor, Biosphere, Environmental Management, and Jewelry

With 168 Figures and 41 Tables

Springer

Prof. Dr. Arnold S. Marfunin

Geological Faculty
University of Moscow
119899 Moscow
Russia, CIS

ISBN 3-540-58245-2 Springer-Verlag Berlin Heidelberg New York

Library of Congress Cataloging-in-Publication Data Advanced mineralogy/Marfunin. Arnold S. (ed.). p. cm. Includes bibliographical references and index. Contents: – v. 3. Mineral Matter in Space, Mantle, Ocean Floor, Biosphere, Environmental Management, and Jewelry ISBN 0-387-58245-2 (v. 3 : U.S.) 1. Mineralogy. 1. Marfunin. Arnol'd Sergeevich. QE363.2.A35 1994 549 – dc20 94-13315

Printed in Germany

Coverdesign: E. Kirchner, Springer-Verlag
Production: ProduServ GmbH Verlagsservice, Berlin
Typesetting: Fotosatz-Service Köhler OHG, Würzburg

SPIN: 10474900 32/3020 – 5 4 3 2 1 0 – Printed on acid-free paper

Preface

This volume of *Advanced Mineralogy* encompasses six different areas having two features in common: they are related to one of the largest enterprises of the second half of this century; and represent the ultimate and final extension of the concept of mineral matter.

– Understanding mineral matter in Space is one of the principal purposes of cosmic exploration. This includes the results of comparative planetology, lunar epopee, sophisticated meteorite studies (now more than 500 meteorite minerals), discovery of the interstellar mineral dust forming some 60 trillion of earth masses in the Galaxy, and terrestrial impact crater studies. It is possible now to speak of mineralogy of the Universum, and the mineralogical type of the states of matter in the Universe.

- Direct samples of mantle xenoliths and ultrahigh pressure-temperature experiments make it possible to consider the mineralogical composition of the Earth as a whole, including the upper an lower mantle and the Earth's core.
- Deep ocean drilling programs, a scientific fleet of hundreds of vessels and several submersibles have brought about great discoveries in the geology, metalogeny, and mineralogy of the ocean floor the largest part of the Earth's surface, in particular revealing new genetic, crystallochemical, and ore types of mineral formation.
- The extraordinary progress in the past three decades in the knowledge of organism-formed minerals, their diversity, phylum distribution, their role in the paleontological picture of extant life and in sedimentary ore formation, with the special features of these "strange minerals" has led to a new significance for minerals as tracers of evolution and the impact of life on the biosphere.
- Environmental mineralogy and radiation mineralogy have a key role in the global problem of the impact of energy production, metals, chemicals, and radionuclides on air, water, and soil pollution.
- Crucial changes in the trade, art, and mineralogy of diamonds, gold, and colored stones have led to the development of the

multisided branch of gemology, which in this Volume ranks as one of the most important areas: with an annual turnover of 47 billion dollars, it gives a special significance to mineral matter: its quality as an artistic value.

The corresponding chapters must be considered as condensed introductions to these wide fields at advanced textbook level. However, they present also the principal results of these most fascinating aspects of mineral matter studies.

At the end of the Volume, the Contents of Volumes 1 and 2 and abstracts of Volumes 4 and 5 are presented. This will help to conceive the whole picture of contemporary understanding of mineral matter as seen by the team of contributors (more than 200 in the first three Volumes) from many countries.

I would like to express my sincere gratitude to many colleagues from different universities, who encouraged me to go through with producing this laborious collective enterprise.

I am grateful to Prof. H.U. Bambauer (Münster) for advice and suggestions, Prof. F. T. Manheim (Woods Hole, Massachusetts), and Dr. M. Rudnicky (Cambridge) for reviewing the chapter on minerals of the ocean floor, and R. Clark and N. Pleasance (De Beers Centenary Company) for reading the chapter on gemology.

Many thanks to Dr. W. Engel (Springer-Verlag) for patience and understanding.

January 1998 A. S. Marfunin

Contents

Chapter 6 Gemology and Jewelry: Scientific and Technological Bases. Artistic Value of Minerals

List of Contributors

ANDERSON, J. R., Dept of Chemistry and Biochemistry, Arizona State University, Tempe, Arizona 85287, USA

BAMBAUER, H. U. , Institut für Mineralogie, Universität Münster, Vogelweide 7, D-48346 Ostbevern-Brock, Germany

BARSKOV, I. S., Department of Geology, Moscow State University, Moscow 119899, Russia

BATURIN, G. N., Institute of Oceanology, Russian Academy of Sciences, Krasnikova 23, Moscow 117218, Russia

BOGDANOV, YU. A., Institute of Oceanology, Russian Academy of Sciences, Krasnikova 23, Moscow 117218, Russia

BUSECK, P. R., Department of Geology, Arizona State University, Tempe, Arizona 85287, USA

CHELISHCHEV, N. F., Institute of Mineralogy, Geochemistry and Crystallochemistry of Rare Elements, Moscow 113035, Sadovnicheskay Nab. 71, Russia

CALMANO, W., Arbeitsbereich Umweltschutztechnik, TU Hamburg-Harburg, Eichendorferstr. 40, D-21073 Hamburg, Germany

DEISSMANN, G., Institut f. Mineralogie u. Lagerstättenlehre, RWTH Aachen, Wüllnerstr. 2, D-52062 Aachen, Germany

DEUTSCH, A., Institut für Planetologie, Universität Münster, Wilhelm-Klemm-Str. 10, D-48149 Münster, Germany

DUBROVINSKAIA, N. A., Institute of Earth Sciences, Uppsala University, S-75236 Uppsala, Sweden

DUBROVINSKY, L. S., Institute of Earth Sciences; Mineralogy-Petrology, Uppsala University, S-75236 Uppsala, Sweden

FISENKO, A.V., Vernadsky Institute of Geochemistry, Vorobjevskoe 10, Moscow 117975, Russia

FÖRSTER, H. G., Institut für Mineralogie und Lagerstättenlehre der RWTH Aachen, Wüllnerstr. 2, D-52062 Aachen, Germany

FÖRSTNER, U., Arbeitsbereich Umweltschutztechnik, TU Hamburg-Harburg, Eichendorferstr. 40, D-21073 Hamburg, Germany

FRIEDRICH, G., Institut für Mineralogie und Lagerstättenlehre der RWTH Aachen, Wüllnerstr. 2, D-52062 Aachen, Germany

GAFT, M. L., Open University of Israel, Geology and Physics Group, 16 Klausner St., P.O.B. 39328 Ramat-Aviv, Tel-Aviv 61392, Israel

HELING, D. F. W., Institut für Umweltgeochemie, Universität Heidelberg, D-69020 Heidelberg, Germany

KASHKAROV, L. L., Vernadsky Institute of Geochemistry, Vorobjevskoe 10, Moscow 117975, Russia

KERSTEN, M., Institut f. Ostseeforschung, Sektion Marine Geologie, Seestr. 15, D-18119 Rostock-Warnemünde, Germany

KHORASANI, R., FH Hamburg, FB Bauingenieurwesen, Hebebrandstr. 1, D-22297 Hamburg, Germany

KOHLER, E., Lehrgebiet Angew. Geologie, Universität Regensburg, Universitätsstr. D-93053 Regensburg, Germany

KURAT, G., Naturhistorisches Museum, Burgring 7, Postfach 417, A-1014 Vienna, Austria

KOZENKO, A.V., Institute of Physics of the Earth, Bolshaya Grusinskaya 10, Moscow 123810, Russia

KUSKOV, O. L., Vernadsky Institute of Geochemistry, Vorobjevskoe 10, Moscow 117975, Russia

LANGENHORST, F., Bayerisches Geoinstitut, Universität Bayreuth, D-95440 Bayreuth, Germany

MARFUNIN, A. S., Department of Mineralogy, Moscow State University, Moscow 119899, Russia

MASAITIS, V. L., Karpinsky Geological Institute, Sredny Prospect 74, St. Petersburg 199026, Russia

NAZAROV, M. N. Vernadsky Institute of Geochemistry, Vorobjevskoe 10, Moscow 117975, Russia

PANCZER, G., Laboratoire de Physico-Chimie des Materiaux Luminescents Universite Claude Bernard-Lyon 1, 69622 Villeurbanne Cedex, France

PENTINGHAUS, H., Kernforschungszentrum Karlsruhe, Institut für Nukleare Entsorgungstechnik. Postfach 3640, D-76021 Karlsruhe, Germany

PHILIPSBORN, H., Abt. Kristallographie, Universitäts Regensburg, Universitätsstr. 31, D-93053 Regensburg, Germany

PÖLLMANN, H., Fachbereich Geowissenschaft, Universität Halle-Wittenberg, Domstr. 5, D-06108 Halle, Germany

RIETMEIJER, F. J. M., Department of Earth and Planetary Sciences, University of New Mexico, Albuquerque, New Mexico 87131-1116, USA

ROZANOV, A. YU., Paleontological Institute, Profsoyznaya 123, Moscow 117868, Russia

RYABCHIKOV, I. D., Institute of Ore Deposit Geology, Mineralogy, Petrology, and Geochemistry, Academia of Sciences, Staromonetny 35, Moscow 109017, Russia

SAXENA, S. K., Institute of Earth Sciences, Uppsala University, S-75236 Uppsala, Sweden

SHARKOV, E. V., Institute of Ore Deposit Geology, Mineralogy, Petrology, and Geochemistry, Academia of Sciences, Staromonetny 35, Moscow 109017, Russia

SHIGLEY, J., Gemological Institute of America, 5355 Armada Drive, Carlsbad, San Diego County, California 92008, USA

SNETHLAGE, R., Bay. Landesamt für Denkmalpflege, Hofgraben 4, D-80539 München, Germany

SNYDER, G. A., Planetary Geosciences Institute, Department of Geological Sciences, University of Tennessee, Knoxville, Tennessee 37996-1410, USA

TAYLOR, L. A., Planetary Geosciences Institute, Department of Geological Sciences, University of Tennessee, Knoxville, Tennessee 37966-1410, USA

ULYANOV, A. A., Department of Mineralogeny, Moscow State University, Moscow 119899, Russia

WYNN-WILLIAMS, G., University of Hawaii at Manoa, Institute for Astronomy 2680 Woodlawn Drive, Honolulu, Hawaii 96822, USA

YAKHONTOVA, L. K., Department of Mineralogy, Moscow State University, Moscow 119899, Russia

ZHARKOV, V. N., Institute of Physics of the Earth, Bolshaya Grusinskaya 10, Moscow, Russia

Abstract of Volume 4.
Processes of Mineral Formation: Frontiers in Experiment and Evolution in Geological History

Theoretical background of the physical chemistry of mineral formation. Thermodynamics and thermochemistry of minerals. Paragenetic analysis of magmatic and metamorphic rocks.
Experimental studies of mineral formation: systematics and overview of principal results.
Thermobarogeometry. List of mineral thermobarometers. Summary of P-T characteristics of principal mineral-forming processes.
Synthesis of minerals.
Geodynamic types of mineral formation in the earth's history.

Abstract of Volume 5.
Minerals as a Source of Metals, Energy, and Materials

All 4500 mineral species are distributed between genetic and economic types of mineral formation with characteristics of their genetic position and industrial importance.

CHAPTER 1

Mineral Matter in Space

1.1 Types of States of Matter in the Universe

A.S. MARFUNIN

The extended scope of mineralogy, including extraterrestrial materials, came first from increasingly sophisticated studies of meteorites, then from direct studies of lunar samples, then from the comparative mineralogy of terrestrial and giant planets, and finally from data on interplanetary dust and interstellar presolar mineral dust.

This is, however, not only an extension by the involvement of new objects. The mineral state of matter is now considered as a natural type of the system of types of matter in the Universe. It is the final extension of the term mineral matter.

The concept of the Universe as a whole represents the largest entity that can be studied or imagined. There are two approaches to understanding it which can be called geometric and evolutionary.

The geometric approach is based on a model of the homogeneous isotropic Universe, the model "dust of galaxies". It considers the geometric future of the Universe, depending on either Riemanian geometry, leading to closed space and the collapse of the Universe, or Lobachevsky (open space) and Euclidian (flat space), leading to infinite expansion.

Most contemporary estimations of the average density of the Universe indicate its geometric future as corresponding to open (or flat) space and infinite expansion.

The evolutionary approach considers the Universe as a whole as an evolutionary series of gravitational (and postgravitational) transformations, as a system of the types of matter; a frame for this Universe is a density-temperature diagram with absolute maxima.

A principal complexity exists in the construction of a whole picture of the Universe from the study of its parts. On the one hand, for each of these parts, summarizing surveys are available which can be used to present the Universe as a whole: general astrophysics and cosmology, particle physics and relativistic cosmology, thermonuclear physics and synthesis of elements, equation of states in astrophysics, cosmochemistry, evolution of stars, protostars, and protoplanets, stellar atmospheres, white dwarfs, neutron stars, black holes, very low mass stars and brown dwarfs, giant planets, Solar System, interstellar dust, meteorites, asteroids, and comets.

On the other hand, each of these fields is self-sufficient and self-contained, operating within the framework of accepted concept, basic equations, formalisms, specialized mathematic and theoretical apparatures, and enormous masses of available data.

To bring together these components of the whole Universe, the component types of the states of matter must be presented in a single system.

Density and temperature are measures of expansion and cooling of the earlier Universe and also of gravitational compression and heating in galaxy formation. They are the coordinates of the antigravitational and gravitational evolution of the Universe. They consider the whole evolution of the Universe in this single coordinate system, without omissions, in a physically and logically exhaustive manner.

The origin of the coordinates can be taken at minimum values of density and temperature of the now existing Universe: less than about 10^{-30} g cm^{-3} and less than 2.7 K, or, extrapolating to time corresponding to proton lifetime (more than 10^{30} years), in the order of 10^{-15} K.

To obtain a closed space of the diagram it is necessary to find the absolute maxima of temperature and density.

In the Universe, the evolution and antropicity of which are determined by fundamental physical constants, for these maxima the values of the Planck density and temperature, which are related only to the fundamental constants, can be taken:

$$\rho_{Pl} = c^5/G^2 2\hbar = 5\times 10^{93}\ \mathrm{g\ cm^{-3}} \quad \text{and} \quad T_{Pl} = 1/k(c^5 \cdot \hbar/G)^{1/2} = 10^{32}\ \mathrm{K}$$

(c – light velocity, G – gravitation constant, ħ – Planck constant, and k – Boltzmann constant).

In the expanding Universe model these maxima correspond to the values of the Big Bang singularity.

The closed space thus obtained offers the possibility of inscribing the whole infinite Universe within the framework defined by these coordinates and absolute values. A model of the Universe including the entire diversity of states, at all distances up to 10–20 billion light years, for all the billions of galaxies and trillions of stars in each galaxy, from all times from the beginning to the end of the Universe, can be constructed based on this big square of the Universe (Fig. 1).

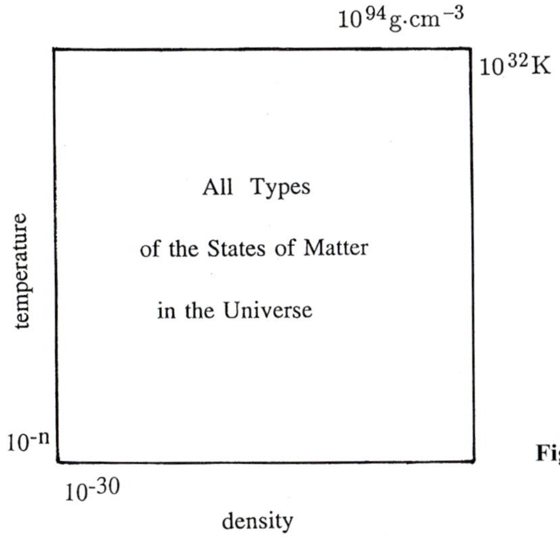

Fig. 1. The frame for the Universe

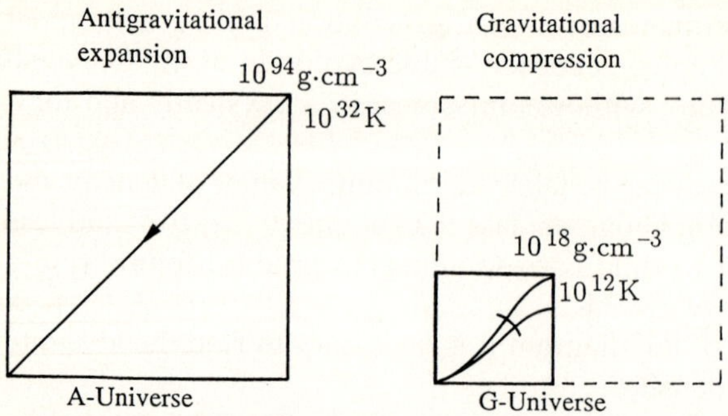

Fig. 2. Division of the evolution of the Universe into A – antigravitational and G – gravitational Universes

- Motive force: Big Bang
- Hot origin
- Linear relation T–t–R
- Pregalactic homogeneous the universe with temporal structure
- From absolute maxima down to ρ, T minima
- First seconds, minutes and up to hundreds of millions and billion years
- Noncreative evolution; dissipation of energy

- Motive force: gravitation; no external force
- Cold origin
- Relations determined by gravitational evolution
- Galactic
- From ρ, T minima up to 10^{12} K and 10^{16} g cm^{-3}
- All next 15–20 billion years and all the future of the Universe
- Creation of galaxies, stars, planets, life

First Fundamental Division in the Evolution of the Universe; Discerning Two Universes. When trying to inscribe into the big square of the Universe all possible states of matter, two groups of states with different trends and times can be distinguished. This corresponds to the most striking initial and fundamental division in the history of the Universe.

Two epochs are distinguished (Fig. 2). whose evolution followed different models and mechanisms, with different origins and opposite directions of motive forces and, different orders of temperature and density:

A: Pregalactic homogeneous initial Universe. Antigravitational expansion and cooling. Hot origin. Motive force: Big Bang. Linear relations $\rho - \mathrm{T} - \mathrm{t} - \mathrm{R}$ (density – temperature – time – radius of the Universe). The Universe with temporal structure. Homogeneous evolution from absolute maxima down to $\rho - \mathrm{T}$ minima. Epoch of the first seconds, minutes, and up to hundreds of millions and the first billion years. This is noncreative evolution; dissipation of energy. Transient states. Extinct Universe.

G: Galactic Universe. Gravitational compression and heating. Cold origin. Motive force: non external force, only gravitation. Fragmentation into galaxies. Evolution from $\rho - \mathrm{T}$ minima up to about 10^{12} K and 10^{16} g cm^{-3}. Epoch of the next 15–20 billion years and all the future of the Universe. Creation of galaxies, stars, planets, life.

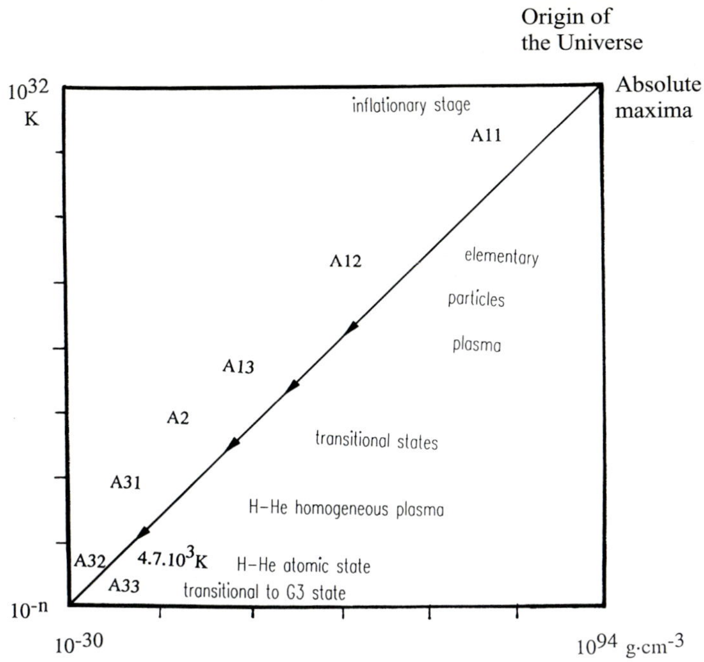

Fig. 3. The types of the states of matter in the A-universe

- Antigravitational expansion and cooling
- Linear relations ρ–T–t–R
- Pregalactic, homogeneous
- Hot origin
- The Universe with temporal structure
- First seconds, minutes up to hundreds of millions and the first billion years
- Extinct Universe

The distributions of the states of mater are completely different for the A- and G-Universes.

General Picture of the A-Universe. Figure 3 describes the behavior with classical simplicity: a straight line drawn by its evolution and traversing all orders of the chosen parameter values:

- from the absolute maxima of the Planck 10^{32} K and 10^{94} g cm^{-3} at the beginning of the Universe (upper right vertex of the square corresponding to the singularity, the Big Bang);
- down to the minima of the parameters near the origin of the coordinates.

All other parts of the diagram are not realizable.

General Picture of the G-Universe (Figs. 4–5). This is not linear, but represented by the entire space limited by about 10^{12} K (i.e., in the stability field of

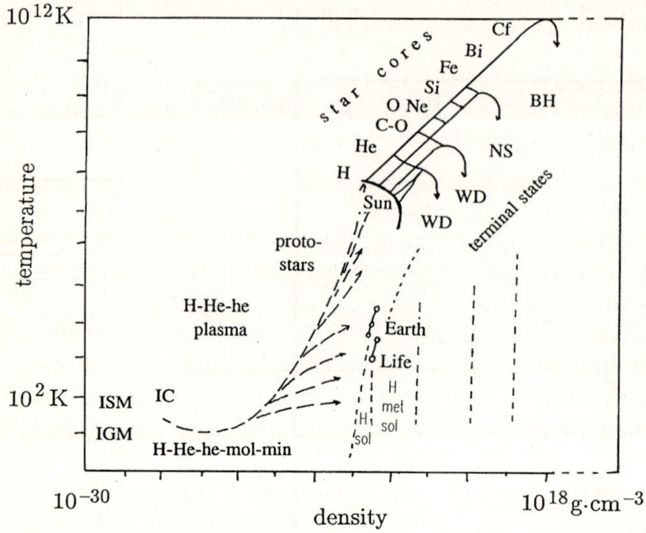

Fig. 4. The G-Universe: ascending states; evolutionary trends

- Interstellar and Intergalactic H-He-he medium
- (Hydrogen-helium-heavier elements)
- Interstellar clouds (molecular-mineral dust)
- Protostars and protoplanets
- Stellar nucleosynthesis
- Terminal states: white dwarfs, neutron stars, black holes
- Mineral states. Stuff of life

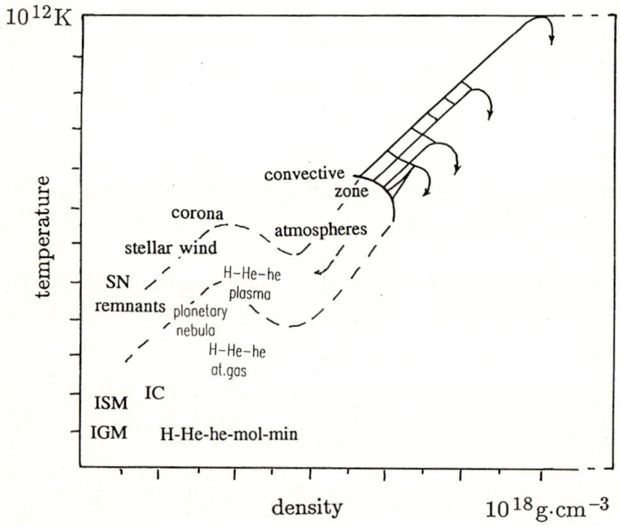

Fig. 5. The G-Universe: descending states

- Star envelopes; convective zone
- Atmospheres, corona
- Stellar wind
- Planetary nebula
- SN remnants

protons and neutrons before transition to quarks) and 10^{16} g cm^{-3} (nuclear matter states).

Thus, the frame of the galactic G-Universe is inscribed in the part of the big square which is lower than the elementary particle field, but covers all combinations of T and ρ in these boundaries.

Temporal Structure of the Pregalactic A-Universe. This results form the preservation of its homogeneity throughout the whole length of its existence; fluctuations do not develop into structural fragmentation. There are series of transient momentary states filling the whole A-Universe in its expanding space. At every moment, the Universe is homogeneous and the momentary states fill it entirely. At any moment in time the whole Universe is described not by a field or line on the $\rho - T$ diagram but by a "running point": there is nothing behind it ("all bridges are burned"), and there are states ahead of it which have never existed before.

This is Universe plasma, Universe light; the Universe a single nonstructured star; at the end of this epoch, there is Universe darkness, nearly emptiness, nearly ultimate coldness, darkness without stars and radiation, without any hint of future, without a sign of a possibility for stars, minerals or intelligence.

Only in the A-Universe do the linear $T-\rho-t-R$ relations exist. The line of its evolution can be identified with the "arrow of time" divided at time-temperature intervals according to the relation: $T(K) = 10^{10} \sqrt{t}\,(s)$.

Directions of Evolution. These are opposite in the A- and G-Universe: in the first, the processes accompany decreasing temperature, expansion, dissipation of energy, and increase in entropy, in the second, gravitational compression, and increase in temperature and energy.

Moving Force. In the A-Universe, this is the inertia of the Big Bang impetus, antigravitational repulsion. This is a single trust in the evolution of the Universe, the only moving force resulting in limitless expansion in time and space. No other forces besides the Big Bang inertia are able to spread the matter of the Universe forever. This inertia of the primordial push, which occurred 15–20 billion years ago, continues to be the single cause of the expansion of the Universe, even after its transition form the A- to the G-epoch.

The moving force in the G-Universe is gravitation, and its action is of two kinds: (1) as the initial cause of the transformation of matter at all stages of the galactic evolution, change in the composition of the Universe from a hydrogen-helium medium to a synthesis of all chemical elements, then to molecules and minerals, to formation of galaxies, stars, planets, and (2) as gravitational mechanisms of celestial mechanics in fragmented hydrogen-helium masses and celestial bodies, media, and particles.

Relations of the A- and G-Universe. These are two separate temporal layers, two epochs in the existence of the Universe. The first epoch elapsed and will never recur in this Universe; this is a transient, extinct Universe. The second exists now and comprises all observable diversity of the Universe from the interstellar medium to stars of all types and planetary systems and life.

All types of states in the Universe can be distributed between three diagrams: for the A-Universe, for the G-Universe ascending states, and for the G-Universe descending states.

A1: The Elementary Particles Universe. One of the fundamental states of matter and the first in the evolution of the Universe. The whole of matter was only in this elementary particles state. There were no compounds, no atoms, no nuclei, no protons or neutrons (and there was no place for thought and word). The entire Universe was composed of this type of matter alone.

- State of elementary matter: more than 300 particles in continuous transformation in other particles, radiations and resonances; annihilation and formation of particle-antiparticle pairs; ephemerality, nondetermination. Unique scale of particle lifetimes and relations of matter – time – space. Total time of existence is about 1 s.
- Represents the largest part of Universe conditions: from 10^{32} to 10^{12} K and from 10^{94} to 10^{14} g cm^{-3}, i.e., 20 orders of temperature and 80 orders of density.
- These particles are not "building stones" (protons, neutrons) for all other types of matter. These elementary particles neither enter into states of the next types of matter, nor compose them. This state was transformed into other types of states and ceased to exist. This is a state of the past, extinct, nonquenchable Universe.
- A1 and A2 are the only states with antiparticles. Antimatter was not separated into a phase; no "antiworlds" existed.
- Limit to subdivision of matter. Limit to infinity in depth.
- First picosecond is the time of the inflationary Universe; inflationary cosmology concerns this first picosecond.
- No primordial chaos; supersymmetry.

There are three subtypes related by phase transitions:

A11: Particles states with supersymmetry interactions (great unification theory for strong and electroweak interactions); 10^{32}–10^{27} K;
A12: Separation of strong and electroweak interactions; 10^{27}–10^{15} K;
A13: Separation of strong, weak, and electromagnetic interactions; 10^{15}–10^{12} K;

A2: Transitional States (from Elementary Particles to Hydrogen-Helium Medium).
A21: Formation of protons and neutrons from quarks; end of free quark state; continuous annihilation and formation of proton-antiproton, neutron-antineutron, and electron-positron, pairs.

A22: Termination of the proton and neutron formation (threshold temperatures 1.0888×10^{10} and 1.0903×10^{10} K; annihilation of equal amounts of protons and antiprotons, neutrons and antineutrons. Proton and neutron surplus (baryon asymmetry) 10^{-9} composes the entire subsequent matter of the Universe.
Initial equal ratio of the survived neutrons and protons changed and stabilized at 5×10^9 K as $n:p \sim 0.2$.
Surviving part of electrons from electron-positron annihilation at a threshold temperature of 5.930×10^{10} K.
At this time, the initial mass of the Universe and the total amount of protons and neutrons with their corresponding electrons were determined. End of antiparticles.

A23: Start of nuclei formation from all neutron and corresponding amount of protons: helium $^4He(2p + 2n)$ with a small admixture of deuterium (2p) and some other isotopes of light elements (Li, Be, B).

Thus, at the end of the transitional states at 5×10^9 K in about 3 min after the Big Bang the final composition of the principal components of the Universe was established: $H_{90}He_{10}$ (atomic ratio) or $H_{75}He_{25}$ (mass ration) with a negligible admixture of some light elements.

A3: Primordial H–He Homogeneous Universe. This was the primitive Universe with the simplest composition, $H_{90}H_{10}$, and traces of light elements with no other elements: without Si, Al, O – elements of minerals and planets, without C, O, P, N – elements of life and without a hundred other elements. This was the pregalactic variant of cosmic space as homogeneous primitive Universe without stars and planets.

A31: H–He plasma.

A32: H–He atomar gas (H and He recombination at 4.7×10^3 K in about 140 000 years from the Big Bang).

A33: Transitional to the galactic Universe state: from A3 to G3. The A3 and G3 states are similar but have two principal distinctions. (1) Relations between the states are determined by the phase supertransition of the entire Universe from homogeneous A-states to inhomogeneous galactic G-states. (2) There are heavier elements in the G-Universe which compose all the forms, bodies, and colors of the world, and all stars and planets, and which are absent in A-states. Duration of the A3 state in determined by (1) the time necessary for the expansion of the Universe to dimensions which permit the arising of the initial galaxys protosuperclusters, (2) the cooling down to temperatures permitting gravitational evolution of inhomogeneities and their fragmentation into galaxies and stars. Hence, uncertainty in duration estimations: 100 K reached in 300 million years, 50 K in 1 billion years. Probably these inhomogeneities in A33 descend genetically from the initial fluctuations already present in the A1 elementary particle state.

G3: cosmic H–He–he Space (Hydrogen-Helium – Heavier Elements).

- On the one hand: it is not emptiness, on the other hand: it is a "vacuum" still many orders higher than the superhighest technical vacuum 10^{-14} g cm^{-3}. It is a "cosmic vacuum": 10^{-30} g cm^{-3}.
- But it is populated (by H–He–he) and this extremely sparse population represents the whole building material for all other states of matter from stars to minerals and life.
- Formula of space composition $H_{90}H_{10}[he_{0.001}]$.
- Always and everywhere in cosmic space, all elements occur.
- The whole volume of space, nearly the whole Universe, is the H–He–he medium. But most of the mass of matter of the Universe (90%) is contained in stars. Second fundamental division the evolution of the Universe: ascending states (evolution trends) and descending states (star outflows).
- Ascending states in the H–He–he space: interstellar clouds-protostars and protoplanets – star envelopes.
- Descending states: star backscattering of H–He–he matter from convective zones, outer envelopes, and atmospheres of stars to corona, stellar wind, planetary nebula, hot ionized gas, cosmic rays, and so on; in all cases it is always H–He–he-plasma.

G31: H–He–he molecular gas and interstellar mineral dust.
G32: H–He–he atomic gas.
G33: H–He–he plasma.

G4: Thermonuclear Stellar States; Synthesis of Elements.

- Star cores: gravitational transformation in the evolution of the Universe occurs only here. Formation of elemental composition of the Universe.
- Star envelopes: backscattering of H–He–he matter, supply space by building material for minerals, planets, and life.

Genetic System of Elements. Consecutive packing of protons and neutrons in nuclei of elements due to gravitational evolution.

Primordial hydrogen; pregalactic helium (and lightest isotopes); main sequence H $\rightarrow$ He; synthesis He–Fe by fusion; synthesis Fe $\rightarrow$ Cf by neutron capture ("slow" and "rapid" processes); "non-Mendeleyevian elements" with excess of neutrons; neutronization.

Gravitational evolution of the Universe represents essentially the synthesis of elements. It covers the whole series from pure proton to pure neutron states.

Genetic Classification of Stars.

1. Core envelope states of stars: main sequence, red giants.
2. Separated cores and envelopes:

- Envelopes $\rightarrow$ coronas, stellar wind, planetary nebulae, dissipation in Space, all H–He–he plasma states.
- Cores $\rightarrow$ gravitational evolution.

G41: Thermonuclear plasma; synthesis from He to Fe.
  G411: Main sequence; H $\rightarrow$ He.
  G412: Synthesis from He to Fe; nondegenerate electrons.
  G413: Degenerate electrons; white dwarfs.
  G414: Plasma crystallization; metallic star cores.
G42: Synthesis of heavy and superheavy elements. Supernovae.
  G421: "Slow" processes; Fe to Bi.
  G422: "Rapid" processes; Fe to Cf.
G43: Neutronization. Neutron stars.
  G431: Neutronized nuclei of "non-Mendeleyevian elements"; neutron nuclei.
  G432: Neutron superfluid state.
G44: Quark-gluon plasma ("quagma"). Black holes.

G5: Mineral State of Matter. The current understanding of matter is related intuitively to matter of the visible world and then transposed to matter in general; and this matter is described as infinitely diverse. However, in the gravitational evolution of the Universe, the whole diversity of terrestrial (planetary) matter, inorganic and organic, crystalline, amorphous, melts, solutions, natural and technogenic, all minerals and materials, constitutes only one type of matter.

Common to all these objects (and distinct from all other states considered above) is the existence in the form of compounds, in a state of chemical bonding, crystal structures, and stereochemistry, of combinations form 104 elements. Their entire diversity is restricted by this single type of matter. Its diversity is the diversity of species and individuals.

It can be named the mineral type of matter, according to the original and predominant interstellar mineral dust and celestial mineral bodies. It constitutes a negligible part of the volume and mass of the Universe, but the whole diversity of matter, its colors and forms, manifest themselves only here.

G51: Interstellar mineral dust and molecules. Largest area of mineral formation; bulk amount of mineral dust in the galaxy is about 60 trillions of Earth masses. Occurs everywhere in all cold regions of all galaxies. The only source of matter for formation of planets and life. Nongravitational formation: mineral condensation from cold H–He–he medium.
G52: Celestial mineral bodies: planets, asteroids, meteorites, comets.
G53: Synthetic compounds.
  G531: Crystalline, amorph, melts, solutions.
  G532: Interactive materials in electronics and technologies, this is a "new race of materials".

G6: Stuff of Life. This can be considered as a carbon-based variety of the mineral type of matter: both are represented by compounds in a state of chemical bonding. Genetically, mineral matter precedes and accompanies the stuff of life; minerals are a substratum of life. Planetary conditions provide only parallel realization of biological and geological evolution.

The principal distinction between stuff of life and minerals is its existence in the form of molecule organizations capable of interaction with the environment, of self-organization and biological evolution. Minerals and materials are compounds, while life is organisms.

There is only one possible composition for the stuff of life: carbon compounds (with hydrogen, oxygen, etc.) because of the existence of the only possible system of elements, and due to the unique ability of carbon to form large C–H–O-based molecules.

Summarizing the foregoing descriptions leads to the "types of matter" concept.

The whole visible world represents a type of matter, called the mineral type, which is only one of the types of matter. Other types of matter exist, fundamental and distinct from each other. The diversity of matter is not boundless; there are a finite number of types (form A1 to G6), together composing matter in general.

The types of matter constitute the evolutionary series of the Universe, complete (without blanks in the series) and with two ends: the beginning of matter (and beginning of time, space, and the Universe) and the terminal states of matter. Each of them has its place and significance in the evolution of the Universe.

The concept "terminal states" designates the end of the evolutionary series of the Universe. States are terminal when they have completed their gravitational transformations due to the exhaustion of a star's gravitational evolutionary resource, limited by the mass of the star.

Depending on the initial mass of a star, a terminal state can be represented by white dwarfs with different elemental composition, by neutron stars with different stages of neutronization, and black holes.

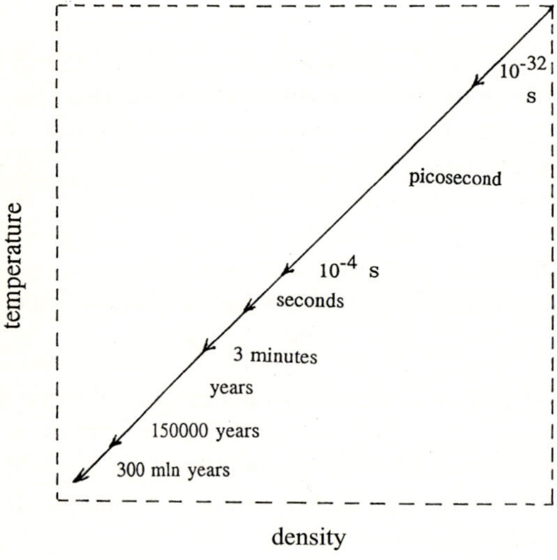

Fig. 6. Arrow of time (time in the pregalactic A-Universe with temporal structure)

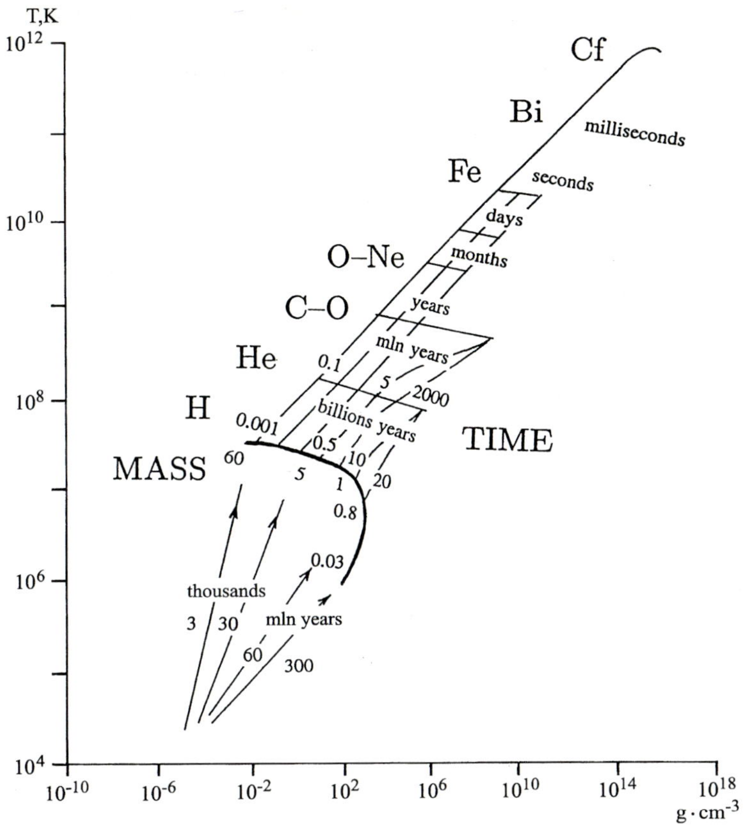

Fig. 7. The well-tempered Universe. (Algorithm of time depending on mass of stars and protostars). This is part of Fig. 4.
Inner side of the arc (main star sequence, corresponding to hydrogen burning): masses of stars are indicated (60; 5; 1 solar mass; 0.8; 0.03). *Outer side of the arc* (Time of stay of stars with different masses in the main sequence (hydrogen burning): 0.001; 0.5; 10; 20 billion years). *Next rows* correspond to times of burning of other elements in stars with different masses: million years, years, months, days, seconds, milliseconds. *Below left* Time of protostars for corresponding star masses

Stars do not disappear, but extinguish and enter terminal states. Terminal states do not evolve further, and are stable and eternal. Mineral states form and exist as terminal states (if not destroyed by external factors). The gravitational evolution of matter in the Universe is finite. The matter of the Universe is eternal in terminal states.

The existence of a frame for the Universe, and a finiteness of the evolutionary series determines the frame and finiteness of time. Time once arose, beginning with the beginning of matter, i.e., with the origin of the Universe. There are amounts of time. In the beginning, the whole of time comprised fractions of a second, now it is 15–20 billion years. The greatest part of time as a whole belongs to the future of the Universe.

Minerals have no proper (evolutionary) time. They form at once in terminal states, which are invariable in time and carried by the flow of geological time.

Time in the Universe is calculable time. The measure of general time is the expansion of the Universe. All times in the evolution of matter are calculated or calculable.

The arrow of time represents this calculation for the pregalactic Universe; it was substantial significance in the density-temperature diagram (Fig. 6). Times of stellar evolution, depending on the mass of stars, compose a temporal algorithm for the galactic Universe.

It can be compared with the structure of a fugue in music. By analogy with the wohltemperiertes Klavier of Bach, it can be described as a well-tempered Universe (Fig. 7).

The calculability of time of the evolution of matter permits the description: "evolutionary determinism" of the Universe.

The suggested ρ–T system arranges events in the whole Universe and coordinates the results of many fields of studies involved in the modeling of the Universe.

Empirical arrangement of the states of matter in the Universe in the density-temperature diagram leads to the fundamental division of Universe evolution, to the system of the types of matter, to the evolutionary model of the Universe, and the definition of time, space, matter, and infinity.

References

Bailey ME, Club SBM, Napter VM (1990) The origin of comets. Pergamon Press, Oxford

Beatty JK, Chaikin A (eds) (1990) The new solar system. Cambridge University Press, Cambridge

Boss AP (1985) Collapse and formation of stars. Sci Am 252:4–11

Burrows A, Hubbard WB, Lunin JI (1989) Theoretical models of very low mass stars and Brown Dwarfs. Astrophys J 345:939–958

Chabrier G, Shatzman E (eds) (1993) The equation of state in astrophysics. Proc IAU Coll No 147, Cambridge University Press, Cambridge

Edmunds ME, Terlevich RJ (eds) (1992) Elements and cosmos. Cambridge University Press, Cambridge

Encrenaz T, Bibring JP (1995) The solar system. Springer, Berlin Heidelberg New York

Gehrels T (eds) (1982) Protostars and planets. Studies of stars formation and of the origin of the solar system. University of Arizona Press, Tucson

Hawking St (1988) A brief history of time. From the Big Bang to Black Holes. Bentam Books

Heide F, Wlotzka F (1995) Meteorites. Springer, Berlin Heidelberg New York

Koester D, Wernez K (1995) White Dwarfs. Springer, Berlin Heidelberg New York

Linde A (1990) Particle physics and inflationary cosmology. Harwood Academic Publishers, Chur

Lipunov VM (1992) Astrophysics of neutron stars. Springer, Berlin Heidelberg New York

Luminet J-P (1992) Black holes, Cambridge University Press, Cambridge

Mihalas D (1978) Stellar atmospheres. WH Freeman, San Francisco

Prantzer NE, Vangioni-Flani M, Casse M (eds) (1992) Origin and evolution of the elements. Cambridge University Press, Cambridge

Silk J (1980) The Big Bang. The creation and evolution of the Universe. WH Freeman, San Francisco

Tenkolsky Sh (1983) Black holes, white dwarfs, and neutron stars. Wiley, New York
Weinberg S (1977) The first three minutes. Basic Books, New York
Whittet DCB (1992) Dust in the galactic environment. Hilger
Winnewisser G, Pelz GC (eds) (1995) The physics and chemistry of interstellar molecular clouds. Springer, Berlin Heidelberg New York
Wynn-Williams S (1992) The fullness of space: nebulae, stardust, and the interstellar medium. Cambridge University Press, Cambridge
Zharkov VN, Gudkova TV (1992) Modern models of giant planets. In: High-pressure research. Terra Sci Publ, Tokyo, pp 393–401

1.2 Interstellar Dust

C. G. Wynn-Williams

Interstellar matter comprises about 10% of the mass of the Milky Way galaxy. About 0.7% of interstellar matter is in the form of microscopic dust grains; most of the remainder is hydrogen or helium gas. As far as we know, there are no liquids in space. Interstellar matter exists both as a diffuse medium that fills the gaps between the stars and as comparatively dense clouds that sometimes take the form of bright or dark nebulae. Under different interstellar conditions the gas may be molecular, atomic, or ionized, with temperatures anywhere from less than 10 to over 10^6 K. In the vicinity of the solar system, gas atoms are separated from each other by distances of the order of 1 cm, while individual dust particles are separated from each other by hundreds of meters. Because interparticle collisions occur so infrequently, the gas and dust are not necessarily in thermal equilibrium with each other. The dust temperature ranges from less than 10 K in dense molecular clouds, to over 1000 K in stellar winds and in the vicinity of newly forming stars.

There is great uncertainty about the physical and chemical nature of interstellar dust. We have clues that indicate that certain materials exist in space, including silicates, graphite, and ice, but many of the data are somewhat ambiguous. No interstellar particles (as opposed to interplanetary particles) have been collected on Earth, although it has been argued that certain species found in meteorites must originally have been interstellar. Current ideas about the nature of interstellar grains come from studying their interactions with starlight, from constraints on the elemental composition of dust based on cosmic abundance arguments, and from infrared and ultraviolet spectral features.

Interactions with Starlight

The most easily observable property of interstellar dust is its ability to absorb and scatter light from background stars. This property is called interstellar extinction (see Mathis 1990 for a review). Short-wavelength light is absorbed and scattered more than long-wavelength light, giving rise to the phenomenon of interstellar reddening of the light from distant stars. If a dense interstellar cloud lies close to a bright star, the scattered light may be observable directly as a reflection nebula. Studies of reflection nebulae indicate that dust grains have albedos of the order of 0.5 at visible wavelengths.

Starlight that is absorbed (as opposed to scattered) by a dust grain is re-radiated into space at infrared wavelengths. The plane of the Milky Way Galaxy is very bright in the 30–300 μm range as a result of dust grains heated to equilibrium temperatures of 10–100 K in the interstellar medium.

Abundance Constraints

Hydrogen and helium cannot solidify under interstellar conditions. Dust must therefore consist mainly of compounds of the more common heavy elements such as oxygen, carbon, and nitrogen, with significant contributions from hydrogen and from the next most common elements: silicon, magnesium, iron, and sulfur. Supporting evidence for this assumption comes from the observation that the relative abundances of some of these elements is much lower in interstellar gas than in the atmospheres of stars – 100 times lower in the case of iron. The widely accepted explanation for these depletions is that the missing elements are locked into dust grains. The most depleted elements are generally those which would condense into dust grains at the highest temperatures, a result that supports the view that grains form in the expanding, cooling winds that blow off the surfaces of giant stars.

Spectral Absorption Features

The strongest interstellar spectral absorption feature lies in the ultraviolet band at 0.22 μm. The most likely carrier is graphite, though other carbonaceous materials have been proposed, including amorphous carbon, quenched carbonaceous composites, and polycyclic aromatic hydrocarbon molecules (see below).

The presence of silicate particles in interstellar dust is deduced from the 9.7 and 18 μm absorption features. The shape of the 9.7-μm feature does not correspond well with any known terrestrial mineral. The discrepancy may be due to the difference in particle sizes between interstellar and laboratory samples, or possibly to the interstellar silicates having an amorphous rather than a crystalline structure. The 11.7-μm feature of silicon carbide has been seen in the immediate surroundings of certain carbon-rich stars, but has not yet been detected in the general interstellar medium.

An absorption feature at 3.4 μm is probably due to C-H stretching in aliphatic hydrocarbons, but a more precise identification cannot be made. A deep absorption feature at 3.1 μm, and a weaker one at 6 μm are attributed to water ice. These features are seen only in dense interstellar clouds, where temperatures are lower and destructive radiation fields are weaker than in the general interstellar medium. Evidence for frozen carbon monoxide and ammonia has also been found.

It is not clear whether the diffuse interstellar absorption bands, which are seen in the visible spectra of reddened stars, have a dust or a molecular origin; in any case, attempts to identify them spectrally have so far met with little success (see Herbig 1995 for a review).

Dust grains that are subject to strong ultraviolet radiation fields display a number of broad emission features at infrared wavelengths. The most prominent are at 3.3, 6.2, 7.7, 8.6, and 11.3 μm. The features have been tentatively identified with polycyclic aromatic hydrocarbon (PAH) molecules, though other identifications have also been suggested. PAHs consist of 10–100 carbon atoms bound into linked benzene-like rings, with hydrogen atoms around the edges. Examples are pyrene and coronene. Emission from PAH molecules is often seen in conjunction with emission from very small grains (see below); PAHs may represent the small end of a continuous size distribution of particles that blur the distinction between molecules and solid grains.

Sizes and Shapes of Dust Grains

Comparison of the observed wavelength dependence of interstellar extinction with theoretical models based on Mie scattering theory indicates that at least some dust grains have diameters of the order of 0.1 μm. A more detailed model by Mathis et al. (1977) posits a power-law distribution of dust sizes with $dn/da \sim a^{-3.5}$ for $0.005 < a < 0.2$ μm. There is some evidence that grains inside dense clouds are larger on average than those in the diffuse interstellar medium; the increase in size may be due to the formation of ice mantles around silicate grains.

There is also evidence for the existence of very small grains (VSGs) consisting of only $10^2 - 10^4$ atoms. The heat capacity of such a grain is so low that a single visible-wave photon will temporarily raise its temperature to values far in excess of thermal equilibrium. VSG dust is detectable by its anomalously strong emission in the 2–20 μm wavelength range.

Almost nothing is known about the shapes of grains. Simple models usually assume sphericity, but needles and flat sheets (as in graphite) have also been considered. The only direct evidence concerning the shapes of dust grains comes from the observation that starlight that has passed through large column densities of interstellar matter usually emerges slightly polarized. The generally accepted explanation for this polarization is alignment of nonspherical grains by interstellar magnetic fields.

Books that provide a starting point for general studies of the interstellar medium include the nontechnical account by Wynn-Williams (1992), the series of lectures by Burton et al. (1992), and the conference proceedings edited by Hollenbach and Thronson (1987) and by Blitz (1990). Useful works that deal with interstellar dust per se are the monograph by Whittet (1992), the conference proceedings edited by Bailey and Williams (1988) and Allamandola and Tielens (1989), and the reviews by Mathis (1990) and Draine (1995).

References

Allamandola LJ, Tielens AGGM (eds) (1989) Interstellar dust. IAU Symp 135. Kluwer, Dordrecht, 525 pp

Bailey ME, Williams DA (eds) (1988) Dust in the universe. Cambridge University Press, Cambridge, 573 pp

Blitz L (ed) (1990) The evolution of the interstellar medium. Astron Soc Pacific, San Francisco, 346 pp

Burton WB, Elmegreen BG, Genzel R (1992) The galactic interstellar medium. Springer, Berlin Heidelberg New York

Draine BT (1995) Properties of dust in diffuse clouds. In: Ferrara A, McKee CF, Heiles C, Shapiro PR (eds) The physics of the interstellar medium and intergalactic medium. Astron Soc Pacific, San Francisco, pp 133–147

Herbig GH (1995) The diffuse interstellar bands. Annu Rev Astron Astrophys 33:19

Hollenbach DJ, Thronson HA (eds) (1987) Interstellar processes. Reidel, Dordrecht, 807 pp

Mathis JS (1990) Interstellar dust and extinction. Annu Rev Astron Astrophys 28:37–70

Mathis JS, Rumpl W, Nordsieck KH (1977) The size distribution of interstellar grains. Astrophys J 217:425–433

Whittet DC (1992) Dust in the galactic environment. Institute of Physics, London, 295 pp

Wynn-Williams CG (1992) The fullness of space. Cambridge University Press, Cambridge, 202 pp

1.3 Interstellar Diamond

A.V. Fisenko

Interstellar diamond (designated C_δ) was discovered in meteoritic matter in 1987.

The diamond was identified as presolar because of its isotopically anomalous xenon composition: it is simultaneously enriched in light isotopes ($^{124.126}$Xe) and heavy isotopes ($^{134.136}$Xe) with respect to solar Xe. This Xe–HL in the diamond phase carrier must have been formed by nucleosynthesis processes.

The grains of diamond are sited in a fine-grained matrix, and their abundance in chondrites reflects mainly the degree of metamorphism undergone. For carbonaceous chondrites the diamond abundance decreases in range CI (Orgueil ~ 0.1 wt.%) > CV (Vigarano ~ 0.5 wt%) ≥ CM (Murray ~ 0.04 wt.%) > CO (Kainsas ~ 0.02 wt.%). For unequilibrium ordinary and enstatite chondrites, the diamond abundance falls within the limit of 0.002–0.013 wt.%.

Interstellar diamond has been isolated from chondrites by corrosive chemical reagents (HCl, HF, $K_2Cr_2O_7$, $HClO_4$). The isolation of most pure diamond grains is based on their tendency to form a colloid at pH ≥ 5 due to the presence of COOH groups at the microdiamond surface. The dry gel represents a brown matter with a density of 2.22–2.33 g/cm^3.

Microdiamond crystals have mainly face-centered cubic lattice structure with the parameter of the unit cell 0.3558 ± 0.0015 nm.

The mean size of diamond grains lies within 10–26 Å. This indicates that the diamond grains were formed as a result of growth and not of fragmentation of their more large predecessors. The kinetic parameters of the oxidation process in air atmosphere differ for interstellar diamond in chondrites of different chemical classes. Thus, for the Murchison CM2 diamond the activation energy is 233.0 ± 2.1 kJ/mol, for the Krymska LL 3.0 diamond 266.1 ± 6.3 kJ/mol.

These distinctions may be due to different PT conditions in the accretion process of the chondrite parent bodies and/or of the metamorphism process.

According to TEM analyses, diamond fractions contain also amorphous material. EELS (electron energy loss spectroscopy) shows the presence of sp^2-bonded atoms in addition to the sp^3-bonded atoms of diamond. The amorphous phase is most probably the hydrogenated surface of diamond grains, but data for the noble gas component in the amorphous phase indicate that it can be the discrete carbonaceous phase (designed C_ζ). The C_ζ-phase should be highly resistant to chemical reagents, but less thermostable than diamond. One candidate for C is the amorphous a–C:H phase.

Carbon in interstellar diamond is "normal", with an average of $\delta^{13}C \sim -33-40‰$. Why the carbonaceous phase carrier of anomalous presolar nucleogenous components has a solar $^{12}C/^{13}C$ ratio is an open question.

The main admixtures in interstellar diamond are hydrogen (10–40 at.%) and nitrogen (up to 1.5 at.%). Hydrogen is most probably sited in the amorphous phase. Hydrogen in diamond-enriched in deuterium relative to SMOW: $\delta D = 245-342‰$. The N abundance in diamond ranges form ~ 0.2 to ~ 0.9 wt.% and depends markedly on the degree of metamorphism of diamond-bearing chondrite: N abundance decreases with an increasing degree of metamorphism. The nitrogen in diamond is depleted in ^{15}N ($\delta^{15}N_{min} \sim -375‰$ relative to atmospheric N). It is possible that N in diamond fractions is either present in two phase carriers, differentialy thermostable to oxidation, or N is a two-component mixture which is heterogeneously distributed within the diamond grains.

Of the trace elements in interstellar diamond, the noble gases are best understood. Their abundances range from 2×10^{-2} to 6×10^{-2} cm^3/g for ^{4}He and from 2×10^{-7} to 5×10^{-7} cm^3/g for ^{132}Xe. Diamond contains almost all trapped He and Ne in chondrites. To the present two components of noble gases in diamond-bearing fractions have been most exactly identified. One of them, with almost "normal" (planetary) isotopic composition, was designated P3 component. Another, containing isotopically anomalous gases, was called HL component. At pyrolyses the P3 component is released at low temperature (<900 °C), the

HL component at 1100–1600 °C. The phase carrier of the P3 component could be phase C_ζ. The HL component is present in diamond grains. In addition to the two components in diamond fractions, there is one further component, released at a temperature higher than the HL component. This third component either has almost normal isotopic composition (called P6) or is an HL subcomponent. The abundance of the P3 component ranges by more than by 2 orders in diamonds isolated from chondrites of different chemical classes and groups. This variation is most probably due to the different degree of metamorphism. At the same time, the abundance of the HL component in the diamond of these chondrites changes by a factor of $\leqslant 2$. The Xe-HL isotopic composition is also similar in all chondrite diamonds. The $^{124}Xe/^{132}Xe$ and $^{136}Xe/^{132}Xe$ ratios are equal to 0.0074 and ~0.642, respectively, almost twice exceeding the ratios for solar Xe. Kr in the HL component is enriched in heavy isotopes $^{84,86}Kr$ (Kr–H), but is depleted in ^{78}Kr. Ar in the HL component is enriched in ^{38}Ar ($^{38}Ar/^{36}Ar = 0.208$), and Diamond grains contain Ne-A2 ($^{20}Ne/^{22}Ne = 8.36 \pm 0.03$, $^{21}Ne/^{22}Ne = 0.035 \pm 0.001$), while the P3 component-Ne-A1 ($^{20}Ne/^{22}Ne = 8.4 \pm 0.08$, $^{21}Ne/^{22}Ne = 0.0299 \pm 0.0003$. Helium in diamond is enriched in ^3He ($^3He/^4He = 1.4 \times 10^{-4}$). Xe-Hl has almost no excess of radiogenetic ^{129}Xe.

At present, there are the following hypotheses for the formation of interstellar diamond:

1. Diamond grain formation by the Chemical Vapor Deposition (CVD)-process. This process is similar to the chemical process of vaporous matter precipitation. Laboratory experiments on epitaxial diamond growth show that the CVD process may occur in a hydrogen-rich medium at relatively low pressures and temperatures ($P < 1$ atm and $T < 1500$ K). Such conditions are present in some astrophysical objects with $C/0 > 1$ (atmosphere of carbon stars and red supergiants, and expanding He shell after a supernova explosion), and therefore the diamond grains can be formed by a CVD mechanism.
2. Diamond grain formation as a result of photolysis of hydrocarbons in the vicinity of stars.
3. Crystallizaton from a melt formed as a result of collision of graphitized carbon particles under the influence of shock waves. The probability of diamond grain formation in this process is estimated at ~5% and therefore diamond in chondrites should be accompanied by considerable amounts of initial carbon grains. These grains, like diamond, should contain anomalous Xe, Kr, and N. However, such carbon grains have not been found in chondrites, and therefore this diamond formation process seems unlikely.
4. Diamond grain formation from chemically connected carbon and/or from amorphous carbon (e.g., soot) in the vicinity of supernova in an extremely unequilibrated process. Shock waves may provide the energy for this process.

References

Allaerts L, Lewis RS, Matsuda J, Anders E (1980) Isotopic anomalies of noble gases in meteorites and their origin. Presolar components in the Murchison C2 chondrite. Geochim Cosmochim Acta 44:189–209

Anders E, Zinner E (1993) Interstellar grains in primitive meteorites: diamond, silicon carbide and graphite. Meteoritics 28:390–514

Bernatowicz TJ, Gibbons PC, Lewis RS (1990) Electron energy loss spectrometry of interstellar diamond. Astrophys J 359:248–255

Blake DF, Freund F, Krishnan K, Echer C, Shiff R, Bunch T, Tielens A, Lupari R, Hetherington C, Chang S (1988) The nature and origin of interstellar diamond. Nature 332:611–623

Buerki PR, Lentwyles S (1991) Homogeneous nucleation of the diamond powder by CO_2-laser-driven gas-phase reactions. J Appl Phys 69:3739–3744

Clayton DD (1989) Origin of heavy xenon in meteorite diamonds. Astrophys J 340:613–619

Fisenko AV, Verchovsky AB, Semjonova LF (1994) Noble gases in the interstellar diamond of Efremovka CV3 chondrite. Geokhimia N7:916–927 (in Russian)

Fisenko AV, Tatsy VF, Semjonova LF, Bolsheva LN (1995) Experimental studies of oxidation kinetics of the meteoritic diamonds by air oxygen. Lunar Planet Sci 26:397–398

Fraundorf P, Fraundorf G, Bernatowicz T, Lewis R, Tang M (1989) Stardust in the TEM. Ultramicroscopy 27:401–412

Huss GR (1990) Ubiquitous interstellar diamond and SiC in primitive chondrites: abundances reflect metamorphism. Nature 347:159–162

Huss GR, Lewis RS (1994) Noble gases in presollar diamonds: Three distinct components and their implications for diamond origins. Meteoritics 29:791–810

Huss GR, Lewis RS (1995) Presolar diamond, SiC, and graphite in primitive chondrites: abundances as a function of meteorite class and petrologic type. Geochim Cosmochim Acta 59:115–160

Jorgensen UG (1988) Formation of Xe-HL-enriched diamond grains in stellar environments. Nature 332:702–705

Lavrukhina AK, Fisenko AV (1994a) On the features and isotopic composition of elements composing the interstellar diamond, SiC and graphite in chrondites. Geokhimia N10: 1372–1382

Lavrukhina AK, Fisenko AV (1994b) On the origin of the interstellar diamonds, SiC and graphite in chondrites. Geokhimia N11:1539–1546

Lewis RS, Anders E (1988) Xenon-HL in diamonds from the Allende meteorite-composite nature. Lunar Planet Sci 19:679–680

Lewis RS, Tang M, Wacker JF, Anders E, Steel E (1987) Interstellar diamonds in meteorites. Nature 326:160–162

Lewis RS, Anders E, Draine BT (1989) Properties, detectability and origin of interstellar diamonds in meteorites. Nature 339:117–121

Nuth JA, Allen JE Jr (1992) Supernovae as sources of interstellar diamonds. Astrophys Space Sci 196:117–123

Ott U (1993) Interstellar grains in meteorites. Nature 364:25–33

Russell SS, Arden JW, Pillinger CT (1991) Evidence for multiple sources of diamond from primitive chondrites. Science 254:1188–1191

Swart PK, Grady MM, Pillinger CT, Anders E (1983) Interstellar carbon in meteorites. Science 220:406–410

Tang M, Anders E (1988) Isotopic anomalies of the Ne, Xe, and C in meteorites. II. Interstellar diamond and SiC: carriers of exotic noble gases. Geochim Cosmochim Acta 52:1235–1244

Tielens AGGM, Seab CG, Hollenback DJ, McKee CF (1987) Shock processing of interstellar dust: diamond in sky. Astrophys 319:L109–L113

Valter AA, Fisenko AV (1993) Diamonds in meteorites. Meteoritika 50:113–122 (in Russian)

Verchovsky AB, Russell SS, Pillenger CT, Fisenko AV, Shukolyukov YuA (1993) Are the C light nitrogen and noble gases located in the same carrier? Lunar Planet Sci 24:1461–1462

Virag A, Zinner E, Lewis RS, Tang M (1989) Isotopic composition of H, C, and N in C diamonds from the Allende and Murray carbonaceous chondrites. Lunar Planet Sci 20: 1158–1159

Wdowiak TJ (1987) Diamond formation in carbon-star atmosphere. Nature 328:385

1.4 Interplanetary Dust

F. J. M. Rietmeijer

Dust particles with diameters between 2–50 µm enter the Earth's atmosphere daily. These nano- to picogram dusts (densities form 0.3–4.2 $g cm^{-3}$) are not affected by gravitational forces, but will spiral towards the sun due to light pressure exerted by the solar wind. As these dusts cross the Earth's orbit, they enter its atmosphere. The interplanetary dust particles [IDPs] collected in the lower stratosphere are derived from solid solar system bodies that cover a wide range of heliocentric distance from the inner asteroid belt to 100s of AU (1 AU = 149600000 km). Both chondritic and nonchondritic particles (i.e., iron-sulfide and Mg-rich olivine grains) could cooccur at the surface of asteroids. Potential IDP sources of most interest are icy protoplanets (not sampled by meteorites) wherein solar nebula dusts were kept at cold storage for 4.56 Ga. They include carbon-rich icy asteroids at ~3.5–5.2 AU, comet nuclei (formed at 20–30 AU) that arrive in the inner Solar System form the Oort cloud, and comets from the Kuiper belt (~40–400? AU). The similarity between chondritic IDP compositions and this composition of CI carbonaceous chondrite meteorites and the solar photosphere show that they were not chemically fractionated during this time. Yet, various alteration processes left their imprint on carbon-rich chondritic IDPs, leaving considerable micrometer-scale heterogeity and nonequilibrium mineral assemblages. It is a challenge to this research to understand the origin(s) of the large numbers of carbonaceous and noncarbonaceous phases alike, and the presence of fully oxidized and reduced carbon, sulfur, and iron, often in the same particle. Another challenge is to recognize the original, or principal, components of these IDPs for a consistent model.

Chondritic Dusts

Original attempts to classify these dusts relied on morphology or mineralogy. That is, aggregate particles, viz. chondritic porous [CP], and less porous chondritic-filled [CF] IDPs, and compact particles, viz. chondritic smooth [CS] and

Advanced Mineralogy, Volume 3
Mineral Matter in Space, Mantle, Ocean Floor, Biosphere, Environmental Management, and Jewelry (ed. by A. S. Marfunin)

chondritic rough [CR] IDPs. The mineralogies allowed division into (1) particles dominated by neso- and ino-silicates plus carbons, and (2) particles dominated by layer silicates plus carbons. Using the *predominant* silicates, the anhydrous particles were subdivided into olivine-rich and pyroxene-rich particles. The hydrated IDPs included smectite-rich and serpentine-rich particles. The mostly anhydrous aggregate IDPs were linked to periodic comets. Compact IDPs that are dominated by phyllosilicates were linked to asteroids. This simple relationship between morphology, dominant silicates, and putative sources is no longer tenable. Using the peak heating temperatures during atmospheric entry (obtained by stepped-heating helium-release profiles) of particles of known size and density, it appears that CS and aggregate IDPs are derived from both asteroids and comets. The carbon-free CR IDPs resemble matrix materials in CI and CM carbonaceous chondrite meteorites and have asteroidal origins.

The typically carbon-rich (2–48 wt.% C) CP, CF, and CS IDPs are unique. They differ significantly in form and texture from components in carbonaceous meteorites and contain some mineral assemblages which do not occur in any meteorite class. These IDPs are nonequilibrium assemblages of (1) ultrafine (2–50 nm) Mg,Fe-olivines, Mg,Fe-pyroxenes, Fe,Ni-sulfides, Fe,Ni-oxides, and Fe,Ni-metals, (2) amorphous materials, and (3) carbonaceous and carbon materials. They may contain small but variable amounts of submicron (~100–1000 nm) minerals. For example, plagioclase-feldspars, K-feldspars, Mg,Fe-smectites, illite, kaolinite, barite, (Mg,Ca,Na)-hydrosulfates, calcite, and brucite in various aggregate IDPs. The CS IDPs contain calcite and various Mg,Fe-carbonates. The predominant Mg,Fe-olivines, Mg,Fe(low-Ca)-pyroxenes, and pyrrhotites show a wide range of grain sizes from 2 to 1000 nm that may correlate with different provenances. The carbonaceous materials also occur in many forms, viz. polycyclic aromatic hydrocarbons, amorphous, turbostratic, and poorly graphitized carbons (300–4000 nm in size), graphite, lonsdaleite, and Fe,Ni-carbides.

IDP Alteration

All particles experience a thermal spike (5–15 s) during deceleration in the atmosphere as a function of size, density, and entry velocity and entry angle, and whereby the entire particle is heated. The ensuing dynamic pyrometamorphic oxidation reactions include laihunitization of Mg,Fe-olivines, and oxidation of common Fe,Ni-sulfides. The resulting magnetite and maghemite often form a thin (<100-nm) rim on the particle. Temperatures in unmelted IDPs can be up to 1050 °C. Other alteration processes affecting the IDPs were recognized. Linear features in olivine and diopside single-crystals are solar flare tracks acquired during Solar System sojourn. Sputter rims on Mg-rich olivines testify that they traveled as small bodies in interstellar space.

The most significant postaccretion alteration of the IDPs is related to residency in a parent body with a heat source that could support the presence of an aqueous fluid. An understanding of the complex mineralogies in chondritic IDPs dawned with the finding that amorphous materials are common in these particles. They are amorphous ferromagnesio-aluminosilica materials [Mg/(Mg+Fe) = 0.23–0.95 (wt.%)], and rare silica-rich and (almost) chondritic materials. Using examples of terrestrial alteration, it is plausible that dry crystallization of amorphous alkali-bearing aluminosilica materials leads to the formation of K-feldspar and plagioclase-feldspars; and aqueous alteration of amorphous materials, or preexisting tectosilicates, yields layer silicates. These secondary minerals, and others, support dry (thermal) and aqueous ($T < 200\,°C$) alteration when these IDPs were resident in icy protoplanets. The scarcity of energy for alteration is an ever present problem. This problem is alleviated by a thin layer of interfacial water that is present on grains in dirty ice. This hydrocryogenic alteration at $T < 0\,°C$ is quite efficient to allow growth layer silicates and salts in IDPs. The least altered chondritic aggregate particles consist of distinct (sub)spherical units, and some amorphous and ultrafine-grained anhydrous units remain in hydrated IDPs.

The Original Dusts

Chondritic aggregates are sediments of (sub)spherical units and rare micrometer-sized ("xenolithic") single-crystal silicates (e.g., enstatite whiskers) and Fe,Ni-sulfides. Local concentrations of carbonaceous (Fig. 8) and carbon-free units (Fig. 9) in almost unaltered IDPs fused into irregular contiguous patches.

The original units, or principal components, include three chemically distinct types (Table 1). The nomenclature is confusing, with different names for the same unit or a general term for different types. Table 1 offers a guide to assess the literature.

Each principal component is a unique chemical end member with a unique (minimum) unit size. Fusion into larger entities may not lead to new chemical types. A recent model for the formation of carbon-bearing chondritic IDPs considers the formation of aggregates that are mixtures of three principal components. The components are the dusts that accreted in the solar nebula. Among others, this model shows evolution of IDP morphology. A high ratio of carbon-free to carbon-bearing ferromagnesio-aluminosilica units in an aggregate particle might, upon hydration and loss of ice, yield a compact CS IDP with relic units. Variable carbon contents of chondritic IDPs reflect the extent of mixing of the carbonaceous component relative to other types during accretion.

Only if the distributions of principal components in the nebula are a function of heliocentric distance, will it be possible to use particle chemistry and morphology as a discriminant of its origin. Many primary and secondary features of chondritic particles remain intimately linked to the proportions of principal components acquired during accretion. These scenarios remain to be proven, but the resulting textures and minerals of carbon-bearing chondritic IDPs may contain information on the scale lengths of accretion in the solar nebula at

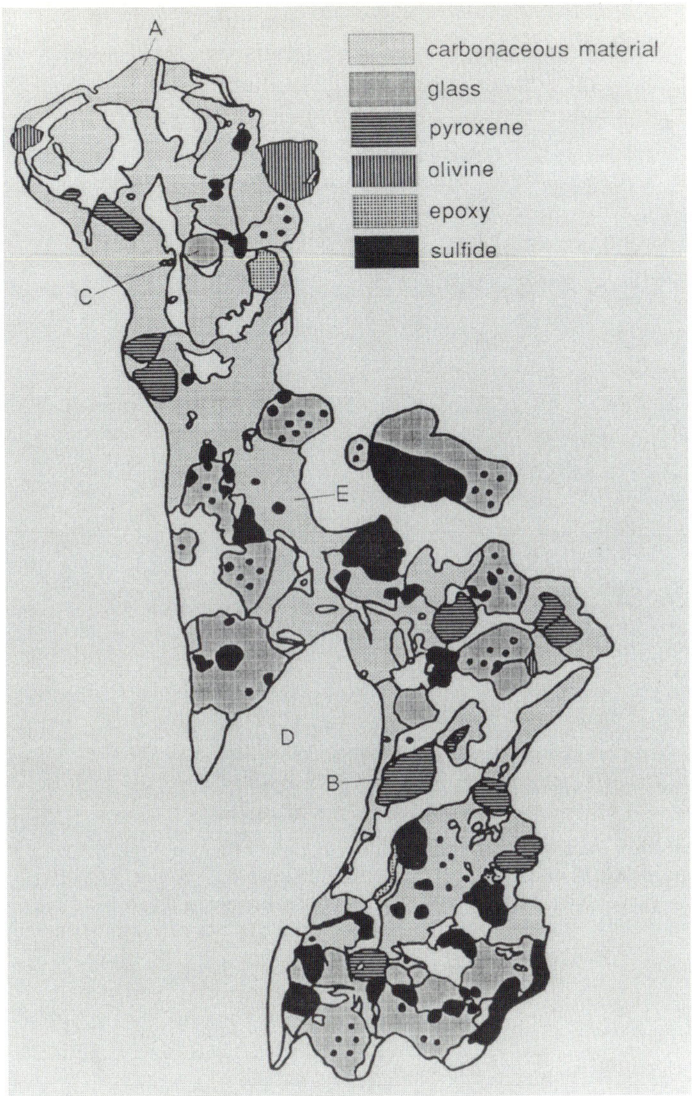

Fig. 8. Mineralogical map based on a transmission electron photomicrograph of an ultrathin section of carbon-rich chondritic IDP W7027H14 showing a contiguous area of carbonaceous material. *A* Carbonaceous material; *B* pyroxene; *C* glass; *D* epoxy, used for embedding of the sample for ultramicrotomy; *E* sulfide. Length of section 5 μm. (Thomas KL et al. 1993)

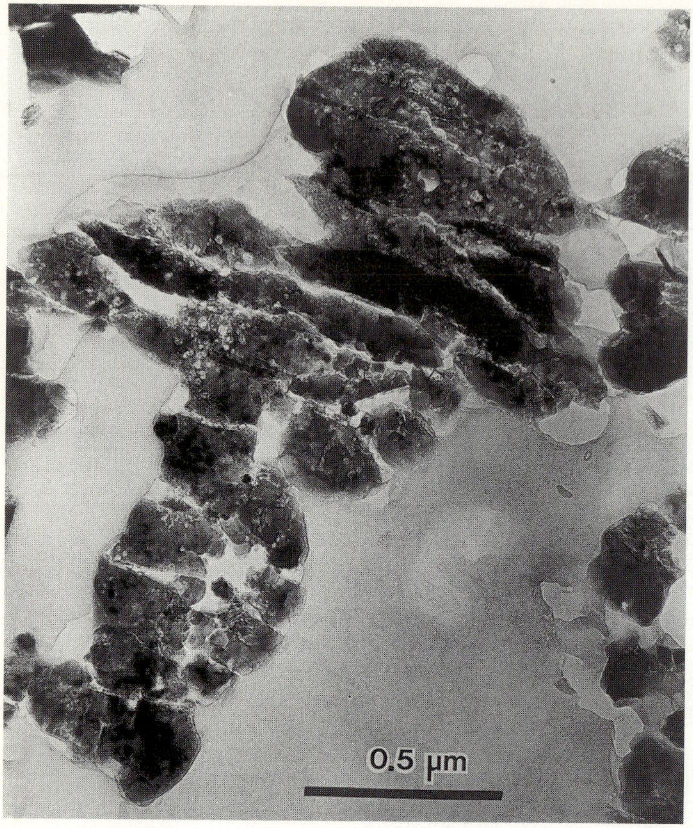

Fig. 9. Transmission electron micrograph of an ultrathin section of chondritic IDP L2011K7 embedded in epoxy (gray background) showing fusion of subspherical (carbon-free) ferromagnesio-aluminosilica (very low Al_2O_3) units with ultrafine-grained Mg, Fe-olivines and pyroxenes and iron oxides. The shattered appearance is an artifact of ultramicrotome sectioning

Table 1. Principal components of chondritic IDPs

Carbonaceous units: refractory hydrocarbons and amorphous carbons Unit size: ~400 nm in diameter
Carbon-bearing ferromagnesio-aluminosilica units Variable Mg/(Mg+Fe) ratios, alkali and Al_2O_3 contents. The units contain ultrafine (2–50 nm) platey Mg,Fe-olivines, Mg,Fe-pyroxenes, Fe,Ni-sulfides, Fe,Ni-metal and Fe,Ni-carbides in a matrix of refractory hydrocarbons and amorphous carbons. Unit size: 100 nm in diameter
(Carbon-free) ferromagnesio-aluminosilica units Variable Mg/(Mg + Fe) ratios, alkali, CaO and Al_2O_3 contents. The units can be amorphous, or holocrystalline (both coarse- and ultrafine-grained) with Mg,Fe-olivines, Mg,Fe-pyroxenes, and Fe,Ni-sulfides. The unit compositions match $(Mg,Fe)_3Si_2O_7$. Units may have an amorphous zone on the periphery. Unit size: 125–1000 nm in diameter (see text)

the sites of their sources. The model received a boost from the 1985/6 encounters with P/comet Halley that recorded an unexpected incidence of light-element, so-called CHON, particles. They could be similar to the carbonaceous units that have since been identified in carbon-rich chondritic IDPs.

Future Research

The minor and trace elements, D/H ratios, and hydrocarbons in carbon-bearing chondritic IDPs also yield new insights. Mineralogical studies of these fascinating dusts remain a prerequisite to constrain dynamic pyrometamorphism and parent body alteration of these IDPs. For example, are oxidation states of carbon, sulfur, an iron correlated, and does S^0 exist in these particles? What are the original (i.e., primary) properties of the principal components? For example, the unit size of the carbon-free component is not yet defined. Their exact bulk compositions also remain uncertain. What is their origin? If they are presolar (or interstellar) dust, the principal components were probably amorphous materials that altered after accretion. Amorphous carbon-free and carbonaceous units occur in several particles but, in others, an amorphous nature of carbon-free and carbon-bearing ferromagnesio-aluminosilica units is inferred from the size distributions of their constituent nanograins. When were the ultrafine silicates and Fe-Ni-sulfides formed? They could be presolar grains that acquired a carbon coating in the solar nebula. In this scenario, efficient evaporation and (re)condensation of (hydro)carbons occurred in the solar nebula at the locations of these IDP sources. This scenario is inconsistent with very high D/H ratios in some of these IDPs, which suggests that carbons were carried to the nebula. It is even possible that the ultrafine minerals are much younger than 4.56 Ga and formed during atmospheric entry. The model creates exciting research opportunities of the complex carbon-bearing chondritic IDPs. The results will constrain the histories of their sources and processes in the solar nebula such as component mixing as a function of heliocentric distance and/or turbulent transport during accretion. The causes of fusion, viz. chemical reactions and/or compaction, of principal components, and associated variations in intraparticle density and porosity), are important. The resulting microtextures, that could determine particle alteration, include partial melting on atmospheric entry.

Conclusions

The Holy Grail of this research would be to find a fluffy, carbon-bearing chondritic aggregate of the three amorphous principal components. It would firmly support the current model of principal component mixing and subsequent parent body alteration and dynamic pyrometamorphism. The carbon-bearing chondritic IDPs are dominated by amorphous materials and (ultra)fine-grained high-temperature minerals.

These properties contribute to a state of high free energy that wild determine particle alteration when energy to activate mineral reactions is scarce. The principal components of chondritic IDPs collected in the lower stratosphere contain a tremendous amount of information on the nature of dusts and processes in the solar nebula. They are a window to the earliest time of our solar system, and to ongoing processes in our galaxy.

References

Bradley JP, Sandford SA, Walker RM (1988) Interplanetary dust particles. In: Kerridge JF, Matthews MS (eds) Meteorites and the early solar system. University Arizona Press, Tucson, pp 861–898

Brownlee DE (1985) Cosmic dust: collection and research. Annu Rev Earth Planet Sci 13:147–173

Mackinnon IDR, Rietmeijer FJM (1987) Mineralogy of chondritic interplanetary dust particles. Rev Geopyhs 25:1527–1553

Rietmeijer FJM (1992) Mineralogy of primitive chondritic protoplanets in the early solar system. Trends Mineral 1:23–41

Sandford SA (1987) The collection and analysis of extraterrestrial dust particles. Fundam Cosmic Phys 12:1–73

Thomas KL et al. (1993) Carbon abundance and silicate mineralogy of anhydrous interplanetary dust particles. Geochim Cosmochim Acta 57:1551–1566

Zolensky ME, Wilson TL, Rietmeijer FJM, Flynn GJ (eds) (1994) Analysis of interplanetary dust. AIP Conf Proc 310, Am Inst Physics, New York, 357 pp

1.5 Cosmogenic Matter in Terrestrial Environments

G. KURAT

The accretionary influx of extraterrestrial matter onto the Earth amounts to about 40000 t/a and is delivered mainly by meteoroids in the size range 50–400 μm (median 220 μm – Love and Brownlee 1993). Not all particles of this size can survive atmospheric entry unaltered. Most are frictionally heated to such an extent that partial to total melting and also partial evaporation takes place (e.g., Kornblum 1969). However, small particles and some of the larger ones penetrating the atmosphere at low velocities and almost tangential entry angles can enter without being melted (Brownlee 1981; Bonny and Balageas 1988; Love and Brownlee 1991).

Melted interplanetary dust particles, cosmic spherules, are omnipresent in sediments throughout geological times (e.g., Taylor and Brownlee 1921). Unmelted interplanetary dust is certainly also present but not easily distinguished from terrestrial dust. It can, therefore, be collected only in places were there are low amounts of terrestrial dust.

Advanced Mineralogy, Volume 3
Mineral Matter in Space, Mantle, Ocean Floor, Biosphere, Environmental Management, and Jewelry (ed. by A. S. Marfunin)

One such place is the stratosphere, where interplanetary dust can be collected utilizing high-flying aircraft or balloons (e.g., Brownlee 1985). However, only small dust particles (average diameter about 10 μm) can be collected in this way, but not the large ones, for several reasons (e.g., Warren and Zolensky 1994). Such particles (stratospheric interplanetary dust particles – SIDPs, unfortunately commonly called IDPs) have been available for study for more than 15 years. Since 1984 also large unmelted interplanetary dust particles – micrometeorites (MMs) – have been available from Greenland ice (Maurette et al. 1986) and subsequently also from Antarctica (Maurette et al. 1991).

In Greenland, micrometeorites can be retrieved from "cryoconite", a dark sediment in melt-water lakes, which consists of dust and cocoons of blue algae and siderobacteria. On average, cryoconite contains about 10 g/kg fine-grained sand and dust, mostly of terrestrial origin, and about 800 cosmic spherules and 200 unmelted to partially melted micrometeorites.

In Antarctica, micrometeorites can be harvested by artificially melting the proper ice. One ton of Antarctic blue ice contains about 100 cosmic spherules with diameters >50 μm and about 500 unmelted to partially melted MMs 50–400 μm in size. Now, fortunately, large amounts of unmelted and almost unaltered samples of the interplanetary dust particles which contribute most to the recent accretion rate on Earth are available for study.

The Nature of Interplanetary Dust

Many micrometeorites have suffered severe alteration by frictional heating in the atmosphere. They are partially to almost totally melted, consisting of foamy glass with variable amounts of unmelted phases. Others have been thermally altered (metamorphosed) but not melted and a few have grossly retained their pristine mineralogy. Some micrometeorites suffered additional alterations in the hostile terrestrial environment.

The pristine mineralogy of micrometeorites is surprisingly simple (e.g., Maurette et al. 1991, 1994; Kurat et al. 1993, 1994a; Klöck and Stadermann 1994). Major minerals are olivine, low-Ca pyroxene, magnetite, and hydrous Mg-Fe silicates (phyllosilicates) like serpentine and saponite. Individual MMs are usually dense, low-porosity mixtures of anhydrous and hydrous phases in proportions ranging from fully anhydrous (coarse-grained "crystalline" micrometeorites) to fully hydrous mineral assemblages (phyllosilicate micrometeorites).

Minor phases comprise Ca-rich pyroxenes, feldspars, Fe-Ni sulfides and metal, Mg-Fe hydroxides, Mg-Al and Fe-Cr spinels, perovskite, ilmenite, hibonite, and others.

The major silicates have highly variable Fe/Mg ratios, even within a given particle (unequilibrated mineral assemblage) and are usually very rich in minor elements as compared to their terrestrial counterparts (Fig. 10).

The hydrous minerals contain some elements in chondritic abundances (e.g., Ti, Al, Cr, Na, K).

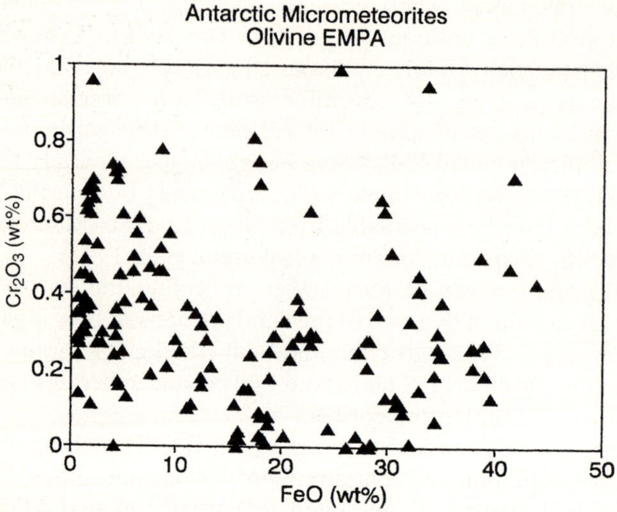

Fig. 10. Plot of Cr_2O_3 vs. FeO contents (wt.%) of olivines from micrometeorites which have varying FeO contents and commonly high Cr_2O_3 contents (terrestrial olivines usually do not contain Cr). Olivines with <0.5 wt% FeO, which are common in CM and CR chondrites, are very rare in MMs

Refractory minerals like Mg-Al spinel are strongly enriched in refractory trace elements (e.g., rare earth elements, Sc, Zr, Hf, etc. – Kurat et al. 1994b) as compared to chondritic rocks.

The mineralogy, mineral chemistry, and the presence of refractory minerals in MMs are similar to those of CM-type (Mighei-type) an CR-type (Renazzo-type) carbonaceous chondrites. However, there are some differences between MMs and CM/CR chondrites, like the presence of abundant Ca-poor pyroxene in MMs (most CM chondrites do not contain such pyroxenes), the lack of very Fe-poor olivines with high Al and Ca contents in MMs (they are common in CM and CR chondrites), and the high abundance of Fe-rich olivines and pyroxenes in MMs.

Bulk major and minor element abundances in phyllosilicate-rich MMs are chondritic, except for Ca, Na, Ni, and S, which are depleted with respect to CI (and CM/CR) chondrites (Fig. 11). Coarse-grained crystalline, anhydrous MMs deviate from the chondritic composition, a feature typical also for anhydrous aggregates and chondrules in carbonaceous chondrites.

Lithophile trace element abundances in phyllosilicate-rich MMs (Fig. 12) follow the abundance pattern of CM chondrites (which is also similar to that of CR chondrites) and deviate from that only in the abundance of K. However, the abundances of siderophile elements in MMs, deviate significantly from those of CI and CM/CR chondrites.

Only the highly refractory elements Os and Ir and the highly volatile Se have abundances similar to those in CI and CM/CR chondrites. The common

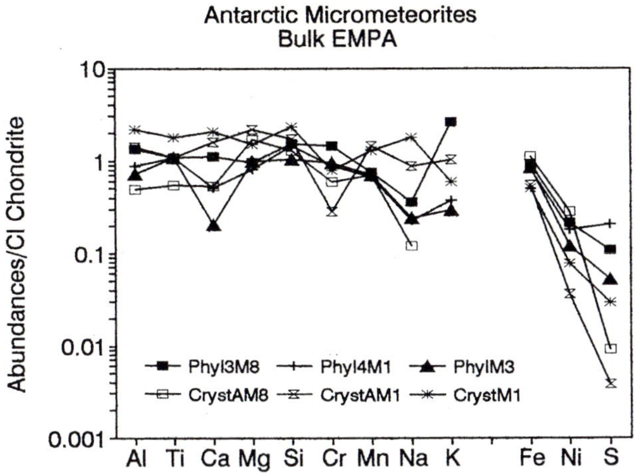

Fig. 11. Chondrite-normalized major and minor element abundances in phyllosilicate-rich and anhydrous crystalline micrometeorites (electron microprobe data from Kurat et al. 1994). Lithophile (*left*) and siderophile (*right*) elements are arranged in order of increasing volatility

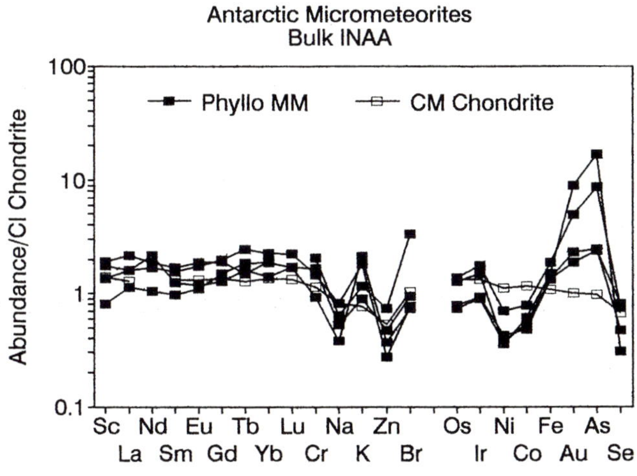

Fig. 12. Chondrite-mormalized abundances of selected trace elements in phyllocilicate-rich micrometeorites (INAA data from Kurat et al. 1994). Lithophile (*left*) and siderophile (*right*) elements are arranged in order of increasing volatility

siderophile elements Ni and Co are depleted compared to chondritic abundances and also fractionated from each other (the Ni/Co ratio is nonchondritic). Iron is enriched over chondritic abundances (moderately), as are Au and As (strongly). The depletion in Ni, Co, and S has been shown (Presper et al. 1993) to be due to terrestrial leaching of Ni-bearing Mg-Fe sulfates from MMs.

Micrometeorites do not contain sulfates which, on the other hand, are abundant in CM and CI chondrites. Large voids present in some MMs may have been occupied by a soluble mineral. Similarly, the depletion of MMs in Ca, as compared to CM chondrites, is possibly due to leaching of carbonates, minerals which are common in CM chondrites but absent from MMs. The enrichments of MMs in Au, As, and K over chondritic abundances must be due to terrestrial contamination. All three elements are strongly enriched in the terrestrial crust as compared to chondrites.

A special compositional feature of MMs (and also SIDPs) is their richness in carbon. Perreau et al. (1993) and Engrand et al. (1994) showed that MMs have C/O ratios which are on average higher than those of CI chondrites, the most C-rich chondrites.. MMs are up to 5x richer in C than CM/CR chondrites.

The small particle portion of the interplanetary dust, the stratospheric interplanetary dust particles – SIDPs, with a typical diameter of about 10 µm, has in principle mineralogical and chemical compositions which are similar to those of micrometeorites. Thus, they consist of either phyllosilicates, anhydrous silicates (olivine or pyroxene), or mixtures thereof (e.g., Brownlee 1985; Bradley 1988; Klöck and Stadermann 1994; Zolensky and Barrett 1994). Also, the chemistry of the major phases is very similar to that of MM phases, i.e., olivine and pyroxene have varying Fe/Mg ratios, and high contents of minor elements and phyllosilicates are mostly saponite with minor serpentine, both rich in minor elements. Magnetite, Ca-rich pyroxene, sulfides, and accessory refractory phases are also present. However, in addition to what has been found in MMs, SIDPs contain sulfates and carbonates of a composition similar to those of carbonaceous chondrites.

Furthermore, there is a class of particles present among SIDPs which has not yet been found among micrometeorites or in meteorites: highly porous aggregates of sub-µm-sized Mg, Fe silicates, Ni, Fe sulfides, and Ni, Fe metal in a glassy and/or carbonaceous matrix (e.g., Bradley 1988; Bradley et al. 1992; Thomas et al. 1994; Rietmeijer 1994).

These aggregates commonly are extremely fluffy and have very high C contents (up to 13×CI, e.g., Thomas et al. 1994).

The chemical bulk composition of SIDPs is chondritic for major, minor, and trace elements (e.g., Jessberger et al. 1992). However, abundances of some elements are nonchondritic due to primary differences (surplus of C) or terrestrial contamination (surplus of K, As, Sb, and Br).

The isotopic composition of several elements has been found to be nonterrestrial and, in some cases, also nonsolar, in both MMs and SIDPs (e.g., McKeegan 1987; Stadermann 1991; Hoppe et al. 1995; Engrand et al. 1996). Specifically, anomalies in isotopic abundances of H/D, C, N, and O have been found.

Interplanetary dust has been exposed to the solar wind and cosmic rays for a sufficiently long time to accumulate large amounts of noble gases and spallogenic isotopes. They have, for example, very high He (up to 10^{-1} cm^{3}g^{-1} at STP) and Ne contents – in excess of 10^{-5} cm^{3}g^{-1} at STP – comparable to only a few very gas-rich chondrites and the lunar soil (e.g., Maurette et al. 1991; Nier 1994).

Helium and neon isotope abundances confirm the extraterrestrial origin of SIDPs and MMs (and some cosmic spherules) as they are comparable to those of solar energetic particles (SEP). In addition, a small contribution from cosmic ray spallation neon was also identified.

Thus, SIDPs and MMs were exposed to cosmic rays and to the solar wind. For the solar wind exposure, the particles must have been of the size recovered. Thus, MMs and SIDPs (e.g., Nier and Schlutter 1990) are true interplanetary dust meteoroids and cannot be products of the breakup of a larger meteoroid in the atmosphere.

Conclusions

Interplanetary dust accreting onto the Earth today bears some similarities to the rare CM/CR carbonaceous chondrites, but differs form them in so many ways that it must be considered a Solar System matter of its own. The features making it different from chondrites are likely to be of primordial origin. These include the mineral abundances, mineral chemistry, the bulk C content, and the presence of unique, fluffy, fine-grained aggregates.

Some deviations in the dust composition from that of chondrites are due to extraction of water-soluble sulfates and carbonates and contamination from the terrestrial environment.

We have good reasons to favor comets as the source of most of the interplanetary dust in the solar system. Therefore, interplanetary dust allows us to study the primordial matter which probably supplied the Earth with volatile elements and the oceans and, possibly, with the ingredients for the development of life.

References

Bonny P, Balageas D (1988) Entry corridor of micrometeorites containing organic material. Lunar Planet Sci Conf XXI: 111–112

Bradley JP (1988) Analysis of chondritic interplanetary dust thin-sections Geochim Cosmochim Acta 52: 889–900

Bradley JP, Humecki HJ, Germani MS (1992) Combined infrared and analytical microscope studies of interplanetary dust particles. Astrophys J 394: 643

Brownlee DE (1981) Extraterrestrial component in deep sea sediments. In: Emiliani C (ed) The sea. vol 7. Wiley & Sons, New York, pp 773–762

Brownlee DE (1985) Cosmic dust: collection and research. Annu Rev Earth Planet Sci 13: 147–173

Engrand C, Christophe Michel-Levy M, Jouret J, Kurat G, Maurette M, Perreau M (1994) Are the most C-rich Antarctic micrometeorites exotic? Meteoritics 29: 464

Engrand C, Deloule E, Hoppe P, Kurat G, Maurette M, Robert F (1996) Water contents of micrometeorites from Antarctica. Lunar Planet Sci Conf XXVII: 337–338

Hoppe P, Kurat G, Walter J, Maurette M (1995) Trace elements and oxygen isotopes in a CAI-bearing micrometeorite from Antarctica. Lunar Planet Sci Conf XXVI: 623–624

Jessberger EK, Bohsung J, Chakaveh S, Traxel K (1992) The volatile element enrichment of chondritic interplanetary dust particles. Earth Planet Sci Lett 112: 91–99

Klöck W, Stadermann FJ (1994) Mineralogical and chemical relationship of interplanetary dust particles, micrometeorites and meteorites. In: Zolensky ME, Wilson TL, Rietmeijer FJM, Flynn GJ (eds) Analysis of interplanetary Dust. Am Inst Physics, New York, pp 51–88

Kornblum JJ (1969) Micrometeoroid interaction with the atmosphere. J Geophys Res 74:1893–1906

Kurat G, Brandstätter F, Presper T, Koeberl C, Maurette M (1993) Micrometeorites. Russ Geol Geophys 34:132–147

Kurat G, Koeberl C, Presper T, Brandstätter F, Maurette M (1994a) Petrology and geochemistry of Antarctic micrometeorites. Geochim Cosmochim Acta 58:3879–3904

Kurat G, Hoppe P, Maurette M (1994b) Preliminary report on spinel-rich CAIs in an Antarctic micrometeorite. Lunar Planet Sci Conf XXV:763–764

Love SG, Brownlee DE (1991) Heating and thermal transformation of micrometeorites entering the Earth's atmosphere. Icarus 89:26–43

Love SG, Brownlee DE (1993) A direct measurement of the terrestrial mass accretion rate of cosmic dust. Science 262:550–553

Maurette M, Hammer C, Brownlee DE, Reeh N, Thomsen HH (1986) Placers of cosmic dust in the blue ice lakes of Greenland. Science 233:869–872

Maurette M, Immel G, Hammer C, Harvey R, Kurat G, Taylor S (1994) Collection and curation of IDPs from the Greenland and Antarctic ice sheets. In: Zolensky ME, Wilson TL, Rietmeijer FJM,, Flynn GJ (eds) Analysis of interplanetary dust. Am Inst Physics, New York, pp 277–289

Maurette M, Olinger C, Christophe Michel-Levy M, Kurat G, Pourchet M, Brandstätter F, Bourot-Denise M (1991) A collection of diverse micrometeorites recovered from 100 tonnes of Antarctic blue ice. Nature 351:44–47

McKeegan KD (1987) Ion microprobe measurements of H, C, O, and Si isotopic abundances in individual interplanetary dust particles. PhD Thesis, Washington Univ, St Louis, Missouri, pp 187

Nier AO (1994) Helium and neon in interplanetary dust particles. In: Zolensky ME, Wilson TL, Rietmeijer FJM, Flynn GJ (eds) Analysis of interplanetary Dust. Am Inst Physics, New York, pp 115–126

Nier AO, Schlutter DJ (1990) He and Ne isotopes in individual stratospheric particles – a further study. Lunar Planet Sci Conf XXI:883–884

Perreau M, Engrand C, Maurette M, Kurat G, Presper T (1993) C/O atomic ratios in micrometer-sized crushed grains from Antarctic micrometeorites and two carbonaceous meteorites. Lunar Planet Sci Conf XXIV:1125–1126

Presper T, Kurat G, Koeberl C, Palme H, Maurette M (1993) Elemental depletions in Antarctic micrometeorites and Arctic cosmic spherules: comparison and relationships. Lunar Planet Sci Conf XXIV:1177–1178

Rietmeijer FJM (1994) A proposal for a petrological classification scheme of carbonaceous chondritic micrometeorites. In: Zolensky ME, Wilson TL, Rietmeijer FJM, Flynn GJ (eds) Analysis of interplanetary dust. Am Inst Physics, New York, pp 231–240

Stadermann FJ (1991) Messung von Isotopen- und Elementhäufigkeiten in einzelnen interplanetaren Staubteilchen mittels Sekundärionen-Massenspektroskopie. PhD Thesis, University of Heidelberg, Germany, pp 97

Taylor S, Brownlee DE (1991) Cosmic spherules in the geologic record. Meteoritics 26: 203–211

Thomas KL, Keller LP, Blanford GE, McKay DS (1994) Quantitative analysis of carbon in anhydrous and hydrated interplanetary dust particles. In: Zolensky ME, Wilson TL, Rietmeijer FJM, Flynn GJ (eds) Analysis of interplanetary dust. Am Inst Physics, New York, pp 165–172

Warren JL, Zolensky ME (1994) Collection and curation of interplanetary dust particles recovered from the stratosphere by NASA. In: Zolensky ME, Wilson TL, Rietmeijer FJM, Flynn GJ (eds) Analysis of interplanetary dust. Am Inst Physics, New York, pp 245–254

Zolensky ME, Barrett R (1994) Olivine and pyroxene compositions of chondritic interplanetary dust particles. In: Zolensky ME, Wilson TL, Rietmeijer FJM, Flynn GJ (eds) Analysis of interplanetary dust. Am Inst Physics, New York, pp 105–114

1.6 Giant Planets

A. V. Kozenko, V. N. Zharkov

Giant planets (GP), brown dwarfs (BD) and very low mass stars (VLM) are objects which have a number of characteristics in common due to the properties of their equations of state. Their luminosities are very low (much less than the solar luminosity, Lo) and their evolutionary time scales are typically measured in Ga.

VLM stars are objects which are more massive than $\sim 10^{-1}$ M⊙ (solar mass).

The term BD has become standard usage for designating a new class of object with the following characteristics: (1) composition similar to the Sun (dominated by hydrogen); (2) masses lie roughly in the range 10^{-2} M⊙ $\leq$ M $\leq 10^{-1}$ M⊙ (i.e., masses about ten times larger than that of Jupiter, but smaller than the critical mass for sustained thermonuclear fusion of hydrogen).

GP are the four largest planets of the Solar System: Jupiter, Saturn, Uranus, and Neptune. Their masses lie in the range 5×10^{-5} M⊙ $\leq$ M $\leq 1\times10^{-3}$ M⊙. The lower mass limit is set by the object's ability to retain hydrogen during accumulation. There is some uncertainty about possible models of origin of objects in the mass range 10^{-3} M⊙ $\leq$ M $\leq 10^{-2}$ M⊙ because the minimum mass for direct formation of a BD from collapse of an interstellar cloud of H and He is ~ 0.02 M⊙. The group of terrestrial planets have no ability to retain hydrogen, due to their being small too masses.

We cannot directly sample the interiors of GP. However, we can use the observational data, experimental data (i.e., equations of state), and physical laws to construct computer models of the interiors.

GPs are gas-liquid, convective, adibatic planets due to:

1. If a GP formed as a hot body, it would not have been able to cool off during its existing time, $t_p \sim 4.5 \times 10^9$ years, because its cooling length $I_{cool} \sim (Xt_p)^{1/2}$ is in the order of 5×10^2 km ($X \sim 10^{-2}-10^{-3}$ cm^2/s is the thermal diffusivity of molecular hydrogen).
2. The heat flow from the planet's interior is calculated as $\sim 10^4$ erg/cm^2/s, that is about 4 orders of magnitude greater than the maximum heat flux that can be transported by molecular conduction.
3. All GPs have instrinsic magnetic fields.
4. It proved that values of the specific dissipative function Q for Jupiter, Saturn, and Uranus are 2.5×10^4, 1.4×10^4, and 5×10^3, respectively and are all about 2 orders of magnitude greater than typical values for matter in the solid state.
5. The planetary gravitational field for GPs corresponds to the field of a planet in hydrostatic equilibrium.

Observational Data

The observational data on cosmochemical abundance and oxygen-carbon ratio give new insight into the internal structure of the GPs. The materials of which the planets consist, are divided into three components according to their volatility: (1) gases (He_2, He, Ne, etc) – the G-component; (2) ices (CH_4, NH_3, N_2O) – the I-component; and (3) rocks and iron-nickel – the R-component.

Depending on (P, T) conditions in the protoplanetary cloud, some compounds of I-component (CH_4 and NH_3) may be present in the G-component. It is assumed that: (1) in the zone of GPs the protosolar cloud contained carbon and nitrogen existing in the forms CH_4 and NH_3, because the pressure in the gaseous phase was significantly larger. Modeling calculations show that the formation of Uranus and Neptune was not accompanied by the formation of gaseous envelopes around them, as in the case of Jupiter and Saturn. Carbon exerts the most noticeable influence on G- and I-components. If it formed CO and was in a gaseous phase, then the mass fraction of the I-component in solid phase (which consists of I and R) and the I/R ratio were strongly decreased.

The Theoretical Background: Equations of State

The equation of state of hydrogen is particularly relevant to the largest two giant planets: Jupiter and Saturn.

Under the circumstances in GP, the relation between pressure, P, and mass density, p, is largely independent of temperature, T, and can be expressed in the form $P \sim p^{\propto}$ with $1.6 < \alpha < 2$. This result is universally true for hydrogen-rich objects in the relevant mass range where the equation of state is mainly governed by the physics of metallic hydrogen (liquid phase) – BD equation of state.

The equation of state for Jupiter follows the BD equation of state most closely, deviating toward a higher-density profile only in a small innermost ice or rock-ice core. By ice, we mean a material composed of the molecules H_2O, CH_3, and NH_3 in solar proportions, not necessarily with intact molecules, and not necessarily in the solid phase.

In Saturn, the deviation from the BD equation of state begins in the metallic-hydrogen region. The hydrogen-helium pressure-density relation can be calculated using purely theoretical results at high pressure and experimental data for effective pair potential at low pressure. The enhanced density of Saturn in the region is interpreted as being caused by enrichment of helium due to immiscibility in metallic hydrogen, and possible enrichment of the ice component.

Uranus and Neptune are composed primarily of ice, a substance which is accessible to laboratory shock compression experiments in the relevant pressure range (typical central pressure reach only a few Mbar). Satisfactory models for both planets have outer hydrogen-helium envelopes which extend to a maximum pressure of ~0.1–0.2 Mbar, below which point the equation of state very closely follows the P(p) curve for ice ($0.1 < P < 8$ Mbar). A rocky core may exist in either Uranus or Neptune.

Models of Giant Planets

Methods for generating the models of GP were developed in early 1970s.

In developing the new models, the abundances of elements were used. Three variants of I+R components were used:

IR1 = (CH_4, NH_3, H_2O + rocks), I/R = 3.1
IR2 = (NH_3, H_2O + rocks), I/R = 2.2,

considering that CH_4 exists in a gaseous phase.

IR3 = (H_2O + rocks), I/R = 1.0,

CH_4 and NH_3 in gaseous phase.

New models of Jupiter and Saturn consist of five layers: the two-layered molecular envelope, an atomic metallic envelope, and a two-layer core. The models with homogeneous helium concentration in the hydrogen envelope or Jupiter ($Y_1 = Y_2 = Y_3$) are models of the first type and do not satisfy the condition of conservation of solar abundance of helium in the planet. In the models of the second type. $Y_2 = Y_3 > Y_1$. The models of Saturn are constructed as those for Jupiter.

The Jupiter-like two-layered molecular envelopes models of Uranus and Neptune are shown in Fig. 13.

The Evolution of Giant Planets

The present general hypothesis is that the embryos of the GPs first formed by accumulation planetesimals. These embryos later gravitationally accreted their gaseous envelopes from the solar nebula. The primary cores are formed by accumulation of rocky and ice planetesimals. If the core mass has reached some critical value, it is capable of concentrating and capturing a gaseous envelope.

The heat flow which corresponds to the observed luminosity, L, of GPs is presumed to be derived from two sources: (1) heat release from the object due to work done on the object's interior, and (2) due to changes in internal energy, E. In GPs the abundance of radioactive high-Z elements such as ^{40}K, ^{232}Th, and ^{238}U is too low for radioactive decay to be significant. The evolutionary age of the planet t_E is defined by the value of t, for which the planets luminosity drops to the present intrinsic luminosities. For the Jupiter, the theory gives $t_E = 5$ Ga, but for Saturn $T_E = 2.5$ Ga. Thus, Jupiter's heat flow is in accordance with the equation of state of hydrogen, but Saturn's is not. In the case of Saturn, it is

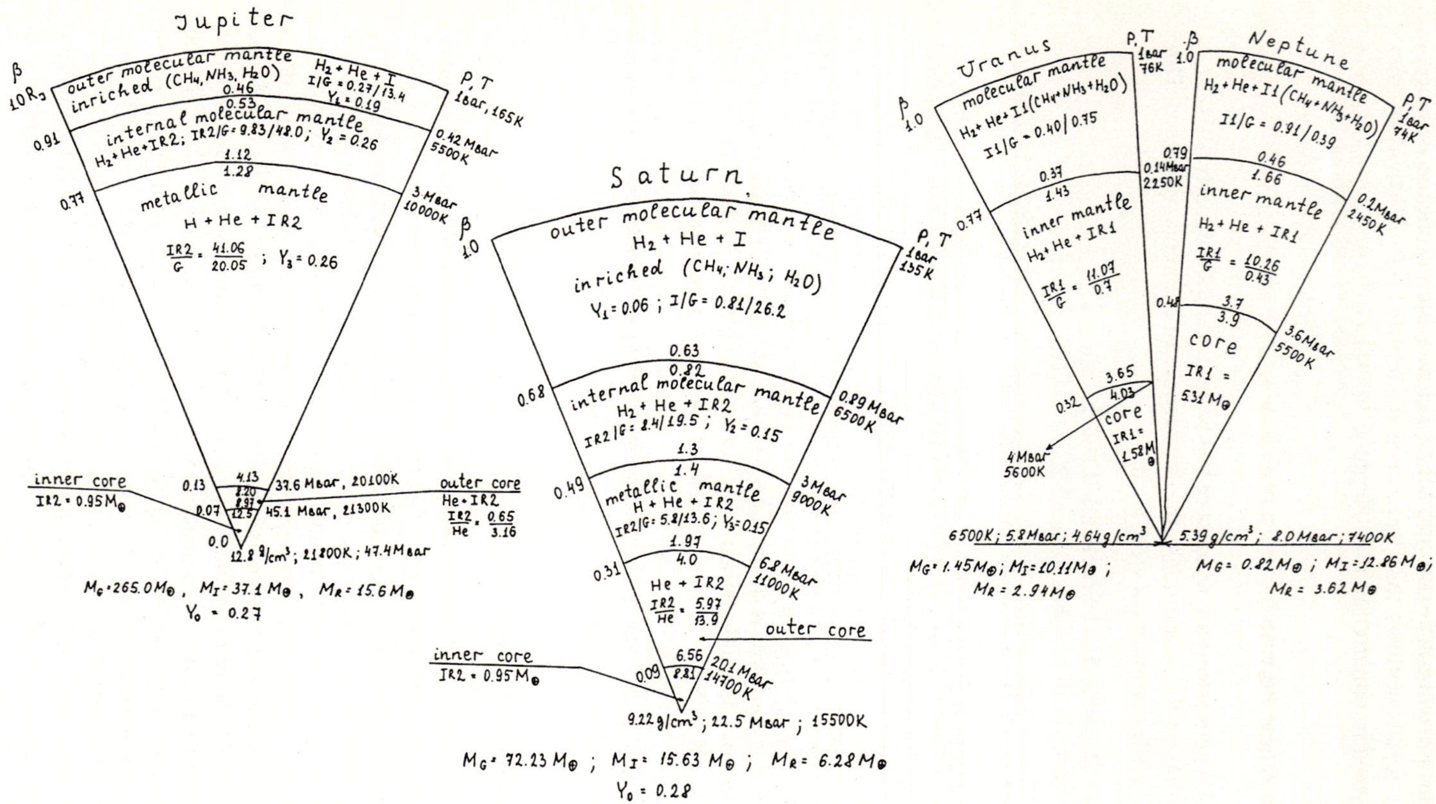

Fig. 13. Schemes of Jupiter, Saturn, Uranus, and Neptune models. Values of the pressure P and temperature T are shown at the interface $\beta(\beta = r/R$ – dimensionless radius) and at the *center*. At the *same levels* in the center of the scheme, the density p (g/cm^3) is shown. The I(IR)/G ratio and helium content in the envelopes are shown. The numerical values of the G, I, and R components are expressed in Earth masses ($M_{\oplus}$). The total values of the masses of the G, I, and R components in Earth masses and total value of helium content (Y_0) are given

necessary to account for the immiscibility of helium in liquid metallic hydrogen and the consequent formation of a helium-enriched core. Due to these processes, the additional gravitational energy is liberated in Saturn's interior. In the case of Uranus and Neptune, the equation of state is dominated by materials heavier than hydrogen and helium. Gravitational energy may be released by a rearrangement of the interior mass distribution in a planet: (1) helium with respect to hydrogen in hydrogen-rich planets (Jupiter and Saturn); (2) I-component with respect to G-component in planets which are rich in both (Uranus and Neptune).

References

Chabrier G, Shatzman E (eds) (1993) The equation of state in astrophysics. Cambridge University Press, Cambridge

Gehrels T, Matthews MS (eds) (1984) Saturn. University of Arizona Press, Tempe

Gudkova TV, Zharkov VN, Leontev VV (1988a) Models of Jupiter and Saturn having a two-layer molecular envelope. Sol Syst Res 22:159

Gudkova TV, Zharkov VN, Leontev VV (1988b) Models of Uranus and Neptune with partially mixed envelopes. Sol Syst Res 22:16

Hubbard WB (1994) Giant planets brown dwarfs, and low-mass star interiors. In: Chabrier G, Shatzman E (eds) The equation of state in astrophysics. Cambridge University Press, Cambridge, pp 443–462

Kafatos MC, Harrington RS, Maran SP (eds) (1986) Astrophysics of brown dwarfs. Cambridge University Press, Cambridge

Weaver HH, Danly L (eds) (1989) The formation and evolution of planetary systems. Cambridge University Press, Cambridge

Zharkov VN (1986) Interior structure of the Earth and planets. Harwood Academic Publ, Chur

Zharkov VN, Kozenko AV (1990) On the role of Jupiter in the formation of giant planets. Sov Astron Lett 16:169

1.7 Constitution of the Terrestrial Planets and the Moon

O. L. Kuskov

The mean densities of the terrestrial planets and the Moon differ. The observed mean density of Mercury is 5.44 g cm^{-3}. The Moon's mean density (3.344 g cm^{-3}) is much lower than that of the Earth (5.52 g cm^{-3}) or the other terrestrial planets (3.934 g cm^{-3} for Mars and 5.245 g cm^{-3} for Venus), probably because of a low metallic iron content. It is quite possible that the differences in mean density and chemical composition between the terrestrial planets and the Moon are explainable by physicochemical mechanisms of Fe/Si and Fe/FeO fractionation.

Moon

Six Apollo missions and three Luna missions provide the background of our knowledge of the Moon. The Moon is depleted in the most volatile elements and iron, and is enriched in refractory elements such as Al, Ca, Ti, and U. However, there is a broad spectrum of the proposed compositional models: from models enriched in Ca and Al to Earth-like compositions in which Ca and Al content is lower; values for the lunar FeO content range between 6 and 18%. The Moon is the only extraterrestrial body for which we have information about seismic velocities as a function of depth. From the point of view of comparative planetology, a major problem to be solved is the determination of the bulk chemical composition of the silicate Moon and the radius of the core.

The mineral composition and seismic structure of the Moon depend strongly on its bulk composition and thermal regime, whereas the bulk chemical composition is controlled by the origin of a planetary body. Although we are not certain what proportions of the major petrogenic oxides were incorporated into the growing Moon, it is almost certain from the seismic observations that their concentrations are different at different depths. The problem arises of how to determine the composition of the lunar mantle depending on the depth. The solution to this problem consists in translating the seismic velocity model into petrological models and vice versa, and comparing the calculated and observed geophysical data, including the mantle seismic velocities, the moment of inertia, and the mass of the Moon. In addition, limits to the composition of the crust, crustal thickness, and mean density, as well as composition and equation of state of the core, must be taken into account.

Geophysical Constraints

Geophysical constraints on the models of the lunar interior are being derived form topographic, gravitational, librational, thermophysical, and seismic data.

Moment of Inertia, Density, and Radius. A value of 0.3905 ± 0.0023 for the normalized moment of inertia of the Moon, as well as a value for the mean density of 3.3437 ± 0.003, are determined. Distribution of pressure can be described by an approximate formula: $P = P_0 [1-((R-H)/R)^2]$, where the pressure in the center of the Moon is $P_0 = 47.1$ kbar, $R = 1738$ km is the radius of the Moon, and H is the depth.

Crust. The mean crustal thickness under different physiographic region are: mascons, 30–35 km; irregular maria, 50–60 km; and highlands, 90–110 km. It is suggested that crustal mean density and thickness change in the range 2.90–3.05 g/cm^3 and 58 ± 8 km.

Mantle. The most recent seismic velocity model based on the complete 5 year data set acquired when four Apollo seismometers were simultaneously operative, consists of three constant-velocity zones with very sharp discontinuous boundaries at depths of 270 and 500 km. The jump in the seismic parameter across the lunar 500-km discontinuity is equal to 20%, which is three to five times greater than that across the 400 or 650 km discontinuity in the Earth's mantle. The jump of velocities and seismic parameter across the lunar 270-km discontinuity is sharp and negative. There is no identical peculiarity for the Earth's mantle. According to this model, the lunar mantle is strongly differentiated and consists of the upper mantle (58–270 km), the middle mantle (270–500 km), and the lower mantle (between 500 km – and a core). Perhaps the most interesting seismic discovery has been the recognition of a widespread low-velocity zone at a depth of 270–500 km (Fig. 14).

Approach and Calculation of Bulk Composition, Mineralogy, and Physical Properties. On the basis of the geophysical constraints, chemical composition of the zoned Moon has been calculated by a method of thermodynamic modeling of phase relations and physical properties in the CaO-FeO-MgO-Al_2O_3-SiO_2 (CFMAS) system including solid solutions. The method involves computation of seismically admissible bulk compositions of the zoned Moon from the given seismic data (inverse problem). The oxide contents and temperature thus determined are adopted as starting approximation, and have been used further in calculations of velocity and density profiles according to the direct procedure.

P- and S-wave velocities have been found from:

$$V_P^2 = \{(K_s + 4/3G)/\rho\}, V_s^2 = \{G/\rho\},$$

where ρ, K_s and G are the density, adiabatic bulk modulus, and shear modulus.

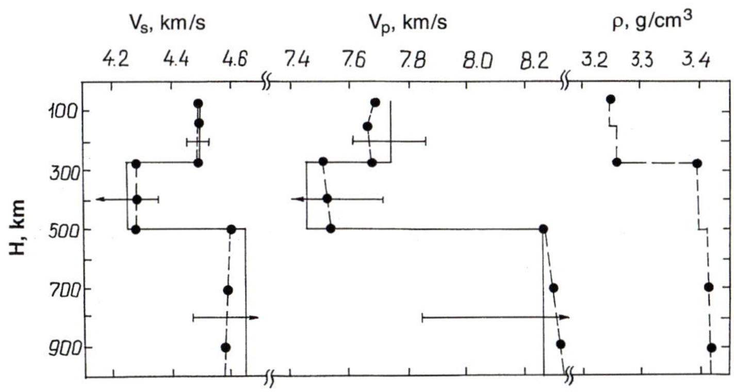

Fig. 14. Calculated velocity and density profiles in the lunar mantle (*dashed lines*) satisfying the moment of inertia and mean density of the Moon in comparison with the observed velocity data (*solid lines*, *error bars* denote seismic uncertainty). Bulk composition models for the upper, middle, and lower mantle are listed in Table 2

Composition of the Moon

The upper mantle of the Moon is enriched in SiO_2 and depleted in FeO relative to the middle mantle, whereas the middle mantle is enriched in SiO_2 and FeO in comparison with the lower mantle; the bulk Moon is enriched in refractories (Table 2). Compositions enriched in refractory elements provide the Al-Ca-rich highland crust.

Comparison of the most probable seismic profiles with those from thermodynamic modeling shows their practical compatibility, whereas the calculated mantle densities satisfy the moment of inertia as well as the mean density constraints (Fig. 14). These results show that there is no reason for density inversion in the Moon. Phase changes from the plagioclase-pyroxenite to the quartz-pyroxenite field taking place at depths of 130–150 km are not able to explain the nature of the 270-km discontinuity, whereas phase changes in any bulk composition model are not able to explain the nature of the 500-km discontinuity. These results lead to the conclusion that the lunar mantle is chemically stratified. At the same time, each separate zone of the mantle is chemically uniform and may be composed of: quartz-bearing pyroxenite containing 0.5–3 mol.% of free silica at depths of 58–270 km (upper mantle); olivine-bearing pyroxenite enriched in FeO at depths of 270–500 km (middle mantle) and olivine-orthopyroxene-clinopyroxene-garnet assemblage enriched in refractory (Al, Ca) elements at depths greater than 500 km (Table 2 and 3).

The preferred solutions for the chemical composition of the middle mantle (an olivine pyroxenite) have an FeO content between 14–18 wt.% (Mg 74–76), indicating that the middle mantle is more iron-rich than the upper and lower mantle. Such an enrichment in FeO may be responsible for the nature of the low-velocity zone at depths of 270–500 km with a density jump from 3.26 to 3.40 g cm^{-3} (Fig. 14). Another explanation is also possible: seismic data give evidence for the occurrence of hydrosilicates in the lunar middle mantle; the mineral assemblage consisting of pyroxene + olivine + garnet + talc and containing hydrosilicate phase (1.5 wt.% H_2O) might be responsible for the low-velocity zone.

The derived composition of the middle mantle is close to that for the source region of Apollo 11 and 12 basalts. Chemical analyses of Apollo 15

Table 2. Chemical composition of the silicate portion of the Moon (wt.%)

Oxide	Crust	Mantle			Entire Mantle	Crust + Mantle
		Upper	Middle	Lower		
SiO_2	45.5	56	52.2	44	50.3	49.9
MgO	7	30.5	24.5	32	29.4	27.5
FeO	6.5	8	17.5	9.5	11.2	10.8
Al_2O_3	25	3	3	8.5	5.2	6.9
CaO	16	2.5	2.5	6	3.9	4.9

Table 3. Phase composition of the lunar mantle (mol.%)[a]

Phase	Upper		Middle		Lower
	58 km	270 km	270 km	500 km	500–1258 km
Anorthite	2.1	0			
Quartz	0	2			
Orthopyroxene	92.8	91.8	94.4	96.8	
Clinopyroxene	5.1	6.2	2.5	0	35.5
Olivine			2.5	2.5	51.5
Garnet			0.6	0.7	13

[a] Solid solutions in the CFMAS system: olivine (binary); garnet (ternary: pyrope, almandine, grossulare); clinopyroxene (five-component solution: diopside, hedenbergite, clinoenstatite, clinoferrosilite, clinocorundum); orthopyroxene (five-component solution: orthodiopside, orthohedenbergite, enstatite, ferrosilite, orthocorundum).
Temperature distribution in the mantle: 58 km 420 °C; 270 km 800 °C; 500 km 1000 °C; 1258 km 1350 °C.

mare basalt samples fall into two distinctive groups, olivine basalts and quartz basalts; it has been suggested that olivine basalts result from magma extraction at depths of 300–400 km, whereas quartz basalts could result from magma segregation at depths of about 200 km. These petrological high-pressure experiments and petrochemical observations are in excellent agreement with the determination of chemical and mineral compositions of the upper and middle mantle from geophysical data.

The Al-Ca-rich model composition (6–9% Al_2O_3, 5–7% CaO) gives the best fit to the lower mantle seismic properties (Fig. 14, Table 3). The lower mantle may be composed of undifferentiated material of bulk lunar composition. In this case, the bulk composition of the crust + upper mantle + middle mantle must correspond to that of the lower mantle.

Core. The radius of the core depends on its composition, thickness of the crust, and distribution of density in the crust and mantle. The seismic constraints, combined with mass and moment of inertia calculations, suggest the presence of a core: 310 km in radius for the γ-Fe(Ni)-core and 480 km for the FeS-core. The existence of a lunar core is consistent with geophysical data (attenuation of seismic velocities, moment of inertia, electrical conductivity, lunar paleomagnetism).

The bulk composition and mineralogies of the silicate portion of the Moon (mantle + crust) generated by the geophysical data show that (Tables 2):

1. The concentrations of FeO, SiO_2, and refractory elements (Ca, Al) are significantly higher than those in the Earth's upper mantle; the bulk Moon is enriched in refractories by a factor of 1.5–2.5 compared with the terrestrial mantle and CI meteorites.
2. The lunar bulk Mg/Si (0.82) and Mg/Al (5.05) atomic ratios are much smaller than those of the Earth and chondrites.

3. The Fe/Si atomic ratio is equal to 0.18 for the silicate portion of the Moon and 0.22 for the Moon as a whole (crust+mantle+core); the latter value is close to the Fe/Si ratio of 0.28 form comet Halley data.
4. The lunar Mg# (Mg/Mg+Fe in the silicate portion) of 0.82 is consistent with some recent estimates for a starting composition for the lunar magma ocean and for the mare basalt source region; the derived lunar Mg# is similar to that for interplanetary dust particles, but is in conflict with that of the Earth's upper mantle (~0.89).

Thus, the bulk compositions of the Earth's upper mantle and the Earth's satellite are different. This conclusion enables us to restrict the number of hypotheses about the origin of the Moon. Fission and coaccretion hypotheses are not able to explain the compositional features of both planetary bodies. The capture event is practically unlikely, and no source region is known to produce bodies of Moon-like composition. The impact hypothesis – collision of a Mars-sized body with the Earth-appears adequate to account for both chemical and dynamical aspects if the protolunar material comes essentially from the mantle of the impactor. However, the same problem as in the capture hypothesis arises of how and where to produce bodies of Moon-like composition? The similarity in oxygen isotopes between Earth and Moon assumes that the impactor formed in the same part of the solar nebula as the Earth. Without specifying a mechanism of the origin of the Moon, we may conclude that the Earth and its satellite formed from compositionally different materials. composition of the Moon remains unusual in comparison with the Earth, the inner planets, chondrites, and achondrites.

Mars

Direct information on the bulk chemistry and structure of Mars is very limited. Recent observations (oxygen isotope ratios, trace element ratios, trapped rare gases, young crystalliaztion ages) have brought evidence that SNC meteorites (four Shergottites, three Nakhlites, and Chassigny) are Martian surface rocks ejected by huge impacts. By assuming that Mars is the parent body of SNC meteorites, the bulk composition of Mars is estimated (wt.%): 44.4% SiO_2, 30.2% MgO, 17.9% FeO, 2.5% CaO, and 3% Al_2O_3. Gross chemical characteristics of Mars are consistent with accretion from more volatile-rich material than the Earth, more extensive oxidation of Fe-metal, and separation of a Fe-Ni-S core.

In the bulk composition models enriched in FeO, the physical property distribution (density and seismic velocities) is essentially different form that in the Earth's mantle. Figure 15 shows that there is no sharp seismic discontinuity in the Martian mantle at 130–140 kbar (1100–1200 km) such as exists in the Earth at the same pressure range (400–420 km depth interval for the Earth). Thus, there is no distinctive separation into upper mantle and transition region. The Martian mantle may consist of three fairly extended zones: crust to

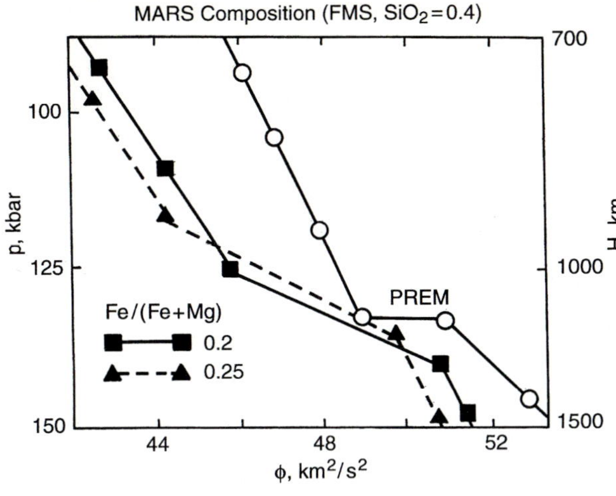

Fig. 15. Seismic parameter profiles ($\phi = V_p^2 - 4/3V_s^2$) in the Martian mantle for two bulk composition models: 40 mol.% SiO_2 and Fe/(Fe+Mg) = 0.2; 40 mol.% SiO_2 and Fe/(Fe+Mg) = 0.25. The PREM model for the Earth with the sharp discontinuous boundary at 134 kbar (400 km for the Earth) is shown for comparison

1000 km, first zone, 1000–1250 km, second (gradient zone); and 1250 km to the lower mantle (if it exists), third zone. Phase transformations are as follows:

$$\text{olivine+pyroxene+garnet} \xrightarrow{1000\text{ km}} \text{olivine+spinel+pyroxene+garnet} \xrightarrow{1250\text{ km}} \text{spinel+garnet.}$$

Models of the internal structure of Mars based on geophysical constraints using the mean density, moment of inertia, and gravitational field suggest that there is a core with radius amounting to from 1000 km for a pure iron core to 2000 km for an iron sulfide one.

Mercury

The observed mean density of Mercury of 5.44 g cm^{-3} implies an iron to silicate ratio in the range from 64 : 34 to 70 : 30, about twice that of any of the other terrestrial planets and the Moon.Three classes of models have been suggested to account for the anomalous composition of Mercury. One class of models employs scenarios which invoke differences in the physical properties (density, ferromagnetism, mechanical strength) of iron and silicates to achieve the required iron/silicate fractionation. A second class of models invokes the differing volatilities of iron and silicates to fractionate the two phases. A third class of models involves a scenario including major planetary collisions, which are suggested to blow off the bulk of the silicate mantle from the original Mercury protoplanet.

Venus

By its mechanical properties (mass $M = 8.815\ M_E$, mean density $\rho = 0.95\,\rho_E$, and mean radius $R = 0.95\ R_E$), Venus is the twin planet of the Earth. This suggests similarity in the bulk composition models, mineralogies, and structure between Venus and Earth. The content of iron in the Earth-like models of Venus is 2% less than that in the Earth. The present temperature of the upper layers of the Venusian mantle is near 1700 K; the temperature at the core-mantle boundary is 3500–4000 K, and is close to that of the Earth.

References

Barsukov VL, Bazilevsky AT, Volkov VP, Zharkov VN (eds) (1992) Venus geology, geochemistry and geophysics. The University of Arizona Press, Tucson
Basaltic volcanism study project (1981). Pergamon Press, New York
Florensky KP, Bazilevsky AT, Burba GA et al. (1981) Essays of comparative planetology. Barsukow VL (ed) Nauka, Moscow
Galimov EM (1995) Problem of the origin of the Moon. In: The main trends in geochemistry. The to 100th Anniversary of Academician AP Vinogradov, Nauka, Moscow, pp 8–43
Hartmann WK, Phillips RJ, Taylor GJ (eds) (1986) Origin of the Moon. LPI, Houston
Hood LL, Jones JH (1987) Geophysical constraints on lunar bulk composition and structure: a reassessment. J Geophys Res 92: E396–E410
Kieffer HH; Jakosky BM, Snyder CW, Matthews MS (eds) (1992) Mars. The University of Arizona Press, Tucson
Kuskov OL (1995) Constitution of the Moon: 3. Composition of the middle mantle form seismic data. Phys Earth Planet Inter 90: 55–74
Kuskov OL, Panferov AB (1993) Thermodynamic models for the structure of the Martian upper mantle. Geochem Int 30 (8): 132–143
Nakamura Y (1983) Seismic velocity structure of the lunar mantle. J Geophys Res 88: 677
Neal CR, Taylor LA (1992) Petrogenesis of mare basalts: a record of lunar volcanism. Geochim Cosmochim Acta 56: 2177–2212
Ringwood AE (1979) Origin of the Earth and Moon. Springer, Berlin Heidelberg New York
Saxena SK (ed) (1986) Chemistry and physics of terrestrial planets. Advances of physical geochemistry. Springer, Berlin Heidelberg New York
Taylor SR (1982) Planetary science: a lunar perspective. LPI, Houston
Taylor SR (1987) The unique lunar composition and its bearing on the origin of the Moon. Geochim Cosmochim Acta 51: 1297–1309

1.8 Mineralogy of Meteorites and Asteroids

A.A. ULYANOV

1.8.1 "Family Tree" of Meteorite Classification

Meteorites are extraterrestrial rocks falling on the surface of our planet, as well as on other bodies of the Solar System. Most of them had elliptical orbits, which suggests that they formed inside our Solar System. Aphelias of the majority of measured and calculated meteorite orbits are in the asteroidal belt, which suggests that the asteroids are a possible source of meteorites.

Upon entry into the atmosphere, the surfaces of the meteorite are heated, melted, and ablated. Only a small fraction of the objects that enter in the atmosphere with appropriate speed and size can pass through the atmosphere to reach the Earth's surface as a meteorite. This raises the question as to whether the meteorites in our museum collections are representative of the objects in interplanetary space.

A meteorite passes through the atmosphere so quickly that the heat is restricted to the outer few millimeters. The interior part of the meteorite is unaltered by atmospheric passage.

Meteorites vary widely in mass and size, from micron-sized particles to large multiton masses. Very large bodies impact at high speeds, are completely destroyed, and may form a meteorite crater.

On the basis of their mineral and chemical composition, meteorites are divided into (Table 4):

1. *Stones*, consisting of magnesium-iron silicates, frequently with troilite and nickel-iron. Stones are further divided into chondrites and achondrites (see below).
2. *Stony-irons*, consisting of an approximately equal mixture of nickel-iron and silicates.
3. *Irons,* consisting essentially of nickel-iron with a small amount of accessory minerals.

There are also some isotopic classifications for meteorites, the most popular being based on oxygen isotopes (e.g., Clayton 1993).

Roughly 4000 meteorites, 1 kg or more in mass, fall to the Earth each year, and only a small percentage of these is recovered. By 1969 the world's meteorite collection comprised just over 3000 meteorites. Since then, more than 10 000 meteorites have been found in the hot and cold deserts of the Antarctic, Australia, and Africa. Some of these meteorites are members of new clans, and about 20 of them are samples from the Moon and Mars.

Figure 16 is a representation of the abundances of various classes of meteorites. The ratio of irons and stones in both finds and falls reflects the very easy identification of irons and their resistance.

Table 4. Classification of meteorites

Class	Subclass	Group	Subgroup	Principal minerals
Stones	Chondrites	Ordinary (OC)	H L LL	Olivine, bronzite, kamasite Olivine, hyperstene kamasite
		Enstatite (E)	EH EL	Enstatite, kamasite, troilite
		Carbonaceous (C)	CI CM CR	Olivine, pyroxene, phyllosilicates
			CV CO CK	Olivine, pyroxene
		Rumurutites (R)		Olivine, pyroxene
	Achondrites	Ca-poor achondrites	Aubrites Diogenites Ureilites	Enstatite Hypersthene Olivine, clynobronzite, FeNi
		Ca-rich achondrites	Angrites Eucrites Howardites	Augite Pigeonite, plagioclase Hypersthene, plagioclase
		Primitive achondrites	Winonaites Acapulcoites Lodranites	Olivine, pyroxene, plagioclase
		Lunar meteorites		Olivine, pyroxene, plagioclase
		Martian meteorites	Shergotites Nakhlites Chassignites	Olivine, pyroxene Diopside, olivine Olivine
Stony-irons	Pallasites			Olivine, kamacite, taenite
	Siderophyres			Orthopyroxene, kamacite, taenite
	Lodranites			Orthopyroxene, olivine, kamacite, taenite
	Mesosiderites			Pyroxene, plagioclase, FeNi
Irons	Hexahedrites			Kamacite
	Octahedrites			Kamacite, taenite
	Ataxites			Taenite, kamacite

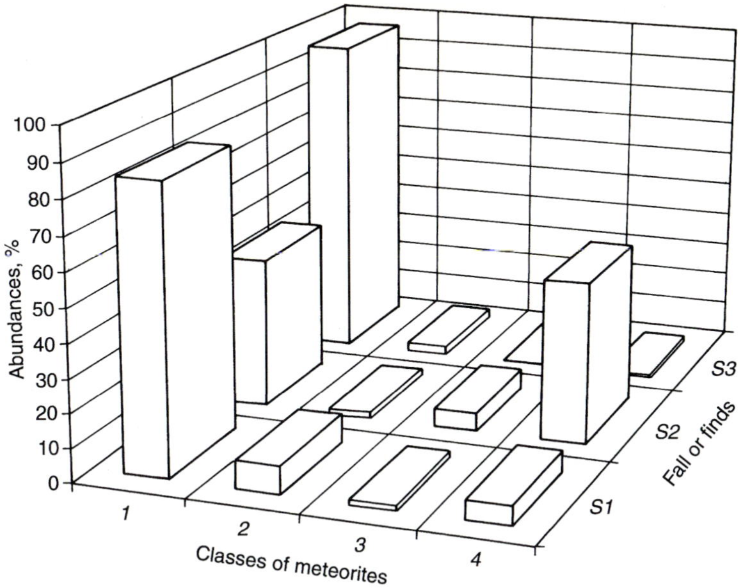

Fig. 16. Abundances of the various classes of meteorites *1* chondrites; *2* achondrites; *3* stony-irons; *4* irons among falls (*S1*) non-Antarctic (*S2*), and Antarctic (*S3*) finds

Table 5. Some mechanical properties of meteorites

Meteorites	Density (g/cm^3)	Porosity (%)
E-chondrites	3.5–3.8	
H-chondrites	3.6–3.8	13.1
L-chondrites	3.4–3.6	10.5
C-chondrites	2.2–3.6	24.4 (for CV)
Aubrites	3.2	4.3–15.1
Diogenites	3.3–3.4	
Ureilites	3.3	
Eucrites	3.1–3.2	14.8
Howardites	3.2–3.3	
Stony-irons	4.3–5.8	
Irons	7.8–8.0	

Meteorites are distinguished also by their main physical mechanical properties, some characteristics of which are summarized in Table 5.

1.8.2 Mineralogical Classification of Meteorites

Stone Meteorites

The prevailing class of meteorites is stones (Fig. 16), the *chondrites* being the most widespread. On some chemical and mineralogical parameters (bulk

MgO/FeO, abundance of metal, Ni/Fe ratio, MgO/FeO in silicates, and abundances of minor elements), textural features, the chondrites are subdivided into: *e*nstatite (E) chondrites, *o*rdinary (OC) chondrites, *c*arbonaceous (C) chondrites, and *r*umurutiites (R-chondrites). Some chondrites cannot be classified into any of these groups, and have been called anomalous, unique, or unclassified. Variations in chemical composition of some chondritic groups are given in Table 6. Van Schmus and Wood (1967) also divided the chondrites into different petrological types based on the degree of recrystallization and equilibration, which they ascribe to thermal metamorphism in the meteorite parent bodies. Chondrites have a peculiar texture, the main feature being the chondrules – predominantly silicate microspherules, the origin of which is widely debated among geologists (e.g., Sears et al., 1995). Chondrules are absent in the rare (but very important) carbonaceous chondrites of the CI clan; these meteorites were classified as chondrites, based on their chemical and mineralogical features. Chondrules may occupy a significant volume of the chondrite matter (≥60%). They show considerable variation in form (difference in spherical degree and degree of recrystallization), sizes, internal texture (porphiritic, radial, barred, and other), and mineral and chemical composition. In the majority of cases, chondrules consist of olivine, pyroxene, plagioclase, and glass. Between the chondrules is a fine-grained matrix, the composition of which is frequently close to the average composition of chondrites. The textural features of different chondrites reflect differences in thermal metamorphism, recrystallization, and shock process (shock deformations, transformation of plagioclase into maskelenite). Thus, most of chondrites are breccias.

A comparison of the composition of chondrites and the solar photosphere (excluding H, He, and some other volatile elements), shows a remarkable similarity in elemental abundances. This distinguishes chondrites from terrestrial rocks. Chondrites contain appreciable quantities of lithophile, chalcophile, and siderophile elements, while terrestrial rocks, as a rule, are predominantly composed of lithophile elements. This suggests that the chondrites are samples of the matter of the early solar nebula, with only small degrees of fractionation of the chemical elements.

In enstatite chondrites almost all iron is in the metallic form, silicates contain very little iron. Variations in the contents of metal iron and other siderophile elements cause the enstatite chondrites to be divided into EH (*h*igh metal contents) and EL (*l*ow metal contents) chondrites. The enstatite chondrites are notable for their low degree of oxidation.

The amounts of metallic iron in ordinary chondrites decrease from H (*h*igh metal contents) through L (*l*ow metal contents) to LL (*l*ow-*l*ow metal contents) chondrites, while the amounts of iron in the silicates increase along the same sequence. Ordinary chondrites are more oxidized than enstatite chondrites.

Of special interest are the carbonaceous chondrites, which have high iron contents, almost entirely in the silicates. The carbonaceous chondrites are characterized by a different degree of oxidation of iron. The carbonaceous chondrites also contain high concentrations of the volatile elements and compounds,

Table 6. Chemical compositions of selected subgroups of meteorites. (After Lavruchina 1973; Yavnel' 1973)

	SiO_2	TiO_2	Al_2O_3	Cr_2O_3	$FeO+Fe_2O_3$	MnO	MgO	CaO	Na_2O	P_2O_5	Σ sil.	FeNi	FeS	Fe_{total}	Σ Fe/SiO_2
Irons	–	–	–	–	–	–	–	–	–	–	0	95–100	0–5	80–93	–
Pallasites	17.0	0.00	0.38	0.68	6.65	0.08	19.8	0.28	0.07	–	45.3	54.04	0.53	54.5	3.22
Mesosiderites	19.5	0.08	4.1	0.36	7.08	0.22	6.4	2.9	0.21	0.2	42.4	50.7	2.83	52.5	2.68
H-chondrites	36.5	0.13	2.43	0.36	8.87	0.25	23.5	1.82	0.85	0.23	75.0	18.9	5.35	27.5	0.755
L-chondrites	39.9	0.13	2.31	0.44	13.1	0.27	25.0	1.9	0.88	0.24	84.3	8.9	6.17	21.8	0.546
Carbonaceous chondrites	33.0	0.11	2.53	0.49	22.0	0.24	23.0	2.32	0.72	0.38	84.8	–	13.6	25.7	0.78
Enstatite chondrites	38.0	0.06	1.87	0.35	1.70	0.14	21.0	0.97	1.00	0.20	65.7	21.6	10.7	7.2	0.72
Aubrites	54.1	0.22	0.91	0.14	1.64	0.06	36.0	1.32	0.12	–	94.4	2.5	1.25	12.1	0.039
Chassinites	37.4	0.07	0.52	0.73	28.0	0.16	32.2	0.11	0.11	–	98.5	0.63	0.88	22.4	0.6
Diogenites	52.1	0.01	1.4	0.32	17.2	0.19	25.9	0.004	0.004	–	98.0	0.82	1.12	14.0	0.27
Ureilites	38.9	0.07	0.80	0.35	13.0	0.09	35.7	0.43	0.05	–	90.0	8.3	–	18.16	0.46
Eucrites	47.6	0.43	13.0	0.36	16.6	0.47	8.46	10.2	0.43	0.09	98.2	1.18	0.56	14.4	0.30
Howardites	49.3	0.1	9.95	0.53	17.1	0.66	11.8	7.7	0.33	0.08	99.0	0.45	0.60	14.0	0.27
Nakhlites	48.8	0.38	1.73	0.33	20.7	0.09	12.0	15.1	0.40	–	99.8	–	0.16	16.2	0.33

including water, carbon, sulfur, rare gases, and considerable amounts of organic substances. Currently, there are six subgroups of carbonaceous chondrites (CI, CM, CO, CV, CR, and CK named after typical members: *I*vuna, *M*urray, *O*rnans, *V*igarano, *R*enazzo, and *K*aroonda, respectively).

The organic compounds of carbonaceous chondrites are of special interest. They include complex organic acids, hydrocarbons, and heterocyclic compounds of C, H, S, N, O, and Cl. Despite their complexity, the organic compounds in carbonaceous chondrites are of abiogenic origin.

Expect for the carbonaceous chondrites, the mineral compositions the members of each chondrite group are mostly uniform. More than 300 minerals are known to be present in chondrites, but many are rare accessory phases (see below). The main minerals of chondrites are olivine and orthopyroxene; small quantities of plagioclase (or/and maskelinite), and the Ca-rich pyroxenes (pigeonite and diopside) are also present. The usual opaque accessories for chondrites are kamacite – α-(Fe, Ni), taenite – γ-(Fe, Ni), and troilite – FeS.

There are many differences between chondritic and terrestrial mineralogy. The common minerals in chondrites are magnesium-iron silicates, while in the Earth's crust the common minerals are quartz and aluminosilicates. Nickel-iron is very rare in terrestrial rocks, and some chondritic minerals have not been found in the Earth's crustal materials. Most chondrites are composed of anhydrous minerals, while hydrated minerals are abundant on Earth.

The mineral composition of different chondritic groups is related to their chemical compositions.

The mineral composition of carbonaceous chondrites (especially the CI-type) also reflects their high degree of oxidation and hydration. CI-chondrites consist mainly of amorphous hydrated silicates, minerals of the serpentine group, and sulfates. In other carbonaceous chondrites the amounts of hydrated silicates are essentially less, and the dominant minerals are olivine and pyroxene.

Characteristic features of carbonaceous chondrites are low abundances of metallic iron and troilite, and large variations in mineral chemistry.

Most carbonaceous chondrites (except the CI-type)) contain white inclusions characterized by high concentrations of calcium, aluminum, and other refractory elements. These inclusions are distinguished by their peculiar textures, anomalies in the isotopic composition of many elements, and complex mineralogical composition (simultaneous presence of both high-temperature minerals – hibonite, spinel, melilite, perovskite, and low-temperature minerals – nepheline, sodalite, and others). High abundances in these Ca, Al-rich inclusions (CAIs) of elements, the condensation temperature of which is very high, provide the foundation for the assumption that these inclusions represent high-temperature mineral assemblages which condensed from high-temperature nebula in the early stages of Solar System evolution (e. g., Wark and Lovering 1977; Grossman 1980). This hypothesis was adopted by many researchers; however, it has not been commonly accepted. It was also suggested that CAIs formed as residues from the selective evaporation of primary silicate materials (e.g., Kurat et al. 1975). More recently, these hypotheses have been modified to include processes involving multistage events of alteration (e.g., MacPherson 1988). However, the question of the sequence of these events is still the subject of considerable debate.

Recent data on several unusual meteorites with affinities to ordinary chondrites made the addition of new chondritic group necessary. Weisberg et al. (1991) termed these unusual meteorites Carlisle Lakes type chondrites. Recently, additional meteorites belonging to this group have been described. The relatively large number of these meteorites justifies the definition of a new group of chondrites. Schulze et al. (1994) suggest naming this new chondritic group R-chondrites (rumurutiites) after Rumuruti, the only fall; all other examples of these meteorites are finds from hot and cold deserts in Antarctica, Australia, and Africa. R-chondrites can be characterized as oxidized, olivine-rich, metal-poor chondrites. The chemical composition and oxygen isotopic composition of these meteorites distinguished them from the other chondrite groups; therefore, these chondrites do not belong to carbonaceous, ordinary, or enstatite chondrites, although they exhibit some similarities to ordinary chondrites.

About 10% of all stony meteorites are *achondrites.* The chemical composition of achondrites is not identical (see Table 6). Usually, achondrites are subdivided into Ca-rich (5–25 wt.% CaO) and Ca-poor (0.1–3 wt.% CaO) achondrites. In Table 4 we presented an additional three groups: primitive achondrites, lunar meteorites, and martian meteorites. The compositions and textures of the two main subgroups of achondrites – eucrites and howardites – are very similar to those of terrestrial igneous rocks. This suggested that achondrites are the products of magmatic differentiation on the meteorite parent bodies. On the other hand, many achondrites have breccia textures similar to those of some chondritic meteorites. Sometimes, plagioclase in achondrites is transformed into maskelenite. Breccia textures and the presence of maskelenite show that some achondrites have undergone impact metamorphic processes. In particular, in ureilites small crystal of diamonds were identified, whose origin may be connected with impacts. Some achondrites with igneous texture may have formed at crystallization of shock melts. Winonaites, lodranites, and acapulcoites are subgroups of the primitive achondrites, that have mineral and bulk compositions similar to those of chondrites, but have nonchondritic textures, and their oxygen isotopic compositions indicate that they were formed on different parent bodies.

Iron Meteorites

The number of iron meteorites takes second place among meteorites. The contents of nickel in them varies widely from 4 to 60%. Expect for Fe and Ni, the iron meteorites contain minor elements such as Co, S, Cu, Cr, and C. Irons are characterized by extremely low abundances of lithophile elements; their composition is rather peculiar and clearly differs from the solar abundances of elements. Probably, the matter of these meteorites was formed either during selective condensation and subsequent accretion processes, or as a result of magmatic differentiation in protoplanetary bodies. The irons consist of two main mineral phases: kamasite (4–7% Ni) and taenite (15–60% Ni). There are some inclusions of shreibersite, troilite, daubreelite, cohenite, graphite, and/or silicates, and phosphates.

The iron meteorites are divided into separate groups, based on the contents of nickel and minor elements (Ga, Ge, Ir, Pt), as well as the structural relationships of kamasite and taenite.

The most widespread group of irons is octahedrites, containing up to 14% Ni. They differ by a peculiar texture, known under the name of Widmanstatten structure: the system of parallel kamasite bands is located in parallel to the sides of the octahedron. The interstition between the plates is filled by taenite and plessite (a fine-grained mixture of kamasite and taenite).

The group of iron meteorites, distinguished by low contents of nickel (less than 6% Ni) and consisting almost wholly of kamasite monocrystals (sometimes more than 0.5 m in size), has received the name hexahedrites.

The third group, ataxites, has no ordered structure. The meteorites of this group differ by their high contents of nickel (more than 14% Ni) and consist of plessite (up to 27% Ni), taenite, and kamasite.

The contents and distribution of nickel in iron meteorites and their peculiar texture testify that this class of meteorite has undergone a melting stage in the history of iron formation. Iron meteorites have undergone numerous intensive shock loads and heating; as a result, there are shock deformations and a number of features of mineral texture in some irons.

On the basis of changes in nickel concentration between taenite and kamasite (on the assumption that the observable picture was formed as a result of nickel diffusion form taenite into kamasite), the cooling speed of iron meteorites can be calculated. From data of Levin and Simonenko (1977), in an interval from 700 up to 500 °C the cooling speed is about 2 °C for 1 Ma: from data of Wood (1971) 1–10 °C for 1 Ma at an interval from 600 to 400 °C. The radiuses of parent bodies are assumed in the order of 50–200 km.

Stony-Iron Meteorites

The class of stony-iron meteorites is an insignificant part of meteorites. Stony-iron meteorites consist of silicates and nickel-iron (in approximately quantities). Meteorites of this class differ by unique textures, representing nickel-iron fragments, cemented together by a silicate matrix, or essentially silicate fragments of different chemical and mineralogical compositions, cemented together by a mixture of nickel-iron and sulfides.

The metal in stony-iron meteorites is kamasite (taenite is rare), having an octahedrite structure. The silicate part basically consists of olivine, orthopyroxene, and plagioclase. Prevailing minerals are in the basis of division of stony-iron meteorites on several subclasses or groups (see Table 4). Some stony-iron meteorites are characterized by breccia texture of silicate parts. The structure and mineral composition of stony-iron meteorites testify that they have passed through a stage of magmatic differentiation.

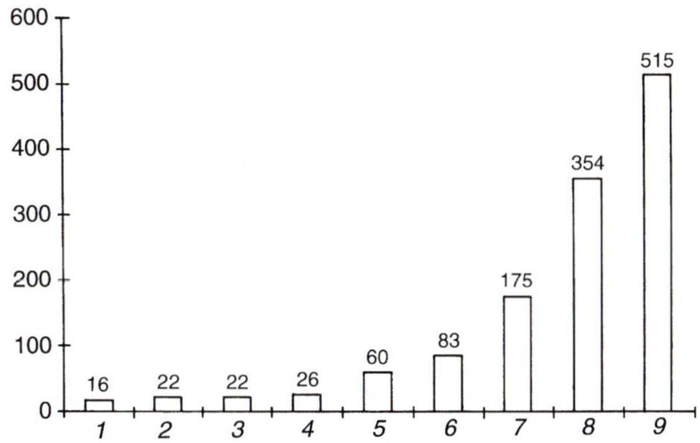

Fig. 17. Number of mineral phases and their varieties recognized in extraterrestrial materials. Data from: *1* Tschermak 1885; *2* Farrington 1915; *3* Heide 1934; *4* Krinov 1955; *5* Mason 1967; *6* Mason 1972; *7* Yudin and Kolomensky 1987; *8* Ulyanov 1991; *9* this chapter

1.8.3 Summary of Extraterrestrial Minerals

Today, more than 500 mineral phases and their varieties are recognized in extraterrestrial materials (meteorites, tectites, interplanetary dust particles, etc.) (Fig. 17).

Reviews of meteorite mineralogy were provided by Mason (1967, 1972), Yudin and Kolomensky (1987), Petaev (1988), Ulyanov (1991), and Yaroshevsky and Ulyanov (1994). These reviews and abstracts of *Lunar and Planetary Science Conferences, Annual Meetings of the Meteoritical Society, Symposiums on Antarctic Meteorites, Russian Meteorite Conferences*, and some publications on meteorite mineralogy in scientific journals (mainly Geochimica et Cosmochimica Acta, Meteoritics, Geochimiya, Meteoritika, etc.) were used to prepare the following list of extraterrestrial minerals. This comprehensive review of space mineralogy included as non-doubtful minerals (normal font in tables) as doubtful minerals, varieties of mineral phases, incompletely described mineral phases (italic font in remarks) in meteorites, tectites, interplanetary dust particles, and cosmic spherules. Table 7 and remarks also include minerals produced by terrestrial weathering of meteorites.

1.8.4 Interstellar Mineral Grains in Meteorites

Some chondrites (so-called primitive meteorites) contain submicroscopic grains of diamond, graphite, SiC, and some other mineral phases. From the extremely anomalous isotopic ratios, originally of noble gases, it is presumed that these grains have an extrasolar system origin and many have formed within stellar atmospheres (e.g. Ott 1993; Zinner 1995). Therefore their formation

Table 7. Native Elements

Mineral	Formula	Crystal system
Kamasite	α-(Fe,Ni)	Cubic
Taenite	γ-(Ni-Fe)	Cubic
Awaruite	Ni_2Fe-Ni_3Fe	Cubic
Wairauite	CoFe	Cubic
Epsilon-iron	ε-(Fe,Ni)	Hexagonal
Tetrataenite	FeNi	Tetragonal
Nickel	Ni	Cubic
Zhanghenite	CuZn	Cubic
Zincum	(Zn, Fe)	Hexagonal
Platinum	Pt	Cubic
Osmium	(Os,Ru)	Hexagonal
Rhenium	(Re, Ru)	Hexagonal
Iridium	Ir	Cubic
Titanium	Ti	Hexagonal
Copper	Cu	Cubic
Sulfur	S	Orthorhombic
Diamond	C	Cubic
Graphite	C	Hexagonal
Lonsdaleite	C	Hexagonal
Chaoite	C	Hexagonal

Fe-metal (Fe,Ni); *α-Fe* (Fe,Ni); *Si-rich Fe,Ni-Metal* (Fe,Ni,Si) (up to 15% Si); *Co-rich kamasite* (up to 7.6% Co; Co > Ni); *Si-rich kamasite* (up to 3.8% Si); *high-Co, low-Ni metal phase* (FeCo-Fe_3Co_2); *Ge-rich awaruite* up to 1.5% Ge; *Ni,Cr-alloy* (Ni,Cr); *Ni,Cu-alloy* (Ni,Cu); *(Zn,Fe)* up to 4% Fe; *$(Fe,Ni)Zn_{13}$, $(Fe,Ni)Zn_{10}$; $(Fe,Ni)Zn_7$, $(Fe,Ni)_3Zn_{10}$; $(Fe,Ni)_3Zn_7$; Fe,Pt-alloy* (Fe,Ni,Pt); *ferroan platinum* PtFe; *Fe,Ni,Ir,Os,Ru-alloy; Pt-alloy* (Pt,Fe,Ni); *Pt,Ir-alloy* (Pt,Ir,Fe,Ni); *Os,Ru-alloy* (Os,Ru); *Ru,Os-alloy* (Ru,Os); *Ru,Os,Fe,Ni,Re-alloy; Os, Ru, Ir, Re, Mo,W-alloy; Ni,Pt-alloy* (Ni,Pt,Fe); *Fe,Ir,As,Mo,Pt-alloy; Mo,Fe,Ni-alloy; native gold* Au; *Au,Ag,Fe-alloy; Mo,Fe,Ni-alloy* (Fe,Ni,Mo); *vanadium* V; *phosphorus* P; *carbon* C; *cliftonite* C, *rustenburgite* Pt_3Sn.

Carbides, nitrides, oxynitrides, phosphides, silicides, and silicophosphides

Cohenite	$(Fe,Ni)_3C$	Orthorhombic
Moissanite	SiC	Hexagonal
Haxonite	$(Fe,Ni)_{23}C_6$	Cubic
Osbornite	TiN	Cubic
Carlsbergite	CrN	Cubic
Roaldite	$(Fe,Ni)_4N$	Cubic
Sinoite	Si_2N_2O	Orthorhombic
Barringerite	$(Fe,Ni)_2P$	Hexagonal
Schreibersite	$(Fe,Ni)_3P$	Tetragonal
Perryite	$(Ni,Fe)_5(Si,P)_2$	Cubic
Suessite	$(Fe,Ni)_3Si$	Cubic

Epsilon-carbide; Fe_3C_2; SiC; TiC; Zr,Mo,Ti-carbide; chalipite Fe_2C; *Ti_2N_2O; Cr-cohenite* $(Fe,Ni,Cr)_3C$; *Fe,Cr-carbide* $(Fe,Cr)_2C$; *Si_3N_4; Fe-barringerite* Fe_2P; *Ni-barringerite* $(Ni,Fe)_2P$; *Cr-barringerite* $(Fe,Cr,Ni)_2P$; *rhabdite* $(Fe,Ni)_3P$; *Fe-Ni-silicide* $(Fe,Ni)_2(Si,P)$; *Fe,Cr,Ni-phosphide* $(Fe,Ni)_5CrP_3$; *Fe,Ti,Ni-phosphide* (Fe,Ni) TiP.

Table 7 (continued)

Sulfides, tellurides, and sulfotellurides		
Pentlandite	$(Fe,NI)_9S_8$	Cubic
Troilite	FeS	Hexagonal
Mackinawite	$(Fe,Ni)_9S_8$	Tetragonal
Pyrrhotite	$Fe_{1-x}S$ (x = 0–0.17)	Monoclinic, hexagonal
Pyrite	FeS_2	Cubic
Bravoite	$(Fe,Ni)S_2$	Cubic
Molybdenite	MoS_2	Hexagonal
Tungstenite	WS_2	Hexagonal
Heazlewoodite	Ni_3S_2	Trigonal
Cinnabar	HgS	Trigonal
Covellite	CuS	Hexagonal
Sphalerite	ZnS	Cubic
Alabandite	MnS	Cubic
Oldhamite	CaS	Cubic
Niningerite	(Mg,Fe)S	Cubic
Chalcopyrite	$CuFeS_2$	Tetragonal
Bornite	Cu_5FeS_4	Cubic
Cubanite	$CuFe_2S_3$	Orthorhombic
Violarite	$FeNi_2S_4$	Cubic
Daubreelite	$FeCr_2S_4$	Cubic
Millerite	NiS	Trigonal
Heideite	$(Fe,Cr)_{1+x}(Ti,Fe)_2S_4$	Monoclinic
Brezinaite	Cr_3S_4	Monoclinic
Caswellsilverite	$NaCrS_2$	Trigonal
Djerfisherite	$K_6(Fe,Cu,Ni)_{24}S_{26}Cl$	Cubic
Cooperite	PtS	Tetragon
Laurite	$(Ru,Os,Ir)S_2$	Cubic
Erlichmanite	$(Os,Ru,Ir)S_2$	Cubic
Chalcosine	Cu_2S	Monoclinic
Moncheite (chengbolite)	$PtTe_2$	Trigonal
Altaite	PbTe	Cubic

Co-pentlandite $(Fe,Ni,Co)_9S_8$ (up to 3.5% Co; Co>Ni); *Cr-pentlandite* $(Fe,Ni,Cr)_9S_8$; *Cu-pentlandite* (up to 13% Cu); *Ti-troilite* (Fe,Ti)S up to 14% Ti; *Mn-troilite* (up to 1.5% Mn); *Zn-troilite* (up to 4.4% Zn); *Ni-troilite* (up to 2.5% Ni);); *Cr-troilite* up to 13% Cr; *Cr,Ni-troilite* (up to 15% Cr and 8% Ni); *Cr,Ni-pyrrhotite* (up to 1.9% Cr and 2.7% Ni); *marcasite* FeS_2; *Fe-sphalerite (= marmatite)* up to 35% Fe; *Ga-sphalerite* (up to 3.7% Ga); *Ferroalabandite* (Mn,Fe)S; *Ag-alabandite* (Mn,Fe, Ag)S (up to 10% Ag); *Cr-alabandite* up to 4% Cr; *ferroniningerite* (Fe,Mg)S; *Mn-niningerite* up to 32% Mn; *chalcopyrrhotite; Cu,Fe-sulfide* 48% Cu, 12% Fe; *Cr,Fe-sulfide* 14–452% Cr and 55–23% Fe; *gentnerite* $Cu_8Fe_3Cr_{11}S_{18}$; *Zn-daubreelite* $(Fe,Zn)Cr_2S_4$ up to 10% Zn; *V-daubreelite* $Fe(Cr,V)_2S_4$ up to 1.2% V; *Ag-daubreelite* up to 0.75% Ag; *Mn-daubreelite* $(Fe,Mn)Cr_2S_4$ up to 3.5% Mn; *CrS; Cr_2S_3; $AgCrS_2$; $AgCr_2S_4$; Ag-sulfide; K,Cr,Ni,Fe sulfide* up to 3% K, 4% Cr, 18% Ni; *Ni_5GeS; Cr-sulfide* up to 30% Cr; *Na,Cr-sulfide; Na,Cu,Cr-sulfide* up to 5% Na, 34% Cr, and 15% Cu; *Na,Cu,Zn,Cr-sulfide* up to 2% Na, 33% Cr, 8% Cu, and 10% Zn; *Mn,Cu-sulfide; V-sulfide; Ti,Cr,Fe-sulfide* up to 40% Ti, 10% Cr, and 10% Fe; *wurtzite* ZnS; *Zn,Fe,Mn-sulfide* (Zn, Fe, Mn)S; *Fe,Zn,Ni,Cr-sulfide* ~50% Fe, 20–25% Zn, 2–3% Ni, and 1–3% Cr; *Fe,Pt,Ni-sulfide* $(Fe,Pt,Ni)_2S_3$ (?); *Pt,Ru,Ir,Os-sulfide* $(Pt,Ru,Ir,Os)_2S$ (?); *Fe,Ru,Ni,Pt,Ir-sulfide* $(Fe, Ru, Ni, Pt, Ir)_9S$ (?); *Mg-sulfide* MgS; *Pt,Au-telluride; Fe,Pt,Cu-sulfutelluride* $Cu(Fe,Pt)_3(S,Te)_3$ (?).

Table 7 (continued)

Oxysulfides and hydrated sulfides		
Tochilinite	$6Fe_{0.9}S*5(Mg,Fe)(OH)_2$	Triclinic
Schollhornite	$Na_{0.3}CrS*2H_2O$	Trigonal

Valleriite $4(Fe,Cu)S*3(Mg,Fe)(OH)_2$; *Fe,Ni-oxysulfide; hydrated Na,Cu,Cr-sulfide* up to 32% Cr, 18% Cu, and 1% Na; *hydrated Fe,Cr-sulfide* $FeCr_2S_4*nH_2O$ (?); *β-phase* $8FeS*9(Ca_{0.8}Fe_{0.2})O*0.3P_2O_5$; *hydrated Ca,Cr-sulfide.*

Oxides		
Cuprite	Cu_2O	Cubic
Periclase	MgO	Cubic
Wustite (= iozite)	FeO	Cubic
Bunsenite	NiO	Cubic
Maghemite	γ-Fe_2O_3	Tetragon
Corundum	Al_2O_3	Trigonal
hematite	α-Fe_2O_3	Trigonal
Escolaite	Cr_2O_3	Trigonal
Baddeleyite	ZrO_2	Monoclinic
α-Quartz	SiO_2	Trigonal
Crystobalite	SiO_2	Tetragonal
Tridimite	SiO_2	Monoclinic, Triclinic
Coesite	SiO_2	Monoclinic
Opal	SiO_2*nH_2O	Amorphous
Rutile	TiO_2	Tetragonal
Brookite	TiO_2	Orthorhombic
Thorianite	$(Th,U)O_2$	Cubic
Spinel	$MgAl_2O_4$	Cubic
Hercynite	$FeAl_2O_4$	Cubic
Coulsonite	FeV_2O_4	Cubic
Magnetite	$FeFe_2O_4$	Cubic
Pleonast	$(Mg,Fe)Al_2O_4$	Cubic
Trevorite	$NiFe_2O_4$	Cubic
Chromite	$(Fe,Mg)Cr_2O_4$	Cubic
Magnesiochromite	$(Mg,Fe)Cr_2O_4$	Cubic
Magnesioferrite	$MgFe_2O_4$	Cubic
Ulvospinel	$TiFe_2O_4$	Cubic
Alumochromite	$(Mg,Fe)(Al,Cr)_2O_4$	Cubic
Perovskite	$CaTiO_3$	Orthorhombic
Pyrophanite	$MnTiO_3$	Trigonal
Ilmenite	$FeTiO_3$	Trigonal
Geikilite	$MgTiO_3$	Trigonal
Grossite	$CaAl_4O_7$	Monoclinic
Hibonite	$CaAl_{12}O_{19}$	Hexagonal
Armalcolite	$(Mg,Fe)Ti_2O_5$	Orthorhombic
Zirconolite (= zirkelite)	$(Ca,Th,Ce)Zr(Ti,Nb)_2O_7$	Monoclinic
Pyrochlor	$(Ca,REE,Th)_2(Nb,Ti,Zr,Al)_2O_7$	Cubic

Fe-rich periclase (Mg,Fe)O; *magnesiowustite* (Fe,Mg)O; *Ni-maghemite* γ-$(Fe,Ni)_2O_3$; *Ti-rich corundum* $(Al,Ti)_2O_3$; *δ-Al_2O_3; V-oxide* V_2O_3; *Bioxide* Bi_2O_3; *pseudobrookite* $(Fe^{3+},Fe^{2+})_2$ $Fe^{3+},Ti)O_5$; *$ZrTiO_4$; Nb-rich rutile* up to 4% Nb_2O_5; *anatase* TiO_2; *cassiterite* SnO_2; *V-spinel $Mg(Al,V)_2O_4$* (up to 3% V_2O_3; *Cr-spinel* $Mg(Al,Cr)_2O_4$ (up to 17% Cr_2O_3); *Cr-hercynite* $Fe(Al,Cr)_2O_4$ (up to 7.7% Cr_2O_3); *Mn-magnetite* $(Fe,Mn)Fe_2O_4$ (up to 4% MnO); *Cr-magnetite* $Fe(Fe,Cr)_2O_4$ (up to 7% Cr_2O_3; *V-magnetite* $Fe(Fe,V)_2O_4$ (up to 22% V_2O_3); *Ti-magnetite* $(Fe,Ti)Fe_2O_4$; *Ni-magnetite* $(Fe,Ti)Fe_2O_4$; (up to 9% NiO); *Cr,V-magnetite* $Fe(Fe,V,Cr)_2O_4$ up to 8% Cr_2O_3 and 23% V_2O_3; *Cr-pleonast* $(Mg,Fe)(Al,Cr)_2O_4$ (up to 18% Cr_2O_3); *V-pleonast* $(Mg,Fe)(Al,V)_2O_4$ (up to 27% V_2O_3); *Zn-pleonast* $(Mg,Fe,Zn)Al_2O_4$ (up to 2.3% ZnO); *ferrochromite* $FeCr_2O_4$; *alumoferrochromite* $Fe(Cr,Al)_2O_4$; *Mn-chromite* $(Fe,Mn)Cr_2O_4$; *Ti-chromite* up to 10% TiO_2; *Ti-alumochromite; Al-magnesiochromite; Mn-magnesiochromite; Zn,Mn-magnesiochromite; ScY,REE-perovskite* $(Ca,Sc,Y,REE)TiO_3$ (up to 6% Y_2O_3 and 1.6% Sc_2O_3); *V-perovskite* $Ca(Ti,V)O_3$ (up to 1.5% V_2O_3); *Nb-rich perovskite* $Ca(Ti,Nb)O_3$; *V-ilmenite* $Fe(Ti,V)O_3$ (up to 2% V_2O_3); *Mn-ilmenite* $(Fe,Mn)TiO_3$ (10–20% MnO); *Mg-ilmenite* $(Fe,Mg)TiO_3$ (7–10% MgO); *Mn,Mg-ilmenite* $(Fe,Mn,Mg)TiO_3$ (up to 19% MnO and 10% MgO); *V-rich hibonite* up to 3% V_2O_3; *$CaAl_2O_4$; $CaAl_{24}O_{37}$; V-rich aluminate* $Ca_4(Al,V)_{20}O_{40}$; *Sc-aluminate; Ti_4O_7; Ti_5O_9; Ti_6O_{13}; Ti_8O_{15}; Ti_9O_{17}; $Ti_{10}O_{19}$; Mg-armalcolite* $(Mg,Fe)Ti_2O_5$; *Cr,Zr -armalcolite* up to 9% Cr_2O_3 and 1.2% ZrO_2; *Al_2TiO_5; zirkelite-like mineral; Ca,Al,Ti,Mg-oxide; Ti,Ca,Mg, Al-oxide; Ti,Ca,Sc,Zr-oxide; Zr,Y-oxide; V-oxide* V_2O_3; *$(Ti,Zr,Si)(Al,Ti)_2O_5$; mineral-T* $(Ca,Mg)Ti_3O_7$; *Ti,Ca,Mg,Al-oxide; Ti,Ca,Sc,Zr-oxide; Ca,Al,Ti,Mg-oxide; Ca,V-oxide.*

Hydrated oxides and hydrooxides

Goethite	α-FeO(OH)	Orthorhombic
Akaganeite	β-FeO(OH,Cl)	Tetragonal
Lepidocrokite	γ-FeO(OH,Cl)	Orthorhombic
Brucite	$Mg(OH)_2$	Tetragonal
Gibbsite	$Al(OH)_3$	Orthorhombic
Ferrihydrite	$5Fe_2O_3*9H_2O$	Hexagonal
Portlandite	$Ca(OH)_2$	Hexagonal

S-Ni-rich ferrhydrite; $Ni(OH)_2$; gell of Fe-hydroxide; hollandite $(Fe_{15}Ni)(O_{12}(OH)_{20})Cl(OH)_2$.

Sulfates, phosphates, carbonates, silicophosphates, wolframates, molybdates, vanadates

Anhydrite	$CaSO_4$	Orthorhombic
Barite	$BaSO_4$	Orthorhombic
Epsomite	$MgSO_4*7H_2O$	Orthorhombic
Gypsum	$CaSO_4*2H_2O$	Monoclinic
Bloedite	$Na_2Mg(SO_4)_2*4H_2O$	Monoclinic
Melanterite	$FeSO_4*7H_2O$	Monoclinic
Copiapite	$FeFe_4(SO_4)_6(OH)_2*20H_2O$	Triclinic
Jarosite	$KFe_3(SO_4)_2(OH)_6$	Triclinic
Hexahydrite	$MgSO_4*6H_2O$	Monoclinic
Buchwaldrite	$NaCaPO_4$	Orthorhombic
Brianite	$Na_2CaMg(PO_4)_2$	Monoclinic
Panetite	$(Na,Ca,K)_2(Mg,Fe,Mn)_2(PO_4)_2$	Monoclinic
Farringtonite	$Mg_3(PO_4)_2$	Monoclinic
Sarcopside	$(Fe,Mn,Mg)_3(PO_4)_2$	Monoclinic
Beusite	$Fe_{1.5}Mn_{1.5}(PO_4)_2$	Monoclinic
Graftonite	$(Fe,Mn,Ca)_3(PO_4)_2$	Monoclinic

Table 7 (continued)

Sulfates, phosphates, carbonates, silicophosphates, wolframates, molybdates, vanadates		
Stanfieldite	$Ca_4Mg_3Fe_2(PO_4)_6$	Monoclinic
Whitlokite (= merrillite)	$Na_2Ca_{18}Mg,Fe)_2(PO_4)_{14}$	Trigonal
Chladniite	$Na_2CaMg_7(PO_4)_6$	Hexagonal
Maricite	$FeNaPO_4$	Orthorhombic
Collinsite	$Ca_2(Mg,Fe,Ni)(PO_4)_2*2H_2O$	Triclinic
Cassidyite	$Ca_2(Ni,Mg)(PO_4)_2*2H_2O$	Triclinic
Lipscombite	$(Fe,Mn)Fe_2(PO_4)_2(OH)_2$	Tetragonal
Vivianite	$Fe_3(PO_4)_2*8H_2O$	Monoclinic
Monazite	$CePO_4$	Monoclinic
Arupite	$Ni_3(PO_4)_2*8H_2O$	Monoclinic
Chlorapatite	$Ca_5(PO_4)_3Cl$	Hexagonal
Fluorapatite	$Ca_5(PO_4)_3F$	Hexagonal
Calcite	$CaCO_3$	Trigonal
Vaterite	$CaCO_3$	Hexagonal
Magnesite	$MgCO_3$	Trigonal
Siderite	$FeCO_3$	Trigonal
Rhodochrosite	$MnCO_3$	Trigonal
Dolomite	$CaMg(CO_3)_2$	Trigonal
Ankerite	$CaFe(CO_3)_2$	Trigonal
Breunnerite	$(Fe,Mg)CO_3$	Monoclinic
Nyerereite	$Na_2Ca(CO_3)_2$	Orthorhombic
Barringtonite	$MgCO_3*2H_2O$	Triclinic
Hydromagnesite	$Mg_5(CO_3)_4(OH)_2*4H_2O$	Monoclinic
Nesquehonite	$Mg(HCO_3)(OH)*2H_2O$	Monoclinic
Reevesite	$Ni_6Fe_2(CO_3)(OH)_{14}*4H_2O$	Trigonal
Sheelite	$CaWO_4$	Tetragonal
Powellite	$CaMoO_4$	Tetragonal

Bassanite $2CaSO_4*H_2O$; *Fe,Ni-sulfate; Mg-sulfate; Fe-merrillite* (Na,Ca) $Ca_{18}(Fe,Mg)_2$-$(PO_4)_{14}$; *Ca-merrillite* $(Ca,Na)Ca_{18}(Mg,Fe)_2(PO_4)_{14}$; *REE-merrillie* Ca_{18} $(Y,REE)_2$ $(Mg,Fe)_2$-$(PO_4)_{14}$; *Na,Fe-phosphate* $(Fe,Na,Mn)_3(PO_4)_2$ (up to 5% Na_2O); *Na,Fe-phosphate* $Na(Mg,Fe)_{2.5}(PO_4)_2$; *Fe-phosphate; Mg,Na,K-phsophate* up to 7% Na_2O and 5% K_2O; *Mg,Ca,Fe,Na-silicophosphate; Fe,Mn,Na,K-phosphate* $(Na,K)_2(Fe,Mn)_8(PO_4)_6$; *Fe,Mn,K,Na-phosphate* $(Na,K)_2(Fe,Mn)_8(PO_4)_6$; *Mg,Ni,K-phosphate; Na,Mg-phosphate* $Na_{1+x}Mg(PO_4)F$; *Na,Ti-phosphate* $Na_2Ti(PO_4)_2$; *Ca-phosphate; hydroxylapatite* $Ca_5(PO_4)_3(OH)$; *aragonite* $CaCO_3$; *Mn-siderite* $(Fe,Mn)CO_3$; *Mn-dolomite* $Ca(Mg,Mn)(CO_3)_2$; *hydrophase II* $Mg(HCO_3)$-$(OH)*6H_2O$; *zaratite* $Ni_3(CO_3)(OH)_4*4H_2O$; *molybdosheelite* $Ca(W,Mo)O_4$; *Ca-vanadate; Ru-rich Ca,Fe-molybdate-phosphate; Fe-molybdate; Fe-molybdate-vanadate; Fe-vanadate.*

Orthosilicates		
Forsterite	$Mg_2[SiO_4]$	Orthorhombic
Olivine	$(Mg,Fe)_2[SiO_4]$	Orthorhombic
Fayalite	$Fe_2[SiO_4]$	Orthorhombic
Montichellite	$CaMg[SiO_4]$	Orthorhombic
Kirschsteinite	$CaFe[SiO_4]$	Orthorhombic
Almandine	$Fe_3Al_2[SiO_4]_3$	Cubic
Grossular	$Ca_3Al_2[SiO_4]_3$	Cubic
Andradite	$Ca_3Fe_2[SiO_4]_3$	Cubic
Majorite	$Mg_3(Fe,Si)_2[SiO_4]_3$	Cubic
Zircon	$Zr[SiO_4]$	Tetragonal

Table 7 (continued)

Ringwoodite	γ-$(Mg,Fe)_2[SiO_4]$	Cubic
Wadsleyite	β-$(Mg,Fe)_2[SiO_4]$	Orthorhombic
Kyanite	$Al_2[SiO_4]O$	Triclinic
Sphene (= titanite)	$CaTi[SiO_4]O$	Monoclinic
Akermanite	$Ca_2Mg[Si_2O_7]$	Tetragonal
Melilite	$Ca_2(Mg,Al)[(Si,Al)SiO_7]$	Tetragonal
Gehlenite	$Ca_2Al[AlSiO_7]$	Tetragonal
Cordierite	$(Mg,Fe)_2Al_3[AlSi_5O_{18}]$	Orthorhombic
Roedderite	$(Na,K)_2(Mg,Fe)_5[Si_{12}O_{30}]$	Hexagonal
Merrihueite	$(Na,K)_2(Fe,Mg)_5[Si_{12}O_{30}]$	Hexagonal
Osumulite	$(K,Na)(Fe,Mg)_2(Al,Fe)_2(Al,Fe)_3[(Si,Al)_{12}O_{30}]$	Hexagonal
Yagiite	$(Na,K)_{1.5}Mg_2(Al,Mg,Fe)_3[(Si,Al)_{12}O_{30}]$	Hexagonal

Mn-forsterite $(Mg,Mn)_2[SiO_4]$ (up to 2.7% MnO); *Cr-forsterite* up to 0.65% Cr_2O_3; *Ca-rich olivine* up to 12% CaO; *Ni-olivine* up to 0.7% NiO; *Cr-olivine* up to 1.4% Cr_2O_3; *P-olivine*-$(Mg,Fe)_2[(Si,P)O_4]$ l(up to 5% P_2O_5); *Fe-montichellite* $Ca(Mg,Fe)[SiO_4]_3$; *Mg-kirschsteinite* $Ca(Fe,Mg)[SiO_4]$; *grossular-andradite* $Ca_3(Al,Fe)_2(SiO_4)$; *hydrogarnet; cocharite* Mg_3Fe_2-$[SiO_4]_3$; *pyrope* $Mg_3Al_2[SiO_4]_3$; *humite* $3Mg_2SiO_4*Mg(F,OH)_2$; *vesuvianite* $Ca_{10}(Mg, Fe)_2 Al_4$-$[SiO_4]_5[Si_2O_7]_2(OH,F)_4$; *pumpellyite* $Ca_2(Al,Fe)_3[SiO_4]$ $[Si_2O_7](OH)_2*nH_2O$; *epidot* Ca_2Al_2-$FeSi_3O_{12}(OH)$; *prehnite* $Ca_2Al_2Si_3O_{10}(OH)$; *Na-melilite* $(Ca,Na)_2(Mg,Al)[(Si,Al)SiO_7]$; *Ca,Ti,Al-silicate* $Ca_3Ti(Al,Ti)_2[(Si,Al)_3O_{14}]$.

Pyroxenes, pyroxenoides and amphiboles

Clinoenstatite	$Mg_2[Si_2O_6]$	Monoclinic
Clinobronsite	$(Mg,Fe)_2[Si_2O_6]$	Monoclinic
Clinohypersthene	$(Fe,Mg)_2[Si_2O_6]$	Monoclinic
Pigeonite	$(Mg,Fe,Ca)(Mg,Fe)$ $[Si_2O_6]$	Monoclinic
Augite	$(Ca,Na)(Mg,Fe,Al,Ti)[(Si,Al)_2O_6]$	Monoclinic
Diopside	$CaMg[Si_2O_6]$	Monoclinic
Omphacite	$(Ca,Na)(Mg,Al)[Si,Al)_2O_6]$	Monoclinic
Salite	$Ca(Mg,Fe)[Si_2O_6]$	Monoclinic
Ferrosalite	$Ca(Fe,Mg)[Si_2O_6]$	Monoclinic
Hedenbergite	$CaFe[Si_2O_6]$	Monoclinic
Fassaite	$(Ca,Mg,Fe)(Mg,Fe,Al)[(Si,Al)_2O_6]$	Monoclinic
Jadeite	$Na(Al,Fe)[Si_2O_6]$	Monoclinic
Kosmochlor (= ureyite)	$NaCr[Si_2O_6]$	Monoclinic
Enstatite	$Mg_2[Si_2O_6]$	Orthorhombic
Bronzite	$(Mg,Fe)_2[Si_2O_6]$	Orthorhombic
Hypersthene	$(Fe,Mg)_2[Si_2O_6]$	Orthorhombic
Ferrohyperstene	$(Fe,Mg)_2[Si_2O_6]$	Orthorhombic
Wollastonite	$Ca_3[Si_3O_9]$	Triclinic
Rhodonite	$CaMn_4[Si_5O_{15}]$	Triclinic
Rhonite	$Ca_2(Mg,Fe)_4FeTiO_2[Si_3Al_3O_{18}]$	Triclinic
Krinovite	$Na_2Mg_4Cr_2O_2[Si_6O_{18}]$	Triclinic
Sapphirine	$(Mg,Al)_7(Mg,Al)O_2[Al,Si)_6O_{18}]$	Triclinic
Pyroxferroite	$(Fe,Mn,Ca)_7[Si_7O_{21}]$	Triclinic
Richterite	$Na_2Ca(Mg,Fe)_5[Si_8O_{22}](F,OH)_2$	Monoclinic
Kaersutite	$NaCa_2(Mg,Fe)_4Ti[Si_6Al_2O_{22}](OH)O$	Monoclinic

Table 7 (continued)

Cr-clinoenstatite up to 0.7% Cr_2O_3; *Al-clinohypersthene; Mn-augite; ferroaugite; Cr,Mn-augite* up to 2.7% Cr_2O_3 and 2.6% MnO; *Mn-enstatite* up to 19% MnO; *Al-enstatite* up to 7.5% Al_2O_3; *Ni-bronsite* up to 0.3% Ni; *Cr-hyperstene* up to 2.5% Cr_2O_3; *Cr-diopside* up to 2.3% Cr_2O_3; *Fe-fassaite; Mg-fassaite; Cr-fassaite* up to 2.4% Cr_2O_3; *Sc-fassaite* up to 18% Sc_2O_3 and 8% ZrO_2; *Al,Ti-fassaite; V-fassaite* up to 10.3% V_2O_3; *Ti-fassaite* up to 18% TiO_2; *acmite* $Na(Al,Fe)[Si_2O_6]$; *aegirine* $NaFe[Si_2O_6]$; *magnesiorichterite* $Na_2Ca(Mg,Fe)_5[Si_8O_{22}]$-$(F,OH)_2$; *ferrokaersutite* $NaCa_2(Fe,Mg)_4Ti[Si_6Al_2O_{22}](OH)O$; *mullite* $Al[Al_xSi_{2-x}O_{5.5-0.5x}]$.

Phyllosilicates

Kaolinite	$Al_4[Si_4O_{10}](OH)_8$	Triclinic
Serpentine (= chrysotile)	$Mg_6[Si_4O_{10}](OH)_8$	Monoclinic
Lizardite	$Mg_6[Si_4O_{10}](OH)_8$	Trigonal and hexagonal
Antigorite	$Mg_6[Si_4O_{10}](OH)_8$	Monoclinic
Clinochrysotile	$Mg_6[Si_4O_{10}](OH)_8$	Monoclinic
Berthierine	$(Fe,Mg)_{4-6}(Si,Al)_4O_{10}(OH)_8$	Monoclinic
Talc	$Mg_3[Si_4O_{10}](OH)_2$	Monoclinic and triclinic
Pyrophyllite	$Al_2[Si_4O_{10}](OH)_8$	Monoclinic and triclinic
Montmorillonite	$(Na,Ca)_{0.3}(Al,Mg)_2[Si_4O_{10}](OH)_2{*}nH_2O$	Monoclinic
Saponite (= kerolite)	$(Ca,Na)_{0.3}(Mg,Fe)_3[Si_4O_{10}](OH)_2{*}nH_2O$	Monoclinic
Clinochlor	$(Mg,Fe)_5Al[AlSi_3O_{10}](OH)_8$	Monoclinic
Chamosite	$(Fe,Mg)_5Al[AlSi_3O_{10}](OH)_8$	Monoclinic
Kronstedtite	$Fe_4Fe_2[Fe_2Si_2O_{10}](OH)_8$	Monoclinic and trigonal
Greenalite	$Fe_6[Si_4O_{10}](OH)_8$	Monoclinic
Clintonite	$Ca(Mg,Al)_3[Al_3SiO_{10}](OH)_2$	Monoclinic
Margarite	$CaAl_2[Al_2Si_2O_{10}](OH,F)_2$	Monoclinic

Alumoserpentine $(Mg,Al)_{6-x}[Si_4O_{10}](OH)_8$; *Ni-serpentine* $(Mg,Ni)_6[Si_4O_{10}](OH)_8$; *pecoraite* $Ni_6[Si_4O_{10}](OH)_8$; *ferroantigorite* $(Mg,Fe)_6[Si_4O_{10}](OH)_8$; *metahalloysite* $Al_4[Si_4O_{10}](OH)_8$; *Na-talc; sericite* $KAl_2[AlSi_3O_{10}](OH,F)_2$; *muscovite* $KAl_2[AlSi_3O_{10}](OH,F)_2$; *phlogopite* $KMg_3[AlSi_3O_{10}](OH,F)_2$; *Na-phlogopite* $(Na,K)Mg_3[AlSi_3O_{10}](OH)_2$; *seladonite* $K(Fe,Al)$-$(Mg,Fe)(OH)_2[Si_4O_{10}]{*}nH_2O$; *vermiculite* $(Mg,Fe,Al)_3[(Al,Si)_4O_{10}](OH)_2{*}4H_2O$; *Na-saponite* $(Na, Ca)_{0,3}(Fe,Mg)_3[Si,Al)_4O_{10}](OH)_2{*}nH_2O$; *ferrosaponite* $(Ca,Na)_{0.3}(Fe,Mg)_3[(Si,Al)_4O_{10}]$-$(OH)_2{*}nH_2O$; *pennine; ripidolite; hysingerite* $Fe_4Si_4O_{10}(OH)_8$ $*4H_2O$; *palygorskite* $(Mg,Al)_2$-$Si_4O_{10}(OH){*}4H_2O$; *sepiolite* $Mg_4Si_6O_{15}(OH)_2{*}6H_2O$.

Alumosilicates of the frame structures

Albite	$Na[AlSi_3O_8]$	Triclinic
Oligoclase	$Ab_{90-70}An_{10-30}$	Triclinic
Andesine	$Ab_{70-50}An_{30-50}$	Triclinic
Labradore	$Ab_{50-30}An_{70-50}$	Triclinic
Bitownite	$Ab_{10-30}An_{90-70}$	Triclinic
Anortite	$Ca[Al_2Si_2O_8]$	Triclinic
Celsian	$Ba[Al_2Si_2O_8]$	Monoclinic
Nepheline	$(Na,K)[AlSiO_4]$	Hexagonal
Sodalite	$Na_8[AlSiO_4]_6Cl_2$	Cubic

Table 7 (continued)

K-feldspar; anortoclase; sanidine $K[AlSi_3O_8]$; *kalsilite* $K[AlSiO_4]$; *maskelenite* (Na,Ca)-$[Al(Al,Si)Si_2O_8]$; *scapolite* $(Na,Ca)_4[Al(Al,Si)Si_2O_8](Cl,CO_3)$; *meionite* $Ca_4(Al_2Si_2O_8)CO_3$; *5-1-zeolite; 4-4-1-zeolite; S6r/D6R-zeolite; chabasite* $(Ca,Na)_2[Al_2Si_4O_{12}]*6H_2O$; *ferrierite* $(Na,K)_2Mg[(Si,Al)_{18}O_{36}](OH)*9H_2O$; *stilbite* $NaCa_2Al_5Si_{13}O_{36}*14H_2O$.

Halides

Halite	NaCl	Cubic
Sylvite	KCl	Cubic
Lawrencite	$(Fe,Ni)Cl_2$	Trigonal
Molysite	$FeCl_3$	Hexagonal

Organic minerals

Weddellite	$CaC_2O_4*H_2O$	Tetragonal

Mineral Phases with Unknown Structures

Ni-rich Mg,Ca,Al,Na,K-silicate; Na,Mg,Ca,Fe,Al-silicate; Hydrated Ni-phillosiliate; Na,-K,Al-silicate $(K,Na)[AlSi_5O_{12}]$; *Fe,Ca,Al-silicate; Cr-silicate* $Ca_{0.5}Mg_{0.5}Fe_{0.2}Cr_{0.5}[SiO_4]$, *Na,Ti-silicate.*

Mineral Mixtures and Glasses

Glasses; lechatelierite; iddingsite; plessit. HAP (= *h*igh *a*luminum *p*hyllosilicate); *LAP* (= *l*ow *a*luminum *p*hyllosilicate); *PCP* [= *p*oor (or *p*artly) *c*haracterized *p*hase].

environment was plasma (Cassidy and Kern 1993) – an electrically neutral gas containing a high proportion of its component molecules in the ionized state, with charge balance satisfied by the presence of free electrons.

Until now diamond, graphite, chaoite, SiC, TiC, corundum, spinel, and Si_3N_4 have been identified as interstellar grains in meteorites. Early work on interstellar material in meteorites has been reviewed by Anders (1988); for recent reviews, see Anders and Zinner (1993) and Ott (1993). Many interstellar grains in meteorites are presolar and have extremely anomalous isotopic compositions which indicate circumstellar origin. Type II supernova have previously been proposed as stellar sources of these presolar grains (Amari et al. 1994a, Zinner et al. 1995). Meyer et al. (1995) report yields for a large number of isotopes in eight distinct zones in the ejecta form a 25 M⊙ supernova.

High $^{12}C/^{13}C$ and $^{16}O/^{17}O$ and low $^{16}O/^{18}O$ ratios in some graphite and SiC grains show the signature of the He/C zone in presupernona stars. The dominant components for isotopically anomalous noble gases in these grains show the signature of the *s*-process (neutron capture on a *s*low time scale) and apparently come from AGB (*A*symptotic *G*iant *B*ranch) stars (Amari et al. 1994b). Moreover, two extreme types of *s*-Kr imply two kinds of AGB-stars with different mean neutron exposures. In additional to neon from AGB-stars, He-burning shells of interstellar graphite grains also contain monoisotopic ^{22}Ne from the β^+-decay of ^{22}Na ($t_{1/2} = 2.58$ a), which was chemically trapped in grains. The short half-life of ^{22}Ne requires a short time scale, and suggests the presence in interstellar graphite grains of material form nova and supernova ejects. Thus, interstellar graphite includes contributions from different stars and their shells.

Additional evidence of the presence of nova and supernova materials in interstellar grains comes from the large variety and unusual isotopic compositions of potassium and calcium, reflecting their stellar sources. Some interstellar grains have ^{41}K excesses due to the decay of ^{41}Ca($t_{1/2} = 1.03 \times 10^5$ a) and ^{44}Ca excesses due to the decay of short-lived ^{44}Ti ($t_{1/2} = 52$ a). Evidence for ^{41}Ca and ^{44}Ti in some interstellar grains strongly suggests mixing between the C-rich and O-rich zones and contributions from the innermost zone to the supernova ejecta from which the grains formed. Thus, the wide range of isotopic ratios for different chemical elements in interstellar grains provides additional evidence for extensive and deep heterogeneous mixing in supernova explosions, confirming astronomical observations and hydrodynamic calculations. Moreover, isotopic data suggest that interstellar grains in primitive meteorites include contributions from different known star sources: AGB-stars, supernovae, novae, and perhaps Wolf-Rayet (WR) stars (Amari et al. 1995).

Isotopic featuries and properties of interstellar grains in different types of primitive chondrites (Cl, CM, CV EH, LL, etc.) with the exception of grain size, are very similar. Differences in grain sizes of interstellar components may represent size sorting or selective destruction of fine-grained material in the nebula (Gao et al. 1995).

Since interstellar grains formed in stellar atmospheres, their investigation provides information on different processes (nuclear, chemical, etc.) in stars. The first kind of information to be obtained from isotopic studies of interstellar grains in meteorites is the number of stellar sources that admitted material into the solar system (e.g., Alexander 1993; Zinner 1995). The second is obtained "from single-grain isotopic data that either have no counterpart in astronomical observations or that cannot be explained by existing models of nucleosynthesis and thus provide stimulation for further theoretical works" (Zinner 1995; e.g., Meyer et al. 1995).

Interstellar grains were found not only in meteorites. Brilliant et al. (1992) studied the isotopic composition of nitrogen in lunar soil separates. They found that the nitrogen release and isotope profile of the major component corresponded exactly to that encountered when interstellar diamonds from primitive meteorites are analyzed before removal of contaminating organic material. Therefore, these authors tentatively interpret isotopic data as the first recognition of interstellar grains in the lunar regolith. Unfortunately, we know of no appropriate investigations to confirm this conclusion.

Isotopic data from Yates et al. (1992) strongly suggest that interstellar grains are in terrestrial sedimentary rocks (e.g., cryoconite – dark sediments from temporary ice lakes on the Greenland ice sheet; it contains both melted extraterrestrial sperules and unmelted micrometeorites). Thus, the sources of interstellar grains for laboratory investigations may be not only primitive chondrites.

1.8.5 Cosmic Spherules in Terrestrial Environments

Cosmic spherules are subspherical to spherical objects produced by melting *i*nterplanetary *d*ust *p*articles (IDPs) and large objects during atmospheric entry. Evidence that precursors of these spherules are extraterrestrial material has been presented by many investigators (see, e.g., Taylor and Brownlee 1991). Most of the cosmic spherules form the Antarctic and Greenland ice caps, deep-sea sediments, ancient oceanic deposits, and salt formations are melted IDPs. On the basis of mineral and chemical composition, IDPs are divided into three main types. ***Iron** spherules* (I-type) consist of magnetite crystal and interstitial wustite. Some of the iron spherules contain Fe-Ni metal beads and small Pt-group nuggets (Brownlee et al. 1984). *High-iron **glassy** spherules* (G-type) compositionally have over 50% FeO and are predominantly dendritic magnetite in a glassy matrix. Some of them also contain Fe-Ni metal beads. ***Stony** spherules* (S-Type) are the most common type and composed of olivine, magnetite, and glass. Bonte et al. (1987) found Pt-group nuggets in them. Work by Taylor and Brownlee (1991) on cosmic spherules with different ages and from a variety of sedimentary environments suggests that proportions of spherule types change as a function of time, with the number of iron spherules increasing with the age of the sample. However, it is not clear, if this variation in spherule types represents a real change in the meteoroid complex or is a result of differential weathering of spherules in the Earth's environment.

1.8.6 Stages in the Complex Processes of Meteorite Origin

A key problem in meteoritics is that of the origin of meteorites. In this question there are two points of view, which however, do not completely exclude one another. According to the first, meteorites are fragments of asteroids; according to the second, fragments of nucleis of "dead" comets.

Both points of view are based on the same facts – parameters of orbits and features of the chemical composition of matter. At present, the most convincing is the hypothesis of an asteroidal origin of meteorites.

One of the important questions in the problem of the origin of meteorites is the question of the character of the protomaterial of their parent bodies. The isotopic composition of bulk meteorites shows minimum differences from the suggested initial protoplanetary substance, and is primary; that is, the distribution of isotopes approximate averages their contents on the Sun and in processes of nucleo synthesis. The analysis of the chemical and isotopic composition of meteorites has allowed many researchers (e.g., Wood 1971; Wasson 1985; Heide Wlotzka 1995) to suggest that the composition of carbonaceous chondrites CI-type (except hydrogen, helium, and some other volatile components), as well as the ratios of major chemical elements (for example, Na/Ca, S/Ca) are close to solar values. This can be specified by genetic relationships between the substance of CI-chondrites and the Sun. The presence of volatile components and hydrous silicates in CI-chondrites permits considering the latter as matter not altered by essential changes and heat for the past 4.5×10^9 years, and reflecting the primary composition of gas-dust nebula. The chondritic meteorites of types other than CI were changed more significantly. Their textures, bearing bright attributes of thermal recrystallization, show that these types of chondritic materials were subjected to essential heating (possibly up to 800 °C) (Wood 1971). However, most chondrites did not pass through the stage of complete melting. On the contrary, the composition and texture of achondrites, stony-iron meteorites, and most irons, specify that they are probably products of a remelting process.

The systematization of the extensive data on meteorite research has allowed a general scheme of origin meteorites (Wood 1971, 1988; McSween 1987). Under this scheme, the parent bodies of meteorites are asteroids, formed as a result of condensation and aggregation of solid materials from gas-dust nebula and having a primary composition close to composition of carbonaceous chondrite.

At the beginning of geological history ($\sim 4.5 \times 10^9$ years ago), the internal parts of asteroids were subjected to strong heating (source of heat could be, for example, decay of ^{26}Al or ^{60}Fe), partial or complete melting, and igneous differentiation of substance. The external parts of parent bodies have not undergone heating. They were considerably heated up, but nonremelting took place in intermediate layers. Fragments of internal layers of asteroids are probably achondrites, stony-iron meteorites, and some irons. The chondritic meteorites represent fragments of an intermediate layer, and carbonaceous chondrites the material of the outside (unaltered) part of asteroids.

The formation of asteroids occurred in various parts of the asteroid belt, characterized by some features of chemical composition. These features of chemical composition can be determined in the observed composition of meteorites. This scheme of formation of meteorites is widely believed, but not unique. There are reasonable representations (Vinogradov 1971; Dodd 1986), that some separate classes and groups of meteorites could be formed directly as a result of condensation and agglomeration from high-temperature nebula.

There are several main stages in the history of formation of meteoritic materials. A complex sequence of events is recorded in the micro- and macrochemical, isotopic, and textural patterns of meteorite minerals. We can resolve the following discrete stages.

The interstellar stage. Atoms had an existence prior to the collapse of the interstellar material which formed the Sun and protosolar nebula. Some refractory objects and the matrix in some types of chondrites appear to preserve presolar grains with unusual isotopic patterns.

Solar nebula stage. The young Sun was surrounded by a rotating disk of gas and dust. Interstellar dust accreting to the young Sun was transformed at high temperature into the chondrules and some kind of inclusions.

Aggregation of chondrite components stage. After the chondrite components were formed, they aggregated into planetosimales. Accreting solids were not uniform in composition.

Stage of metamorphic alteration and igneous melting. After the accretion stage was finished, chondrite planetosimales were heated internally. Two types of alterations have been observed in different chondrites (Wood 1988): anhydrous metamorphism at relatively high temperature (700–1200 K) and low-temperature hydrothermal metamorphysm that altered mafic mineral phases into clay minerals. In the interior parts of planetosimales, solid materials may be melted. As a result of this process, achondritic magmas and other silicate-metal-sulfide melts may be generated.

Collisions of planetosimale stage. As a result of the collision of planetosimales, the latter grew in size and became asteroids.

Ejection stage. Some planetosimales were disturbed by shock events and small fragments were ejected into space.

Orbital evolution stage. Initially near-circular heliocentric orbits of these fragments were perturbed by the gravity of large planets, into elliptical orbits and were captured by Earth. The time during which meteorites existed as small fragments in space is typically a few tens of millions of years.

Fall to the Earth stage. During the atmospheric flight, the fragments were friction-heated and their surfaces were melted, but melted materials is ablated as fast as it is heated.

After falling to the Earth. The "terrestrial life" of meteorites begins. Less than 1% of fall meteorites have been collected. Most fall meteorites weather and become the substance of the Earth's crust.

Different minerals in space bodies are indicators of all of these stages in the history of the formation materials that we now call meteorites.

1.8.7 Meteorite Ages

The absolute ages of meteorites, dating the time of their chemical fractionation by melting in the parent bodies determined by ^{207}Pb-^{206}Pb, ^{87}Rb-^{87}Sr, or ^{187}Re-^{187}Os methods [4.4–4.7×10^9 years (Birk et al. 1975; Church 1975; Minster and Allegre 1976)], correspond to (at present) the accepted time of age of the Solar System, and are one of the proofs of the simultaneous origin of the Earth and the parent bodies of meteorites.

The ages of retaining rare gases in meteorites (or ages of cooling and crystallization of molten material), determined by $^{40}K/^{40}Ar$- and U-Th-He-methods, give various interpretations, which change from 0.5 to 4.5 × 10^9 years (Dymek et al. 1976; Jessberger et al. 1977; Schultz and Signer 1977; Wasserburg and Burnett 1978).

These variations reflect the effects of collisions of meteorite parent bodies, the heating-up of these bodies, with their partial or complete melting, and loss of gases. In general, in the interpretation of data on meteorite ages, it is always necessary to take into account, that the geological history of meteorites can be rather complex. Secondary processes (especially shock metamorphism) frequently yield rejuvenation values for the absolute age of meteorites. Moreover, it is necessary to consider that some meteorites are breccia and consist of separate clasts with a different igneous history.

In addition to the ages of differentiation and crystallization of meteorites, there are cosmic-ray ages, which characterize the duration of irradiation after the reduction of the meteorite to meter-sized fragments. These ages determine the time since the moment of breakup of the meteorite parent bodies (as a result of shock processes); that is the time of the independent existence of a meteorite in space. Cosmic-ray irradiation ages of meteorites are much shorter than the ages of differentiation and gas retention. The space ages are different for the various groups of meteorites.

1.8.8 Isotopic Composition and Isotopic Anomalies of Elements in Meteorites

Studies of the isotopic composition of Mg, Cr, Ag, Tl, U, and Xe in samples from primitive meteorites have established the presence of several short-lived nuclides with a mean life ≥~1 Ma in the early Solar System (Wasserburg 1985; Cameron 1993). The presence of these nuclides in the early Solar System solids manifests itself in an excess in their daughter nuclide concentrations (e.g., ^{26}Mg in the case of ^{26}Al). A list of some radioactive nuclei and their decay products is given in Table 8. Although a "fossil" origin for such excesses was also suggested (see, e.g., Clayton 1986), the experimental data, particularly the correlation between the excess in the daughter nuclide concentration and abundance of the stable isotope in the parent elements (e.g., correlation of excess ^{26}Mg in CAIs with their ^{27}Al content), argues against such a proposition. These stu-

Table 8. Short-lived nuclides

Parent	Product	$T_{1/2}$Ma
^{41}Ca	^{41}K	0.15
^{99}Tc	^{99}Ru	0.21
^{60}Fe	^{60}Ni	0.3
^{36}Cl	^{36}Ar	0.31
^{26}Al	^{26}Mg	0.72
^{53}Mn	^{53}Cr	3.7
^{98}Tc	^{98}Ru	4.2
^{107}Pd	^{107}Ag	6.5
^{182}Hf	^{182}W	9
^{205}Pb	^{205}Tl	14
^{247}Cm	^{235}U	16
^{129}I	^{129}Xe	17
^{244}Pu	^{232}Th, SF	82
^{146}Sm	^{142}Nd	103

dies have also suggested possible heterogeneities in the distribution of ^{26}Al in the solar nebula, and have helped in elucidating the processes leading to the formation of the CAIs themselves (e.g., MacPherson et al. 1988). The studies of coarse-grained CAIs, in particular, showed relationsships specific between their magnesium isotopic compositions and petrological types. For example, some type B1 CAIs are characterized by the presence of ^{26}Mg excess due to decay of extinct ^{26}Al, and yield well-behaved Mg-Al isochrons with $(^{26}Al/^{27}Al)_9$ at the time of formation of these inclusions, clustering around the value 5×10^{-5}. The Mg-Al systematics in the type B2 CAIs, on the other hand, are often disturbed, and are characterized by a lower value for $(^{26}Al/^{27}Al)_0$, indicating late disturbances in the magnesium isotopic systematics due to reequilibrium of magnesium isotopes and/or heterogeneity in the distribution of ^{26}Al in the solar nebula (Goswami et al. 1994).

The presence of short-lived nuclides in early Solar System solids limits the time interval between the last addition of freshly synthesized material to the solar nebula and the formation of these solids. Obviously, this time interval can be better limited through the observation of radionuclides with the shortest mean life. Until very recent times, ^{26}Al, with a mean life of 1 Ma, was the shortest-lived radionuclide whose presence in the early Solar System was conclusively established. Recently, the search for other shorter-lived nuclides (e.g., ^{36}Cl, ^{41}Ca, ^{60}Fe, ^{99}Tc, etc.) has yielded conclusive results (e.g., Srinivasan et al. 1994). These studies have presented strong evidence in favor of the presence of short-lived nuclides with a mean life of <1 Ma in the early Solar System.

Another type of isotopic effect found in meteorites also appears to call for some memory of the chemical state in which nuclei were carried into the solar nebula (Meyer et al. 1995). These effects include ^{16}O excesses (Clayton et al. 1973) and correlated ^{48}Ca, ^{50}Ti, and ^{54}Cr isotopic anomalies (e.g., Ireland 1990) in CAIs from carbonaceous chondrites. Then understanding of these isotopic effects would aid in constructing the chemical history of matter in the interstellar medium.

1.8.9 Links Between Asteroids and Meteorites

Almost all meteorites are derived from the asteroidal belt. A comparison of asteroid and meteorite spectra should identify the possible meteorite parent bodies. However, only a few asteroids with unique spectral characteristics have been linked with some meteorite types. They include: (1) asteroid 4 Vesta, which for more than 20 years has been debated as the parent body for the HED (howardite, eucrite, and diogenite) basaltic achondrites; (2) asteroids 19 Fortuna and 13 Egeria appear to be possible CM chondrite parent bodies; (3) asteroid 3103 Eger has spectral characteristics very similar to aubrites, and (4) small asteroid 3628 (estimated diameter 7 km) could be the source body for ordinary chondrites (e.g., Binzel et al. 1993, Burbine and Binzel 1995). A discrepancy between the observed flux of ordinary chondrite meteorites and the virtual lack of observed ordinary chondrite asteroids may be due to two processes (Binzel et al. 1993): (1) space weathering alters the surface of asteroids and, therefore, changes the spectral characteristics; (2) ordinary chondrite bodies exist only among small asteroids. Continued spectroscopic surveys of asteroids hold exciting promise for further success.

References

Alexander CMO'D (1993) Presolar SiC in chondrites: how variable and how many sources. Geochim Cosmochim Acta 57:2869–2888

Amari S, Lewis RS, Anders E (1994a) Interstellar grains in meteorites: III. Graphite and its noble gases. Geochim Cosmochim Acta 59:1411–1426

Amari S, Zinner E, Lewis RS (1994b) C-, N-, O-, Si-, and Ti-isotopic ratios of low density graphite grains from Murchison indicate a supernova origin. Lunar Planet Sci 25: 2728

Anders E (1988) Circumstellar material in meteorites, noble gases, carbon, and nitrogen. In: Kerridge JF, Matthews MS (eds) Meteorites and the early solar systm. University of Arizona Press, Tucson, pp 927–955

Anders E, Zinner E (1993) Interstellar grains in primitive meteorites: Diamond, silicon carbide, and graphite. Meteoritics 28:490–514

Binzel RP, Xu S, Bus SJ, Skrutskie MF, Meyer M, Knezek P, Barker ES (1993) The asteroid-meteorite connection: the discovery of a main belt ordinary chondrite asteroid. Meteoritics 28:324

Birk JL, Minster JF, Allegre CJ (1975) ^{87}Rb-^{87}Sr chronology of achondrites. Meteoritics 10:364–365

Bonte Ph, Jehanno C, Maurette M, Brownlee DE (1987) Platinum metals and microstructure in magnetic deep sea cosmic sperules. Proc Lunar Planet Sci Conf 17:E641–E648

Brilliant DR, Franchi IA, Arden JW, Pillinger CT (1992) An interstellar component in the lunar regolith. Meteoritics 27:206–207

Brownlee DE, Bates BA, Wheelock MM (1984) Extraterrestrial platinum group nuggets in deep-sea sediments. Nature 309:693–695

Burbine TH, Binzel RP (1995) Could G asteroids be the parent bodies of the CM chondrites? Meteoritic 30:494

Cameron AGW (1993) Nucleosynthesis and star formation. In: Levy EH, Lunine J (eds) Protostars and planets III. University of Arizona Press, Tucson

Cassidy WA, Kern CM (1993) Primordial mineral growth in a plasma. Meteoritics 28:335–336

Church SE (1975) Radiogenic isotope research. Rev Geophys Space Phys 13:98–101

Clayton DD (1986) Interstellar fossil ^{26}Mg and its possible relationship to excess meteoritic ^{26}Mg. Astrophys J 310:490–198

Clayton RN (1993) Oxygen isotopes in meteorites. Ann Rev Earth Planet Sci 21:115–119

Clayton RN, Grossman L, Mageda TK (1973) A component of primitive nuclear composition in carbonaceous meteorites. Science 182:485–488

Dodd RT (1986) Meteorites. Petrology and geochemistry. Moscow, Mir, 384 pp (in Russian)

Dymek RF, Albee AL, Chodos AA, Wasserburg GJ (1976) Petrography of isotopically dated clasts in the Kapoeta howardite and petrologic constraints on the evolution of its parent body. Geochim Cosmochim Acta 40:1115–1130

Farrington OS (1915) Meteorites, Published by the author, Chicago, Illinois

Gao X, Nittler LR, Swan PD, Walter RM (1995) Presolar grains in Indarch. Meteoritics 30:508

Goswami JN, Srinivasan G, Ulyanov AA (1994) Ion microprobe studies of Efremovka CAIs: I. Magnesium isotope composition. Geochim Cosmochim Acta 58:431–447

Grossman L (1980) Refractory inclusions in the Allende meteorite. Annu Rev Earth Planet Sci 8:559–608

Heide F (1934) Kleine Meteoritenkunde. Springer, Berlin Heidelberg New York

Heide F, Wlotzka F (1995) Meteorites. Messengers from space. Springer, Berlin Heidelberg New York

Ireland T (1990) Presolar isotopic and chemical signatures in hibonite-bearing refractory inclusions from the Murchison carbonaceous chondrite. Geochim Cosmochim Acta 54:3219–3237

Jessberger EK, Standacher Th, Dominik B, Herzog GF (1977) Ar^{40}-Ar^{39} dating of the Puebleto de Allende meteorite. Meteoritics 12:266–269

Krinov EL (1955) The basis of meteoritika. Moscow, Gostechizdat, 392 pp (in Russian)

Kurat G, Hoinkes G, Fredriksson K (1975) Zoned Ca-Al-rich chondrule in Bali: new evidence against the primordial condensation model. Earth Planet Sci Lett 26:140–144

Lavruchina AK (1973) Differenciation of the chemical elements in protoplanetary nebula. Meteoritika 32:7–24 (in Russian)

Levin BYu, Simonenko AN (1977) Unresolved problems in origin of meteorites. Meteoritika 36:3–22 (in Russian)

MacPherson GD, Wark DA, Armstrong JT (1988) Primitive material surviving in chondrites: refractory inclusions. In: Kerridge JF, Matthews MS (eds) Meteorites and the early solar system. University of Arizona Press, Tucson, pp 746–807

Mason B (1967) Extraterrestrial mineralogy. Am Mineral 52:307–325

Mason B (1972) The mineralogy of meteorites. Meteoritics 7:309–326

McSween HY (1987) Meteorites and their parent planets. Cambridge University Press, Cambridge

Meyer BS, Waver TA, Woosley SE (1995) Isotope source table for a $25M_{\odot}$ supernova. Meteoritics 30:325–334

Minster JF, Allegre CJ (1976) ^{87}Rb-^{87}Sr history of Norton County enstatite achondrite. Earth Planet Sci Lett 32:191–198

Ott U (1993) Interstellar grains in meteorites. Nature 364:25–33

Petaev MI (1988) List of meteorite minerals. Meteoritika 47:156–166 (in Russian)

Schultz L, Signer P (1977) Noble gases in the St. Mesmin chondrite: implications to the irradiation history of a brecciated meyeorite. Earth Planet Sci Lett 36:363–371

Schulze H, Bischoff A, Palme H, Spettel B, Dreibus G, Otto J (1994) Mineralogy and chemistry of Rumuruti: the first meteorite fall of the new R chondrite group. Meteoritics 29:275–286

Sears DWG, Shaoiong H, Benoit PH (1995) Chondrule formation, metamorphism, breccia-tion, an important new primary chondrule group, and the classification of chondrules. Earth Planet Sci Lett 131:27–39

Srinivasan G, Ulyanov AA, Goswami JN (1994) ^{41}Ca in the early solar system. Astrophys J Lett 431:L67–L70

Taylor S, Brownlee DE (1991) Cosmic spherules in the geologic record. Meteoritics 26:203–211

Tshermak G (1885) Die mikroskopische Beschaffenheit der Meteoriten. Stuttgart. (Facsimile reprint, with English translation in Smithson Contrib Astrophys 1964, 4:239 p)

Ulyanov AA (1991) The meteorite minerals. In: 14th Brown-Vernadsky Microsymposium on Comparative Planetology, Moscow, 20 pp

Van Schmus WR, Wood JA (1967) A chemical-petrologic classification for the chondritic meteorites. Geochim Cosmochim Acta 31:747–765

Vdovykin GP (1967) Carbonaceous matter in meteorites. Moscow, Nauka, 269 pp (in Russian)

Vinogradov AP (1971) High-temperature protoplanetary processes. Geohimiya 11:1283–1296 (in Russian)

Wark DA, Lovering JF (1977) Marker events in the early evolution of the solar system: evidence from rims on Ca-Al-rich inclusions in carbonaceous chondrites. Proc 8th Lunar Sci Conf 1:95–112 Pergamon Press, Oxford

Wasserburg GJ (1985) Short-lived nucli in the early solar system. In: Black DC, Matthews MS (eds) Protostar and planets II. University of Arizona Press, Tucson, pp 703–773

Wasserburg GJ, Burnett DS (1978) The status of isotopic age determinations on iron and sto-ne meteorites. Proc Symp Meteorite Research, Dordrecht, pp 467–479

Wasson JT (1985) Meteorites. Their record of early solar system history. WH Freeman, New York

Weisberg MK, Prinz M, Kojima H, Yanai K, Clayton RN, Mayeda TK (1991) The Carlisle Lakes-type chondrites: a new grouplet with high $\Delta^{18}O$ and evidence for nebular oxidation. Geochim Cosmochim Acta 55:2657–2669

Wood JA (1971) Meteorites and origin of the solar system. Moscow, Mir, 172 pp (in Russian)

Wood JA (1988) Chondritic meteorites and the solar nebula. Annu Rev Earth Planet Sci 16:53–72

Yaroshevsky AA, Ulyanov AA (1994) Catalogue of meteorite minerals. Meteor Conf 22: 91–92 (in Russian)

Yavnel' AA (1973) Classification of meteorites and its role in the problem of the origin of meteorites. Meteoritika 32:25–36 (in Russian)

Yates PD, Arden JW, Wright IP, Pillinger CT, Hutchison R (1992) A search for pre-solar material within an acid-resistant residue of Greenland cryoconite. Meteoritics 27: 309–310

Yudin IA, Kolomensky BD (1987) Mineralogy of meteorites. Sverdlovsk, Ural Sci Centre USSR Acad Sci, 200 pp (in Russian)

Zinner E (1995) Astronomy by mass spectrometry: interstellar grains in meteorite. Meteori-tics 29:555

Zinner E, Amari S, Ravaglio C, Gallino R, Busso M, Woosley S (1995) The isotopic compo-sition of interstellar graphite from the Murchison meteorite: evidence for supernova mixing. Lunar Planet Sci 26:1561–1562

1.9 Irradiation Effects on the Lunar Solids and Meteorites: Solar Wind, Solar Flares, and Galactic Cosmic Ray Records in the Lunar Minerals; Ion Implantation

L. L. KASHKAROV

Cosmic materials such as interplanetary dust particles, micrometeorites, meteorites, and the regolith matter of the Moon and asteroids constantly undergo irradiation by cosmic ray particles. Depending on the flux value, composition, energy spectrum, spatial variation, and angular distribution of the interplanetary charged particles in the inner Solar System, there are four basic constituents of the particle irradiation: solar wind (SW), suprathermal ions (STI), solar flares (SF), and galactic cosmic rays (GCR).

Fundamental components in cosmic radiation are protons (hydrogen nuclei, ~90%) and α-particles (nuclei of helium ~7%). The iron-group nuclei $(23 < Z < 28)$ account for ~10^{-3}% of the total cosmic particle flux only. These three groups of particles are the most important from the point of view of their interaction with exposed material that gives specific quantitatively measured radiation effects.

In total, irradiation effects are the results of the energy loss process (ionization, nuclear excitation, nuclear spallation reactions, etc.) by the charged particles during their penetration through material under exposure; the depth of these effects is due to a particle kinetic energy (Fig. 18). The differential energy spectrum for the total Solar System charged particle flux is shown in Fig. 19.

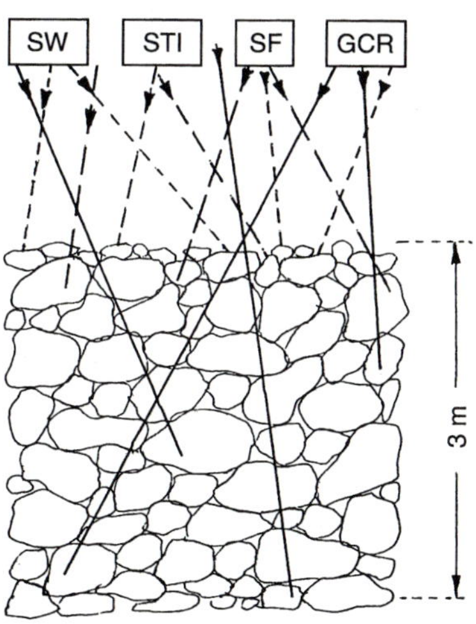

Fig. 18. Schematic sketch through the surface layer of the lunar regolith or the stone meteorite breccia illustrating the depth penetrating for the SW, STI, SF and GCR charged particles

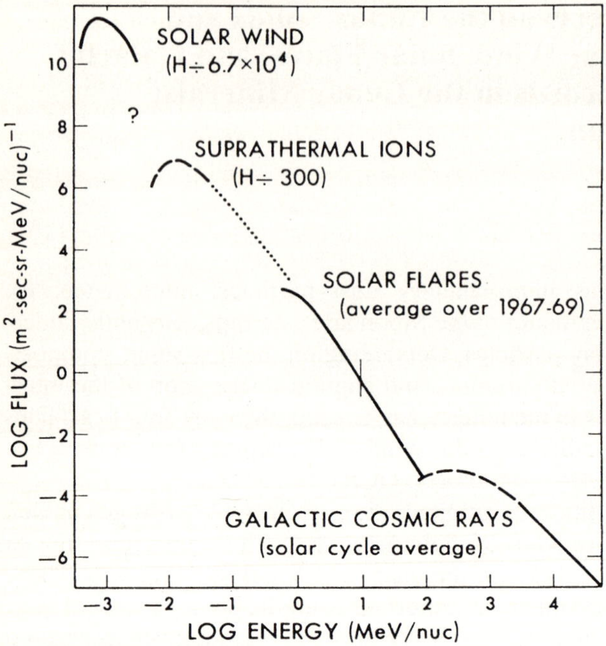

Fig. 19. Average present-day energy spectrum of the iron group nuclei and protons (H-nuclei) in the interplanetary space

The SW-particles with E ~ 1 keV are very easily absorbed in the very thin (up to ~0.1 μm) surface layer of crystals. Thus, the radiation effects induced by SW-ions can be preciseley determined and measured only in the case of the thin-crushed micron-sized dust grains and microcrystal material under exposure. These effects are: (1) very high concentrations of solar-type rare gases (^{4}He, ^{20}Ne, ^{38}Ar, etc.) implanted in the exposed material, and (2) a superficial layer of amorphous coatings on the crystalline grains, whose thickness is about 0.05 nm.

The penetration depth in crystals for the STI with kinetic energy E ~ (0.01–0.1) MeV/nucleon is up to ~1 μm. As the result of stopping such intense STI flux in the thin surface layer of crystals and the SW-ions, the implantation of solar-type rare gases the formation of deeper amorphous layers are observed. In addition, the following STI irradiation effects were identified. With the help of high-voltage electron microscopy, in the micron-sized crystal grains nonetched latent nuclear particle tracks were observed – the linear trails of radiation-damaged material with a diameter of ~100 Å and track-density values (number of tracks measured on square unity of the search crystal surface) equal to 10^{10}–10^{12} track/cm^2. Upon heating (thermal annealing), these tracks act as nucleation sites and are transformed into track crystallites.

One of the very important irradiation effects of iron metal blebs accumulation in the lunar soils can be considered. The production of the iron metal occurs in a complex process, including: (1) implantation of SW and STI hydrogen in soil microcrystals; (2) micrometeorite impact events; (3) production in the small fuse volumes of the glass phase; and (4) near-instantaneous reduction of Fe^{2+} in the melt material. So-called agglutinated glasses containing iron metal are widespread in lunar soils. The ratio of extremely fine-grained iron metal particles (<30 nm) to total iron oxide (I_s/FeO) is an accepted measure of the soil's "maturity" – the characteristic of their residence times on the lunar surface. The estimated effective times or the surface exposure ages for lunar regolith samples generally vary from 100 to 10000 times longer than for chondrite breccias.

The SF-charged particles, due to their higher kinetic energy in interaction with the silicate materials, penetrate in the material to a depth of up to some millimeters. However, in comparison with SW and STI particles, the SF have an essentially lesser degree of intensity and correspondly higher exposure times on the regolith surface of the cosmic objects are needed to obtain quantitatively measured radiation effects. The main radiation effects from SF particles consist of: (1) the formation of tracks from the iron-group nuclei characterized by specific track-density depth gradient inside the first 100–200-μm layer from the individual crystal surface, with the track-density values from 10^6–10^9 track/cm^2 interval; (2) disordered crystal structure up to the amorphous state, analogous to a metamictization process under high-fluency α-irradiation in zircons and other minerals with high U and Th concentration.

The GCR high energy p- and α-particles (E > 100 MeV/nucleon) give rise to a number of spallation nuclear reactions, for example: ^{26}Mg(p,n)^{26}Al, ^{25}Mg(p,α)^{22}Na, etc. The depth of penetration in silicate matter of protons with E > 1000 MeV is equal to 300–400 g/cm^2, but the integral flux in these depths is strongly decreased. Thus, the products of p-induced nuclear reactions can be measured in the lunar regolith soil at different depths, but the total output of this radiation effect is decreased from the upper several centimeters down to some meters by ~100 times. Besides primary p-and α-particles in the nuclear interactions, a great number of secondary nuclear-active (neutrons, π-mesons, etc.) and X- and γ-ray radiation are generated. Under the influence of these secondary particles in matter, different nuclear reactions occur, one of which (ca. some MeV energy) is the formation of recoil nuclei, as a result of which process a great number of dislocations in the crystalline lattice are seen.

Chiefly under X- and γ-ray radiation in silicate crystals, the process of thermoluminescence storage occurs.

Neutrons of moderate energy give rise to fission of heavy nuclei (Th, U), and then, by thermalized (E > 0.01 eV) neutrons with high probability, fission of ^{235}U isotope nuclei occurs.

At the end of the total path of the GCR iron-group nuclei in silicate crystals tracks are formed with a length 10–20 μm, and track-density values 10^5–10^7 track/cm^2.

References

Allen CC, Morris RV, Lauer HV Jr, McKay DS (1993) Microscopic iron metal on glass and minerals – a tool for studying regolith maturity. Icarus 104:291–300

Crabb J, Schultz L (1981) cosmic ray exposure ages of ordinary chondrites and their significance for parent body stratigraphy. Geochim Cosmochim Acta 45:2151–2160

Durrani SA, Bull RK (1987) Solid State nuclear track detection principles. Pergamon Press, Harwell, UK

Fleischer RL (1981) Nuclear track production in solids. In: Progress in materials science. Chalmers Anniversary Volume. Pergamon Press, New York, pp 97–123

Hartmann WK (1983) Moons and planets. Wadsworth Publishing, Belmont, California

Housen KR, Wilkening LL (1982) Regoliths on small bodies in the solar system. Annu Rev Earth Planet Sci 10:355–376

Kashkarov LL (1988) High-energy cosmic ray VH-nuclei in the early solar system. Izv Acad Nauk USSR Ser Phys 52:2321–2324

Kashkarov LL (1990) Low-energy VH-nuclei cosmic ray tracks in meteorites. In: Int Worksh on solid state track detectors and their applications. Dubna, Russia, pp 87–91

Kashkarov LL (1995) VH-nuclei cosmic-ray tracks in chondrites as indicators for radiation-thermal history of the meteorite matter at the early stage of Solar system primary body formation. Radiat Measurements 25:311–314

Kashkarov LL, Genaeva LI, Kalinia GV, Lavrukhina AK (1988) Irradiation effects for the ordinary chondrites in the early stage of the Solar system body formation. Meteoritika 47:113–122

McSween HY Jr (1987) Meteroites and their parent planets. Cambridge University Press, New York

Morris RV (1980) Origins and size distribution of metallic iron particles in the lunar regolith. Proc Lunar Planet Sci Conf 11th:1697–1712

Perron C, Naury M (1986) Very heavy ion track etching in olivine. Nucl Tracks 11:73–80

Price PB (1982) Applications of nuclear track-recording solids to high-energy phenomena. Philos Mag 45:331–346

Wasson JT (1985) Meteroites, their record of early solar system history. WH Freeman, New York

1.10 Mineralogy of Astroblems – Terrestrial Impact Craters

1.10.1 Introduction

A. Deutsch and F. Langenhorst

Hypervelocity collision of extraterrestrial projectiles with the Earth (impact process) has played a crucial role in the evolution of the atmosphere and life, and still represents a certain danger for man. The importance of terrestrial cratering was acknowledged by the geoscience community only in 1980, when the

Advanced Mineralogy, Volume 3
Mineral Matter in Space, Mantle, Ocean Floor, Biosphere, Environmental Management, and Jewelry (ed. by A. S. Marfunin)

impact of a roughly 10 km projectile was proposed as the cause for the long-known mass extinction at the Cretaceous-Tertiary (K/T) boundary. From that time research on all aspects of cratering started to flourish. The discovery of the 65 Ma Chicxulub impact structure, Mexico, as source crater of the ejecta layer at the K/T boundary, and the impact of the comet Shoemaker-Levi 9 on Jupiter in summer 1994 finally convinced a broader audience of the fundamental importance of hypervelocity impact as a geological process.

This chapter aims to review current knowledge of minerals which occur either in direct spatial relation to terrestrial impact structures or as constituents of distant ejecta deposits. In addition, a few specific, well-investigated structures of proved or suggested impact origin will be discussed in the light of recent mineralogical and geochemical research. As a preface to these topics, concise sections cover basic aspects of shock processes, cratering, and the general geology of impact craters and ejecta layers. For a comprehensive introduction to impact processes with particular emphasis on physical conditions, we refer the reader to Melosh (1989). More special aspects, such as geology, geophysical properties, and dating of terrestrial impact structures, mineralogical and geological shock effects, are reviewed by, e.g., Masaitis et al. (1980), Grieve (1991), Pilkington and Grieve (1992), Deutsch and Schärer (1994), Stöffler and Langenhorst (1994), Grieve and Shoemaker (1994), and Grieve et al. (1996).

Beyond phenomenological descriptions of minerals produced in or affected by impact process, this chapter focuses on the characteristics of these minerals on the submicroscopic scale, as well as on the processes and physical conditions which cause shock effects. This basic information is necessary to combine observational data at natural impact craters and shock experiments with theoretically derived models of cratering mechanics. Instructive examples are given on the basis of the most detailed investigated minerals in the environment of impact craters and ejecta layers. The important contribution of modern geochemical and mineralogical research to a better understanding of impact processes will be shown in the context of the impact structures reviewed, Nördlinger Ries, Sudbury, Vredefort, and Popigai.

1.10.2 Cratering and Shock Metamorphism

A. Deutsch and F. Langenhorst

Cratering and Shock Physics

Impact craters on Earth are produced by the hypervelocity impact of asteroids and comets at velocities between 11 and 72 km s^{-1}, with a mean relative collision velocity of around 20 km s^{-1}. Crater formation is subdivided into three subsequent but overlapping stages: (1) Contact of the projectile with the target

Advanced Mineralogy, Volume 3
Mineral Matter in Space, Mantle, Ocean Floor, Biosphere, Environmental Management, and Jewelry (ed. by A. S. Marfunin)

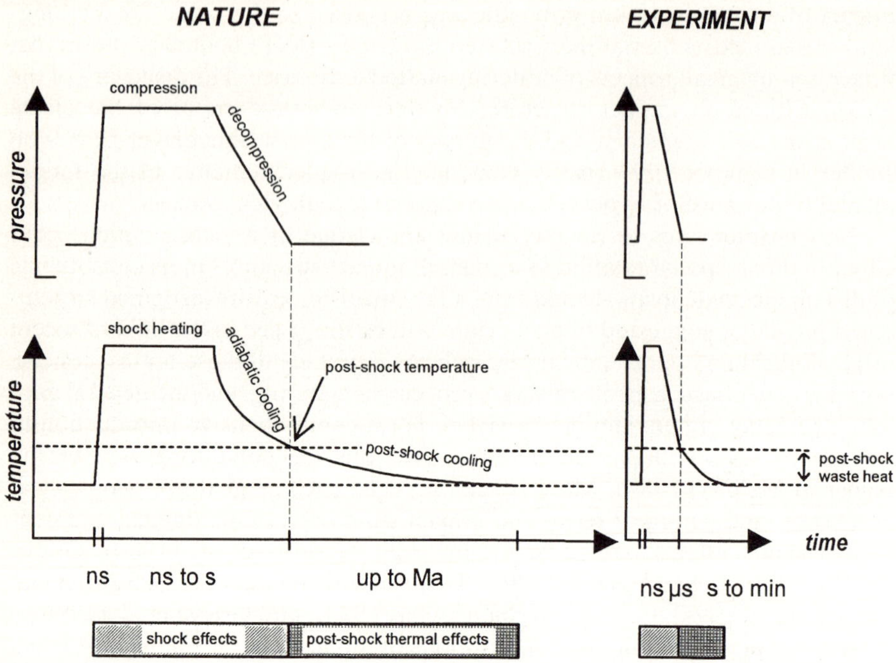

Fig. 20. Schematic pressure and temperature vs. time profiles for a shock event in nature (*left*) and experiment (*right*). Note change in time scale between *nature* and *experiment.* Shock and post-shock thermal effects are related to different stages of the process

surface, followed by compression and vaporization of the projectile and parts of the target, (2) excavation, and (3) modification of the crater. During the initial contact of the projectile with the surface, impact-melted target material is ejected as a primary jet at supersonic speed at low angles; this material forms tektites. Strong compression waves (shock waves) propagate with supersonic velocity into both projectile and target as most of the impactor's kinetic energy is transferred to the target. Shock waves result in drastic changes in the physical parameters of the affected material, which are idealized as discontinuities in pressure, temperature, density, and internal energy. This is shown schematically in Fig. 20. An even faster travelling rarefraction wave follows, resulting in decompression of the compressed material. In contrast to "normal" compression waves, shock waves are accompanied by a high-speed material transport behind the shock front giving rise to particle motion and ejection. As a result of the nonisentropic nature of shock compression, residual heat remains in the shocked material after pressure release (post-shock waste heat in Fig. 20). The projectile and parts of the target rocks are vaporized at pressures, which may reach several 100 GPa (Gigapascal), and temperatures of a few 10^4 Kelvin, forming an expanding vapor plume above the growing crater. In the target, the hemispherical shock wave and the rarefraction wave lead to shock metamorphism, material excavation, and ejection, until a bowl-shaped "transient" crater is formed within seconds. Due to gravitational forces, the transient cavity collapses rapidly during the modification stage of the crater, leading to turbulent mixing of shocked and unshocked materials, and slumping of material along the crater walls.

The cratering process results in different, characteristic geological formations. Figure 21 illustrates the distribution of these impact formations in and around a complex crater formed in a crystalline target. About 80 vol.% of the target affected by the shock wave consists of undisplaced or slightly rotated target rocks underneath and around the crater ([par-]authochtonous or authigenic breccias). The remaining 20% is formed by displaced, allochthonous (allogenic) breccias showing various degrees of shock metamorphic overprint, and impact melt rocks, with the relative proportion of melt rocks increasing with crater size. Clastic matrix breccias with melt particles (suevitic breccias) compose up to 10% of the displaced masses, but the dominant material are breccias devoid of impact melt.

Shock Metamorphism

Shock metamorphism is a fundamental and common process in the Solar System. It is caused by the short pulse of high dynamic pressure, which occurs exclusively in the context of natural impact events, nuclear or chemical explosions, or shock experiments. In shock metamorphism, pressure, temperature, as well as strain and quench rates, exceed by orders of magnitude the conditions existing during endogenic processes. This fundamental difference is illustrated in Fig. 22 for pressure and temperature. Typical strain rates acting during shock compression range from 10^6 to 10^9 s^{-1}, roughly 20 orders of magnitude higher than typical strain rates of endogenic metamorphism. However, these exceptional physical conditions prevail at the maximum for a few seconds only, even in the case of very large impact events (Fig. 20). The extreme and ultra-short shock stress causes unique physical and chemical disequilibrium changes in minerals and rocks. Rocks are the subject of textural changes, ranging from brecciation, through melting, to vaporization, i. e., ceasing of target and projectile material as a physical entity. Minerals show a variety of microscopic to sub-microscopic changes, e. g., formation of lattice defects or high-pressure polymorphs and dissociation, the so-called (residual) shock effects. These different categories of change and formation in rocks and minerals have been traditionally summarized as shock or impact metamorphism, although these terms are not synonymous in a strict sense. The latter term has to be preferred because several phenomena identified in the context of impact craters and ejecta layers are not directly related to shock compression and decompression. Impact metamorphism covers, in addition, a variety of processes confined by a geological "normal" p-T-time space. For example, post-shock cooling, differentiation, and crystallization of a large impact melt pool resemble the formation of a suite of

IMPACTITES IN A COMPLEX CRATER

coherent impact melt layer ①
- clast-free melt rock
- polymict impact melt breccia
  with glassy or crystalline matrix

post-shock T ≥ 1500° C; variable cooling rates depending on fragment content and thickness of the melt sheet

allochthonous breccia deposits ②
- polymict breccias with melt: *suevitic breccia*
  melt absent: *fragmental breccia*

variable shock pressure; variable formation T; variable cooling rates

crater basement ③
- monomict brecciated megablocks
- parauthochthonous polymict breccias
- pseudotachylites (3a)
- breccia dikes (3b)
- impact melt dikes (3c)

variable shock pressure but generally ≤ 45 GPa; variable formation T (inverse T) profile); low, crater-size depending cooling rates

distant ejecta ④
- tektites
- differently shocked rock and mineral clasts
- microtektites, microkrystites, impact melt glass

variable shock pressure; very high cooling rates

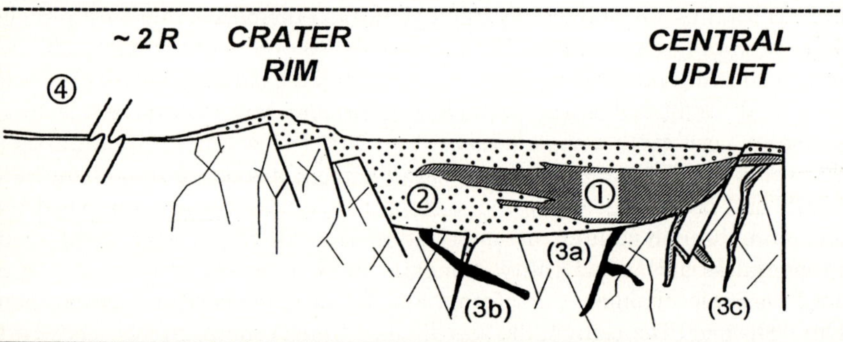

Fig. 21. Schematic cross-section of a complex crater with a central uplift showing locations of the different impact formations, arranged according to their variable pressure-temperature history. *R* Radius of the final (modified) crater. Not to scale

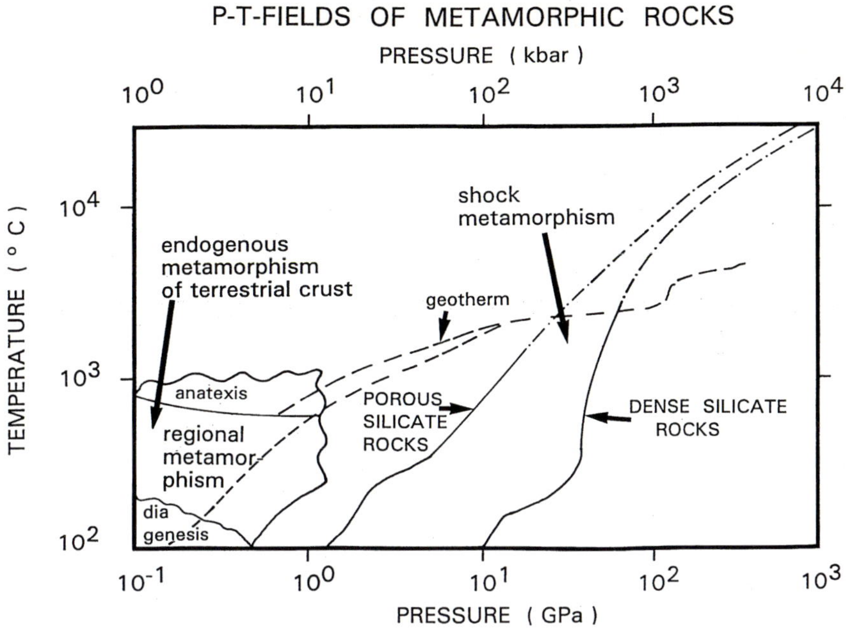

Fig. 22. Schematic pressure vs. temperature diagram, delineating fields for endogenic metamorphism of the terrestrial crust and for shock metamorphism

endogenic magmatic rocks, but cannot be considered as metamorphic processes. Rocks being affected by impact metamorphism are summarized under the collective term impactite.

In this chapter, a clear distinction is drawn between primary shock effects related to shock compression and pressure release (shock metamorphism in the strict sense) and post-shock thermal effects, which occur after decompression (Fig. 20). We will concentrate exclusively on such shock and post-shock phenomena that are characteristic of impact. The current interpretation and understanding of shock effects in minerals relies mainly on careful transmission electron microscope (TEM) studies as well as on systematic shock experiments. A recent example of the use and combination of both techniques is given in Langenhorst and Deutsch (1994) and Langenhorst (1994).

Shock and Cratering Experiments

Shock recovery and cratering experiments have yielded important findings for the understanding of impact processes, including, for example, the calibration of threshold pressures for particular shock features or ejecta mechanisms (e.g., Boslough 1991). The two most commonly used set-ups for the acceleration of projectiles in experiments are light-gas guns (compressed air, one- and two-stage argon or helium devices) and high-explosive devices.

Recently, powerful lasers and electric discharge techniques have gained in importance, and shock pressures in excess of 100 GPa can be reached with coherent waves. In addition, investigations at nuclear test-sites have resulted in significant scientific contributions to cratering research.

Distinct differences between shock waves generated in the laboratory compared to nature have to be considered in interpreting experimental data. For example, and as illustrated in Fig. 20, the duration of the peak pressure plateau lasts for orders of magnitude longer in natural impacts than in experiments. This shorter pressure pulse is only a consequence of the small thickness of projectiles in all types of experiments. Other differences originate from the geometry of the shock waves, the energy density at the projectile's stagnation point compared to chemical or nuclear charges, and the loading path to reach the peak pressure. Moreover, shock and post-shock temperatures are significantly lowered compared to the natural case if techniques with multiple shock reverberations are used. Major advantages of shock recovery experiments, which were pioneered by Milton and DeCarli (1959), are (1) the precise control of physical and chemical properties of the specimen, (2) direct measurements of physical parameters and their change during shock compression by sophisticated VISAR methods (*V*elocity *I*nterferometer *S*ystem for *A*ny *R*eflector), and (3) separation of effects which originate directly during shock compression from those, caused by post-shock annealing after pressure release. The latter point makes recovery experiments indispensable in unraveling the original nature of shock effects. Numerous investigations have shown that, despite the above-named caveats, minerals display identical shock features in both experiment and nature down to the atomic scale as long as they are related only to the magnitude of the shock pressure. Therefore, shock and cratering experiments are established as important tools to investigate the physics of impact metamorphism and cratering, and to calibrate shock effects in minerals.

1.10.3 Characteristics of Terrestrial Impact Structures

A. DEUTSCH and F. LANGENHORST

Recognition

Distinctive evidence for confirming the impact origin of a geological structure is established by the occurrence of (1) residual shock effects in minerals and/or certain high-pressure polymorphs, (2) shatter cones, and, in special cases, (3) projectile remnants adjacent to the crater (Small Meteorite Craters, p 86). Second-order diagnostic tools are (4) breccia sequences and breccia dikes,

Advanced Mineralogy, Volume 3
Mineral Matter in Space, Mantle, Ocean Floor, Biosphere, Environmental Management, and Jewelry (ed. by A. S. Marfunin)

including melt bodies with rock and mineral fragments and pseudotachylites (frictional melt rocks), (5) circular geophysical anomalies, (6) topographic features imaged by aerial photographs or remote sensing techniques, (7) overturned stratigraphy in the breccias of the ejecta deposits, and (8) geochemical signatures, such as enhancement of certain siderophile elements (e.g., Ni, Co, Cr, and PGEs), or isotope ratios, for example, a low $^{187}Os/^{188}Os$ ratio. (9) Exotic chemical compounds such as fullerenes have been detected only in the context of the Sudbury impact structure, Ontario, Canada, and in K/T ejecta deposits; their diagnostic value is unknown so far. None of the features listed under (4) to (8), however, provides alone ultimate proof for an impact origin of a geological structure or a distinct layer embedded in a sedimentary sequence. Criteria for recognizing deeply eroded impact structures have yet to be developed.

Temporal, Spatial, and Size Distribution of Impact Craters

By the end of 1996, 156 terrestrial impact structures were known, ranging in diameter (D) from 15 m (Haviland, Kansas, USA) to about 350 km; the oldest structures have an age of about 2 Ga (billion years), the youngest were formed in historical times: Kaalijärvi, Estonia, for example, is dated at about 4000 B.C. A crater list containing information on size, age, and projectile type (if known) was published by Grieve and Shoemaker (1994). R.A.F. Grieve at the Canadian Geological Survey, Ottawa, regularly presents an updated report of confirmed craters, including the newly discovered impact structures – at present, at a rate of up to five craters per year (http://gdcinfo.agg.emr.ca/crater/world_craters.html).

The craters on land show concentrations in North America (especially Canada), Australia, parts of the former Soviet Union, and northern Europe. This distribution reflects active crater-search programs in some countries and the fact that these regions are old cratonic areas with small time-integrated erosion rates, long exposed to the cosmic bombardment, and, therefore, with a larger cumulative number of craters. Other stable shields, e.g., in Africa, S. America, or China, have not yet been searched systematically for impact structures, but recent discoveries indicate the presence of a significant number of additional craters in these areas as well. On land, terrestrial craters show varying stages of preservation and exposure, ranging from deeply eroded, e.g., Vredefort, South Africa, to examples with well-preserved ejecta deposits outside the rim of the final crater, e.g., Nördlinger Ries, Germany. About 30% of the known crater population are completely buried by sedimentary rocks, and, in rare cases, tectonism resulted in re-shaping of the circular form as exemplified by Beaverhead, Montana, USA, or the Sudbury Structure, Ontario, Canada. In general, however, impact structures are circular, except that projectiles with a very low impact angle produce extended scars, such as the Spider Structure, W.A., Australia. If the size-frequency distribution of terrestrial craters is compared with that on other planetary bodies, a distinct deficiency for craters with $D \leq 20$ km is obvious due to preferred removal of smaller craters by endogenic and exogenic forces.

About 60% of the known terrestrial craters are younger than 200 Ma (million years), reflecting again the highly dynamic nature of the crust. For about one third of the total crater population, isotopic ages are available, and for about 30 craters, the relative precision on the age is better than 10%. However, even for some of these, the age significance is under dispute. Other methods of dating are biostratigraphy in post-crater sediments and paleomagnetic methods; under favorable conditions, both methods can yield an age with high precision and accuracy. Age data and size distribution indicate quite clearly that our knowledge of the terrestrial cratering record is strongly biased towards younger and larger craters. This was taken into account in an estimation of the recent terrestrial cratering rate, which is $5.6 \pm 2.8 \times 10^{-15}$ $km^{-2} a^{-1}$, using only craters with $D \geq 20$ km and younger than 120 Ma (Grieve and Shoemaker 1994). Despite the inherent bias in the cratering record, a periodicity in cratering during Phanerozoic times with a period of 26–33 Ma has been claimed. The insufficient number of high-quality crater ages limits any analysis of whether this assumption is statistically valid (see discussion in Grieve and Shoemaker 1994).

The morphology of relatively well-preserved impact craters on Earth compares, in principle, well with crater forms on other planetary bodies ranging with increasing diameter from simple, bowl-shaped depressions, to complex craters with a central peak, a peak-ring, a ring basin, to multi-ring basins. The relations of the morphometric parameters and their definition have been reviewed in detail by Grieve (1987). Differences in the morphometric parameters to, for example, the lunar data, are due to gravitational effects, which influence crater modification and erosion. In the following, simple and complex craters are characterized, and special cases of terrestrial cratering are outlined.

Historic Collisions

Tunguska. The great fireball near the Podkamennaya-Tunguska river, Siberia, on the morning of June 30, 1908, is the only historically well-documented encounter of a high-energetic extraterrestrial body with planet Earth: The body, probably a 50-m fragment of a stony meteoroid, did not strike the surface, but disintegrated at an estimated altitude of about 8.5 km, causing atmospheric over-pressure pulses, and, by coupling to the ground, seismic waves, which both were recorded as far away as Greenwich, Jena, or Tiflis. Analysis of barograms, seismic records, and scaling from nuclear airburst yields an estimated kinetic energy of 12 up to 48 megatons TNT equivalent ($\approx 10^{17}$ J) for the Tunguska bolide. The event resulted in devastation of more than 2000 km^2 of Siberian taiga. Despite intense research, only minute traces of cosmic material in the form of glass and magnetite spheres have been recovered on the ground. Peat at the impact site contains a weak Ir signal (17.2 pg g^{-1}), whereas, Crête Site B ice core, drilled in the ice sheet at Central Greenland, lacks an Ir anomaly related to the Tunguska event (Rasmussen et al. 1995).

Besides Tunguska, two other collisional events occurred in this century. In February 1947, a bolide with an estimated kinetic energy of 1.7 kt TNT equivalent ($\approx 10^{14}$ J) broke up at an altitude of about 6 km over Shikote-Aline, Siberia, yielding more than 100 craters with a diameter ranging from 0.5 up to 26.5 m. More than 23 tons of the iron meteorite fragments have been recovered. A similar, although much smaller and about 4 ka (10^3 years) old meteorite-strewn field is known in South Australia (the Henbury craters). In March 1965, a bright fireball was observed over southern British Columbia, Canada, due to the passage of an object with an estimated kinetic energy of 20 ktTNT equivalent. Near Revelstoke, tiny fragments of a carbonaceous chondrite with a total mass of 1 g were discovered and, in addition, airborne collectors over the United States gathered a large number of silicate glass and magnetite spherules.

Simple Impact Craters

The basic form of a simple crater is a shallow bowl surrounded by a structurally uplifted rim. Beneath the floor of this so-called apparent crater, a lens of allochthonous impact breccias lies on top of the brecciated, authochthonous crater basement. The interface between breccia lens and authochthonous rocks defines the true crater, which shows a roughly parabolic cross-section. The typical depth/diameter ratio of the true crater is 1 : 3. Several simple craters have been investigated in great detail, using geological and geophysical exploration techniques as well as shallow and deep drilling. The most outstanding examples are the 49.7 ka Barringer (Meteor) crater, Arizona, USA, the 450 Ma Brent crater, Ontario, Canada, and the 455 Ma Kärdla crater, Estonia.

Barringer crater, with a diameter of 1.2 km, still displays morphological features nearly unchanged by erosional forces. The apparent crater is about 200 m deep. The rim uplifted about 50 m above the original ground surface, is covered by an overturned flap of ejecta material, which continues into the hummocky outer ejecta blanket. The exceptional stratigraphic control of the sedimentary target and the wealth of observational data allowed to constrain all parameters relevant for calculating impact conditions for the Barringer crater (Roddy et al. 1975). Brent with D = 3.8 km is the largest simple crater in a crystalline target (Grieve 1978). This crater owes its excellent preservation to a 260 m thick cover of post-impact sediments, which sealed the up to 630 m thick breccia lens. Brent has a thin basal zone of impact melt rocks directly overlying a parauthochtonous breccia of the brecciated crater floor. Mineralogical data form Brent have served to define the radial attenuation of shock pressure in simple craters. Kärdla, with D = 4 km, has a stratified target of Ordovician and Lower Cambrian sand- and siltstones on top of Proterozoic crystalline rocks

(Plado et al. 1996). More than 300 drill holes penetrated the breccia lens, the rim as well as the brecciated basement. Remnants of the ejecta occur in the sediments up to 50 km away from the crater rim. A very small central uplift indicates that Kärdla is close to the transition from simple to complex crater forms.

Small Meteorite Craters Formed by Iron Meteorites. Meteorite craters are considered a special group of prehistoric impact scars because – under favorable climatic conditions – remnants of the projectile are partly conserved. Under morphological aspects, however, this type of crater belongs to the simple bowl-shaped craters. The projectiles are high-density iron or stony-iron meteorites less than 150 m in diameter. They can penetrate the atmosphere, whereas, weaker stony bodies of similar dimension are disrupted by atmospheric break-up. The only known exception is the 100 ka Rio Cuarto cratering field, Argentina, where a low impact angle resulted in ricochet of the chondritic projectile, remnants of which have been recovered at the impact site. The largest meteorite crater is the 49.70±0.85 ka Barringer Crater. This age is based on different techniques, i.e. surface dating of boulders in the ejecta blanket using ^{10}Be and ^{26}Al exposure ages, ^{14}C-dating of organic material in rock varnish, and thermoluminescence dating of quartz in shocked sandstone (e.g., Nishiizumi et al. 1991; Phillips et al. 1991).

In the surroundings of meteorite craters, several types of impactor-related particles are recognized. They include ablation spherules produced during atmospheric entry of the projectile, shattered fragments of the meteorite itself and metal oxide fragments with meteoritic Fe/Ni ratios, glassy spherules composed of target and impactor material, and melted target rock particles with vesicular texture. Holohyaline spherules (impactite lapilli) at the Barringer Crater contain up 24 wt.% meteoritic material, and carry metal spherules with up to 90 wt.% Ni, interpreted as partially oxidized, molten meteoritic material (Kargel et al. 1996). At the Berringer Crater, geochemical properties of the projectile (Canyon Diablo IA iron meteorite, original diameter of the projectile 30 m) and target formations are well constrained, allowing a mass balance for several elements in minerals and rocks of the ejecta blanket. At the elliptical-shaped 310-ka Wolfe Creek crater, W.A., Australia (D = 300 m), remnants of the III AB iron projectile are altered to laminated shale balls consisting of Ni-serpentine and Ni-carbonate.

Complex Impact Structures

Transition from simple to complex craterform structures occurs on Earth at diameters above 2 km in sedimentary targets, and 4 km in crystalline targets. The most characteristic features of complex craters are (1) modified rim areas, and (2) uplifted areas expressed as central peak and/or topographic rings; moreover, complex impact structures are shallower than simple craters with respect to their depth/diameter ratio. The presence of near-surface lithologies of the pre-impact target stratigraphy within the rim and annular trough indicates that the maximum radial excavation by the cratering flow field is restricted to half the distance of the final structural rim. The depth of excavation is on the order of 0.05 times the final rim diameter. The rim area itself, often displaying terrace-like morphology, consist of blocks faulted inwards towards the center of the structure. In complex impact structures, annular troughs and central basins are filled with allochthonous impact breccias; if crystalline rocks dominate among target lithologies, a coherent impact melt layer is present. Examples for complex impact structures are illustrated below.

Rocks in the central hump are uplifted from maximum depths on the order of one tenth of the diameter of the final crater rim. According to geomechanical models using hydrocodes, some overshooting may occur during formation of this central peak (Ivanov 1994). Good examples for this structural uplift are found at the 15 Ma Steinheim Basin, Germany (D = 3.8 km), which most probably represents a twin crater of the Nördlinger Ries, or at the about 175 Ma Puchez-Katunki, Russia (D = 80 km). The central uplift of this exceptionally well-preserved structure is drilled to a depth of 5374 km (Vorotilov deep borehole). Down to about 3 km it consist of 100 m sized megablocks. Impact metamorphic overprint in the central uplift decreases with depth and is apparently concentrated at interblock boundaries (Masaitis et al. 1995).

It is important to note that erosion may preferentially remove brecciated material, resulting in a significantly modified ore even inverse topographical expression of the final crater. An excellent example of this is the 142.5 Ma Gosses Bluff, WA, Australia, showing only a 6 km wide structural uplift as topographic relic in the center of the structure with an original diameter of 22 km. Erosional effects also reshaped the morphology of the 23 Ma Haughton Dome, N.W.T., Canada (D = 24 km), and of the 280 Ma Clearwater West crater, Quebec, Canada (D = 36 km). Both impact structures display prominent topographic rings, which are formed not by structurally uplifted material, but simply by the preservation of more erosion-resistant rocks.

Submarine Impact Structures

Although submarine craters cannot be considered as a distinct group with regard to size or age, they are a specific problem because of their small number. Only a few confirmed underwater craters are known: Montagnais located offshore Nova Scotia, Canada (50.5 Ma, D ≈ 45 km), Tvären, Sweden (455 Ma, D = 2 km), Mjølnir in the Barents Sea (Gudlaugsson 1993), north of Norway (Jurassic; D ≈ 40 km), and Chesapeake Bay offshore the Atlantic coast of Virginia, USA (≈ 35 Ma; D ≈ 90 km). Additionally, half the Chicxulub structure is buried offshore the Yucatan platform below thick post-impact sediments, and the Kara Sea, Russia, contains relics of a crater representing either a twin structure of the 71 Ma Kara crater, named Ust-Kara, or part of the Kara crater itself. A Late Pliocene impact is recorded in an about 300 000 km² area of the South Pacific (Eltanin Sea Mt.) by the occurrence of microtektites and an iridium anomaly in absyssal sediments, yet the small-sized projectile (∅ ≤ 0.5 km) probably did not reach the ocean floor (≈ 5000 m) there (Margolis et al. 1991; Gersonde et al. 1997). The debris layer is conserved in sediment cores from the Bellinghausen Sea; so far, vesicular impact melt, glassy spherules and unmelted meteorite clasts have been recovered in the cored material. Craters on oceanic crust are unknown to date, reflecting not only the young mean age of the oceanic crust, but also our relatively poor knowledge of two thirds of the Earth's solid surface.

Projectile Identification

Larger impacting bodies penetrate the atmosphere and collide with cosmic velocities, yielding shock pressures in the MPa range and extremely high shock and post-shock temperatures. These extreme conditions cause melting and vaporization of both the projectile and the target. Recondensation and admixture of the projectile material to impactites in the crater, especially impact melt material, and to ejected impact glass, produce a geochemical signature, allowing, in principle, identification of the projectile type by analyzing platinum group elements (PGEs), nickel, cobalt, and chromium, as well as Re-Os isotope systematics (e.g., Evans et al. 1993; Koeberl and Shirey 1997). These trace elements, which are depleted to various degrees in the Earth's crust, occur in high abundances and with distinctive interelement ratios (like ruthenium/iridium or rhenium/iridium) in certain groups of meteorites. For example, Irons show a pronounced enrichment in rhodium, while all types of chondrites have a flat C1 normalized distribution pattern limiting the use of PGEs for fine classification. Achondrites are enriched in Cr compared to most terrestrial rocks; in contrast, other types of impactors, like iridium- and osmium-poor Stony-Irons, are chemically inconspicuous and will not leave traces in impact melt rocks. Using the characteristic meteoritic signature to identify the projectile type and evaluate the projectile's contribution to the geochemical composition of an impactite, the so-called meteoritic component, is hampered by two facts. Constraining the net amount of the meteoritic component requires subtraction of the indigenous component which is derived from the target rocks. If the target has a heterogeneous composition with presence of mafic to ultramafic rocks, a precise knowledge of the relative proportions of the different lithologies in the crater area is required to avoid overestimation of the indigenous component. The second impediment in projectile identification via PGEs results from the fact that fractionation of the siderophile elements can occur, either directly during the impact process or by post-impact alteration processes. Ir proved to be the most diagnostic element, while Re, for example, shows some mobility in weathered material.

Despite these complications, the projectile type has been identified for 27 larger impact craters (Grieve and Shoemaker 1994). In some cases, however, specification relies on only a few elements, requiring reanalysis with more sophisticated methods. The analyzed impact melt rocks usually contain 1 wt.% projectile material, but knowledge of the three-dimensional distribution of the meteoritic component within an impact melt sheet is poor. PGE data indicate a uniquely high amount of 7.4 wt.% chondritic component in impact melt rocks of the 280 Ma East Clearwater structure, Quebec, Canada (Palme et al. 1979). This observation is explained by a low impact velocity (≤ 17 km s^{-1}). It is assumed that the metal component of the projectile was only melted but not vaporized, and directly incorporated down into the vaporized and melted target rocks. The principal reliability of PGE data for determining projectile contribution has gained enormous importance in constraining the general influx of extraterrestrial material, and, of even greater interest, in the search for distal ejecta layers.

1.10.4 Geological Formations in and around Impact Structures

A. DEUTSCH and F. LANGENHORST

Based on their regular distribution, up to four different geological formations can be distinguished in and around an impact crater: (1) crater basement, (2) allochthonous breccia deposits, (3) coherent impact melt layer, and (4) distant ejecta. Distribution of these formations is shown in Fig. 21, giving a schematic cross-section through a complex impact structure with a central uplift. The geological formations are characterized by certain impact metamorphic features, which are discussed below; in part, these characteristics reflect the distinct post-shock thermal histories. The three main textural types of polymict impact breccias which occur as essential lithologies in the context of an impact structure and their characteristics are explained in Fig. 23. It is important to note that no general consensus about the terminology for imapctites has been reached yet, and local terms or names comprising both genetic and descriptive attributes are still in use. A straightforward rock classification, allowing identification of impactites in meteorites, the lunar collection, and terrestrial impact craters, is not available to date. A draft of a consistent nomenclature is currently being reviewed by the IUGS subcommission on the systematics of metamorphic rocks, study group for impactites (Stöffler and Grieve 1994).

Crater Basement

The crater basement consists of monomict brecciated megablocks and parauthochthonous and authochthonous (authigenic) breccias, which are either polymict or monomict, depending on the target stratigraphy. Dikes of fragmental breccias, pseudotachylites, and impact melt rocks transect the basement

Advanced Mineralogy, Volume 3
Mineral Matter in Space, Mantle, Ocean Floor, Biosphere, Environmental Management, and Jewelry (ed. by A. S. Marfunin)

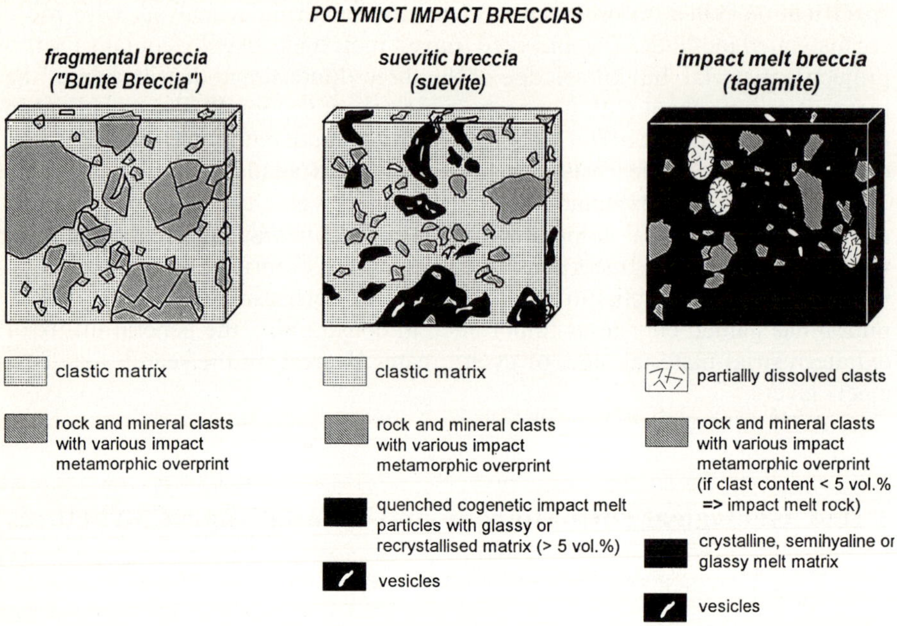

Fig. 23. Schematic drawing of the three main textural types of polymict impact breccias. The size of clasts and cogenetic melt particles (in suevitic breccias) may vary for orders of magnitudes. Rock fragments may display internal brecciation.

(Fig. 21); contact relationships, matrix type, and clast population suggest that the dike breccias originate during compression and decompression as well as in the crater's modification stage. Knowledge of this impact formation relies on research in deeply eroded craters or in the central uplift of complex structures as well as on drilling and geophysical measurements. Maximum shock pressures recorded in the crater basement are about 45 GPa; above this threshold, partial melting starts. Usually, shock features in quartz and feldspar are used for shock wave barometry. Isobars indicate a systematic decay of the maximum recorded shock pressure from the stagnation point of the projectile towards the crater rim and with depth, and trace the upwelling in the central mount, e.g. at Puchez-Katunki. The regular distribution of isobars has been used successfully to reconstruct the original dimension of now deeply eroded impact structures. Shatter cones corresponding to a shock pressure in the 4 ± 2 GPa range, are the diagnostic shock effects which occur radially most outwards from the crater center. There are good reasons to believe that the formation of shatter cones fades out at about 0.6 to 1 radii of the outer rim of the transient cavity. Shocked minerals are unknown from regions outside the rim of the transient cavity, where the rocks only show various kinds of brecciation.

Post-impact thermal history in the crater basement is determined by post-shock waste heat in rocks of this basement and annealing by the overlying crater fill, which may comprise impact melt. A voluminous melt sheet can cause a reverse temperature gradient in the crater floor, and produce a thermal contact aureole in the footwall. In large impact structures, such as Sudbury, where the transient cavity reaches deep crustal levels, cooling rates in the basement also depend on the ambient geothermal situation. Cooling is, therefore, related to the crater size, and temperature decay is slow compared to all other impact formations. A special thermal situation is documented at some impact structures, e.g., at Puchez-Katunki, where shear heating due to differential movement of large blocks has produced zones of coptoblastolithes, i.e., impact metamorphosed rocks, which have undergone high-temperature solid state recrystallization (Masaitis and Mashchak 1996).

Allochthonous Breccia Deposits

Allochthonous breccia deposits occur in the crater fill as breccia lens, and, up to two to three crater radii away from the tectonic rim of an impact structure (Fig. 22), as part of the continuos ejecta blanket which can contain large amounts of local, unshocked surface material. Mineral and rock fragments in the polymict breccias display a wide range of impact metamorphic features; melt particles may be present (Fig. 23). The breccia lens is formed by material which remained inside the hemispherically growing transient crater, slumping back from the crater walls during the modification, and by fallback clasts of all size. Late fallback material from the vapor cloud, rich in glassy particles, may form a thin layer on top of these deposits. Ballistic sedimentation is the process forming the breccia deposits outside the crater: debris formed by the passage of the shock wave are ejected ballistically in an expanding ejecta curtain during crater growth at predominantly subsonic speed, followed by radially outward flow at the surface. This flow erodes the surface layer and incorporates significant amounts of local material. Part of the debris found outside the crater rim is laid down by a base surge, which is a gravity-driven density current descending from the expanding vapor cloud above the crater. The observed polymict nature of the breccia deposits is due to the complex material movement in both types of processes.

In allochthonous breccia formations, thermal equilibrium is reached quickly after deposition. Depending on the abundance of melt particles, the equilibrium temperature may exceed 600 °C in suevitic breccias. Observations at the Ries crater indicate that cooling of the about 400 m thick suevite layer to ambient conditions occurred on the order of months to a few hundred years. Hot gases and fluids, rising form deeper levels of the crater, result in more complex thermal histories of allochthonous breccias and can cause partial obliteration of shock effects.

Coherent Impact Melt Layers

The coherent impact melt layer ranges from clast-free melt rocks to polymict breccias with a melt matrix that contain shocked and unshocked rock and mineral fragments in varying abundance (Fig. 23). The melt matrix is holohyaline to holocrystalline, occasionally showing a coarse-grained igneous texture. Massive impact melt rocks are called tagamites after the Tagamy hills in the Popigai crater, Russia. Size and thickness of the coherent impact melt sheet is scaled with the crater dimension, and the relative amount of melt increases with the crater size (Cintala and Grieve 1994). Coherent melt layers are extremely rare to absent in craters with sedimentary target materials, especially when carbonates dominate. Melting on the whole rock scale is confined to a hemispherical zone shocked at pressures in excess of about 60 to 80 GPa. Partial melting is restricted to highly shocked material, which is engulfed by the impact melt, and to contact zones, where it may occur after deposition of the coherent melt layer in the crater. Time scales for cooling depend on fragment content and thickness of the melt layer, and are variable, with the center of the melt sheet remaining hot much longer than the clast-rich marginal facies. Good examples of coherent impact melt sheets exist in the 100 km sized 210 Ma Manicouagan structure, Quebec, Canada, where up to 230 m of the melt layer is preserved, at Popigai with an up to 600 m thick melt layer, and at Sudbury.

Independent of crater size and target material, unaltered impact melt rocks display common characteristics. Geochemically, the melt rocks duplicate the composition of target rock mixture, except in certain cases, when enhanced abundances of some siderophile elements and Cr occur due to projectile contamination. In coherent impact melt layers, most major and minor elements are well homogenized with respect to the precursor material. This characteristic homogenization is a corollary of material mixing during the turbulent, high-velocity flow of the melt within the expanding transient cavity. It can be estimated from the Manicouagan melt sheet that chemical homogeneity reaches at least over horizontal distances of tens of kilometers and vertical distances exceeding 200 m, although local irregularities exist (Floran et al. 1978). Commonly observed variations in alkali elements may be due to preferred assimilation of country rock material with a low melting point or to secondary processes, documented for a number of craters. Excess of calcium and, to a lesser degree, of magnesium may indicate a cover of carbonatic sediments in the target area at the time of the impact, which has now been eroded. Within a crater, the observed variation in REE abundances between different melt rock samples is less than that for the target lithologies, whose REE distribution patterns bracket those of the impactites. On the level of isotopes, complete homogenization is rarely reported. Variable initial Sr isotopic ratios are an important primary characteristic of coherent impact melt layers (Jahn et al. 1978), and the Rb-Sr data for whole rock samples define mixing lines or cluster in the field delimited by target lithologies. Samarium-neodymium isotope systematics seem to be determined by the presence of inherited refractory minerals – at least at the kg scale of whole rock samples, and, hence, Nd model ages help to delimit precursor material of impact melt rocks. If sufficient geochemical and geological data are available, the relative proportions of target lithologies contributing to the impact melt rocks can be modeled as mixtures of particular basement lithologies by means of least-squares mixing programs. This has be done for several craters, and usually silica, aluminum, and iron yield the best fits. It should be noted, however, that hydrothermal processes initiated by, e.g., circulating fluids driven off crater basement lithologies by the hot overlying melt sheet, may obliterate some of the given clear geochemical features.

At small craters with limited impact melt layers, differentiation, if it occurs at all, is certainly restricted, due to the extreme sheet-like character of the melt body and the high viscosity of the clast-laden melt during crystallization. At Sudbury, however, the more than 2.5 km thick impact melt (so-called Sudbury Igneous Complex) differentiated like a classical endogenic magma (see below). So far, other examples for differentiated impact melt sheets are unknown from terrestrial impact structures.

Distant Ejecta

Distant ejecta comprise variably shocked mineral and rock fragments, high-pressure phases, impact melt glass, including tektites, as well as spherules of widely varying chemical and textural composition and morphology, amongst them so-called microkrystites, and exotic spinel crystals; the latter probably represent high-temperature condensates from the vapor plume. Distant ejecta material may be heavily contaminated by constituents of the impactor, resulting in enhanced abundances of siderophile elements (mostly PGEs). Exotic chemical components such as fullerenes and extraterrestrial amino acids (α-amino-isobutric acid, and racemic isovalin) have been reported only from the K/T boundary ejecta layer and are still a subject of discussion. Geochemical signals like sharp excursions in the stable isotope record may characterize marine sediments on top of distant ejecta deposits. However, these "anomalies" are secondary consequences of impact, and can only be related to the short- and long-term corollaries of a cratering event if unambiguous mineralogical evidence for impact metamorphism is documented (e.g., planar deformation features in quartz). Distant ejecta can occur up to thousands of kilometers away from the crater. The material, having suffered rapid quenching, is either ejected ballistically or suspended and transported by large-scale motions in the stratosphere.

Tektites (derived from *τεκτοσ* = molten) are a subgroup of impact glasses characterized by their chemical homogeneity and an H_2O content below 0.02 wt.% (Koeberl 1994; Beran and Koeberl 1997). Compared to volcanic glasses of similar silica-rich composition, tektites are depleted in elements of different volatility (e.g., Cu, Ce, Sn, Pb), whereas, enrichments in Cr, Ni, and Co, as well as low $^{187}Os/^{188}Os$ isotope ratios, reflect incorporation of vaporized projectile material. Tektites may contain inclusions of lechatelierite (SiO_2

glass), coesite, relics of shocked minerals, and baddeleyite as decomposition product after zircon. The subgroup of Moung Nong-type tektites displays alternating light- and dark-colored layers and carries submicron-sized crystals of corund, zircon, rutile, and chromite. Tektites originate extremely early in a cratering event by impact melting of the uppermost target rock layer, followed by jet ejection of the melt at low angles with supersonic velocity. The up to several centimeters large objects often display spherical symmetry like teardrops or dumbbells due to quenching of rotating liquids during transport. The aerodynamically shaped flange-button Australites provide evidence for re-melting in the course of re-entering the atmosphere. Moung-Nong-type tektites have a blocky appearance and the mass of single objects may exceed 24 kg. Geochemical characteristics of tektites, e.g., ^{10}Be abundances, reflect the origin from near-surface material, and their Sr-Nd isotope systematics have been used successfully in some cases to delimit the source region or to identify the precursor lithologies (Blum et al. 1992).

Tektites, which are ideal objects for Ar-Ar dating, occur in four major strewn fields (e.g., Koeberl 1994). These are the Australasian, the Ivory Coast, the Central European (= Moldavites), and the North American strewn fields. Isolated discoveries are, e.g., the 24 Ma Urengonites in Siberia (Deutsch et al. 1997), tektite-like objects near the Frasnien-Fammenian boundary in Belgium and China, and microtektites with a diameter of generally less than 1 mm in deep-sea drill cores, for example, the ca. 35.5 Ma microtektites detected in Caribbean Sea cores close to the Eocene-Oligocene boundary. Impact melt glasses display shapes similar to tektites, but differ by a slightly higher H_2O content (0.02 to 0.06 wt.%) and more heterogeneous chemistries. This has been documented, e.g., for impact melt glass droplets at the K/T boundary covering a broad range from calcium to silica-rich compositions due to the incorporation of varying proportions of different sedimentary precursor rocks. Apparently, impact melt glasses originate from deeper levels of a growing impact crater than tektites.

The Createaceous-Tertiary boundary stands as uniquely documented example for the so far only known global occurrence of distant ejecta (Smit 1994). This may be due to the fact that the knife-sharp extinction horizon in the marine realm, and the well-known fern-spore peak in continental sediments of North America facilitate the precise location of the sometimes very thin ejecta layer (< mm) in undisturbed sections. Mineralogical, sedimentological, and geochemical anomalies characterize this ejecta layer, which is linked to the Chicxulub impact structure with a diameter on the order of 200 km (Morgan et al. 1997). Compared to the number of proven craters, impact debris is very scarce in the stratigraphic column (Grieve 1997). Only in three other cases, the source crater for distant ejecta material impact has been identified unambiguously. These are (1) the 1.03 Ma Ivory Coast tektites with a total mass of 20×10^9 g, which are connected with the Bosumtwi Structure (D = 10.5 km) in Ghana, (2) Moldavites with a total mass of 3×10^9 g, which are correlated with the 15 Ma Nördlinger Ries, and (3) the ejecta horizon of the approximately

590 Ma Lake Acraman Structure (D = 90 km), which has been traced over an area of 20000 km^2 in the monotonous Late Proterozoic shaly sediments of the Adelaide Geosyncline, South Australia. This ejecta layer consists of up to fist-sized rock fragments, shocked minerals, spherules, and shards pseudomorphosed after impact glass, and shows a pronounced anomaly of siderophile elements (Gostin et al. 1989). In addition, based on magnetostratigraphic, chemostratigraphic, and biostratigraphic arguments, Popigai (D = 100 km) is suggested as source crater for the PGE-enriched layer at the end of the Eocene (Langenhorst 1996), where shocked mineral grains and microtektites occur, but confirmation by high-precision ages is so far lacking. Most recently, the Chesapeake Bay impact structure has been proposed as source crater for tektites of the 35.5 Ma North American strewn field (Poag 1996).

Systematic search of the sedimentary record for distant ejecta layers by, for example, examination of quartz for shock features has proved to be painstaking and unsuccessful due to dilution by undeformed quartz grains of sedimentary origin (e.g., Schmitz et al. 1994). So far, distant ejecta deposits have been recognized first by Ir anomalies, followed by very detailed mineralogical analysis to prove the impact origin. Still controversially interpreted are unusual spherule deposits in the Early Archenan Barberton greenstone belt of South Africa (for a recent critical discussion, see Koeberl and Reimold 1995). The two beds, up to 1.5 m thick, are enriched in Ir and contain spinels of peculiar morphology, composed dominantly of chalcophile elements and with high Ni and ferric iron contents. The spinels are seen as products of condensation from a mixed target bolide vapor. If this interpretation is correct, these layers, and a recently proposed horizon in the around 2.5 Ga Hamersley group, W.A., Australia (Simonson et al. 1996), would represent the Earth's oldest distant ejecta deposits. The high cratering rate in the Early Archean implies that more distant ejecta layers should exist in such old sediments; they are, however, not yet identified.

1.10.5 Minerals in Terrestrial Impact Structures and their Characteristic Features

F. Langenhorst and A. Deutsch

In the context of impact events, minerals can be affected or formed by a large variety of processes. From the genetic and chronological point of view, a distinction must be made between (1) primary minerals displaying shock-metamorphic overprint (including formation of glass and high-pressure polymorphs), and (2) new minerals formed by post-shock crystallization of impact melt or post-shock condensation of vaporized material. Shocked minerals are unequivocal indicators of impact events. They are, therefore, not only indispensable to prove the impact origin of a suspected crater-form structure, but also to finally confirm certain sedimentary features as distal ejecta layers (Bohor et al.

Advanced Mineralogy, Volume 3
Mineral Matter in Space, Mantle, Ocean Floor, Biosphere, Environmental Management, and Jewelry (ed. by A. S. Marfunin)

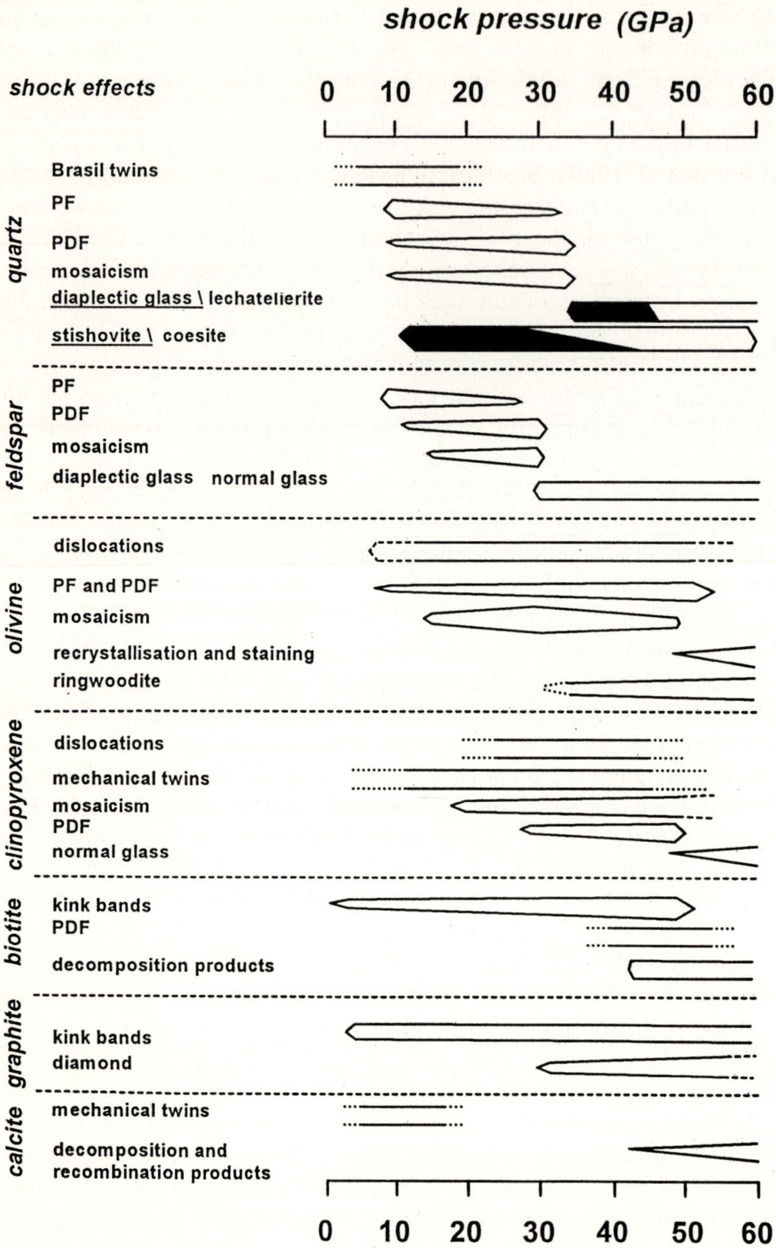

Fig. 24. Occurrence of specific shock effects in rock-forming minerals as function of shock pressure. Data obtained by shock experiments on compact samples.

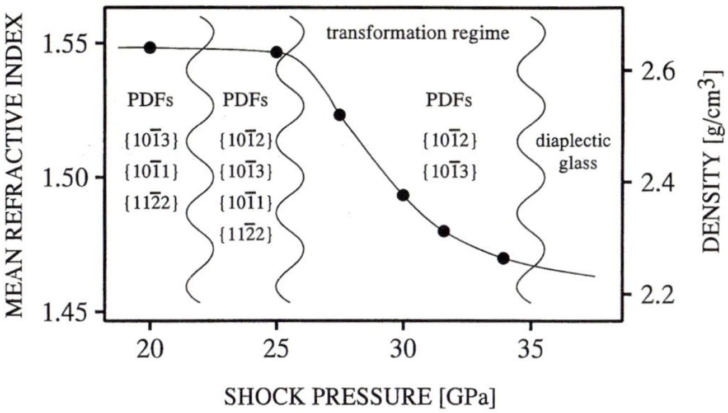

Fig. 25. Mean refractive index, density, and combinations of PDF orientations of experimentally shocked quartz as function of shock pressure. These properties serve as shock barometers (Langenhorst and Deutsch 1994).

1984). In addition, shock signatures of primary minerals provide important clues to shock conditions. Calibration data on the formation of specific shock effects in certain pressure regimes are available from shock recovery experiments on rocks or minerals (Figs. 24, 25). The knowledge of shock pressure and its attenuation rate is crucial for scaling impact craters and understanding cratering mechanics. Among newly formed minerals, spinels in distant ejecta deposits are of special interest, as their characteristics could help to better model physical and chemical processes taking place in the expanding vapor plume, although, so far, knowledge in this field is rather limited. Characteristics and conditions of formation of both shocked and newly formed minerals and their significance in impact research are the subjects of this chapter. Both types of minerals are often altered by post-shock annealing and/or hydrothermal overprinting. In as far as these processes yield secondary features or minerals, which are still suggestive of impact, they will also be discussed.

Shock Deformation and Transformation of Minerals

During shock compression and decompression, unique physical and chemical changes in minerals and rocks, generally referred to as shock effects, are induced. Shock effect is a collective and neutral term covering different categories of shock-induced changes, such as formation of lattice defects, high-pressure polymorphs, and diaplectic glasses, as well as changes in physical properties. Changes in physical properties are considered as indirect shock effects simply reflecting shock-induced phase transformations. In general, the term shock effect should be used only for changes originating during shock compression and decompression. Post-shock thermal effects, such as dissociation of minerals in a superheated (≥2000 °C) impact melt, can be typical for the impact environment, but are not directly related to the shock process (Fig. 20). As a consequence of their short formation time (<1 s in nature), shock effects represent disequilibrium changes that are, to some extent, unknown for minerals deformed in endogenic regimes.

Although a basic knowledge of some shock phenomena was already available from the petrographic study of meteorites (e.g., "maskelynite", a diaplectic plagioclase glass, was first described by Tschermak in 1872), most shock effects were discovered only in the late 1950 and early 1960 in either experimentally or naturally shocked material. At that time, shock effects were characterized by optical microscopy, which, due to limited resolution, prevents a clear recognition of the physical nature of shock effects. Recent transmission electron microscope (TEM) observations have largely improved knowledge and understanding of shock effects and have even led to the discovery of new submicroscopic shock effects, such as dislocations and microtwins; this field of research established a link to the material sciences. In this chapter, we concentrate on the new information which is now available for the rock-forming minerals, quartz, feldspar, olivine, and pyroxene. Characterization and definition of the most important shock effects are summarized in Table 9.

Table 9. Shock effects in minerals

Characteristic shock effects in minerals	Short definition
1. Formation of crystal defects	
a) Dislocations	Linear lattice defect
b) Planar microstructures	
– Planar fractures	Crystallographically oriented cleavage planes activated under dynamic compression
– Planar deformation features	Crystallographically oriented sets of amorphous lamellae (undecorated PDFs) and their post-shock modifications (decorated PDFs)
c) Mechanical twins	Crystal domains related by a point symmetry element (mirror, rotation or inversion axis), produced by deformation
d) Kink bands	Externally rotated crystal domains without crystallographic relationship to the host lattice, produced by deformation
e) Mosaicism	Internal blocky structure of shocked crystals
2. Transformations to	
a) High-pressure polymorphs	Densely packed modifications of minerals
b) Diaplectic glass	Quenched high-pressure melt preserving shape, internal textures, and composition of the precursor crystal
3. Decomposition	Dissociation of minerals into new solid phases (and gaseous species)
4. Melting and vaporization	Production of rock and mineral melt, and vapor

Dislocations

In its simplest form, a dislocation is a narrow line defect in which a plane of atoms stops within a crystal. Formation and migration of dislocations is the mechanism of deforming crystals under high stress. TEM analyses indicate that dynamic (shock) deformation is able to produce a large number of dislocations in certain minerals. Shocked olivine and clinopyroxene stand as instructive examples for minerals with high dislocation densities. On the other hand, shock cannot activate dislocations in minerals like quartz, in which the known formation mechanisms are simply too slow for activation of dislocations within the μs to s shock pulse.

Quartz. The lack of perfect dislocations in shocked quartz is due to its special mechanical behavior (Stöffler and Langenhorst 1994). Static deformation experiments have shown that the strength of dry quartz is comparable to the theoretical elastic limit of a few GPa. In the presence of water, however, quartz becomes ductile, a phenomenon known as hydrolytic weakening. Water in quartz crystals is the source for grown-in point defects. Diffusion of these water-related defects controls formation and mobility of dislocations. This is an extremely slow process compared to the time scale of shock compression, impeding formation and multiplication of dislocations in dynamically deformed quartz. When naturally shocked quartz contains numerous dislocations, these may have been formed by pre- or post-shock tectonic deformation.

Olivine. Dislocations are highly concentrated in olivine crystals from shock experiments and heavily shocked meteorites with densities up to 2×10^{14} m^{-2} (Fig. 26a). Since dislocations in olivine generally occur in the immediate vicinity of planar fractures, fracturing is interpreted as the cause of their formation (Langenhorst et al. 1995). The dislocations are almost exclusively of pure screw character and have the Burgers vector [001]. If present, edge segments are very short, enabling determination of the shock-induced slip systems, which are (100)[001], {110}[001], and (010)[001]. Apparently, dislocations with the Burgers vector [001] are quicker to activate than the energetically favorable dislocations with the shorter Burgers vector [100]. This consideration is consistent with results for olivine of static deformation experiments, revealing the dominance of dislocations with the Burgers vector [001] at relatively high strain rates (Carter and Avé Lallemant 1970).

Clinopyroxene. Shocked clinopyroxene is dominated by edge dislocations occurring with densities of up to 10^{14} m^{-2}. The predominant slip systems of dislocations are (100)[001] and, to a lesser extent, {110}[001]. Instead of a random distribution as in olivine, dislocations in clinopyroxene are often organized in the form of bands. As revealed by shock experiments (Leroux et al. 1994a), these dislocations originate at Frank-Read sources (Fig. 26b). A migration velocity of 500 ms^{-1} has been calculated from the distance to Frank-Read

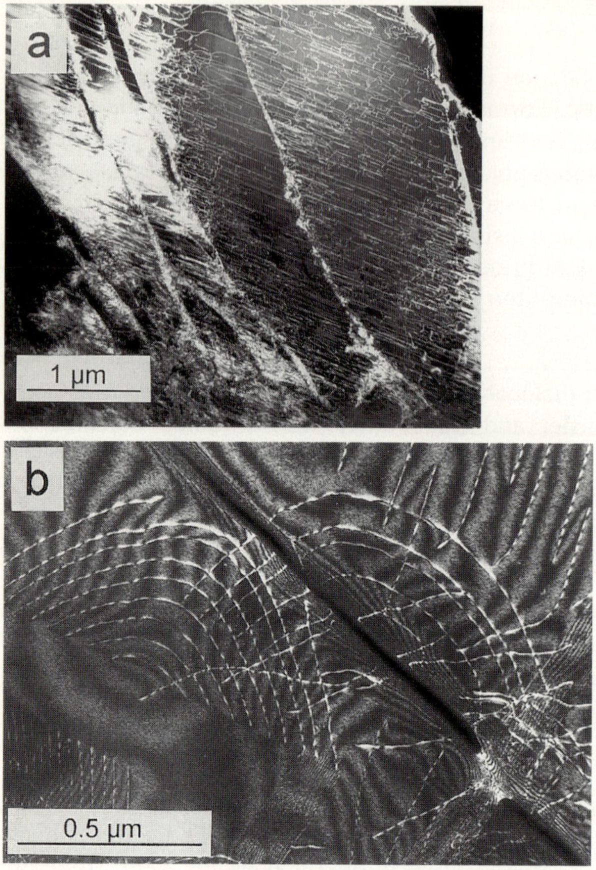

Fig. 26 a, b. TEM dark field images of one-dimensional shock defects (dislocations) in olivine and pyroxene. **a** Dislocations with Burgers vector [001] and irregular fractures in olivine from the Kernouvé chondrite, which has been experimentally shocked at 35 GPa. **b** Frank-Read source in diopside experimentally shocked at 45 GPa (Courtesy H. Leroux).

sources and the length of the shock pulse. Due to the significantly longer shock pulse in natural impact events, Frank-Read sources are more difficult to detect in naturally shocked clinopyroxene, since dislocations have migrated far away from their source.

Planar Microstructures

Many rock-forming minerals, e.g., quartz, feldspars, and mafics fail mechanically upon shock compression in two basic modes: by formation of planar fractures and planar deformation features. "Planar microstructures" as defined by Stöffler and Langenhorst (1994) is a collective term for these two different types of microstructures, which are both oriented parallel to rational crystallographic planes of low Miller indices.

Planar Fractures (PFs). PFs usually form sets of parallel open fissures with a spacing of about 20–30 μm, allowing unambiguous recognition with the optical microscope, and occur in nearly all rock-forming silicates such as quartz, pyroxene, olivine, feldspar, amphibole, and garnet. PFs are unequivocal shock indicators if they are present (1) in minerals without normal cleavage, or (2) along planes which are not parallel to the normal cleavage of a mineral. Quartz, for example, known for brittle and irregular fracturing when statically deformed, develops under shock compression distinct PFs parallel to low index planes such as (0001) and $\{10\bar{1}1\}$ (Fig. 27a). Olivine, displaying in the unshocked state only indistinct cleavage, fractures under shock compression preferentially parallel to the crystallographic planes (100), (010), (001), (110), (130), and (111). Orthopyroxene stands as example of a mineral, in which the shock-developed PF orientations such as (001) and (010) differ from the normal cleavage, which is parallel to (210) and (100). In some minerals, PFs commonly act as grain boundaries for the much more closely spaced planar deformation features.

Planar Deformation Features (PDFs). PDFs are multiple sets of parallel, planar optical discontinuities with a typical spacing <5 μm (Grieve et al. 1996). PDFs of the undecorated type ("fresh" PDFs) are resolvable by the optical microscope as unsharp, thin (<1–2 μm) lamellae. PDFs become clearly visible if, as an effect of post-shock annealing, their planes are decorated with tiny vugs (decorated PDFs). Decoration of PDFs is well known for quartz. At the TEM scale, fresh PDFs represent thin glass lamellae with a composition identical to the host crystal. These glass lamellae have been detected by TEM in quartz, feldspar, and clinopyroxene (e.g., Kitamura et al. 1977; Gratz et al. 1992; Leroux et al. 1994a; Fig. 27b, c). Based on optical observations, "PDFs" have been reported for a number of other minerals, e.g., olivine, amphibole, scapolite, biotite, sillimanite, apatite, and zircon (Stöffler 1972; Bohor et al. 1993), yet confirmation of the glassy nature of these planar microstructures by TEM is lacking. Minerals such as quartz and feldspars with a relatively low density, i.e., $<3\ g\ cm^{-3}$, and thus high compressibility, tend to develop more and thicker PDFs than mafics such as pyroxene and olivine.

Quartz. Shocked quartz displays multiple sets of well-developed PDFs pervading the entire host crystal. Single quartz crystals may contain up to 15 differently oriented sets of PDFs. Displacements are not observed at the intercepts of PDF sets, indicating a shearless formation. The recent model of PDF formation assumes an origin directly in the shock front, which separates as sharp boundary uncompressed and compressed parts of the affected crystal (Goltrant et al. 1992). The lattice misfit between the two parts is compensated by the formation of densely packed amorphous lamellae, the PDFs. Predominant PDF

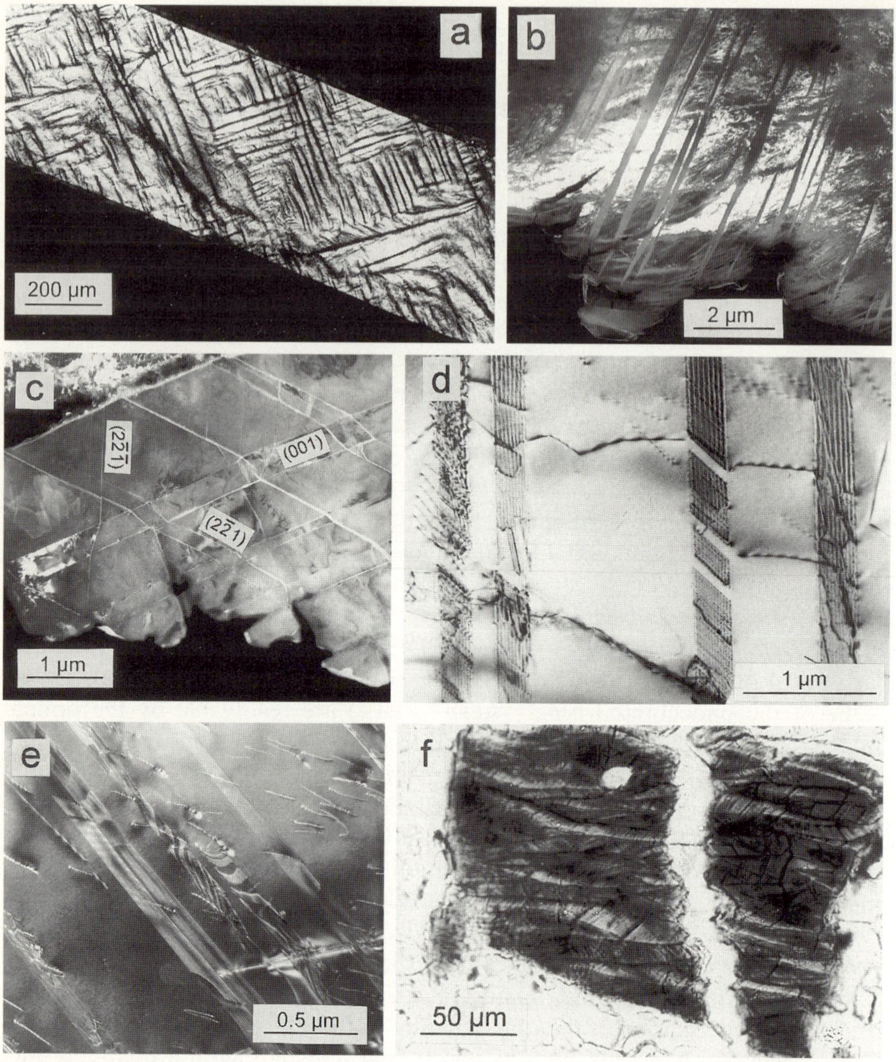

Fig. 27a–f. Two-dimensional shock effects (planar microstructures, mechanical twins, kink bands) in minerals. **a** Optical micrograph of planar fractures (PFs) in quartz experimentally shocked at 27.5 GPa (Langenhorst and Deutsch 1994). PFs are oriented parallel to $\{10\bar{1}1\}$. **b** TEM bright field image of amorphous lamellae, the PDFs, in quartz experimentally shocked at 30 GPa (Langenhorst 1994). **c** TEM dark field image of two PDF sets parallel to {221} in shocked diopside from the ordinary chondrite Tenham (Langenhorst et al. 1995). PDFs are deflected by mechanical twins parallel to (001). **d** TEM bright field image of basal Brazil twins in shocked quartz from the collar region of the Vredefort impact structure. (Courtesy H. Leroux). Twins are inclined to the electron beam giving rise to the formation of the fringe pattern. Twin boundaries contain numerous partial dislocations. **e** TEM dark field image of shocked diopside in the ordinary chondrite Tenham. Fringe patterns are inclined twins parallel to (100). Besides perfect dislocations, partial dislocations are visible in the twin boundaries. **f** Optical micrograph of kink bands in biotite in a suevitic granite clast from Seelbronn, Ries crater.

orientations in quartz are rhombohedral forms $\{10\bar{1}n\}$ with n = 1, 2, and 3, and trigonal bipyramids $\{11\bar{2}n\}$with n = 1 and 2. Identification of other reported orientations, e.g., $\{21\bar{3}1\}$, $\{22\bar{4}1\}$, $\{51\bar{6}1\}$, $\{10\bar{1}0\}$, and $\{11\bar{2}0\}$ is based only on optical measurements. The frequently reported orientation (0001), known as basal deformation features (Carter 1965), does not belong to the PDF type of microstructures but is a fundamentally different type of defect, namely Brazil twins.

Universal stage measurements on PDFs in naturally shocked quartz have revealed that specific crystallographic PDF orientations or combinations thereof vary systematically with increasing shock intensity (von Engelhardt and Bertsch 1969; Robertson and Grieve 1977; Gurov and Gurova 1991; see literature in Stöffler and Langenhorst 1994). This pressure dependence of PDF orientations has been calibrated experimentally (Hörz 1968; Langenhorst and Deutsch 1994), and serves as a reliable barometer for estimating shock pressures in impactites; a compilation of PDF orientation data combined with refractivity and density data is given in Fig. 25. TEM studies on experimentally shocked quartz indicate that the mean thickness of PDFs correlates with pressure: below 25 GPa, PDFs are 20 to 30 nm thick, whereas those produced above 25 GPa are about 200 nm thick (Ashworth and Schneider 1985; Langenhorst 1994).

Physical properties of shocked quartz are distinctly influenced by the amount of PDFs. Larger amounts of these amorphous lamellae cause a reduction in refractivity and density. Enhanced lattice parameters of experimentally shocked quartz have also been explained by the presence of PDFs, which produce a distortion of the crystal lattice (Langenhorst 1994).

Feldspars. PDFs in feldspars also consist of amorphous material; they closely resemble those in quartz, and occur in a similar pressure range from 10 to 30 GPa. Their thickness may amount to 300 nm. Due to the variable composition and low symmetry of feldspars, especially plagioclase, it is difficult to reliably index PDF planes with stereographic projection. Dominant PDF orientations in andesine and labradorite are (001), (010), and (100); other reported orientations are (101), (111), $(1\bar{2}0)$, (130), and (203) (Stöffler 1972).

Clinopyroxene. The occurrence of PDFs in clinopyroxene was noted long ago by optical microscopy, but conclusive evidence for their presence has been provided only recently by TEM studies on clinopyroxene, either experimentally shocked or occurring in heavily shocked ordinary chondrites (Leroux et al. 1994a; Langenhorst et al. 1995; Fig. 27c). PDFs in clinopyroxene start to appear above 30 GPa and are thin (~50 nm) glass lamellae with a typical spacing <1 μm. The monoclinic prisms {221} and {331} are the dominant PDF forms in clinopyroxene; they generally coexist with mechanical twins. In contrast to quartz, shearing seems to play an important role in PDF formation in clinopyroxene because PDF sets are displaced if they intersect each other or mechanical twins.

Mechanical Twins

Twinning is a common reaction of minerals of low symmetry to shock deformation. Mechanical twins in shocked minerals are visible under the optical microscope as narrow (<10 μm) polysynthetic bands; due to this appearance, the twins could be mistaken as exsolution lamellae. Extended and massive twin areas, however, are unknown in shocked minerals. Shock-induced mechanical twin lamellae have been reported for quartz, clinopyroxene, amphibole, sphene, ilmenite, and calcite.

Quartz. TEM studies revealed that one characteristic feature of weakly shocked quartz is mechanical Brazil twins with a mean thickness of about 50 nm; they are oriented exclusively parallel to (0001) (Kieffer et al. 1976; Goltrant et al. 1992; Leroux et al. 1994a). The twin boundaries contain numerous partial dislocations, indicating the mechanical nature of the Brazil twins (Fig. 27d). Identification of this feature by optical methods is complicated in thin sections of normal thickness if the twin boundaries are free of tiny fluid inclusions due to secondary alteration (decorated Brazil twins). The similar appearance of decorated PDFs and decorated Brazil twin boundaries was the reason for the erroneous description of the mechanical twins as PDFs.

Brazil twinning involves a change form left- to right-handed quartz or vice versa. It is known as either deformation or growth defect. The deformation Brazil twins are exclusively oriented parallel to (0001) and decorated with partial dislocations. In contrast, growth Brazil twins lack such dislocations and occur parallel to $\{10\bar{1}1\}$ and $\{11\bar{2}0\}$. Mechanical Brazil twins have not yet been reported for quartz shocked experimentally in the critical pressure range up to 15–20 GPa for probably two reasons: (1) thorough TEM search for Brazil twins in such material is still lacking, and (2) shock experiments were mainly performed with unfavorable quartz orientations [e.g., shock front parallel to (0001)], which impede high shear forces on (0001) and, thus, the production of Brazil twins. In static deformation experiments, however, thin mechanical Brazil twin lamellae develop if high (>1.5 GPa) and rapid shear stress is applied on (0001) (McLaren and Phakey 1966). Such conditions are absent in endogenous environments, and, hence, the occurrence of mechanical Brazil twins parallel to the basal plane is diagnostic for shock deformation in nature. The presence of Brazil twins proved to be an especially useful shock indicator when post-shock annealing has caused disappearance of the less robust PDFs.

Clinopyroxene. Clinopyroxene shocked in experiments, nuclear explosions, and natural impact events shows two types of mechanical twins with twin boundaries parallel to (100) and (001) (Müller 1993; Leroux et al. 1994a; Langenhorst et al. 1995; Fig. 27c, e). Both twin operations involve a change in the crystal orientation. The (100) twins are distinctly thinner (<10 nm) than the (001) twins (<0.5 μm). Therefore, only the latter can be identified by optical microscopy on the basis of alternating extinction. The (100) twin boundaries contain concentric partial dislocations which are absent in the (001) boundaries. Thus, (100) twins are interpreted to form by gliding of partial dislocations, whereas (001) twins might originate from a rapid kinking process. If this explanation is correct, (001) twins represent a specific type of kink bands with a crystallographic relationship to the host crystal. Both types of twins have been produced in static experiments with high strain rates (10^{-4} s^{-1}; Kirby and Christie 1977); however, boundaries of the statically produced (001) twins contain partial dislocations, indicating a formation by gliding. Tectonically deformed diopside shows only (100) twins (Skrotzki 1994). The occurrence of dislocation-free (001) twins in diopside is therefore considered as reliable shock indicator.

Other Minerals. Little is known of the twinning behavior of other minerals. Amphibole has been reported to develop mechanical twins parallel to (001) and (100). In ilmenite, twins occur parallel to $\{10\bar{1}1\}$ and (0001), and sphene displays twins parallel to {221}. Mechanical twins were assumed to be rarely present in plagioclase, but proof for their mechanical nature is difficult because this mineral shows a large variety of polysynthetic growth twins. Shock-induced mechanical twinning probably takes place in carbonates. Calcite in limestones with shatter cones from impact structures shows intense microtwinning parallel to $\{01\bar{1}2\}$ and $\{10\bar{1}1\}$ (Robertson and Mason 1975). Similar twins form also in static deformation experiments (Barber and Wenk 1979).

Kink Bands

Kinking is a deformation mode commonly found in minerals with a sheet structure, e.g., micas and graphite, which exhibit only one dominant glide plane. Kinking develops if stress is applied in such a manner that extensive gliding is geometrically impossible, i.e., if the shock front is approximately perpendicular to the glide plane. In contrast to deformation twins, kink bands are not oriented parallel to rational crystallographic planes and display variable disorientation with respect to the host lattice.

Biotite. Kink bands have been studied most intensely in shocked biotite from the Ries crater, shock experiments, and nuclear test sites (Cummings 1964; Schneider 1972; Lambert and MacKinnon 1984; Fig. 27f). Compared to tectonically deformed biotite, shocked biotite shows a strong asymmetry of kink bands. Degree of asymmetry, frequency, and width of kink bands are correlated to shock pressure. In the case of the Ries crater, the orientation of kink bands has been used successfully to reconstruct the original position of ejected and overturned megablocks of the crystalline basement (Graup 1978).

Mosaicism

The term mosaicism describes the internal fragmentation of a single crystal into a mosaic of slightly disoriented crystal domains. Mosaicism is a common shock-wave damage in minerals, often accompanied by other shock effects such as planar microstructures. In thin sections, mosaicism appears as a highly irregular, "mottled" extinction pattern, which is distinctly different from undulatory extinction. The internal fragmentation causes a reduction of the long-range order of a crystal, allowing unambiguous detection by various X-ray and electron diffraction techniques (see references in Stöffler and Langenhorst 1994). In single-crystal diffraction patterns, mosaicism results in streakiness and broadening of the diffraction spots. In powder diffraction patterns, line broadening and a loss of higher-order reflections is observed, indicating domain sizes of <200 nm and the presence of internal strain. Successful attempts were made to quantitatively correlate domain size and internal strain with shock pressure using quartz, pyroxene, and carbonates (e.g., Dachille et al. 1968; Hörz and Quaide 1972; Hanss et al. 1978; Schneider et al. 1984; Ashworth and Schneider 1985; Langenhorst 1994; Martinez et al. 1995).

Diaplectic Glass

Minerals known to develop PDFs show at enhanced shock pressures an isochemical transformation to the amorphous state. This shock-produced glass is defined as diaplectic (from the Greek *διαπλεσσοσ* = to destroy by striking; von Engelhardt et al. 1967) or thetomorphic glass, yet the latter term has not gained acceptance, and is considered obsolete. Diaplectic glass is characterized by the preservation of morphological and textural features of the former crystal, e.g., crystal shape and twin boundaries, and by the absence of flow structures and vesicles. These features allow a clear distinction between diaplectic and shock-induced melt glass in thin sections. Diaplectic glass displays higher refractive index and density than synthetic glass of identical chemical composition. The formation of diaplectic glass, one of the most characteristic features of shock metamorphism, is mainly restricted to quartz and feldspars, i.e., minerals with open crystal structures containing three-dimensionally linked $[SiO_4]$ tetrahedra. Transformation to diaplectic glass is strongly related to the formation of PDFs (Langenhorst 1994).

Quartz. In shocked quartz, amount and width of amorphous lamellae (PDFs) increase gradually with shock pressure until the PDFs coagulate at about 35 GPa to an entirely amorphous phase, the diaplectic glass. Calculation of shock temperatures and TEM observations indicate that PDFs formed at pressures >25 GPa represent, in compressed state, molten bands of silica. Therefore, diaplectic quartz glass has been interpreted as a high-pressure melt phase quenched upon decompression (Langenhorst 1994). Diaplectic quartz glass may hence have maintained the densified structure of the melt, resulting in enhanced refractivity and density; its refractive index ranges up to 1.468, in contrast to 1.459 for synthetic silica glass (Stöffler and Langenhorst 1994). In nature, diaplectic quartz glass often coexists with the high-pressure polymorphs coesite and, to a minor extent, stishovite.

Due to the continuous nature of the transformation from shocked quartz with PDFs to diaplectic glass, refractive indices, birefringence, and density change continuously in the pressure range from 25 to 35 GPa (Fig. 25). These properties have been precisely calibrated by shock experiments on single crystal quartz, and serve as the most accurate indicator in shock barometry of terrestrial impactites. Careful application of the calibration data is required because pre-shock properties of the rock such as porosity, temperature, and polycrystallinity of the quartz influence onset and completion of the transformation, causing a shift in the pressure reading.

Feldspars. Diaplectic feldspar glass is a common constituent of terrestrial, lunar, martian impact lithologies, and ordinary chondrites (Stöffler et al. 1991). Maskelynite is a synonym for diaplectic glass of plagioclase composition. Formation of diaplectic feldspar glass is assumed to equal the transformation of shocked quartz to diaplectic quartz glass. Shocked feldspars also develop broad PDFs (although not to the same amount as quartz), which coagulate at enhanced pressures to an entirely amorphous phase. Chemical composition of unaltered diaplectic feldspar glass resembles that of the former feldspar; substantial loss of alkalies has not been observed. Systematic shock experiments have shown that the change in optical properties of alkali feldspars and plagioclases correlates with shock pressure (Ostertag 1983). According to the incipient decrease of refractive indices, transformation to diaplectic plagioclase glass begins in the pressure range of 22 to 25 GPa, and is complete between 26 and 30 GPa, dependent on the chemical composition. Albite transforms at higher pressures, and shows a stronger decrease in refractivity than anorthite. The gradual change in physical properties of feldspars also serves as a shock barometer, although it is less sensitive to pressure than in the case of quartz. This is due to the fact that the drop in refractivity is less significant, and, moreover, variable chemical compositions yield additional uncertainties. A peculiarity of diaplectic feldspar glasses is that they can have refractive indices and densities comparable to those of the unshocked original crystals.

High-Pressure Polymorphs

The presence of high-pressure polymorphs is striking evidence for natural impact events and, in some cases, has provided the ultimate proof for the impact origin of controversially interpreted geological structures such as the Ries and Vredefort. Quartz, graphite, orthopyroxene, and olivine are known to undergo partial or complete phase transitions to high-pressure polymorphs. Quartz transforms to coesite and stishovite, graphite to diamond and lonsdaleite, orthopyroxene to majorite, ilmenite, and perovskite, and olivine to wadsleyite and ringwoodite. Except the solid-state transitions in the carbon system, these transformations are not direct and require, in an intermediate state, the formation of a high-pressure melt from which the high-pressure polymorphs crystallize. On the basis of Hugoniot data, the presence of high-pressure polymorphs has been proposed for zircon, rutile, and plagioclase, yet such phases have never been observed in naturally or experimentally shocked material.

Coesite and Stishovite. Both silica high-pressure polymorphs occur in dense crystalline impactites, shocked sandstone with high porosity, and thin pseudotachylite veins in quartzite country rocks (Stöffler 1971; Kieffer et al. 1976; Martini 1991; White 1993; Leroux et al. 1994b). In dense crystalline rocks, up to 40% of the host quartz is transformed to coesite, and up to 5% to stishovite. Coesite is found in quartz with PDFs and reduced refractive index, and, more frequently, in diaplectic glass, whereas stishovite occurs commonly in quartz with PDFs and less frequently in diaplectic glass. The estimated pressure ranges for the formation of coesite and stishovite are 30–60 GPa, and 12–45 GPa, respectively (Stöffler and Langenhorst 1994). Significant amounts of high-pressure polymorphs have been found in the porous Coconino sandstone of the Barringer crater but, in this case, coesite coexists with more than 80% of quartz, displaying PFs and diaplectic quartz glass. Stishovite occurs in Coconino sandstone at a higher shock level in an assemblage consisting of fractured quartz, amorphous silica, and coesite. The coexistence of the high-pressure polymorphs with fractured quartz implies formation at rather low shock pressures on the order of about 10 GPa. Thin pseudotachylites veins at the Vredefort Structure contain coesite and stishovite in paragenesis with silica glass and recrystallized quartz (Martini 1991); the immediate wall rocks carry quartz with mechanical Brazil twins.

According to TEM studies, coesite and stishovite form fine-grained (<3 μm), polycrystalline aggregates (White 1993; Leroux et al. 1994b). Coesite shows intense growth twinning parallel to (010) (Fig. 28a). These characteristics, as well as the general association of both high-pressure phases with diaplectic quartz glass or lechatelierite, provide evidence for a formation by rapid crystallization from a high-pressure silica melt.

Stishovite has been synthesized in shock experiments on quartz but, due to the extremely short pressure pulse of experiments (cf. Fig. 20), to a distinctly lesser extent ($\leqslant$1%) than in nature. This observation supports the outlined crystallization model. Formation of coesite has been reported in a shock experiment with quartz powder (Deribas et al. 1966) but could not be duplicated since.

Coesite, interpreted as an unambiguous shock indicator, has also been discovered in highly tectonized slabs of subduction zones and volcanic diatremes as inclusions in high-pressure phases such as diamond and pyrope. Mineral assemblages, texture, and crystal appearance, however, allow a clear distinction between endogenic and shock-produced coesite.

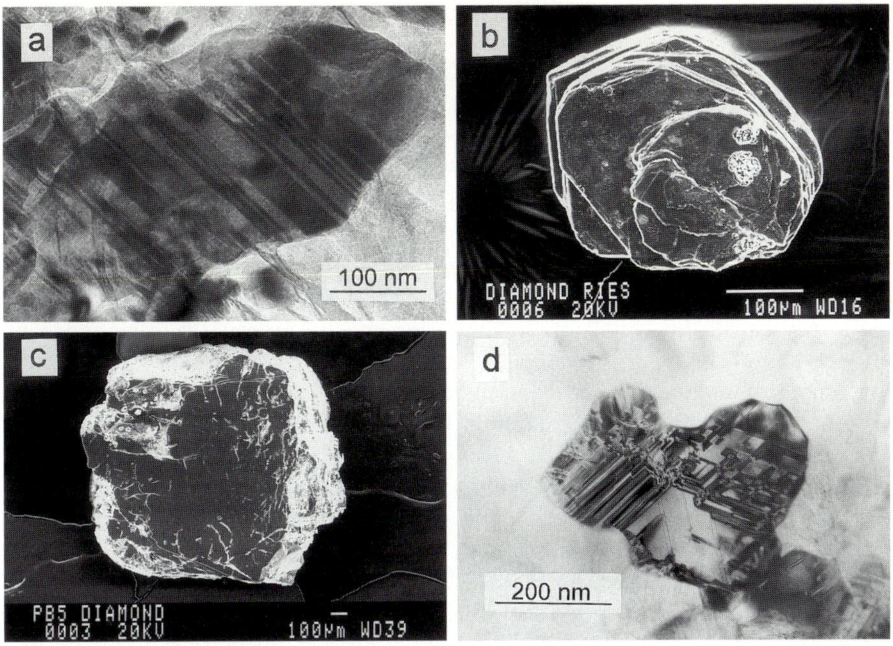

Fig. 28 a–d. High-pressure phases found in terrestrial and extraterrestrial impactites. **a** TEM bright field image of a small coesite grain within a polycrystalline coesite aggregate in a suevitic granite clast from Bollstadt, Ries crater. The grain is pervaded by numerous growth twins parallel to (010). **b** Secondary electron image of an idiomorphic diamond platelet extracted from a suevite of the Otting quarry, Ries. **c** Secondary electron image of a black, polycrystalline diamond from the Popigai impact crater, Russia. **d** TEM dark field image of a ringwoodite grain in a shock vein of the ordinary chondrite Tenham. Planar defects are stacking faults parallel to {110} planes (Langenhorst et al. 1995).

Diamond and Lonsdaleite. Shock-produced diamond and the hexagonal carbon polymorph, lonsdaleite, were first synthesized in shock experiments (DeCarli and Jamieson 1961; Trueb 1971). On the basis of these experimental results it has then been realized that the long known diamonds in ureilites and the iron meteorite Canyon Diablo are of impact origin (Lipschutz 1964; Hannemann et al. 1967; Frondel and Marvin 1967; Clarke et al. 1981). Subsequently, impact diamonds have been detected in terrestrial impact structures, e.g., Popigai (Masaitis et al. 1972; Valter et al. 1992), several other craters of the former USSR territory (Gurov et al. 1985; Masaitis 1993), the Ries (Rost et al. 1978; Hough et al. 1995), and at two sites of the Cretaceous/Tertiary ejecta layer (Carlisle and Braman 1991; Gilmour et al. 1992). Shock-produced diamonds show frequently anomalous birefringence, and various colors, ranging from colorless, yellow, gray, to black, and are characterized by light carbon isotopes. They commonly occur as idiomorphic platelets or fragments of platelets. This morphology and other structural features (twins, layered structure) are inherited from the precursor mineral graphite, which has undergone a shock-induced, martensitic solid-state transformation. Due to the inherited properties, platy impact diamonds are also called apographitic diamonds or paramorphs. X-ray diffraction experiments revealed the polycrystallinity of the paramorphs and an orientation relationship between diamond and graphite with (111) parallel to (0001), respectively. Diamond paramorphs are found as inclusions in strongly shocked (>30 GPa) crystalline clasts (mostly gneisses) of suevites and tagamites. In most cases, diamond paramorphs coexist with the hexagonal carbon polymorph, lonsdaleite, which, according to X-ray diffraction may amount up to 50–60% of the paragenesis. Lonsdaleite could not yet be imaged with TEM in these impact diamonds (Hough et al. 1995). Variable lonsdaleite contents may cause the variations in the color and the unusual birefringence of impact diamonds.

Impact diamonds formed by transformation of coal in sedimentary rocks present a peculiarity at the Kara crater. This type of impact diamonds is porous, has variable color (white, brown, black), and shows textural relics of the coal.

In suevites of the Ries, tiny skeletal diamonds intergrown with hexagonal and cubic SiC, phases forming at ambient pressure and high temperature, have recently been reported by Hough et al. (1995). Since epitaxial intergrowth of these phases is known from chemical vapor deposition (CVD) experiments, Hough et al. (1995) concluded that the skeletal Ries diamonds originated from vapor condensation. If this interpretation is correct, formation of CVD diamonds is not a primary shock effect, but related to vapor plume cooling.

Carbonados occurring exclusively in placers and low-grade metamorphic rocks of Precambrian age show similarities to impact diamonds (Grieve and Masaitis 1994). They also are irregular polycrystalline aggregates; their light $\delta^{13}C$ values and noble gas contents indicate trapped atmosphere, and rare earth element abundances point to a crustal source. This overall signature could indicate that carbonados are remnants of large Precambrian impact events (Smith and Dawson 1985; Hough et al. 1995). However, recent TEM studies revealed distinctly different microstructures in carbonados and impact diamonds (Langenhorst 1997). Impact diamonds are pervaded by planar defects (stacking faults) parallel to {111}, whereas carbonados display perfect and helically shaped dislocations in the {111} glide planes, pointing to plastic deformation in the deeper crust or upper mantle.

Majorite, Wadsleyite, Ringwoodite, Ilmenite, and Perovskite. These high-pressure phases are major phases of the Earth's lower mantle and transition zone, and have been thoroughly studied in static high-pressure experiments (e.g., Manghnani and Syono 1987). These phases have also been detected as components of melt veins in heavily shocked ordinary chondrites (Putnis and Price 1979; Madon and Poirier 1983; Mori 1984; Langenhorst et al. 1995; Chen et al. 1996; Sharp et al. 1997). On Earth, however, these high-pressure polymorphs have not yet been identified in impactites, but may occur in pseudotachylites transecting ultramafic lithologies such as peridotites. Among these minerals, only ringwoodite can be unambiguously identified on the optical scale by its distinct violet color. Majorite, ilmenite, and perovskite have pyroxene composition $(Mg,Fe)SiO_3$ and crystallize in the garnet, ilmenite, and perovskite structures, respectively. Wadsleyite and ringwoodite are $(Mg,Fe)_2SiO_4$ high-pressure polymorphs and crystallize in the β- and γ-spinel structures, respectively. Characterization of lattice defects revealed the presence of stacking faults in the spinels, occurring parallel to (100) in wadsleyite, and parallel to {110} in ringwoodite (Fig. 28d).

Recent TEM studies of heavily shocked (shock stage S5 or S6) ordinary chondrites have revealed a large variety of disequilibrium high-pressure mineral assemblages occurring as fine-grained, polycrystalline aggregates in thin shock dikes. Majorite has been found in the Sixangkou L6 chondrite in high-pressure assemblages with magnesiowüstite or ringwoodite (Chen et al. 1996). The majorite, which is cogenetic with magnesiowüstite, shows enhanced contents of minor elements such as, for example, Na, Ca, Al, Cr, and it thus represents a majorite-pyrope solid solution. The difference in composition to enstatite in the host meteorite suggests crystallization from a high-pressure melt as formation mechanism. On the other hand, the majorite (low Ca) plus ringwoodite assemblage shows no significant compositional deviation from the host enstatite and olivine, respectively. This assemblage has been interpreted as having formed by a solid-state transformation.

Mori (1994) reports a high-pressure assemblage of amorphous $(Mg,Fe)SiO_3$ grains plus magnesiowüstite. It is assumed that perovskite was the precursor phase of the $(Mg,Fe)SiO_3$ glass since perovskite, at ambient pressure and moderate temperature (>150 °C), is unstable. A new mineral assemblage consisting of ringwoodite, $(Mg,Fe)SiO_3$ ilmenite, and amorphous $(Mg,Fe)SiO_3$ grains, which are also interpreted to be the result of perovskite amorphization, has been discovered in the L5-6 chondrite Acfer 040 (Sharp et al. 1997). It is concluded that crystallization from the high-pressure melt started with perovskite formation and was followed by growth of ilmenite and ringwoodite. Recently, crystalline $(Mg,Fe)SiO_3$ perovskite and ilmenite have also been identified in topotaxial relationship with pyroxene in the Tenham meteorite (Tomioka and Fujino 1997).

Shock experiments on olivine and enstatite generally failed to synthesize high-pressure Mg-silicates, with one exception. Jakubith and Hornemann (1981) report the production of majorite at 40 and 45 GPa. The detection of majorite was based on X-ray photoelectron spectroscopy (XPS) and X-ray diffraction techniques. Conclusive evidence is, however, lacking because imaging techniques have not been employed, and other attempts to synthesize majorite failed (Ahrens and Gaffney 1971).

Decomposition

Shock-induced decomposition of volatile-bearing minerals, e.g., sulfates, carbonates, and hydrous silicates (micas, serpentine, amphibole) played an important role in the formation of the early atmosphere. In the context of the mass extinction at the K/T boundary, the amount of CO_2 and SO_2 shock-released from the sediments in the Chicxulub target area is crucial for evaluating possible atmospheric perturbation (Brett 1992; Pope et al. 1994). Experimental data for the threshold pressure of outgassing are inconclusive and lower than those estimated from thermodynamical equilibrium calculations (Martinez et al. 1995). Grain size and sample porosity have strong effects on lowering the pressure for incipient volatilization, the experimental set-up also seems to influence the result.

Sulfates. Sulfates are highly resistant to shock loading. The shock pressure for incipient vaporization of anhydrite is about 80 GPa, whereas an anhydrite-quartz mixture starts to devolatilize in experiments already at about 30 to 40 GPa. The mineralogical proof for the reaction, for example, in the form of a solid reaction product such as $CaSiO_3$, has neither been reported from the recovered experimental material nor discovered in naturally shocked sulfates (Chen et al. 1994).

Carbonates. CO_2 release from porous chalk has been observed in experiments at particularly low pressures around 10 GPa, which is attributed to the formation of localized hot spots. Recent experiments indicate that devolatilization of non-porous dolomite and calcite starts at 55 to 65, and 35 to 45 GPa, respectively, which is in good accordance with thermodynamical calculations (Martinez et al. 1995). Mineralogical corroboration for shock-induced degassing of carbonates by the occurrence of newly formed minerals requires TEM techniques. So far, solid reaction products are documented in only two cases. Dolomite, shocked experimentally at 60 GPa, contains 20-nm-sized MgO crystals. At the Haughton crater, Ca-Mg-rich melt glass, larnite, and augite crystals occur in highly shocked clasts of carbonate sediments. One likely reason for the apparent lack of calcium-magnesium oxides, hydroxides, and silicates in shocked carbonates is the immediate back reaction of CO_2 with the very fine-grained CaO/MgO crystals, known for their efficiency in trapping carbon dioxide (Agrinier et al. 1995).This process has been documented via stable isotope measurements in the allochthonous breccia deposits at the Haughton Crater (Martinez et al. 1994). Most recently, up to 2 cm sized dolomite spheroids in the ca. 15 m thick K/T ejecta deposits in Belize have been interpreted by Pope et al. (1996) as recondensates of the Chicxulub vapor cloud; yet these spheroids may represent an alteration feature (caliche).

Hydrous Silicates suffer under shock compression substantial loss of water. Dehydration of shocked hydrous silicates is often evidenced by darkening and pervasive vesiculation of the mineral grains. In some cases, shock experiments provide data on the amount of dehydration as function of shock pressure. Incipient and complete dehydration of serpentine is observed at 20 and 65 GPa, respectively (Lange et al. 1985). Shock experiments on gneisses reveal that the water loss in micas is accompanied with shock-melting (Lambert and Mackinnon 1984). Complete water loss in biotite has been noted at 70 GPa.

Shock-Fused Glass

Shock-induced melting and vaporization represents the highest degrees of shock. Calculations of shock temperatures indicate that highly shocked (>50 GPa) minerals melt already under compression, whereas vaporization takes place during decompression when the liquid-vapor boundary is approached (Melosh and Vickery 1991).

Shock-Fused Glasses with Mineral Composition. Threshold pressures for shock melting of minerals are variable and depend largely on the compressibility of the respective mineral as well as on pre-shock temperature, and porosity of the host rock. Strongly shocked, polymineralic rocks undergo selective melting. Post-shock quenching of the melts results in formation of glasses with mineral composition, which, in contrast to diaplectic glasses, show flow structures and vesicles, and obliteration of the original crystal shape. This observation indicates that shock-fused glasses were, for a short period of time after decompression, still molten. The occurrence of these glasses has been reported for leucocratic minerals and a few mafics. Glasses of plagioclase or alkali feldspar compositions are widely observed in crystalline rocks shocked at pressures >45 GPa. Silica glass with flow structures, lechatelierite, quite commonly occurs in impact craters formed in sedimentary rocks or loose quartz-rich sediments (French and Short 1968; Kieffer et al. 1976; Hörz et al. 1989; Stöffler and Langenhorst 1994). Formation of lechatelierite in dense rocks requires pressures in excess of 50 GPa.

Mafic minerals such as (OH)-bearing phases (micas, amphiboles) either decompose incongruently or form glass matching the starting material in composition, except that crystal water is quantitatively driven off (Lambert and McKinnon 1984). Despite quenching, nm-sized spinels of unusual composition may crystallize in such shock-fused glasses (Martinez et al. 1993; Schrand et al. 1996). Monomineralic melts of other mafics, such as olivine, are not capable of being quenched to glass, but rapidly crystallize in the form of fine-grained polycrystalline aggregates. This phenomenon is erroneously referred to as recrystallization although the term is restricted to solid state processes.

Rock Glasses. Melting at the whole-rock scale occurs for most dense rocks at pressures exceeding 80 to 100 GPa. The resulting melt is deposited as splash-form glass particles and "bombs" in suevitic breccias or as coherent impact melt body. Small melt volumes might be quenched to rock glasses, whereas in the case of large coherent impact melt sheets such as observed at Sudbury, rapid quenching to glass is impossible and the melt crystallizes as observed in endogenic magmatic complexes. Initial post-shock temperatures of coherent melt bodies are on the order of 2000 °C. Another type of impact melt is produced very early in cratering from the uppermost target layers and ejected ballistically at high velocities; it forms the tektites (see p. 93).

Formation of New Minerals in Impact Melt and Vapor

Crystallization of impact melt and condensation of the vapor result in formation of new minerals. In the case of large impact melt bodies, e.g., at Sudbury, the slow cooling allows differentiation, yielding a "normal" igneous crystallization sequence. Minerals suggestive of impact, however, form in smaller impact melt volumes due to quenching with distinct undercooling. The solid state can already be achieved during the ballistic transport or immediately after deposition in close contact with cold neighboring lithologies. In both cases, quenching is accelerated by the admixture of cold solid clasts, and results in the formation of minerals with exotic crystal shapes. For example, plagioclase whiskers with hollow crystal morphologies and forked faces are a typical constituent in parts of impact melt sheets with glassy matrix and perlitic textures, e.g., at the Scandinavian craters Dellen, Mien, and Lappajärvi (Carstens 1975; Deutsch et al. 1992). This plagioclase shows oscillating zoning and the rectangular holes are filled with glass.

Metal Spherules. Impact glasses are important carriers of metallic spherules with diameters ranging from a few μm up to 100 μm (El Goresy et al. 1968; Fregerslev and Carstens 1976). Metal spherules are usually found at craters formed by impact of iron meteorites, e.g, Barringer, Bosumtwi, Henbury, Montaruqui, Chile (D = 460 m), and Wabar, Saudi Arabia (D = 97 m); such spherules occur also in craters produced by a stony projectile, e.g., at Lappajärvi. The metallic spherules consist of an FeNi metal core surrounded by pyrrhotite. Other sulfides, e.g., chalcopyrite, pentlandite, and millerite, are present to a minor extent.

Spinels in Microtektites. Spinels are important new mineral phases in distal ejecta horizons such as the K/T and Eocene/Oligocene boundaries and the Late Pliocene sediments of the South Pacific close to the Eltanin Mt. impact (Smit and Kyte 1984; Robin et al. 1992; Kyte and Bostwick 1995). The spinels are constituents of quenched impact melt droplets, less than 1 mm in diameter (microtektites), which are assumed to form in the ejecta plume. Small grain sizes (<10–20 μm) and skeletal/dendritic crystal morphologies suggest an origin of the spinels by rapid crystallization from the melt. The impact spinels are almost exclusively magnesioferrites, or magnetites with low Ti and Cr, but high Ni, Mg, and Al contents, making them distinctly different to all other terrestrial spinels. Due to this peculiar chemical composition, the spinels serve as good indicators for large-scale impact events. Their high Fe^{3+}/Fe^{2+} ratio reflects the highly oxidizing conditions during formation in the atmosphere. Recently, Ni-rich magnesiowüstite, a phase crystallizing only from ultrarefractory MgO-rich liquids, has also been detected in K/T microtektites (Kyte and Bohor 1995). Although the glassy groundmass of these K/T microtektites is generally replaced by sanidine and glauconite, the spinels survived this alteration. In the case of the Eltanin Mt. impact, the impact glass is still preserved (Margolis et al. 1991; Fig. 29a, b).

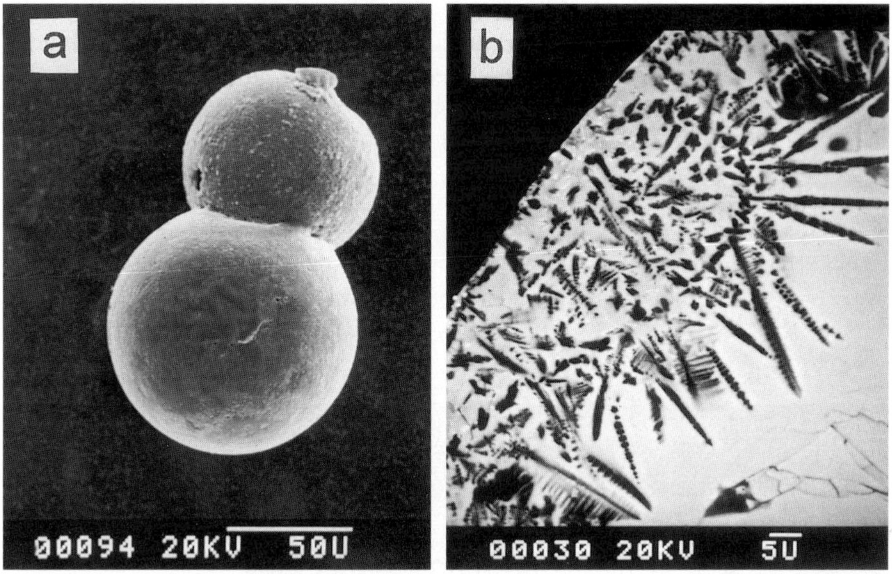

Fig. 29 a, b. Secondary electron images of microtektites found in Late Pliocene sediments containing debris from a small asteroid impact in the Southern Ocean. (Courtesy P. Claeys). **a** Two spherules sticking together. **b** Cross-section of a glass spherule with dendritic spinels

Decomposition Products. Along with decomposition and devolatilization of minerals under shock and decompression, decomposition also occurs as a corollary to annealing if minerals are incorporated in superheated impact melts. Decay of zircon to baddeleyite and silica glass is a common phenomenon in impact glasses, e. g., at Bosumtwi, Henbury, and the Ries, and is also known for Libyan desert glass (El Goresy et al. 1968; Kleinmann 1969). The reaction starts above 1676 °C, which is well above the temperature regime in melts of endogenous origin. A further example for post-shock decomposition is the breakdown of Cr-rich spinels to eskolaite (Cr_2O_3) and spinel ($MgAl_2O_4$) (El Goresy et al. 1968).

Post-shock Annealing and Alteration of Shocked Minerals

After deposition of impact debris and melt in and around the modified crater, shocked minerals are exposed to annealing and hydrothermal alteration. Annealing effects and alteration products develop particularly in impact melt rocks and suevitic breccias because of their high residual temperatures; circulating hydrothermal fluids are the second fundamental factor in these post-shock processes. Annealing is most important for the modification of impactites, and finally can result in total recovery of primary shock effects (Grieve et al. 1996) and devitrification of impact glasses.

Annealing Effects in Shocked Minerals. Characteristic annealing effects, including recovery of primary shock features, develop if clasts of shocked minerals and rocks are incorporated in superheated impact melt. The effects are less pronounced in suevites or clastic matrix breccia due to their lower formation temperatures. Regional metamorphic and hydrothermal activity may enhance annealing.

Clasts of shocked plagioclase and microcline acquire in impact melts so-called checkerboard and ghost-like textures (Bischoff and Stöffler 1984). Checkerboard-feldspars consist of about 10 μm subgrains (Fig. 30 a) which are equally oriented and surrounded by an interstitial mesostasis of different

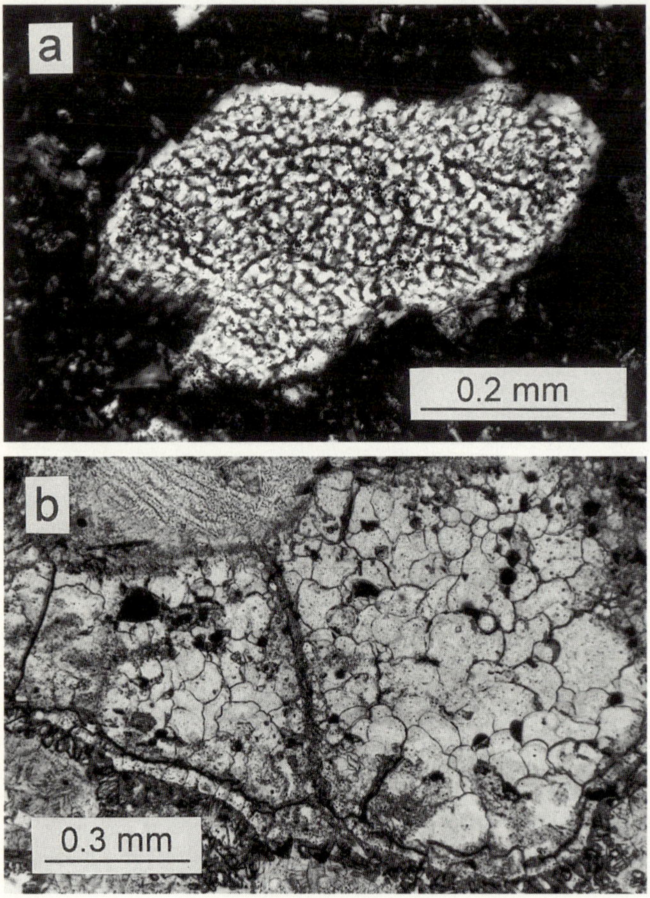

Fig. 30 a, b. Annealing effects in mineral clasts found in Kärnäite (impact melt breccia) of the Lapajärvi impact structure, Finland. **a** Plagioclase fragment with checkerboard texture within a fine-grained crystalline matrix. **b** Quartz fragment with ballen texture and reaction zone embedded in the crystalline matrix.

composition. The uniform orientation is interpreted as memory effect to the crystal orientation of the precursor feldspar, which was either weakly shocked (<30 GPa) or a diaplectic glass (30–45 GPa). For example, in impact melt rocks at Lappajärvi, weakly shocked microcline clasts are converted into An-rich plagioclase subgrains rimmed by sanidine and a mesostasis consisting of tiny sanidine and quartz grains. These characteristics indicate heating of the shocked microcline above the liquidus temperature for a short period of time. Subsequent quenching caused fractional crystallization of the melt of microcline composition, which, according to the phase diagram, started with crystallization of plagioclase and completed with eutectic crystallization of sanidine and quartz. Ghost-like feldspar clasts have blurred contours and consist of randomly oriented feldspar laths with minor amounts of mesostasis. These ghosts are interpreted as having been fused at shock pressures in excess to 45 GPa prior to incorporation in impact melt. Compared to the weakly shocked feldspar clasts developing the checkerboard texture, the shock-fused feldspars were sustained for a longer period of time above liquidus temperature and the new feldspars crystallized with random orientation from the molten clast.

Shocked quartz incorporated in impact melt may develop a "ballen" texture (Carstens 1975; Grieve et al. 1996; Fig. 30b), which resembles a cauliflower-like fracture pattern. It is assumed that ballen quartz represents a paramorph of quartz after lechatelierite. Due to the high temperatures of impact melts, quartz has, however, not directly formed by devitrification of lechatelierite. In a transient state, high-cristobalite recrystallized and transformed to low-cristobalite, causing a volume contraction which subsequently results in the typical fracture pattern. Annealing of experimentally produced diaplectic quartz glass at temperatures above 1200 to 1300 °C yields a similar ballen texture. At these temperatures, diaplectic quartz glass recrystallizes to a mixture of crystalline α-quartz and cristobalite, whereas synthetic silica glass always transforms to pure cristobalite under identical conditions.

As illustrated in Fig. 31 for shocked quartz, a combination of aqueous alteration and annealing cause progressive modification of PDFs, leading finally to their decoration with tiny bubbles (Grieve et al. 1996). The aqueous fluids dissolve predominantly in PDFs because of the high solubility of water in amorphous silica, which is up to 6 wt.% H_2O, whereas, only <100 ppm are soluble in quartz (Cordier and Doukhan 1989). Annealing of the shocked quartz starts with recrystallization of the amorphous material, which itself is assisted by the presence of water. This leads in the first stage to a mixture of amorphous silica and minute quartz crystallites (<10 nm; Fig. 31b), as observed by Goltrant et al. (1991, 1992). Progressive recrystallization causes coarsening of the quartz crystallites and exsolution of water within the former PDFs in the form of bubbles. This results simply from the fact that the newly formed quartz crystallites are extremely oversaturated in water. The internal pressure increases with increasing size of the bubbles. Relaxation of the overpressure is achieved by incorporating SiO_2 from the bubble surface as extra half planes into the surrounding quartz lattice. A high density of sessile dislocation loops is formed

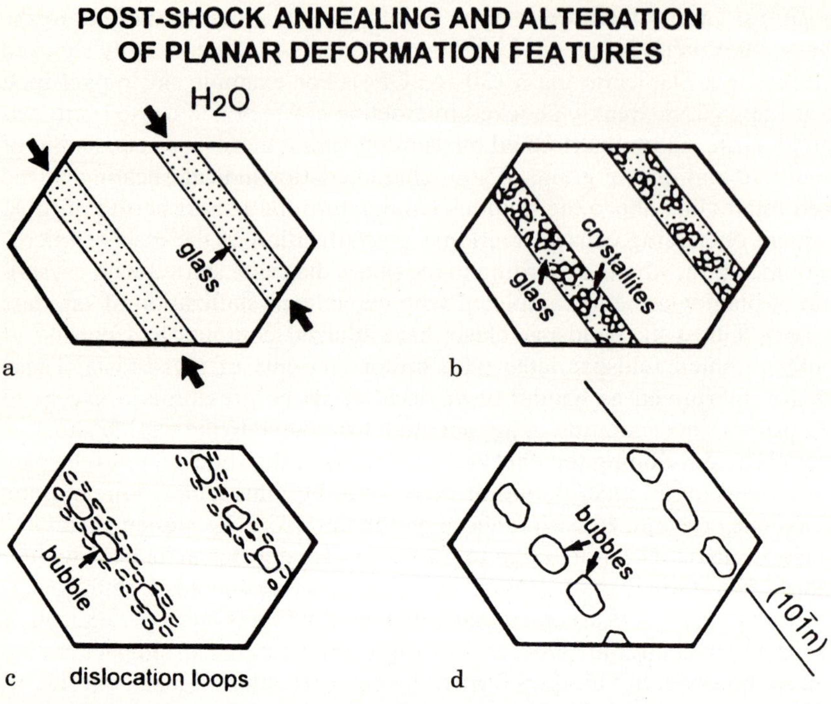

Fig. 31a–d. Schematic drawings of progressive post-shock modification of planar deformation features in quartz. **a** Initial state of glassy PDFs affected by hydrous fluids. **b** Incipient recrystallization of the glass to tiny quartz crystallites. **c** Progressive recrystallization leading to the disappearance of glass, exsolution of water, and incorporation of dislocation loops in the surrounding lattice. **d** Final state of a PDF consisting of tiny bubbles aligned along the planes of the original PDFs

decorating, together with the bubbles, the original PDF (Fig. 31 c). Such PDFs, consisting of a large density of dislocations and bubbles, are particularly observed in quartz grains from K/T sites and relatively young, Cenozoic impact craters (Goltrant et al. 1991, 1992). If annealing is sustained much longer, as in the case of the 2 Ga Vredefort structure, the only remnants of PDFs are aligned bubbles still showing the orientation of the former amorphous lamellae (Fig. 31 d). The dislocations have been transported out of the crystal or are reorganized in the form of subgrain boundaries.

Knowledge of post-shock annealing of shocked mafic minerals is relatively limited (Feldmann 1994). Recrystallization of shock-damaged biotite from the Janisjärvi and Popigai craters led to an assemblage of K-feldspar, orthopyroxene, and ilmenite. Strongly shocked garnets from the Popigai and Ries craters recrystallized to an assemblage consisting of fine-grained hercynite, hypersthene, and alkali feldspar. Annealing of strongly shocked amphiboles from the Puchezh-Katunky crater resulted in the formation of fine-grained aggregates composed of plagioclase, amphibole, clinopyroxene, magnetite, and, rarely, K-feldspar.

Devitrification of Impact Glasses. Due to rapid quenching of impact melts, the resulting glasses are often oversaturated in water and exhibit high internal strain. Therefore, post-shock annealing easily initiates devitrification of impact glasses. Pyroxene, plagioclase, and subordinate magnetite have been identified as devitrification products in impact glass of the Ries suevite (von Engelhardt et al. 1995). Ortho- and clinopyroxene trichites occur in the form of hairs or have worm-like shapes ranging in thickness from 1 to 5 μm or as minute star-like and point-like crystallites. Clinopyroxenes show remarkably high Al_2O_3 contents of up to 10 wt.%, which has been detected so far only in lunar impact melt rocks of basaltic composition. Plagioclases form laths or fan-like to spherulitic aggregates.

Alteration Products. Mobilization of water and other volatiles results from shock-induced decomposition itself and, more importantly, post-shock waste heat in the deposited breccias, which expel volatiles from underlying colder lithologies. Vertical degassing pipes with a diameter of 2 to 40 cm in suevites at the Ries illustrate the importance and extent of volatile mobilization in impact deposits (von Engelhardt 1972). Alteration products in terrestrial impact formations are dominated by clay minerals (smectite, chlorite, montmorillionite, illite), zeolites (analcite, phillipsite, chabasite, etc.), K-feldspar, and quartz; to a minor extent, carbonates (calcite, siderite) and sulfates (barite) are present (Allen et al. 1982). Clay minerals such as Fe-chlorites and smectites reflect formation temperatures on the order of 100 to 300 °C. Suevites of the Ries contain up to 20 wt.% of Fe-rich montmorillionite occurring as coating of vesicles with constant thickness of about 10 μm. Their honeycomb texture, lack of illite interlayers, and presence of zeolites indicate temperatures around 100 °C for the hydrothermal alteration (Newsom et al. 1986), even though the deposition temperatures are estimated to be about 750 °C (von Engelhardt et al. 1995).

1.10.6 Examples for Terrestrial Impact Structures

A. Deutsch

The Nördlinger Ries – an Excellently Preserved Complex Impact Structure

Nördlinger Ries. The 15.0 Ma Ries (N 48°53′ E 10°37′) is a complex crater with a rim-to-rim diameter of about 26 km (Pohl et al. 1977). Due to the quite

Advanced Mineralogy, Volume 3
Mineral Matter in Space, Mantle, Ocean Floor, Biosphere, Environmental Management, and Jewelry (ed. by A. S. Marfunin)

complete preservation of the continuous ejecta blanket outside, and the breccia lens inside the crater, this complex impact structure is one of the most remarkable documents of impact cratering on Earth.

The Ries crater has a stratified target comprising crystalline basement of Hercynian age, covered by up to 700 m thick sediments which range from Permian to Miocene; among the sediments, limestones and marls dominate over shales, clays, and sandstone. The Ries consists of an almost circular, central basin with a diameter of about 12 km, whose flat surface is formed by 250 to 350 m of post-impact lake sediments. This central basin is bordered in the W, SW, and E by a horseshoe-shaped chain of isolated hills, the so-called inner ring, standing about 50 m above the central plain and consisting of large monomict brecciated, uplifted blocks of the crystalline basement with breccia dikes and shatter cones. Displaced megablocks, covered in part by lake sediments, make up the about 6 km wide zone between the inner ring and the tectonically modified crater rim, reaching up to 150 m above the plain. Differential erosion has removed part of this rim to the north. A unique feature of the Ries as compared to other terrestrial complex crater, is the excellent preservation of an appreciable part of the continuous ejecta blanket, extending mainly to the south and east up to about 45 km from the crater center, and forming a hummocky terrane.

This ejecta deposit is an up to 200 m thick clastic matrix breccia – the local term Bunte Breccia is now used to characterize similar rocks at other impact sites, composed of unshocked or only moderately shocked rock and mineral fragments of all stratigraphic units in the target. Up to 95 vol.% of the fragments are sedimentary, the remainder crystalline rocks; clast range in particle size from μm to nearly 1 km sized blocks. Beyond the continuous ejecta deposit, isolated fist-sized and brecciated fragments of Malmian limestone (Reuter blocks) occur within a layer of Miocene sediments in the Molasse basin south of the river Danube at distances up to 70 km. Distant ejecta are represented by the Moldavite tektite strewn field found at a range of more than 350 km in the Czech Republic, and at discrete localities in East Germany (Lausitz area) and Austria (villages of Stainz and Horn).

The Bunte Breccia is locally overlain by an up to 25 m thick layer of suevitic breccia (fallout suevite) occurring inside a radius of about 22 km as patches of several hundred m^2 to some km^2 extension. Impact melt particles in this suevite – their local name is Flädle – range in size between sub-mm and 50 cm and show signs of aerodynamical transport (glass bombs). A recent petrofabric examination indicate that transport of the fallout suevite occurred in part by horizontal movement, similar to a pyroclastic flow (Bringemeier 1994). Amongst the fragments, the up to some decimeters sized lithologies of the crystalline basement dominate over the much smaller sedimentary clasts. Crystalline rock fragments display all degrees of impact metamorphism. Impact diamonds occur in the suevite at the Otting quarry. From this locality, coesite and stishovite have been reported, providing in the 1960s the final proof for an impact origin of the Ries. Deposition temperature of the fallout suevite was below 750 °C. Chilled zones are known from the base (~1 m) and the top (~10 m) of the suevite deposits. Emanating gases and hot vapor caused large-scale montmorillonitization of the matrix and formation of degassing pipes.

Coherent, clast-rich impact melt, the so-called red suevite, occurs at the eastern crater rim in several large masses extending to 50 m^2. The once glassy, vesicular matrix of this massive rock is completely recrystallized to a fine-grained aggregate of feldspar, pyroxen, hematite, and cristobalite. The fragment load of this melt consists nearly exclusively of crystalline lithologies displaying all stages of impact metamorphism.

The crater cavity is filled with a breccia lens, consisting from top to bottom of several m of sedimentary reworked suevitic breccia, about 15 m of graded suevite, and an up to 400 m thick suevite layer which, in its lower section, is gradually poorer in melt particles. At a depth of about 500 to 550 m from the present surface, drilling reached the brecciated crystalline basement, which is penetrated by dikes of suevitic breccia. Below about 650 m, the crystalline rocks still display a weak impact metamorphic overprint, decreasing with depth, as well as shatter cones, and numerous up to cm-thick dike breccias. The absence of a thick coherent layer of impact melt at the Ries is ascribed to the predominantly sedimentary target rocks, which, at high shock and post-shock temperatures, may decompose into gas and a solid residue but not melt.

Extensive geophysical measurements have revealed details of the subsurface structure of the Ries. Strong and irregular negative magnetic anomalies with an inverted field vector have been determined for the central cavity; they reflect the magnetization of the suevite layer. The area within the crater rim is characterized by a centrosymmetrical gravity low of up to 18 mgal. The hemispherical zone of brecciated basement with reduced density and seismic wave velocities extends to a depth of about 6 km and a radius of about 13 km.

Research on the Ries, in particular drilling, e.g., the Forschungsbohrung Nördlingen 1973 with 1206 m core recovery, and minute petrographic work, yielded some basic facts for a better understanding of the cratering process. For example, analyses of nine shallow drill cores, penetrating through the Bunte Breccia of the continuous ejecta blanket outside the crater rim into the uppermost ground zero surface, produced first evidence for regularities in the composition of such ejecta deposits (Hörz et al. 1977). Grain size of this breccia, as well as the size of megablocks, and the abundance of stratigraphically older target rocks (i.e., from a deeper level of the transient cavity) decrease radially from the crater center, while the mass ratio of locally derived material vs. primary ejecta increase. In the size fraction <10 mm, up to 90 vol.% local materials – Tertiary sands, shales, pebbles, and coal – has been observed. This relationship plays an important role in interpreting and understanding the material sampled on the Moon during the Apollo missions.

Large Proterozoic Impact Structures (Sudbury, Vredefort), and the Bushveld Enigma

Compared to other solid bodies in the solar system, the Earth comprises surprisingly few (<10) "old" (pre-Cambrian) impact structures. This is explained by the inherent dynamic nature – with both exogenic and endogenic forces – of our planet. Two of the preserved craters, Sudbury and Vredefort, however, are the largest terrestrial impact structures, and both are associated with world-class ore deposits.

Sudbury. The 1850 Ma Sudbury Structure (N 46°36′ W81°11′) which has been recently reviewed by Deutsch et al. (1995), represents the best example of a terrestrial multi-ring impact structure. Sudbury is superimposed on the Proterozoic Huronian Supergroup of the Southern Province and Archean basement rocks of the Superior Province. As depicted in Fig. 32, the structure covers an area of >15000 km^2. Proceeding inwards and stratigraphically upwards, three main geological formations directly related to the impact event have been identified at Sudbury: the brecciated crater floor, extending up to a distance of 80 km from the outer margin of the Sudbury Igneous Complex (SIC), which is part of the coherent impact melt sheet in the central inner ring of the structure, and the allochthonous breccia lens.

The crater floor consists of the Footwall rocks, transected irregularly by dikes of the so-called Sudbury Breccia. Four types of breccia dikes can be distinguished, differing in matrix characteristics (melt vs. clastic matrix), thickness, clast content, and contact relationship to the country rock. They are most abundant in an approximately 13 km wide zone adjacent to the SIC and in concentric zones at 20 to 33, 42 to 48, and 78 to 80 km distance from the North Range of the SIC. The size of some pseudotachylite dikes indicates an amount of displacement in the order of 1 km between individual blocks (Spray and Thompson 1995). Erosional remnants of the Huronian Supergroup are preserved in a ring-like graben zone parallel to the outer margin of the SIC at a distance of 20 to 25 km. At the basal contact of the SIC with the footwall, the polymict parautochthonous Footwall Breccia, which includes clasts of the Sudbury Breccia and contains sulfide ores, forms a narrow, irregular zone. Annealing of the breccia by the overlying impact melt layer resulted in partial melting of the fine-grained clastic material, which, according to orthopyroxene-clinopyroxene geothermometry, crystallized below 1040 °C.

The coherent impact melt sheet ranges from a clast-rich, discontinuous impact melt breccia, the Contact Sublayer, through the dominantly inclusion-free magmatic-textured Main Mass of the SIC, to a capping, extremely inclusion-rich impact melt breccia, the Basal Member of the Onaping Formation. Apophyses of the melt complex, the Offset Dikes, either emanate radially from the outer margin of the SIC up to a distance of at least 30 km, or form dikes concentric

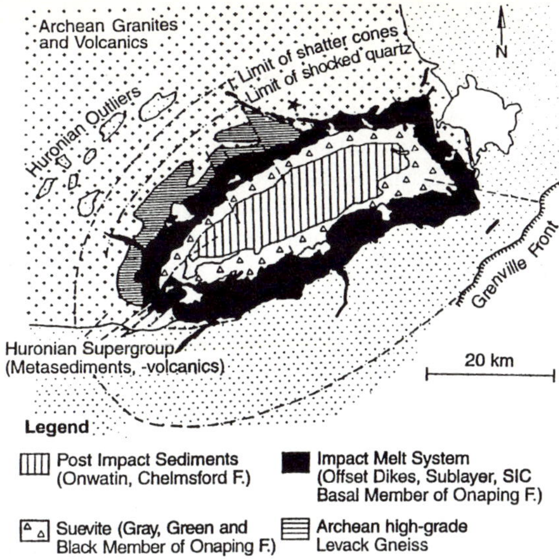

a

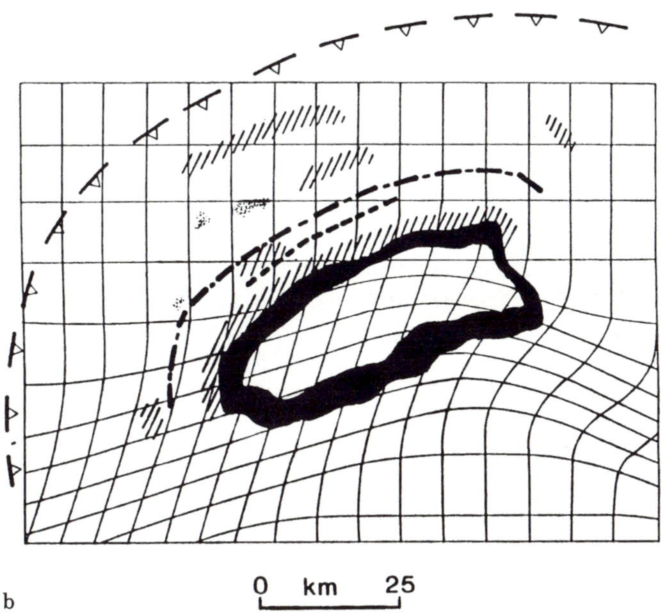

b

Fig. 32. a Geological sketch map showing the main lithological and stratigraphic units of the Sudbury Structure; ★ Foy Offset Dike; *stippled area* Southern Province; *crosses* Superior Province. The outer limits of occurrence shatter cones and decorated planar features in quartz ("shocked quartz") correspond to a shock pressure of about 2, and 6 to 7 GPa, respectively. Huronian outliers are downfaulted remnants of pre-impact cover rocks. **b** Sketch map of the Sudbury area showing radial limits north of the outer margin of the Sudbury Igneous Complex (SIC; *dark gray*) for decorated PDFs in quartz (*dashed lines*), shatter cones (*point-dash line*), the relic occurrence of downfaulted Huronian cover rocks (*light gray*) and zones with abundant Sudbury Breccia (*hatched*), and an estimate of the original outer rim of the final modified crater (*broken line with triangles*). The strain pattern, as estimated by Roest and Pilkington (1994), is given in *solid lines* which indicate the paleo-latitude and longitude lines.

to the SIC in the crater floor. The crystalline impact melt breccias of both the Contact Sublayer and the Offset Dikes contain abundant fragments of local material, as well as mafic to ultramafic inclusions, which are otherwise unknown at Sudbury. The up to 3.5 km thick Main Mass of the SIC consist of norites, a quartz-gabbro, and granophyres; it represents the central, differentiated part of the impact melt complex with an estimated volume of 1.0 to 2.5×10^4 km^3. The Basal Member contains up to 80 vol.% clasts of mainly Huronian metasediments in a crystalline matrix, which is the result of annealing by the underlying SIC.

The allochthonous breccia lens comprises fallback material, the Gray and the Green Member of the Onaping Formation (OF), whereas breccias of the topping, up to 1200-m-thick Black Member (OF) mark the onset of post-impact redepositional processes. These breccias are covered by the Onwatin and Chelmsford Formations, representing the sedimentary infill of the crater. The Gray Member is a suevite-like clastic matrix breccia with variably shocked clasts and irregularly shaped melt fragments, which displays flow structures, vesiculation, and a fine-grained crystalline texture. Huronian materials dominate amongst the fragments. The 5 to 70 m thick Green Member consists of fine-grained, often chloritized, clastic material and some larger lithic clasts embedded in a microcrystalline matrix. The unique texture of this impact melt breccia may have resulted from an agglomeration of melt particles and clastic debris deposited as fallback material.

The impact origin of the Sudbury Structure is substantiated by the occurrence of diagnostic shock indicators. These are Brazil twins and decorated PDFs in quartz, planar fractures in zircon of the Onaping breccias and footwall rocks, and shatter cones. Second-order features are huge pseudotachylite dikes and thick breccia sequences as well as fullerenes (C_{60} and C_{170}) occurring with an abundance of up to 10 ppm in Onaping breccias; these fullerenes were probably synthesized within the vapor plume (Becker et al. 1994). The present spatial distribution of impact-related formations and shock metamorphic effects, sketched in Fig. 32, has been used to constrain the original size of the Sudbury impact structure to approximately 250 km. This multi-ring impact basin was tectonized immediately after the impact and is now heavily eroded. Only the central depression inside the first ring is preserved, but remnants of additional rings are probably delineated by the four roughly semi-circular zones containing large bodies of Sudbury Breccia and pseudotachylites.

Solid arguments in support of the above outlined impact model are provided by structural, geophysical, geochemical, and geochronological investigations. High-precision U-Pb dating yields, within the given precision of +3.5/–2.6 Ma (2σ), an identical 1.85 Ga age for zircon and baddeleyite extracted from the Offset Dikes, the included ultramafic lithologies, and the SIC (e.g., Ostermann et al. 1996). These formations carry only newly crystallized grains, totally lacking shock features and inherited cores. The 1.85 Ga age is also recorded in heavily shocked zircons of the Onaping formation, titanite in the footwall south, and newly grown zircons in the basement north of the SIC, showing that all formations, from the footwall up to the breccia layers, were either the product of or affected by the impact event.

All impact melt lithologies (coherent melt sheet, melt particles in the suevitic breccias) as well as mafic to ultramafic clasts in the Sublayer and the Offset Dikes display a strong geochemical affinity to crustal material. Slopes to the REE distribution patterns, which are bracketed by REE patterns for basement and cover rocks now exposed in the footwall, are steeply negative. Isotopic characteristics of the impact melt lithologies, with highly radiogenic initial Sr ratios, and $\varepsilon_{Nd}^{T=1.85\,Ga}$ ranging from −5 to about −9, correspond to those of local country, and, hence, target rocks. Nd model ages, relative to a depleted mantle source, of 2.7 Ga are identical with the true formation age of the crust in the Sudbury target area. In addition, osmium and lead isotope compositions of Cu-Ni ores in the Sublayer and Offset Dikes resemble values determined for local country rocks (e.g., Dickin et al. 1996). The geochemical features can be interpreted without particular requirement for an additional, 1.85 Ga mantle-derived component, and, therefore, strongly support an origin of the SIC by impact melting of crustal target lithologies. Further evolution of the SIC into a differentiated impact melt complex has not yet been modeled in detail. The present distribution with about 27 vol.% norites (mean SiO_2 content 56 wt.%), quartz-diorite (54 wt.% SiO_2), 63 vol.% of granophyric material (68 wt.% SiO_2), and the apparently abrupt geochemical transitions between the units, is currently best explained by the double diffusive convection system in the impact melt. The ores may have originated as immiscible sulfide droplets resulting from a decrease in sulfur solubility in the cooling siliceous impact melt sheet. Additional support for this simple model is provided by mass balance calculations, indicating that for the South Range a remixing of known ore bodies into the norite would raise its nickel content by only 10 ppm to about 90 ppm, which is well within the known range of nickel contents for Huronian metavolcanics in that area.

According to surface structural observations and a detailed geophysical survey within the frame of the LITHOPROBE transect, the present elliptical pattern of the SIC is the result of northwest-directed tectonic compression, which started immediately after the impact event and outlasted cooling of some impact formations. Deformation and tectonic displacement resulted in horizontal bulk shortening of the Sudbury Structure and exhumation of a lower level of the SIC in the south. After removing the effects of a simple NW-SE stress field, which is illustrated in Fig. 32b, the SIC could well have been originally circular with an estimated NW-SE diameter of about 60 to 65 km for the outer margin of the SIC at the present erosional level (Roest and Pilkington 1994). Seismic reflection data have revealed that the Levack Gneiss Complex and SIC layers of the North Range dip gently southward at 20 to 30° and appear continuous under the Sudbury Basin, with gradually increasing thickness towards the center. Indications for a magmatic plumbing system or feeder dikes are totally lacking. The now highly asymmetric SIC is rootless at depth, providing a very strong argument for the origin of the SIC by impact melting.

Vredefort. The circular 2023 Ma (Kamo et al. 1996) Vredefort structure (S 27°0′ E 27°30′) is the largest known terrestrial impact structure; for reviews see Grieve and Masaitis (1994) and Reimold and Koeberl (1994). As shown in Fig. 33, Vredefort consists of a central core of updomed Archean basement rocks with a diameter of about 36 km, rimmed by an approximately 18 km wide collar of steeply dipping to overturned Archean to Proterozoic sequences of the Witwatersrand and Ventersdorp Supergroups, and a roughly 28 km concentric synclinorium of the Proterozoic Transversal sequence. The southeastern half of the structure is covered by the Karoo sequence. The so-called Vredefort Dome is considered as the deeply eroded remnant of a large, complex impact structure with only the crater floor preserved. Taking an erosion of about 8 km into account, and using well-established relationships of the diameter of the uplifted area to the original crater diameter, the original rim-to-rim diameter of the Vredefort impact structure is estimated to be between 300 and 335 km (Therriault et al. 1996).

A variety of first- and second-order impact-related effects have been observed at Vredefort, e.g., extraordinary occurrences of pseudotachylites (Vredefort is the type locality of "pseudotachylytes"), shatter cones, planar microstructures in quartz, and coesite and stishovite (Martini 1991). Planar microstructures, which have been the key issue on the controversy of endogenic vs. exogenic origin for Vredefort, represent closely spaced sets of mechanical Brazil twins (Leroux et al. 1994b). Compared to other complex impact structures, where residual shock effects in quartz document increasing shock levels towards the center, shocked quartz at Vredefort seemingly shows an irregular distribution, and lowest pressures have been calculated for the center. This particular observation is ascribed to post-impact thermal metamorphism in the uplifted core, which was caused by intrinsic heat plus residual thermal energy, deposited by the shock event; in addition, a significant heat transfer from the hot overlying melt sheet could be expected. The planar deformation features, i.e., amorphous lamellae, in quartz responded to this annealing with formation of decorated planar elements and, finally, total recrystallization obliterated the shock effects. Therefore, PDFs reflecting high shock pressures disappeared, whereas the more robust mechanical Brazil twins survived.

Pseudotachylites are concentrated in the collar rocks surrounding the central uplifted area, but also occur in this core, in the Witwatersrand Basin at a distance of about 80 km from the presumed crater center, as well as around Johannesburg. ^{39}Ar-^{40}Ar dating resulted in uniform 2 Ga ages for all pseudotachylite occurrences, although some gas loss due to later low-temperature annealing is recorded. The age identity serves as an independent argument for the large size of the Vredefort impact structure.

Impact melt lithologies at Vredefort are preserved only in the form of dike-like bodies of the so-called 2023±4 Ma (2 σ) bronzite granophyre, representing apophyses of the originally overlying massive impact melt layer into the crater basement. These dikes resemble the Offset Dikes at Sudbury. The geochemical composition of the granophyre has been successfully modeled by mixing of

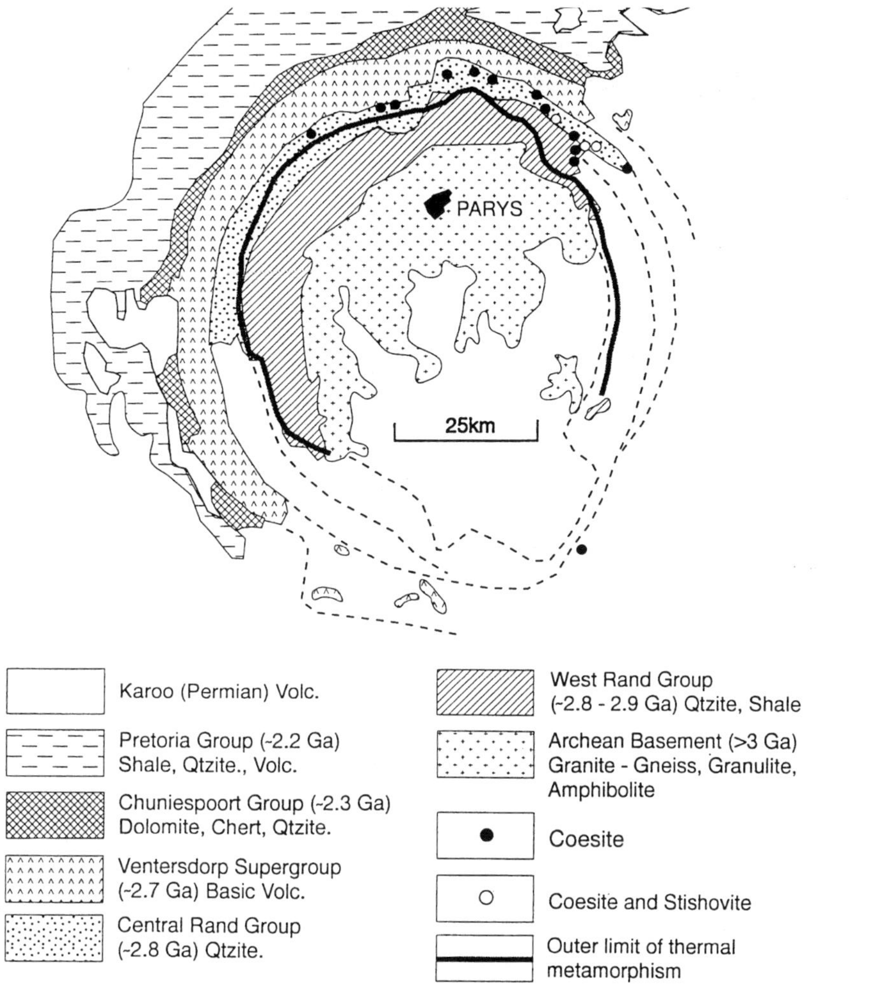

Fig. 33. Geological sketch map of the central part of the Vredefort impact structure. Qtzite = quorzite, Volc = volcanics. (After Martini 1991 and Grieve et al. 1996)

lithologies which constitute the lower part of the Vredefort target stratigraphy. In addition, Re-Os systematics indicate the presence of a meteoritic component.

The incredibly rich gold fields of the Witwatersrand Basin, accompanied by extremely valuable deposits of detrital uranite, occur in an 180° arc within the estimated original diameter of the Vredefort structure. Preservation of these progenetic sedimentary deposits is due to downdropping of higher stratigraphic levels of the target within an annulus surrounding the central uplift (Grieve and Masaitis 1994).

Bushveld Complex. Geological setting, geochemical and petrographic characteristics of the 2055±4 Ma (2 σ) (Kamo et al. 1996) Bushveld Complex, South Africa, have been reviewed recently by von Gruenewaldt and Harmer (1992). The extended and ore-rich Bushveld Complex covers an area of 65 000 km^2. It consists of the mafic to ultramafic, up to 7.5 km thick Rustenberg layered sequence, which is overlain by large volumes of granites and granophyres forming the up to 5 km thick Lebowa sheeted granite suite. Divided into a western and eastern compartment, the mafic series, with an estimated volume of 500 000 km^3, crop out in up to 40 km wide and at least 400 km striking arcuate belts. On both lobes, the layered sequence is slipping at low angels towards the center of the whole complex. The Bushveld Complex is topped by the Rooiberg Group, ranking amongst the largest known accumulations of felsites, with a volume of about 300 000 km^3. This siliceous, massive, and quite uniform sequence shows features suggestive of high emplacement temperatures (>850 °C). Stratigraphically, the Rooiberg Group is considered equivalent to the Stavoren granophyre of the Dullstrooom Formation, forming locally the floor of the layered suite, and, hence, considered as pre-dating the Bushveld Complex, though perhaps separated by only a very short time span. Both the granophyres, the Rustenberg and the Lebowa Suites, are not dated with the required precision to resolve the chronostratigraphic framework in detail. Published ages overlap within given errors, being typically on the order of 10 to 30 M (2 σ).

Since 1956, it has been repeatedly suggested that the Bushveld Complex could have originated due to simultaneous multiple impact events, although a self-consisting impact model taking into account and explaining all observational features has not yet been outlined. Main arguments in support of this presumption rest so far on quite speculative morphological interpretations, as well as on textural and geochemical properties of the Rooiberg felsites, which have been interpreted as impact melt (Elston 1992). Quartz grains throughout the Rooiberg Group display slightly curved microstructures with a typical spacing of >15 μm, which microscopically resemble deformation structures in quartz of tectonic shear zones. Despite intense field work and elaborate petrographic investigations, conclusive shock indicators such as residual shock features in quartz have not been reported from the Bushveld area (Joreau et al. 1997).

The impact hypothesis is at odds with the widely accepted interpretation that the Bushveld Complex represents the world's largest layered intrusion with an intracratonic setting, possibly emplaced in a rifted environment. Several geochemical data, however, are not well understood at the moment. Amongst these, highly radiogenic Sr, Pb, Os isotope ratios, and heavy oxygen isotope compositions have to be mentioned. For example, $^{86}Sr/^{86}Sr$ ratios recalculated to 2.05 Ga, range for most Bushveld lithologies from 0.7056 to about 0.709, compared to 0.7022 for a "normal mantle" having bulk Earth character. The initial ε_{Nd} value for the layered sequence is highly negative with −6, and the Nd model age relative to CHUR is about 3.5 Ga, corresponding to the true age of the oldest basement rocks of the northeastern Kapvaal craton. For one

Rooiberg high-magnesian felsite, a Nd model age of 3.4 Ga relative to a depleted mantle has been published. The data exclude a direct origin of the named elements from the contemporaneous bulk mantle, and are currently interpreted as evidence for (1) mixing of magmas derived from at least two mantle domanes with a highly complicated multistage pre-Bushveld history, and (2) contamination of the Bushveld material by crustal lithologies. Calculated amounts of assimilation vary between 10% for oxygen and up to 60% for Os (Hart and Kinloch 1989).

To conclude, an impact origin of the Bushveld Complex cannot be totally ruled out, but solid evidence for this assumption is lacking (e.g., French 1990). Taking the lessons from the development of the impact concept at the Sudbury Structure into account, a thorough reassessment of the vaguely defined impact hypothesis for the Bushveld Complex requires more elaborate geochemical work, e.g., systematic Nd isotope analyses, and high-precision dating. A detailed geochronological frame is one important step to foster or discard any impact model for the Bushveld area or part of it. Field and labor work should concentrate on the discovery of diagnostic residual shock effects, truly related to the Bushveld Complex. It must be taken into account, however, that the volume of impact melt increases relative to the volume of the transient cavity, causing obliteration shock effects which are documented at smaller terrestrial impact structures (Grieve and Cintala 1992).

Impact diamonds at the Popigai Impact Structure

V. L. Masaitis

The Popigai impact structure is located in northwest Siberia on the northern edge of the Archean Anabar shield, which is, in part, covered by Upper Proterozoic to Cretaceous sedimentary rocks. The Popigai impact event occurred about 35.7 ± 0.2 Ma (2 σ) ago (Bottomley et al. 1997). Popigai may be regarded as a multi-ring impact structure with a diameter of about 100 km (Masaitis et al. 1975; Masaitis 1994). It consists of an about 45 km wide central depression, encircled by an uplifted ring of shocked gneisses, which, in turn, is surrounded by the ring through. This through and the central depression are filled by allogenic (allochthonous) impact breccias and impact melt rocks, reaching a maximum thickness of about 2.5 km.

The allogenic breccia consists of large blocks and fragments of crystalline and sedimentary rocks, and occupies the lower part of the crater filling. Many fragments display shatter cones and petrographical features of shock metamorphism, including the presence of coesite and stishovite. Coherent, up to some 100 m thick sheets of massive tagamites overlap the breccia. Tagamites also occur as irregular bodies in suevitic breccias (suevites), which are widely distributed in the upper part of the crater fill. Small patches of breccias and tagamites are preserved as relics of the ejecta blanket outside the crater depression. Tagemites and suevitic breccias originated by impact melting and partial dissipation of melt have a chemical composition identical to that of basement gneisses. Tagamites containing numerous clasts of shocked rocks and minerals show either a glassy or a crystalline matrix with plagioclase and pyroxene as main constituents. Suevites are made up of fragments and bombs of chilled impact glass with an admixture of rock debris. The impact melt rocks are enriched by Ni, Co, and Ir due to contamination of projectile material presumed to be composed of ordinary chondrite.

The Popigai crater is the place where, at the beginning of the 1970s, impact diamonds have been found for the first time directly in shocked and impact-melted rocks (Masaitis et al. 1972). Impact diamonds originated from martensitic transformation of graphite, which is a rock-forming mineral in garnet-biotite gneisses of the Popigai target lithologies. Impact diamonds differ from the well-known diamonds in kimberlites, and occur as irregular grains and tabular paracrystals; they are very rarely colorless, and mostly yellow, gray, or black. The largest grains may reach 10 mm, but the average dimension of the diamonds is 0.5–2 mm. Tabular paracrystals inherit the crystallographic shape of the precursor graphite, especially the pinacoid faces $\{0001\}$. However, the angles between the prismatic $\{10\bar{1}0\}$ faces of the paracrystals are distorted due to the anisotropy of compressibility of graphite. The inherited twinning striation is common for pinacoid faces, but some planar fissures observed on them may be caused by shock deformation. Grains and paracrystals of impact diamonds display dissolution patterns, the honeycomb microrelief on their surfaces being caused by phase inhomogeneities.

X-ray studies showed that paracrystals mainly consist of two microcrystalline phases: the spinel law cubic phase (diamond) and the hexagonal high-pressure phase (lonsdaleite), whose content in paracrystals may reach 50%, some remnant or secondary graphite may be present as well. The size of the single microcrystals of these high-pressure phases is about 5×10^{-4} cm or less. The polycrystalline aggregate is structured: the pinacoid of paracrystals, i.e., the (0001) plane of the precursor graphite, corresponds to the (111) plane of the cubic phase and is perpendicular to the (0001) plane of lonsdaleite. According to the optical properties, paramorphs imitate monocrystals of non-cubic symmetry. In some cases, the paracrystals display transparency, strong birefringence (up to 0.02) and usually total straight extinction. Sometimes, the twinning of paracrystals is inherited from Veselovsky twins of the precursor graphite.

The density of impact diamonds varies due to the admixture of graphite and minerals and glasses from the host rocks, but some paracrystals are characterized by anomalous density (up to 3.61 g cm^{-3}) which is higher than the X-ray density of cubic diamond. Inclusions in impact diamonds are graphite, alumosilicat glasses, silica-glass, and Fe-oxide; diamond paramorphs exhibit a minor admixture of Si, Ti, Al, Na, Ca, similar to that observed in the precursor graphite.

Spectra of combinative dispersion display intensive widening of the Raman line at 1332 cm^{-1} (up to 30 times that of endogenic diamonds), which reflects the degree of defectivity of polycrystals. Photoluminescence spectra are characterized by systems of non-phonon lines in the interval 650–780 nm and the absence of the defect represented by the 640-nm line. These data and the study of IR spectra may indicate the presence of N-centers of absorption as well as centers caused by the presence of CO_2 and H_2O. In UV light, impact diamonds usually have orange-red luminescence. The carbon isotope composition of the precursor graphite and the impact diamonds is identical, ranging from –12 to –17 ‰ $\delta^{13}C$.

Impact diamonds occur as accessories in tagamites and suevites, as well as in fragments of strongly shocked gneisses (protoimpactites). In these graphite-bearing rocks, all minerals are diaplectic or transformed into shock-fused monomineralic glasses; diamonds included is such lithologies may be regarded as authigenic. The study of shock features in all silicate minerals of these protoimpactites indicates a pressure of about 34–36 GPa for the onset of the transformation from graphite to diamond; the transformation is complete at a pressure of about 60 GPa. The degree of transfer is presumed to be controlled not only by the maximum shock pressure but also by the duration of the compression as well as the crystal orientation relative to the front of the shock wave.

There are some regularities in the distribution of diamonds in the impactites within the Popigai crater. Some radial zones are enriched with diamonds due to azimuthal inhomogeneities in the distribution of graphite in target rocks, superimposed by the concentric zoning of shock metamorphism. The primary distribution of shock-induced high-pressure carbon phases was disturbed by the radial ejection of melted material with a different diamond content. The preservation of diamonds from oxidation and graphitization depended on the post-shock temperature of the ejected material and its cooling rate after deposition. The overheated fractions of melt appear to be virtually barren of diamonds after chilling.

Modern placers of impact diamonds adjacent to Popigai originated as a result of impactite disintegration and redeposition. Single grains of impact diamonds were found at a distance of about 100–150 km from the crater rim, which suggested that they are relics of totally rewashed distal ejecta.

In terms of energy source and ore formation mechanism, the primary deposits of impact diamonds belong to an absolutely new category of economic minerals and are basically different from sedimentogenic, magmatogenic, metamorphogenic, and other concentrations of mineral raw materials generated in the Earth's crust and on its surface. At present, similar impact diamonds have been found in other impact structures originating in graphite-bearing target rocks (Ries, Kara, Puchezh-Katunki, Terny, Zapadnaya astroblemes). Thus, impact diamonds may be widely distributed, and finds of this material in placer deposits indicate the rewashing of impactites of some buried or strongly eroded craters. Although the lonsdaleite phase has not yet been established in carbonados occurring in placer deposits in Central Africa and Brazil, there is a supposition that this type is polycrystalline diamond originated in the Precambrian by large impacts on a carbon-bearing target. However, preliminary TEM observations seem to disprove this hypothesis (see above: High-Pressure Polymorphs).

Impact diamonds may be regarded as a criterion for the impact origin of lava-like or tuff-like rocks, where they can be found similarly to the high-pressure phases, coesite and stishovite. Their presence in any thin layer in the sedimentary sequence may be caused by dust-cloud transport, indicating a synchronous impact event to have occurred in some distant region.

Special investigations and commercial tests showed that polycrystalline impact diamonds from the Popigai crater are occasionally superior to the kimberlite ones in a number of properties, in particular in resistance to destructing load. Light-colored impact diamonds are the strongest, which allows them to be used for drills and other abrasive tools. Apographitic impact diamonds represent a new type of industrial diamond raw material, which offers a number of advantages over the well-known types of natural and synthetic diamonds. The economic value of this industrial impact diamond deposit may be significant, taking into account the high content of this mineral in impactites, and the very large resources of ores ensured by the widespread occurrences of diamond-bearing rocks at Popigai.

References

Agrinier P, Boyd SR, Martinez I, Schärer U, Javoy M, Deutsch A (1995) On the kinetics of $CaO + CO_2 \Rightarrow CaCO_3$ and CO_2 released during impact processes. Ann Geophys 13 Suppl III C:738

Ahrens TJ, Gaffney ES (1971) Dynamic compression of enstatite. J Geophys Res 76:5504–5514

Allen CC, Gooding JL, Keil K (1982) Hydrothermally altered impact melt rock and breccia: contributions to the soil of Mars. J Geophys Res 87:10,083–10,101

Ashworth JR, Schneider H (1985) Deformation and transformation in experimentally shock-loaded quartz. Phys Chem Min 11:241–249

Barber DJ, Wenk HR (1979) Deformation twinning in calcite, dolomite, and other rhombohedral carbonates. Phys Chem Min 5:141–165

Becker L, Bada JL, Winans RE, Hunt JE, Bunch TE, French BM (1994) Fullerenes in the 1.85 billion-year-old Sudbury Impact Structure. Science 265:642–645

Beran A, Koeberl C (1996) Water in tektites and impact glasses by FTIR Spectrometry. Meteoritics Planet Sci 32:211–216

Bischoff A, Stöffler D (1984) Chemical and structural changes induced by thermal annealing of shocked feldspar inclusions in impact melt rocks from Lappajärvi crater, Finland. J Geophys Res 89:B645–B656

Blum JD, Papanastassiou DA, Koeberl C, Wasserburg GJ (1992) Neodymium and strontium isotopic study of Asutralasian tektites: new constraints on the provenance and age of target materials. Geochim Cosmochim Acta 56:483–492

Bohor BF, Foord EE, Modreski PJ, Triplehorn DM (1984) Mineralogical evidence for an impact event at the Cretaceous-Tertiary boundary. Sience 224:867–869

Bohor BF, Betterton WJ, Krogh TE (1993) Impact shocked zircons: discovery of shock-induced textures reflecting increasing degrees of shock metamorphism. Earth Planet Sci Lett 119:419–424

Boslough MB (1991) Shock metamorphism and chemistry and planetary geologic processes. Ann Rev Earth Planet Sci 19:101–130

Bottomley RJ, Grieve RAF, York D, Masaitis VL (1997) The age of the Popigai impact event and its relation to events at the Eocene/Oligocene boundary. Nature 388:365–368
Brett R (1992) The Cretaceous-Tertiary extinction: a lethal mechanism involving anhydrite target rocks. Geochim Cosmochim Acta 56:3603–3606
Bringemeier D (1994) Petrofabric examination of the main suevite of the Otting quarry, Nördlinger Ries, Germany. Meteoritics 29:417–422
Buchanan PC, Reimold WU (1996) Analysis of deformation lamellae in quartz grains from the Rooiberg felsite, Bushveld Complex, South Africa, and associated rocks. Lunar Planet Sci Conf XXVII: 175–176
Carlisle DB, Braman DR (1991) Nonometre-size diamonds in the Cretaceous/Tertiary boundary clay of Alberta. Nature 352:708–709
Carstens H (1975) Thermal history of impact melt rocks in the Fennoscandian shield. Contrib Mineral Petrol 50:145–155
Carter NL (1965) Basal quartz deformation lamellae, a criterion for recognition of impactites. Am J Sci 263:786–806
Carter NL, Avé Lallemant HG (1970) High temperature flow of dunite and peridotite. Geol Soc Am Bull 81:2181–2202
Chen G, Tyburczy JA, Ahrens TJ (1994) Shock-induced devolatilization of calcium sulfate and implications for K-T extinctions. Earth Planet Sci Lett 128:615–628
Chen M, Sharp TG, El Goresy A, Wopenka B, Xie X (1996) The majorite-pyrope + magnesio-wüstite assemblage: constraints on the history of shock veins in chondrites. Science 1271:1570–1573
Cintala MJ, Grieve RAF (1994) The effects of differential scaling of impact melt and crater dimensions on lunar and terrestrial craters: some brief examples. In: Dressler BO, Grieve RAF, Sharpton VL (eds) Large meteorite impacts and planetary evolution. Geol Soc Am Spec Pap 293:51–59
Clarke RS, Appleman DE, Ross DR (1981) An antarctic iron meteorite contains preterrestrial impact-produced diamond and Ionsdaleite. Nature 291:396–398
Cordier P, Doukhan JC (1989) Water solubility in quartz and its influence on ductility. Eur J Mineral 1:221–237
Cummings D (1964) Kink-bands: shock deformation of biotite resulting from a nuclear explosion. Science 148:950–952
Dachille F, Gigl P, Simons PY (1968) Experimental and analytical studies of crystalline damage useful for the recognition if impact structures. In: French BM, Short NM (eds) shock metamorphism of natural materials. Mono Book Corp, Baltimore, Maryland, pp 555–570
DeCarli PS, Jamieson JC (1961) Formation of diamond by explosive shock. Science 133:1821–1822
Deribas AA, Dobretsov NL, Kudinov VM, Zyuzin NI (1966) Shock compression of SiO_2 powders. Dokl Akad Nauk USSR 168:127–130
Deutsch A, Schärer U (1994) Dating terrestrial impact events (invited review). Meteoritics 29: 301–322
Deutsch A, Buhl D, Langenhorst F (1992) On the significance of crater ages – new ages for Dellen (Sweden) and Araguainha (Brazil). Tectonophysics 216:205–218
Deutsch A, Grieve RAF, Avermann M, Bishoff L, Brockmeyer P, Buhl D, Lakomy R, Müller-Mohr V, Ostermann M, Stöffler D (1995) The Sudbury structure (Ontario, Canada): a tectonically deformed multi-ring impact basin. Geol Rundsch 84:697–709
Deutsch A, Ostermann M, Masaitis VL (1997) Geochemistry and Nd-Sr isotope signature of tektite-like objects (Urengoites, South-Ural glass). Meteoritics Planet Sci 32:679–686
Dickin AP, Artan MA, Crocket JH (1996) Isotopic evidence for distinct crustal sources of North and South Range ores, Sudbury Igneous Complex. Geochim Cosmochim Acta 60:1605–1613
El Goresy A, Fechtig H, Ottermann J (1968) The opaque minerals in impactite glasses. In: French BM, Short NM (eds) Shock metamorphism of natural materials. Mono Book Corp, Baltimore, Maryland, pp 531–553

Elston WE (1992) Does the Bushveld-Vredefort system (South Africa) record the largest known terrestrial impact catastrophe? Int Conf on Large meteorite impacts and planetary evolution. Sudbury LPI Contrib No 790:23–24

Evans NJ, Gregoire DC, Grieve RAF, Goodfellow WD, Veizer J (1993) Use of platinum-group elements for impactor identification: terrestrial impact craters and Cretaceous-Tertiary boundary. Geochim Cosmochim Acta 57:3737–3748

Feldman VI (1994) The conditions of shock metamorphism. In: Dressler BO, Grieve RAF, Sharpton VL (eds) Large meteorite impacts and planetary evolution. Geol Soc Am Spec Pap 293:121–132

Floran RJ, Grieve RAF, Phinney WC, Warner JL, Simonds CH, Blanchard DP, Dence MR (1978) Manicouagan impact melt Quebec 1 stratigraphy petrology and chemistry. J Geophys Res 83:2737–2759

Fregerslev S, Carstens H (1976) FeNi metal in impact melt rocks of Lake Lappajärvi, Finland, Contrib Mineral Petrol 55:255–263

French BM, Short NM (1968) Shock metamorphism of natural materials. Mono Book Corp, Baltimore, Maryland, 644 pp

French BM (1990) Absence of shock-metamorphic effects in the Bushveld Complex, South Africa: results of an intensive search. Tectonophysics 171:287–301

Frondel C, Marvin UB (1967) Lonsdaleite, a hexagonal polymorph of diamond. Nature 214:587

Gersonde R, Kyte FT, Bleil U, Diekmann B, Flores JA, Gohl K, Grahl G, Hagen R, Kuhn G, Sierro FJ, Voelker D, Abelmann A, Bostwick JA (1997) Geological record and reconstruction of the late Pliocere impact of the Eltanin asteroid in the Southern Ocean. Nature 390:357–363

Gilmour I, Russell SS, Arden JW, Lee MR, Franchi IA, Pillinger CT (1992) Terrestrial carbon and nitrogen isotopic ratios from Cretaceous-Tertiary boundary nanodiamonds. Science 258:1624–1626

Goltrant O, Cordier P, Doukhan JC (1991) Planar deformation features in shocked quartz: a transmission electron microscopy investigation. Earth Planet Sci Lett 106:103–115

Goltrant O, Leroux H, Doukhan JC, Cordier P (1992) Formation mechanism of planar deformation features in naturally shocked quartz. Phys Earth Planet Int 74:219–240

Gostin VA, Keays RR, Wallace MW (1989) Iridium anomaly from the Acraman impact ejecta horizon: impacts can produce sedimentary iridium peaks. Nature 340:542–544

Gratz AJ, Nellis WJ, Christie JM, Brocious W, Swegle J, Cordier P (1992) Shock metamorphism of quartz with initial temperatures −170 to +1000 °C. Phys Chem Min 19:267–288

Graup G (1978) Das Kristallin im Nördlinger Ries. Petrographische Zusammensetzung und Auswurfsmechanismus der kristallinen Trümmermassen, Struktur des kristallinen Untergrundes und Beziehungen zum Moldanubikum. Enke, Stuttgart, 190 pp

Grieve RAF (1978) The melt rocks at Brent crater, Ontario, Canada. Proc Lunar Planet Sci Conf 9th: 2579–2608

Grieve RAF (1987) Terrestrial impact structures. Annu Rev Earth Planet Sci 15:245–270

Grieve RAF (1991) Terrestrial impact: the record in the rocks. Meteoritics 26:175–194

Grieve RAF (1997) Extraterrestrial impact events: the record in the rocks and the stratigraphic column. Palaeogeo. Paleoclimat. Palaeoecol. 132, 5–23

Grieve RAF, Cintala MJ (1992) An analysis of differential impact melt-crater scaling and implications for the terrestrial impact record. Meteoritics 27:526–538

Grieve RAF, Masaitis VL (1994) The economic potential of terrestrial impact craters. Int Geol Rev 36:105–151

Grieve RAF, Shoemaker EM (1994) The record of past impacts on Earth. In: Gehrels T (ed) Hazards due to comets and asteroids. University of Arizona Press, Tucson, pp 417–462

Grieve RAF, Langenhorst F, Stöffler D (1996) Shock metamorphism of quartz in nature and experiment: II. Significance in geoscience. Meteoritics Planet Sci 31:6–35

Gudlaugsson ST (1993) Large impact crater in the Barents Sea. Geology 21:291–294

Gurov EP, Gurova EP (1991) Geological structure and composition of rocks in impact craters. Nauka Press, Kiev (in Russian)

Gurov EP, Melnychuk EV, Metalidi SV, Ryabenko VA, Gurova EP (1985) The characteristics of the geological structure of the eroded astrobleme in the western part of the Ukrainian Shield. Dopovidi Akad Nauk Ukrainskoi Radyanskoi Sotsialisychnoi Republiky, Seriya B: 8–11 (in Russian)

Hannemann RE, Strong HM, Bundy FP (1967) Hexagonal diamonds in meteorites: implications. Science 155: 995–997

Hanss RE, Montaque BR, Davis MK, Galindo C, Hörz F (1978) X-ray diffractometer studies of shocked materials, Proc Lunar Planet Sci Conf IX: 2773–2787

Hart SR, Kinloch ED (1989) Osmium isotope systematics in Witwatersrand and Bushveld ore deposits. Econ Geol 84: 1651–1655

Hörz F (1968) Statistical measurements of deformation structures and refractive indices in experimentally shock-loaded quartz. In: French BM, Short NM (eds) Shock metamorphism of natural materials. Mono Book Corp, Baltimore, Maryland, pp 243–254

Hörz F, Quaide WL (1972) Debye-Scherrer investigations of experimentally shocked silicates. Moon 6: 45–82

Hörz F, Gall H, Hüttner R, Oberbeck VR (1977) Shallow drilling in the Bunte Breccia impact deposits, Ries Crater, Germany. In: Roddy DJ, Pepin RO, Merrill RB (eds) Impact and explosion cratering. Pergamon Press, New York, pp 425–448

Hörz F, See TH, Murali AV, Blauchard DP (1989) Heterogeneous dissemination of projectile materials in the impact melts from Wabar Crater, Saudi Arabia. Proc Lunar Sci Conf IXX: 697–709

Ivanov BA (1995) Geomechanical models of impact cratering: Puchezh-Katunki structure. In: Dressler BO, Grieve RAF, Sharpton VL (eds) Large meteorite impacts and planetary evolution. Geol Soc Am Sec Pap 293: 81–91

Jahn BM, Floran RJ, Simonds CH (1978) Rb-Sr isochron age of the Manicouagan melt sheet Quebec, Canada. J Geophys Res 83: 2799–2803

Jakubith M, Hornemann U (1981) Majorite formation from enstatite by experimental shock-loading. Phys Earth Planet Int 27: 95–99

Joreau P, Reimold WU, Robb LJ, Doukhan JC (1997) A TEM study of deformed quartz grains from volcaniclastic sediments associated with the Bushveld Complex, South Africa. Eur J Mineral 9: 393–401

Kamo SL, Reimold WU, Krogh TE, Colliston WP (1996) A 2.023-Ga age for the Vredefort impact event and a first report of shock-metamorphosed zircons in pseudotachylitic breccias and granophyre. Earth Planet Sci Lett 144: 369–387

Kargel JS, Coffin P, Kraft M, Lewis JS, Moore C, Roddy D, Shoemaker EM, Wittke JH (1996) Systematic collection and analysis of meteoritic materials from Meteor Crater, Arizona. Lunar Planet Sci Conf XXVII: 645–646

Kieffer SW, Phakey PP, Christie JM (1976) Shock processes in porous quarzite: transmission electron microscope observations and theory. Contrib Mineral Petrol 59: 41–93

Kirby SH, Christie JM (1977) Mechanical twinning in diopside Ca(Mg,Fe)Si_2O_6: structural mechanism and associated crystal defects. Phys Chem Min 1: 137–163

Kitamura M, Goto T, Syono Y (1977) Intergrowth textures of diaplectic glass and crystal in shock-loaded P-anorthite. Contr Mineral Petrol 61: 299–304

Kleinmann B (1969) The breakdown of zircon observed in the Libyan desert glass as evidence of its impact origin. Earth Planet Sci Lett 5: 497–501

Koeberl C (1990) The geochemistry of tektites: an overview. Tectonophysics 171: 405–422

Koeberl C, Reimold WU (1995) Early Archaean spherule beds in the Barberton Mountain Land, South Africa: no evidence for impact origin. Precambrian Res 74: 1–33

Koeberl C, Shirey SB (1996) Re-Os systematics as a diagnostic tool for the study of impact craters and distal ejecta. Palaeogeo. Paleoclimat. Palaeoecol. 132, 25–46

Kyte FT, Bohor BF (1995) Nickel-rich magnesiowüstite in Cretaceous/Tertiary boundary spherules crystallized from ultramafic, refractory silicate liquids. Geochim Cosmochim Acta 59: 4967–4974

Kyte FT, Bostwick JA (1995) Magnesioferrite spinel in Cretaceous/Tertiary boundary sediments of the Pacific basin: remnants of hot, early ejecta from the Chicxulub impact? Earth Planet Sci Lett 132:113–127

Lambert P, Mackinnon IDR (1984) Micas in experimentally shocked gneiss. J Geophys Res 89:B685–B699

Lange MA, Lambert P, Ahrens TJ 81985) Shock effects on hydrous minerals and implications for carbonaceous meteorites. Geochim Cosmochim Acta 49:1715–1726

Langenhorst F (1989) Experimentally shocked plagioclase: changes of refractive indices and optic axial angle in the 10–30 GPa range. Meteoritics 24:291

Langenhorst F (1994) Shock experiments on α- and β-quartz: II. X-ray investigations. Earth Planet Sci Lett 128:683–698

Langenhorst F (1996) Characteristics of shocked quartz in late Eocene impact ejecta from Massignano (Ancona, Italy): Clues to shock conditions and source crater. Geology 24: 487–490

Langenhorst F (1997) Impaktdiamanten als Zeugen des Bombardements aus dem All. Humboldt-Spektrum, Humboldt-Universität zu Berlin, Heft 2/97, 32–37

Langenhorst F, Deutsch A (1994) Shock experiments on preheated α- and β-quartz: I. Optical and density data. Earth Planet Sci Lett 125:407–420

Langenhorst F, Joreau P, Doukhan JC (1995) Thermal and shock metamorphism of the Tenham meteorite: a TEM examination. Geochim Cosmochim Acta 59:1835–1845

Leroux H, Doukhan JC, Langenhorst F (1994a) Microstructural defects in experimentally shocked diopside: a TEM characterization. Phys Chem Min 20:521–530

Leroux H, Reimold WU, Doukhan JC (1994b) A TEM investigation of shock metamorphism in quartz from the Vredefort dome, South Africa. Tectonophysics 230:223–239

Lipschutz ME (1964) Origin of diamonds in the ureilites. Science 143:1431–1434

Madon M, Poirer JP (1983) Transmission electron microscope observation of α, β and $(Mg,Fe)_2SiO_4$ in shocked meteorites: planar defects and polymorphic transitions. Phys Earth Planet Int 33:31–44

Manghnani MH, Syono Y (1987) High-pressure research in mineral physics. Geophysical Monograph 39, American Geophysical Union

Margolis S, Claeys Ph, Kyte FT (1991) Microtektites, microcrystites and spinels from a late Pliocene asteroid impact in the Southern Ocean. Science 251:1594–1597

Martinez I, Schärer U, Guyot F (1993) Impact-induced phase transformations at 50–60 GPa in continental crust: an EPMA and ATEM study. Earth Planet Sci Lett 119:207–223

Martinez I, Agrinier P, Schärer U, Javoy M (1994) CO_2 production by impact into carbonates? A SEM-ATEM and stable isotope (^{18}O, ^{13}C) study of carbonates from the Haughton impact crater. Earth Planet Sci Lett 121:559–574

Martinez I, Deutsch A, Schärer U, Ildefonse Ph, Guyot F, Agrinier P (1995) Shock recovery experiments on dolomite and thermodynamical modeling of impact-induced decarbonation. J Geophys Res 100, B8:15,465–15,476

Martini JEJ (1991) The nature, distribution and genesis of the coesite and stishovite associated with the pseudotachylite of the Vredefort Dome, South Africa. Earth Planet Sci Lett 103:285–300

Masaitis VL (1993) Origin of the Sudbury Structure from the points of new petrographic mineralogical and geochemical data. Trans All-Russian Mineral Soc 122:1–17 (in Russian)

Masaitis VL (1994) Impactites from the Popigai crater. In: Dressler BO, Grieve RAF, Sharpton VL (eds) Large meteorite impacts and planetary evolution. Geol Soc Am Spec Pap 293:153–162

Masaitis VL, Mashchak MS (1996) Recrystallization and blastesis of shock-metamorphozed rocks in impact structures. Proc Russ Mineral Soc CXXV:1–18

Masaitis VL, Futergendler DI, Gnevushev MA (1972) Diamonds in impactites of the Popigay meteorite crater. Zap Vsesoyuznogo Mineralogicheskogo Obshchestva 101:108–112

Masaitis VL, Mikhailov MV, Selivanovskaya TV (1975) Popigai meteorite crater. Nauka, Moscow, 124 pp

Masaitis VL, Danilin AI, Mashchak MS, Raikhlin AI, Selivanovskaya TV, Shadenkov EM (1980) The geology of astroblemes. Nedra Press, St Petersburg, USSR, Russia, 231 pp (in Russian)

Masaitis VL, Mashchak MS, Naumol MV, Orlovo JV, Selivanovskaya TV (1995) Puchezh-Katunki impact crater: main features of geological structure. Doklady Ac Si 342:358–360

McLaren AC, Phakey PP (1966) Electron microscope study of Brazil twin boundaries in amethyst quartz. Phys Stat Sol 13:413–422

Melosh HJ (1989) Impact cratering. A geological process. Oxford University Press, New York, 245 pp

Melosh HJ, Vickery AM (1991) Melt droplet formation in energetic impact events. Nature 350:494–497

Milton DJ, De Carli P (1963) Maskelynite: formation by explosive shock. Science 140:670–671

Morgan J, Warner M and the Chicxulub working group (1997) Size and morphology of the Chicxulub impact crater. Nature 390, 472–476

Mori H (1994) Shock-induced phase transformations of the Earth and planetary materials. J Mineral Soc Jpn 23:171–178

Müller WF (1993) Thermal and deformation history of the Shergotty meteorite deduced from clinopyroxene microstructure. Geochim Cosmochim Acta 57:4311–4322

Müller-Mohr V (1992) Breccias in the basement of a deeply erode impact structure Sudbury Canada. Tectonophysics 216:219–226

Newsom HE, Graup G, Sewards T, Keil K (1986) Fluidization and hydrothermal alteration of the suevite deposit at the Ries crater, West Germany, and implications for Mars. J Geophys Res 91:E239–E251

Nishiizumi K, Kohl CP, Shoemaker EM, Arnold JR, Klein J, Fink D, Middleton R (1991) In situ ^{10}Be-^{26}Al exposure ages at meteor Crater, Arizona. Geochim Cosmochim Acta 55:2699–2703

Ostermann M, Deutsch A, Schärer U (1996) Impact melt dikes in the Sudbury multi-ring basin (Canada): Implications from U-Pb geochronology on the Foy Offset. Meteor Planet Sci 31, 494–501

Ostertag R (1983) Shock experiments on feldspar crystals. J Geophys Res 88:B364–B376

Palme H, Goebel E, Grieve RAF (1979) The distribution of volatile and siderophile elements in the impact melt of East Clearwater (Quebec) Proc Lunar Planet Sci Conf 10th:2465–2495

Pevzner LA, Masaitis VL (1996) (in Russian)

Phillips F, Zreda MG, Smith SS, Elmore D, Kubik PW, Dorn RI, Roddy DJ (1991) Age and geomorphic history of Meteor Crater Arizona from cosmogenic Cl and C in rock varnish. Geochim Cosmochim Acta 55:2695–2698

Pilkington M, Grieve RAF (1992) The geophysical signature of terrestrial impact craters. Rev Geophys 30:161–181

Plado J, Pesonen LJ, Elo S, Puura V, Suuroja K (1996) Geophysical research on the Kärdla impact structure, Hiiumaa Island, Estonia. Meteor Planet Sci 31:289–298

Poag CW (1996) Structural outer rim of Chesapeake Bay impact crater: seismic and bore hole evidence. Meteor Planet Sci 31:218–226

Pohl J, Stöffler D, Gall H, Ernstson K (1977) The Ries impact crater. In: Roddy DJ, Pepin RO, Merill RB (eds) Impact and explosion cratering. Pergamon Press, New York, pp 343–404

Pope KO, Baines KH, Ocampo AC, Ivanov BA (1994) Impact winter and the Cretaceous/Tertiary extinctions: results of a Chicxulub asteroid impact model. Earth Planet Sci Lett 128:719–725

Pope KO, Ocampo AC, Fisher AG, Morrison J, Sharp Z (1996) Carbonate condensates in the Chicxulub ejecta depositis from Belize. Lunar Planet Sci Conf XXVII: 1045–1046

Putnis A, Price GD (1979) High pressure $(Mg,Fe)_2SiO_4$ phases in the Tenham chondritic meteorite. Nature 280:217–218

Rasmussen KL, Clausen HB, Kallemeyn GW (1995) No iridium anomaly after the 1908 Tunguska impact evidence from a Greenland ice core. Meteoritics 30:634–638

Reimold WU, Koeberl C (1994) About the impact origin of the Vredefort structure. Geobull Geol Soc South Africa 37:3–7

Robertson PB, Grieve RAF (1977) Shock attenuation at terrestrial impact structures. In: Roddy DJ, Pepin RO, Merill RB (eds) Impact and explosion cratering. Pergamon Press, New York, pp 687–702

Robertson PB, Mason GD (1975) Shatter cones from Haughton dome, Devon Island, Canada. Nature 255:393–394

Robin E, Bonté Ph, Froget L, Jéhanno C, Rocchia R (1992) Formation of spinels in cosmic objects during atmospheric entry: a clue to the Cretaceous-Tertiary boundary event. Earth Planet Sci Lett 108:181–190

Roddy DJ, Boyce JM, Colton GW, Dial AL Jr (1975) Meteor crater, Arizona, rim drilling with thickness, structural uplift, diameter, depth, volume, and mass-balance calculations. Proc Lunar Sci Conf 6th:2621–2644

Roest WR, Pilkington M (1994) Restoring post-impact deformation at Sudbury: a circular argument. Geophys Res Lett 21:959–962

Rost R, Dolgov YA, Vishnevskiy SA (1978) Gases in inclusions of impact glass in the Ries crater, West Germany, and finds of high-pressure carbon polymorphs. Dokl Acad Nauk USSR 241:165–168 (in Russian)

Schmitz B, Jeppsson L, Ekvall J (1994) A search for shocked quartz grains and impact ejecta in early Silurian sediments on Gotland. Sweden. Geol Mag 131:361–367

Schneider H (1972) Shock-induced mechanical deformations in biotites from crystalline rocks of the Ries crater (Southern Germany). Contrib Mineral Petrol 37:75–85

Schneider H, Vasudevan R, Hornemann U (1984) Deformation of experimentally shock-loaded quartz powders: X-ray line broadening studies. Phys Chem Min 10:142–147

Schrand Ch, Deutsch A (1996) Phase transformations in pre-heated granitic rock samples in shock recovery experiments at 85 GPa: formation of shock-generated crystals. Meteor Planet Sci 31, A124–A125

Sharp TG, Lingemann CM, Dupas C, Stöffler D (1997) Natural occurrence of $MgSiO_3$-ilmenite and evidence for $MgSiO_3$-perovskite in a shocked L chondrite. Science 277: 352–355

Simonson BE, Davies D, Wallace M, Reeves S (1996) PGEs and quartz grains in a resedimented Late Archean impact horizon in the Hamersley Group of Western Australia. Lunar Planet Sci Conf XXVII: 1203–1204

Skrotzki W (1994) Defect structure and deformation mechanisms in naturally deformed augite and enstatite. Tectonophysics 229:43–68

Smit J (1994) Extinctions at the Cretaceous-Tertiary boundary. The link to the Chicxulub impact. In: Gehrels T (ed) Hazards due to comets and asteroids. University of Arizona Press, Tucson, pp 859–878

Smit J, Kyte FT (1984) Siderophile-rich magnetic spheroids from the Cretaceous-Tertiary boundary in Umbria, Italy. Nature 310:403–405

Smith JV, Dawson JB (1985) Carbonado: diamond aggregates from early impacts of crustal rocks? Geology 13:342–343

Spray JG, Thompson LM (1995) Friction melt distribution in a multi-ring impact basin. Nature 373:130–132

Stöffler D (1971) Coesite and stishovite: Identification and formation conditions in shock-metamorphosed rocks. J Geophys Res 76:5474–5488

Stöffler D (1972) Deformation and transformation of rock-forming minerals by natural and experimental shock processes: I. Behavior of minerals under shock compression. Fortschr Mineral 49:50–113

Stöffler D, Grieve RAF (1994) Classification and nomenclature of impact metamorphic rocks: a proposal to the IUGS subcommission on the systematics of metamorphic rocks. Lunar Planet Sci Conf XXV: 1347–1348

Stöffler D, Langenhorst F (1994) Shock metamorphism of quartz in nature and experiment: I. Basic observation and theory. Meteoritics 29:155–181

Stöffler D, Keil K, Scott ERD (1991) Shock metamorphism of ordinary chondrites. Geochim Cosmochim Acta 55:3845–3867

Therriault AM, Reid AM, Reimold WU (1993) Original size of the Vredefort Structure, South Africa. Lunar Planet Sci Conf XXIV: 1419–1420

Thompson LM, Spray JG (1994) Pseudotachylytic rock distribution and genesis within the Sudbury impact structure. In: Dressler BO, Grieve RAF, Sharpton VL (eds) Large meteorite impacts and planetary evolution. Geol Soc Am Sec Pap 293:275–287

Tomioka N, Fujino K (1997) Natural (Mg, Fe)SiO_3-ilmenite and -perovskite in the Tenham meteorite. Science 277, 1084–1086

Trueb LF (1971) Microstructural study of diamonds synthesized under conditions of high temperature and moderate explosive shock pressure. J Appl Phys 42:503–510

Tschermak G (1872) Die Meteoriten von Shergotty and Goalpur. Sitzungsber Akad Wiss Wien Math-Naturwiss Kl 65 Teil 1:122–145

Turner FJ (1964) Analysis of kinks in micas of an Innsbruck mica schist. N Jb Miner Mh 9:51–83

Valter AA, Yeremenko GK, Kwasnitsa VN, Polkanov YA (1992) Shock-metamorphosed carbon minerals. Nauka Press, Kiev (in Russian)

von Gruenewaldt G, Harmer RE (1992) Tectonic setting of proterozoic layered intrusions with special reference to the Bushveld Complex. Proterozoic crustal evolution (ed KC Condie). Developments in Precambrian geology 10: Chapter 5. Elsevier, Amsterdam, pp 181–213

von Engelhardt W (1972) Shock-produced rock glasses from the Ries Crater. Contrib Mineral Petrol 36:265–292

von Engelhardt W, Bertsch W (1969) Shock-induced planar deformation structures in quartz from the Ries crater, Germany. Contrib Mineral Petrol 20:203–234

von Engelhardt W, Arndt J, Stöffler D, Müller WF, Jeziorkowski H, Gubser RA (1967) Diaplektische Gläser in den Breccien des Ries von Nördlingen als Anzeichen für Stoßwellenmetamorphose. Contrib Mineral Petrol 15:91–100

von Engelhardt W, Arndt J, Fecker B, Pankau HG (1995) Suevite breccia from the Ries crater, Germany: origin, cooling history and devitrification of impact glass. Meteoritics 30:279–293

White JC (1993) Shock-induced melting and silica polymorph formation, Vredefort structure, South Africa. In: Boland JA, FitzGerald JD (eds) Defects and processes in solid state: geoscience applications. Elsevier, New York, pp 69–84

1.10.7 The Cretaceous-Tertiary Boundary Impact Event

M. A. Nazarov

The Cretaceous period was terminated by a massive extinction of major groups of animals and plants including dinosaurs, the great marine reptiles, the flying reptiles, the ammonites and belemnites, many bivalve groups, and major groups of marine phyto- and zooplankton. This is one of the most dramatic mass extinction events ever recorded. It has been estimated that as many as 75% of Cretaceous species may have been eliminated. Numerous theories have been proposed to explain the nature of the biotic crisis. However, the problem with almost all these theories was that they rested on very little physical data. Physical evidence for extinction mechanisms has largely come from the 1-30-cm-thick clay layer marking the Cretaceous-Tertiary (K-T) boundary. This layer, deposited during the K-T transition approximately 65 Ma ago, contains strong evidence for an impact of an extraterrestrial body that had struck the Earth at the end of the Cretaceous, causing the massive extinction.

The K-T boundary clay layer is enriched relative to the crustal abundance in Ir and other siderophile elements (Ni, Co, Ru Rh, Pd, Os, Pt, Au) by factors of 5–4000. The enrichment is global in extent and has been documented at more than 75 K-T boundary sites throughout the world. It means that the siderophile anomaly should have been produced by a global-scale event for a short time. The catastrophic event could not be related to an abrupt shift in ocean chemistry because the anomaly is found in both marine and nonmarine sediments. The very low $^{187}Os/^{188}Os$ ratio of the K-T clay (up to 0.135) points to either the mantle or an extraterrestrial source for the K-T siderophiles, whereas the trace element abundance pattern of the clay is compatible only with an extraterrestrial chondritic source. Thus, geochemical data indicate firmly a catastrophic accretionary event at the end of the Cretaceous. However, no extraterrestrial particles have been detected at the K-T boundary by mineralogical methods. Only siderophile-rich, magnetite spherules and Ni-rich magnesioferrite crystals found in the K-T layer are suggested to be derived from extraterrestrial matter.

Possible accretionary mechanisms include: (1) accretion of extraterrestrial dust or a crushed material of an accreting object undergone by tidal and atmospheric disruption; (2) the impact of compact extraterrestrial objects. If the K-T accretionary event is an impact, then an impact ejecta should be present in the K-T boundary layer. In fact, this material has been recognized. It has been demonstrated that the K-T clay contains quartz grains exhibiting features characteristic of shock metamorphism. The grains show multiple sets of planar features precisely oriented with respect to the crystal structure. The only known way to produce such features is by a high-velocity impact, and quartz grains exhibiting such lamellae are commonly present in terrestrial impact craters and are produced by laboratory shock experiments. At the K-T boundary shocked quartz grains have been detected worldwide. Besides individual shocked quartz grains, composite shock-deformed grains of quartz-quartz and quartz-feldspar occur globally in the K-T layer. Shocked grains are most abundant and largest in size in K-T sections of the USA and Mexico. Here, stishovite, a high-pressure phase, was also identified. Tektitelike glasses discovered in the K-T sites of Haiti and Mexico provide further strong evidence for the presence of an ejecta component in the K-T layer. The glasses have mainly dacitic to andesitic compositions. However, some of the glasses are Ca-rich and can have resulted from mixing of impact-melted basement rocks with limestones and/or anhydrites.

The diameter of the K-T object estimated from the observed noble metal concentrations in about 10–20 km. The projectile could form an impact crater of more than 200 km in diameter. However, the K-T object could be disrupted by tidal forces and, therefore, a few craters of a smaller size could be formed. The presence of the shocked quartz grains and the andesitic composition of the tektitelike glasses shows that the K-T object, or at least its fragment, impacted the continental crust. It has been demonstrated that the 180-km-diameter buried circular structure centered at Chicxulub, on the north coast of Yucatan, Mexico, could be a continental K-T boundary impact crater. Now the structure is under detailed investigation. another possible K-T crater is the Kara impact structure located on the Kara Sea shore, near to the Ural Mountains in Russia. The structure has a diameter of 120 km, but it may be a little older than the K-T boundary. If the Chicxulub and Kara are indeed of different ages, then the cratering rate in the Cretaccous should be much higher than the normal rate.

The kinetic energy associated with the K-T object is $>10^{30}$ erg. The transfer of the energy to the environment should lead to significant physical and chemical changes inducing massive extinction of species. Possible environmental stresses caused by the impact include: short-term heating of the atmosphere, storm winds and tsunamis, global fires, shielding of the Sun by lofted fine impact ejecta and a subsequent sharp cooling of the oceans and atmospheres, a decline in photosynthesis rates, a CO_2 and/or H_2O-greenhouse, shock production of NO giving acid rains, and/or enhanced transmission of UV light due to destruction of the ozone layer. Geochemcial studies of the K-T clay do show evidence for some environmental changes. This clay is globally enriched in soot, indicating major wildfires, whereas C, O, and N isotopic compositions of the clay record significant temperature changes, decline of bioproduction, and precipitation of acid rains.

The impact scenario of the K-T transition suggests that other mass extinctions in geological history could also have been caused by impact events. In fact, the extinctions appear to have been periodic, with a 26-Ma cycle, and terrestrial impact-crater age distributions show a similar periodicity. Therefore, it has been suggested that impacts from cyclic swarms of comets in the inner Solar System were responsible for the periodic extinctions. However, geochemical studies across main bioevent horizons in the fossil record have not provided clear evidence for the hypothesis.

References

Alvarez LW, Alvarez W, Asaro F, Michel H (1980) Extraterrestrial cause for the Cretaceous-Tertiary extinction. Science 208:1095–1108

Glen W (ed) (1994) the mass extinction debates: how science works in a crisis. Stanford University Press, Stanford, 370 pp

Sharpton VL, Ward PD (eds) (1990) Global catastrophes in Earth history. Geol Soc Am Spec Pap 247:631

Silver LT, Shultz PH (eds) (1982) Geological implications of impacts of large asteroids and comets on the Earth. Geol Soc Am Spec Pap 190:528

CHAPTER 2

Mineralogy of the Mantle and Core

2.1 Mineralogical Structure of the Earth: Earth Geoscience Transect

O. L. KUSKOV

The determination of the bulk chemical composition of the Earth and its seismic structure is one of the major goals in the Earth Sciences. Cosmochemical and meteoritic studies suggest that the composition of the bulk Earth is essentially close to chondritic in terms of the major lithophile and refractory elements. According to the chondritic concept, a sufficiently adequate description of mantle rocks proves to be the system CaO-FeO-MgO-Al_2O_3-SiO_2 (CFMAS), where the ratios of Mg, Si, Ca, and Al are chondritic (cosmic) and Fe/(Fe+Mg)=0.1 (atomic ratio). Modeling of cosmochemical condensation-evaporation processes showed the stability of a condensation product, which presumably could be considered as corresponding to the average chemical composition of the terrestrial planets. This equilibrium mineral assemblage was designated solar chondrite. The composition of the solar chondrite assemblage is in agreement with the main petrological models of the Earth's upper mantle whereas its seismic properties are in agreement with the seismological models of the mantle. Another possibility for the primitive mantle composition has been derived from petrological considerations: pyrolite, which can yield basal-

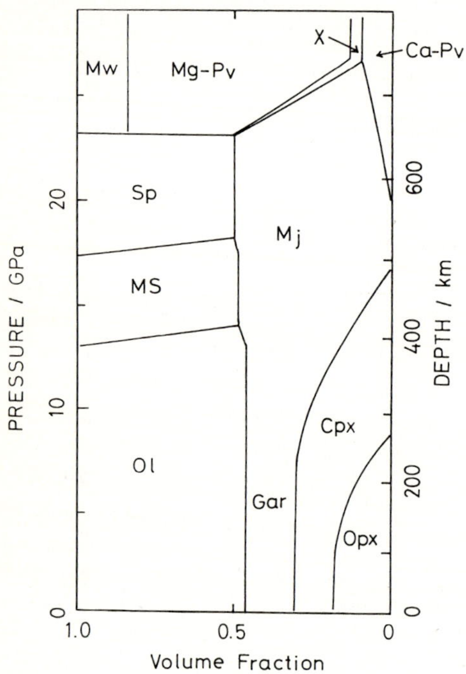

Fig. 34. Mineral constitution of pyrolite in the Earth's mantle. *Ol* Olivine; *MS* modified spinel (β-phase); *Sp* γ-spinel; *Opx* Ca-poor pyroxene; *Cpx* Ca-rich pyroxene; *Gar* garnet; *Mj* majorite; *Mg-Pv* $MgSiO_3$-rich perovskite; *Ca-Pv* $CaSiO_3$-rich perovskite; *Mw* magnesiowustite; *x* Al-rich phase

Table 10. Chemical compositions (wt.%) proposed for the Earth's mantle

Oxide	Chondritic mantle	Pyrolite	Solar chondrite
SiO_2	50.2	45.2	48.3
Al_2O_3	3.6	4.4	3.4
FeO	7.2	8.0	10.7
MgO	36.2	39.0	34.7
CaO	2.8	3.4	2.9

tic magma. The pyrolite composition is (Fig. 34) depleted in silica compared to the chondritic models (Table 10).

Compositional and mineralogical models of the upper mantle (~40–400 km), transition zone (400–650 km) and lower mantle (650–2900 km), as well as the chemical composition of the entire mantle and the nature of the mantle discontinuities are uncertain and have been a matter of intense discussion. Phase diagrams of mineral systems provide the basis for modeling and understanding the constitution of the Earth's mantle, but they tell us nothing about the elastic properties. In the filed of Earth sciences, only seismology provides direct measurements of the mantle properties. An important task is to convert the phase diagram into velocity and density profiles, connecting seismology and mineral physics. Through experimental and theoretical studies, it has been possible to determine the effects of the elements on phase equilibria and crystal chemistry of minerals as well as on physical properties (elastic properties and density) of mineral assemblages and compare them with seismic data.

Phase Transformations, Composition, and the Nature of the Mantle Discontinuities

Phase Transformations

Olivine, pyroxene, and garnet are the major minerals of the upper mantle. The olivine (α)-modified spinel (β)-spinel (γ) transitions in the system Mg_2SiO_4-Fe_2SiO_4 were studied and equilibrium boundaries of the phase transitions were performed up to 21 GPa and 1600 °C. The pyroxene-garnet phase transition, the stability fields of majorite garnet, ilmenite, and perovskite were determined in the system $Mg_4Si_4O_{12}$(enstatite)-$Mg_3Al_2Si_3O_{12}$ (pyrope) at pressures and temperatures of the transition region and lower mantle. The results show that the pyroxene-garnet transition occurs gradually with increasing pressure and is unlikely to yield sharp velocity and density jumps in the mantle. At pressures above 17 GPa, diopside dissociates into $CaSiO_3$ perovskite and $MgSiO_3$ component in the ilmenite or perovskite structure. Phase relations in the FeO-MgO-SiO_2 system above 21–23 GPa are characterized by formation of (Mg,Fe)SiO_3 ilmenite or perovskite with negative slopes for the ilmenite-perovskite transition in $MgSiO_3$ and for the decomposition of γ-spinel to perovskite+periclase in Mg_2SiO_4. Experiments show that the capacity of perovskite to take Al_2O_3 into solid solution is enhanced with increasing pressure; stishovite may transform to the $CaCl_2$ structure, which might be stable in the lowermost mantle.

Elasticity data and thermal expansion were measured at high pressures and temperatures for a number of mantle minerals to evaluate petrological models and to understand the nature of the mantle discontinuities, and to compare the obtained theoretical results with the observed seismic velocity profiles.

400-km Discontinuity

The boundary between the upper mantle and transition zone does not exceed 8–15 km in thickness with a temperature of 1400–1600 °C. The experimentally determined and calculated widths of the α-β transition loop for the upper mantle olivine [Mg/(Mg+Fe) ~ 0.89] are consistent with that observed for the 400-km discontinuity. The observed jumps of the seismic parameter $\Delta\Phi/\Phi_s$ ($\Phi_s = V_P^2 - 4/3\, V_S^2 = K_s/\rho$, where V_P and V_s are the seismic velocities, K_s is the adiabatic bulk modules and ρ is the density) and density ($\Delta\rho/\rho$) across the discontinuity in the PREM model are equal to 4.1 and 5.1%, respectively. The calculated density jump in a pyrolite model is slightly insufficient as compared with the PREM data, while changes in Φ_s exceed seismic ones after the PREM data but are less pronounced than in the regional models. Although the calculated values of Φ_s and ρ and their jumps in a pyrolite model are not in complete harmony with the global and regional models, they agree with seismic estimates within the uncertainties of both thermodynamic data and seismological models. Thus, the 400-km discontinuity may represent the isochemical phase transformations in the olivine component of pyrolite. However, the chemical nature of the boundary between the upper mantle and transition zone cannot be ruled out.

520-km Discontinuity

In the peridotitic mantle the β-phase-γ-spinel transition is commonly invoked to explain the nature of a seismic discontinuity at depth of around 520 km. Other explanations suggest the dissolution of garnet to Ca-perovskite or chemical change in composition. The seismic data are controversial and probably do not require the presence of a 520-km discontinuity in the mantle.

650-km Discontinuity

The boundary region between the transition zone and the lower mantle ("650"-km discontinuity, T ~ 1600–1800 °C) is marked by the presence of a very sharp velocity and density jump (in the PREM model: $\Delta\Phi/\Phi_s = 6.9\%$ and $\Delta\rho/\rho = 9.7\%$) and a high velocity gradient at depths of 600–800 km. The 650-km seismic discontinuity is associated either with phase transition form spinel to perovskite plus magnesiowustite or with change in chemistry. Phase diagrams constrain this phase transition to be extremely sharp (2–4 km wide); therefore they could act as an effective reflector of seismic waves. On the other hand, it is difficult to match the seismic properties simultaneously observed at the depths of 600–650 km and across the seismic discontinuity, and in the lower mantle with any isochemical model considered in the FMS system. Thermodynamic calculations show that the jump of the seismic parameter $\Delta\Phi/\Phi_s$ across the discontinuity in the global and regional models is two to three times greater than in the pyrolite model. From the comparison of the calculated and observed seismic parameter profiles, it is likely that the upper and lower mantle are chemically distinct. Phase changes in the pyrolite model are not able to explain the nature of the 650-km discontinuity. However, a combination of chemical and phase transitions can easily explain this seismic discontinuity. The results of seismic studies of P-to-S converted phases give evidence for a locally sharp 650-km discontinuity; however, recent estimates show that the discontinuity is better modeled by a linear gradient zone 20–30 km thick.

In a chemically homogenous mantle model, the upper and lower mantles are assumed to have the same pyrolitic composition, and a whole mantle convection is expected to occur. However, the thermodynamic calculations indicate that the lower mantle is more iron-rich or silica-rich or simultaneously more silica-iron-rich than the upper mantle. In such a chemically stratified mantle model, two-layer convection is likely to operate.

The lack of seismic activity below 650–700 km has a natural relation to the boundary between the transition zone and the lower mantle, thus suggesting that lithospheric plates penetrate through the upper mantle but not into the lower mantle. Seismic activity ceases in the Bennioff zones at these depths, which is ascribed to a change in composition and absence of physicochemical transformations at depths of the lower mantle. The lower mantle is, in essence, dead therefore in a seismic respect. Seismology and phase-transition theory are in agreement with independent isotopic models in which the depleted upper mantle and the undepleted lower mantle are chemically distinct.

Another important problem is the negative Clapeyron slope of perovskite-forming transitions, which has significant implication because if the slope is sufficiently negative it could inhibit mixing of the upper and lower mantle. The penetrability of the 650-km discontinuity by mantle flow has been the subject of debate. Tomographic images indicate that slab penetration is not universal. Some descending slabs appear to penetrate the discontinuity, while other slabs appear to deflect along the boundary.

Water-Bearing Phases in the Earth's Mantle

Water is present in crustal and mantle rocks. It is accepted that small amounts of water in the shallow levels of the upper mantle are stored in hydrous silicate minerals such as amphibole and phlogopite. A few members of humite group minerals such as clinohumite and chondrodite were also suggested to be stable in the upper mantle. Trace quantities of hydroxyl (OH) have been detected spectroscopically in many nominally anhydrous minerals. For example, pyroxenes contain 200–500 ppm water, and β-Mg_2SiO_4 has been found to contain up to 4000 ppm (0.4 wt.%) water. However, the deep mantle is generally considered to be dry. This suggestion rules out a possible role of dense hydrous magnesium silicates However, a whole family of water-bearing magnesian-silicate mantle minerals was discovered recently in the MgO-SiO_2H_2O system. High pressure-temperature experiments showed that phase A is not stable above 1000 °C from 6–15 GPa; chondrodite is not stable above 900–1000 °C at 100 kbar, and phlogopite is not stable above 1300 °C at 4 GPa in a peridotitic mantle. It has been suggested that the most dense phase B ($Mg_{24}Si_8O_{38}(OH)_4$, $\rho = 3.368$ g cm^{-3}) might be a host mineral for water and might be stable at transition zone pressures, whereas phase D is probably stable in subducted slabs near the 650-km discontinuity. Phases E and F were observed to be stable at 13–17 GPa and 800–1000 °C. An upper temperature stability limit of 1000 °C for the hydrous mantle minerals implies that these phases are able to store water in deep subducted slabs but not in the surrounding mantle. Superhydrous phase B ($Mg_{10}Si_3O_{14}(OH)_4$, $\rho = 3.21$ g cm^{-3}), stabilized by fluorine, has the highest-temperature stability limit and may potentially exist in the transition zone. The assemblage superphase B+stishovite was observed at 16–23 GPa and 1200–1400 °C. The dehydration of this assemblage produces β-phase + stishovite at 16–21 GPa, and ilmenite or perovskite at higher pressures. Dense hydrous magnesium phases and nominally anhydrous minerals are the most likely agents for transporting water to the transition zone. The presence of water in the Earth's interior has been invoked to explain a number of phenomena such as the origin of hydrosphere, the occurrence of deep earthquakes by martensitic phase transformations accompanying dehydration of subducted material, and the occurrence of a strong low shear wave velocity anomaly in the upper region of the transition zone.

Core-Mantle Boundary

The core-mantle boundary is a fundamental compositional discontinuity in the Earth. The D″ region is a layer of low seismic gradient at the base of the lower mantle. Records of P and S waves diffracted at the core-mantle boundary indicate on anisotropy in the lowermost mantle with a thickness of 200–300 km. Anisotropy in D″ could be caused by lattice-preferred orientation, which may arise within the convective boundary layer. An experimental study revealed that molten iron reacts with $(Mg,Fe)SiO_3$ perovskite to form stishovite and an iron alloy containing Si and O at about 3000 K and 25–70 GPa. Reactions between molten iron and silicate melt show that a certain amount of Si and O dissolves in molten iron. No evidence for dissolution of Mg was obtained. Heat flow from the core to the mantle creates a thermal boundary layer, at the same time, chemical reactions may create a layer of different composition and density than the overlying mantle. If such a material is less than 3–6% denser than the overlying mantle, it will be swept away by upwelling plumes.

Core

It is generally accepted that the Earth has a solid inner core of radius 1220–1230 km surrounded by a liquid outer core of radius 3480–3490 km. The density of the outer liquid core is about 10% lower whereas the seismic parameter (Φ_s) is higher than those of iron. This implies that some light elements should enter the outer core. A number of elements – Si, C, S, O, and H – were considered as the light alloying elements. Iron is known to occur in four different crystal structural forms: α, δ, γ, and ε. The phase diagram of iron is greatly complicated at high pressures and temperatures and was recently modified due to new experimental diamond anvil cell data; a new phase was discovered in the megabar pressure range. The ε-iron phase transforms to this lower density phase before melting and this new phase may probably form a part of the Earth's outer core.

The nature of the inner core boundary has remained controversial since its discovery in 1936. A velocity model of the inner core boundary region is developed form broadband waveform modeling. It contains a relatively low velocity gradient above the boundary, a sharp jump at the boundary (~0.78 km/s), and a relatively large gradient underneath. The Earth's inner core is solid iron alloyed with nickel which has a very large Poisson's ratio and probably elastic anisotropy. At the same time, the inner core (as well as the outer core) may contain some amount of light component (for example, several weight present of sulfur or other element). The presently best estimate of temperature at the inner-core boundary is 4000–5000 °C at 330 GPa.

The incorporation of light components into the Earth's core has a direct bearing on the thermal regime of the core and the lower mantle, and on the driving mechanism for the geodynamo. The two most likely power sources for the dynamo are thermal convection resulting form heat loss at the core-mantle boundary and compositional convection arising from fractional crystallization at the outer-inner core boundary.

References

Akaogi M (1993) Mineral physics pertinent to constitution of the mantle and core. J Phys Earth Spec Seismology in Japan

Akimoto S, Akaogi M (1984) Possible hydrous magnesian silicates in the mantle transition zone. In: Sunegawa I (ed) Materials science of the Earth's interior. Terrapub, Tokyo

Anderson DL (1989) theory of the Earth. Blackwell, Boston

Bell DR, Rossman G (19929 Water in the Earth's mantle: the role of nominally anhydrous minerals. Science 255:1391–1397

Bercovici D, Schubert G, Tackley PJ (1995) On the penetration of the 660-km phase change by mantle downflows. Geophys Res Lett 20:2599–2602

Bock G (1994) Synthetic seismogram images of upper mantle structure: no evidence for a 520-km discontinuity. J Geophys Res 99B:15843–15851

Boehler R (1993) Temperatures in the Earth's core form melting point measurements of iron at high static pressures. Nature 363:534–536

Dziewonski A, Anderson DL (1981) Preliminary reference Earth model. Phys Earth Planet Inter 25:297–356

Gasparik T (1993) The role of volatiles in the transition zone. J Geophys Res 98B:4287–4299

Hemley RJ, Cohen R (1992) Silicate perovskite. Annu Rev Earth Planet Sci 20:553–600

High-pressure research in mineral physics (1987) Manghnani MH, Syono Y (eds). Terrapub, Tokyo/AGU, Washington

Ito E, Morooka K, Ujike O, Katsura T (1995) Reactions between molten iron and silicate melts at high pressure: implications for the chemical evolution of Earth's core. J Geophys Res 100B:5901–5910

Jeanloz R, Thompson AB (1983) Phase transitions and mantle discontinuities. Rev Geophys Space Phys 21:51–74

Jephcoat A, Olson P (1987) Is the inner core of the Earth pure iron? Nature 325:332–335

Kellogg LH, King SD (1993) Effect of mantle plumes on the growth of D″ by reaction between the core and mantle. Geophys Res Lett 20:3179–382

Kesson SE, Fitz Gerald JD, Shelley JMG, Withers RL (1995) Phase relations, structure and crystal chemistry of some aluminous silicate perovskites. Earth Planet Sci Lett 134: 187–201

Knittle E, Jeanloz R (1989) Simulating the core-mantle boundary: an experimental study of high pressure reactions between silicate and liquid iron. Geophys Res Lett 6:609–612

Kuskov OL, Panferov AB (1991) Phase diagrams of the FeO-MgO-SiO_2 system and the structure of the mantle discontinuities. Phys Chem Min 17:642–653

Kuskov OL, Sidorov YuI, Shapkin AI (1995) Condensation model of the Earth: solar chondrite. The main trends in geochemistry. To the 100th Anniversary of Academician AP Vinogradov. Nauka Moscow, pp 58–69

Meade C, Jeanloz R (1991) Deep-focus earthquakes and recycling of water into the Earth's mantle. Science 252:68–72

Nolet G, Zielhuis A (1994) Low S velocities under the Tornquist-Teisseyre zone: evidence for water injection into the transition zone by subduction. J Geophys Res 99B:15813–15820

Paulssen H (1988) Evidence for a sharp 670-km discontinuity as inferred from P-to-S converted waves. J Geophys Res 93:10489–10500

Petersen N, Vinnik L, Kosarev G, Kind R, Oreshin S, Stammler K (1993) Sharpness of the mantle discontinuities. Geophys Res Lett 20:859–862

Poirier J-P (1994) Light elements in the Earth's outer core: a critical review. Phys Earth Planet Inter 85:319–337

Ringwood AE (1979) Origin of the Earth and Moon. Springer, Berlin Heidelberg New York

Saxena SK, Shen G, Lazor P (1993) Experimental evidence for a new iron phase and implications for Earth's core. Science 260:1312–1313

Song X, Helmberger DV (1992) Velocity structure near the inner core boundary from waveform modeling. J Geophys Res 97:6573–6586

Thompson AB (1992) Water in the Earth's upper mantle. Nature 358:295–302

Woodhouse JH, Giardini D, Li X-D (1986) Evidence for inner core anisotropy from free oscillations. Geophys Res Lett 13:1549–1552

2.2 The Upper Mantle

2.2.1 General Characteristics of the Upper Mantle

I. D. RYABCHIKOV

Upper mantle composition is better characterized than that of other inner shells, partly because upper mantle material is brought to the surface by tectonic and magmatic processes and is therefore available for direct investigation. The following sources of information are used for this purpose:

1. Xenoliths and xenocrysts of upper mantle rocks brought to the surface by alkaline basaltoids, kimberlites, and lamproites.
2. Tectonic slices of mantle material: abyssal peridotites, mantle patrs of ophiolitic complexes, massifs of orogenic lherzolites.
3. Mantle-derived magmatic rocks.
4. Restrictions in geophysical information in comparison with properties of phases stable at high pressures.
5. Comparison with cosmic abundances.

The reliability of these sources of information may differ. Some authors maintain that xenoliths of mantle rocks reflect in many cases the action of magmatic and metasomatic processes, and in this respect are inferior to tectonically transported material. On the other hand, kimberlites bring material from the deepest parts of mantle undisturbed by retrograde alterations. In the majority of cases, xenoliths and xenocrists in kimberlites are brought from the continental lithosphere including the parts close to the lithosphere's lower boundary, but in exceptional cases minerals and mineral assemblages are found transported from the transition zone (majoritic garnets with the substantial proportion of pyroxenelike components in solid solution – Moore et al. 1991; Sautter et al. 1991) and possibly even from the lower mantle (ferropericlase and orthopyroxene found as mineral inclusions in some diamonds, which are thought to represent silicate perovskite + ferropericlase paragenesis typical for lower mantle with silicate perovskite converted to pyroxene – Kesson and Gerald 1991).

Prevailing rocks are peridotites ranging from fertile lherzolites to harzburgites depleted in magmaphile elements with regular covariation of major elements. Eclogites (subducted oceanic crust), pyroxenites (in many cases magmas solidified in the mantle), subducted continental crustal rocks (coesite-bearing rocks, diamonds in "gneisses"), modally metasomatized rocks with amphibole, phlogopite, richterite, wehrlites, often with apatite, etc.

Given the predominance of peridotitic material, we may conclude that the prevailing mineral in the upper mantle is olivine: the boundary with the transition zone corresponds to the transition of olivine into spinel-like orthosilicate structures, while the crust-mantle boundary corresponds to transition form olivine-rich peridotites to basic and intermediate rocks of the lower crust.

Petrographical investigation of mantle rocks and experimental studies of phase equilibria reveal changes in phase assemblages in the mantle with increasing pressure, which are reflected in the main in the nature of alumina-rich minerals: at the lowest pressures it is represented by plagioclase, which, with increasing pressure and temperature, disappears principally due to the formation of tschermakitic and jadeitic components dissolved in pyroxenes with the simultaneous formation of $MgAl_2O_4$-rich spinel in fertile lherzolites. This may be shown in an example in the following schematic reactions:

$$CaAl_2Si_2O_8 + 2\ Mg_2SiO_4 = MgAl_2O_4 + CaMgSi_2O_6 + 2\ MgSiO_3$$
$$CaAl_2Si_2O_8 + Mg_2SiO_4 = CaAl_2SiO_6 + 2\ MgSiO_3$$
$$NaAlSi_3O_8 + Mg_2SiO_4 = NaAlSi_2O_6 + 2\ MgSiO_3.$$

The transition form plagioclase lherzolites to spinel lherzolites takes place at approximately 10 kbar (at the depth of ca. 30 km), which implies that the uppermost part of suboceanic mantle contains plagioclase-bearing lherzolites, whereas under the continents the crust-mantlc boundary marks the appearance of the rocks belonging to spinel lherzolite facies.

With further rise in pressure, highly aluminous pyroxenes become unstable, and excess alumina enters pyrope-rich garnets, which are present in the upper mantle form the depths of 50–60 km (depending upon the temperature gradient) down to its lower boundary. The income of magnesian garnets is accompanied by the disappearance of aluminous spinel (chrome-rich spinels are stable down to much greater depths). These changes may be illustrated by the following simplified reactions:

$$MgAl_2O_4 + 4\ MgSiO_3 = Mg_3Al_2Si_3O_{12} + Mg_2SiO_4$$
$$MgAl_2SiO_6 + 2\ MgSiO_3 = Mg_3A_{12}Si_3O_{12}.$$

Due to reactions of this type the amount of garnet rises with increasing pressure and decreasing temperature. A further significant increase in the proportion of garnet at the expense of pyroxenes takes place at pressures above ca. 100 kbar due to the entry of the excess of silicon into garnet, which replaces part of the aluminum in the octahedral position (major component in garnet):

$$2\ Mg_2Si_2O_6(Opx) = Mg_3[MgSi]Si_3O_{12}(Garn).$$

Phase assemblage olivine + orthopyroxene + clinopyroxene + garnet remains stable down to the lower boundary of the upper mantle. Complete dissolution of pyroxenes in garnet is attained within the transition zone.

An important role in the geochemistry of the mantle is played by the concept of the primitive mantle, which is understood as a silicate shell formed on the Earth after accretion and formation of metallic core prior to magmatic and other types of geological differentiation. This composition corresponds also to bulk silicate Earth. The diagnostic features of such material are chondritic ratios among nonvolatile lithophile elements, which, in contrast to siderophile and volatile elements, are not changed during processes of condensation and evaporation in protoplanetary nebula or during the segregation of metallic phases forming the core. Refractory ltihophile elements include Ca, Al, Mg, Ti, and many trace elements such as REE, U, Th, Ba, Sr, Zr, Hf, Ta, Nb, etc. Because both parent and daughter elements in Sm-Nd and Lu-Hf isotopic systems belong to the group of refractory lithophile components, Nd and Hf isotope compositions should be the same as in chondritic meteorites. The materials which meet the requirements of chondritic ratios among refractory lithophile elements are found among mantle xenoliths from various locations (Jagoutz et al. 1979). These rocks are represented by fertile lherzolites with a maximum content of clinopyroxene, high concentrations of CaO, Al_2O_3 and TiO_2 and flat chondrite-normalized REE patterns.

The overall compositions of two major units of the upper mantle (rigid lithospheric plates and plastic asthenosphere or sublithospheric convecting upper mantle) differ from the primitive mantle. Judging form the geochemical characteristics of mid-ocean ridge basalts (MORB), whose source lies in the asthenosphere, the sublithospheric upper mantle is characterized by the contents of magmaphile major elements (Ca, Al, Ti, Na) and moderately incompatible trace elements (including HREE) similar to the primitive mantle, whereas strongly incompatible trace elements (LREE etc) are substantially depleted. The isotope composition of MORBs (more radiogenic Nd and less radiogenic Sr and Pb as compared to the bulk silicate Earth) demonstrates that the relative depletion with respect to the less compatible elements (Nd, Rb, U, and Th as compared to Sm, Sr and Pb) is a long-living feature of the sublithospheric upper mantle.

The composition of the lithosphere is markedly heterogeneous because its rigid plates include frozen blocks with diverse petrochemical and geochemical characteristics. The average composition of peridotitic nodules form alkali basaltic rocks and kimberlites, which probably reflects correctly the overall chemistry of the subcontinental lithosphere, is characterized by reduced concentrations of moderately incompatible elements (e.g., HREE) and magmaphile major components (like Ca and Al) as compared to the primitive mantle approximately by a factor of 2, and at the same time it exhibits enrichment in strongly incompatible elements (LREE) relative to the primitive mantle. This probably reflects the complicated history of the lithospheric rocks. Many of them are represented by the residua from the generation of basaltic and more magnesian mantle (harzburgites and low-Ca lherzolites which contain smaller amounts of dense garnet and therefore are more buoyant than more fertile asthenospheric rocks) which were enriched in strongly incompatible elements by low fraction melts and fluids penetrating into the lithosphere form the underlying asthenosphere.

Major sites of magma-generating processes are situated in the upper mantle. The most voluminous magmas, like MORBs are the result of decompression melting of asthenospheric material, but the lithosphere may also suffer partial melting due to its heating by underplating asthenospheric plumes or diapirs. It is also very likely that various types of lithospheric material are subducted below the upper mantle and are subsequently involved into plumes of deep origin producing melts on adiabatic decompression.

A number of components enrich crystalline restite during the magma-generating processes. They include Cr, which is retrained by clinopyroxene and/or chromite, Ni preferentially entering olivine, PGE, Au, and Cu, which are extracted by the heavy sulfide liquid remaining in restite. Large-scale mantle melts may be relatively enriched in these elements, and after intrusion into the crust, they may produce economically important concentrations of these ore components.

References

Jagoutz E, Palme H, Baddenhausen H, Blum K, Cendales M, Dreibus G, Spettel B, Lorenz V, Waenke H (1979) The abundance of major, minor and trace elements in the Earth's mantle as derived from primitive ultramafic nodules. Proc Lunar Planet Sci Conf 10th:2031–2050

Kesson SE, Gerald JDF (1991) Partitioning of MgO, FeO, NiO, MnO and Cr_2O_3 between magnesian silicate perovskite and magnesiowustite: implications for the origin of inclusions in diamond and the composition of the lower mantle. Earth Planet Sci Lett 111:229–240

McDonough WF (1990) Constraints on the composition of the continental lithospheric mantle. Earth Planet Sci Lett 101:1–18

Moore RO, Gurney JJ, Griffin WL, Shimizu N (1991) Ultra-high pressure garnet inclusions in Monastery diamonds: trace element abundance patterns and conditions of origin. Eur J Mineral 3:213–230

Ringwood AE (1991) Phase transformations and their bearing on the constitution and dynamics of the mantle. Geochim Cosmochim Acta 55:2083–2110

Sautter V, Haggerty SE, Field S (1991) Ultra-deep (>300 km), ultramafic xenoliths: new petrologic evidence form the transition zone. Science, 252:827–830

2.2.2 Direct Samples of the Upper Mantle

E. V. Sharkov, G. A. Snyder, and L. A. Taylor

Most of our information about upper mantle rocks and their mineralogy is obtained from three groups of rocks: (1) xenoliths (nodules), xenocrysts, and megacrysts in basalts, kimberlites, and lamproites; (2) ultramafic complexes of the ophiolite associations and oceanic ultramafic rocks from fractures zones in the ocean floor; and (3) ultramafic rocks from tectonic blocks in folded belts. Our interpretation of these rocks leads to the conclusion that the upper mantle is composed mainly of spinel and garnet peridotites and pyroxenites and their metasomatized equivalents. A subordinate, albeit economically important, role is played by eclogites and other high-pressure rocks, in that they are often diamondiferous.

Mantle Xenoliths in Basalts, Ophiolite Complexes, and Ocean-Floor Basalts

Our present understanding of xenoliths and xenocrysts in alkali basalts allows for their interpretation mainly as the material from asthenospheric diapirs. The mantle material hosted by oceanic basalts was most likely derived from both continental and oceanic environments, whereas xenoliths in kimberlites and lamproites represent material derived from two distinct mantle sources – ancient continental lithosphere and deep underlying asthenosphere (Boyd 1987; Pearson et al. 1995a,b).

Mantle Xenoliths (Nodules) in Alkaline Basalts

Mantle nodules are widespread in alkaline basalts (mainly Cenozoic in age) and related alkaline rocks – nephelinites, melilitites, phonolites, ugandites, mafurites, etc. Commonly, the xenoliths are found in vesicular basalts, scorias, and pyroclactics, but rarely in lavas. These nodules have been subdivided into a predominant "green" series and a subordinate "black" one (Wilshire and Shervais 1975).

"Green", or Cr-diopside series xenoliths are represented by spinel- or garnet-peridotites and pyroxenites. They are characterized by the presence of bright green, Cr-bearing, clinopyroxene and very uniform, high-magnesian olivines, and pyroxenes (Mg 88–92). For this group, typical rock types are (in order of decreasing abundance): lherzolites, harzburgites, dunites, websterites, clinopyroxenites, and orthopyroxenites.

Average mineral proportions in the spinel peridotites from different occurrences yield a rock type similar to lherzolite, with olivine contents ranging from 62 to 75 modal%, orthopyroxene 16 to 25%, clinopyroxene 7 to 12% and high-Al spinel 1 to 2%. Nodules of deep-seated garnet-bearing peridotites are rare in alkali basalts. In some nodules, calcic plagioclase coexists with spinel, representing more shallow, transitional environments. However, plagioclase in these rocks is not a primary mineral and is present only in fine-grained reaction rims on other minerals.

The structure of the mantle xenoliths is commonly massive, although, rarely, they may be weakly banded due to thin layers or lenses of pyroxenes or spinels. The textures of the xenoliths are rather similar to those reported in ophiolite complexes (see below) and vary form protogranular to porphyroclastic and granoblastic (Nicolas 1989).

Hydrous minerals are also common and include: pargasitic amphibole, kaersutite, and phlogopite. Usually these minerals form disseminated, individual grains in mantle peridotites. The disseminated hydrous minerals either locally replace spinel, garnet, and pyroxene (in this case they are commonly described as interstitial or poikilitic), or they appear to be in textural equilibrium with other minerals in the peridotite (Dawson and Smith 1982). They are believed to be produced by a fluid percolation and metasomatic reactions in the mantle within originally dry peridotites.

In contrast to those in the "green" series, hydrous minerals play an essential role in the "black" or Al-Ti-augite series. Clinopyroxenes are represented here by black, high-Al, and high-Ti augites. Minerals of this series are more iron-rich (Mg < 85), and their compositions vary over a wide range. This group includes (in order of decreasing abundance) clinopyroxenites, hornblende clinopyroxenites, hornblendites, wehrlites, dunites, websterites, and glimmerites.

Unique nodule types include the complex xenoliths (usually lherzolites of the Cr-diopside series) which contain veins or thin layers of "green" or "black" pyroxenites or hornblendites or glimmerites. Currently accepted interpretations of the "green" pyroxenites consider them a result of partial crystallization of picritic basalt melts. In contrast, amphibole- and phlogopite-bearing veins are thought to form by fractional crystallization of incompatible-element-enriched fluids generated from melts in upper mantle conduits. "Black" series rocks, in this case, are treated as fragments of relatively thick veins.

As a rule, xenoliths in basalts are accompanied by megacrysts – fragments of rather large crystals of clinopyroxene, hornblende, phlogopite, sanidine, and more rarely garnet, ilmenite, olivine, and orthopyroxene, which exist in various combinations and proportions. Sometimes intergrowths of two or more of these minerals occur (e.g., Snyder et al. 1993a). The chemical compositions of these minerals are similar to those of the "black" series. Their Mg's vary from 60 to 80, Cr content is not more than 0.01%, and they commonly have high Ti contents (often 9–12% in phlogopites). The distinctive features of megacrysts are their high homogenity, and lack of zoning and exsolution features. This suggests that they were formed as a result of disintegration of coarse-grained vein rocks of the "black" series (Sharkov et al. 1990), or are near-liquidus phases precipitated from host basalts at elevated pressure (Irving and Frey 1984).

Ultramafic Complexes in Ophiolites and Other Oceanic Occurrences

Another type of mantle rock occurrence is represented by ultramafic complexes of ophiolites and ocean floor basalts. It is now universally accepted that ophiolite associations in Phanerozoic fold belts represent fragments of ancient oceanic lithosphere which were obducted onto continental margins after ocean closure, and that their lower, ultramafic parts are direct remnants of former oceanic mantle (Coleman 1977; Laz'ko and Sharkov 1988; Nicolas 1989, etc.).

Ophiolites occur mainly in belts or elongated bodies, with the most significant part (for our purposes) consisting off mantle-derived ultramafic rocks (so-called Alpine-type peridotites). Individual bodies vary in size over a wide range; from small lenses tens of meters across to large massifs thousands of square kilometers in surface expression. In all cases, they have the form of thrust nappes, with thicknesses up to 5–6 km.

The massifs of Alpine-type peridotites are heterogeneous due to mineralogic layering of the rocks and their internal structures. The most common rocks in these massifs arc harzburgites with 15–20 modal% orthopyroxene; dunites and pyroxenites are less common. Rock-forming silicates are characterized by uniformly high-magnesium compositions worldwide. Lherzolites are uncommon and the proportion of clinopyroxene in them is typically small. The only exceptions are the massifs of the western Mediterranean, where lherzolites predominate above harzburgites. Among peridotites, only Cr-spinel (or chromite) predominates as the aluminous phase, although plagioclase does occur less commonly. The diverse combinations of these rocks in cross-sections and the variable distribution of the major minerals (olivine, ortho- and clinopyxenes, Cr-spinel, and plagioclase) define implicated inner structures in the massifs. One of the main features of these internal structures is layering or banding, caused by the regular alternation of mineral proportions, from olivine pyroxenites to enstatite dunites; however, orthopyroxene nowhere exceeds 15–20 modal%. Large-scale stratification is absent in most ultramafic complexes, and the composition of the minerals is uniform over several kilometers of stratigraphy in many harzburgite massifs (Laz'ko and Sharkov 1988).

Veins of dunites, websterites, ortho- and clinopyroxenites, and hornblendites also cross-cut the layering. The thickness of these veins vary from centimeters to tens and even hundreds of meters.

Most of the rocks in ultramafic complexes have undergone solid-state, plastic deformation under mantle conditions and exhibit varying degrees of deformation and recrystallization. The textures of these rocks are commonly protogranular, porphyroclastic, and granoblastic (Nicolas 1989). The least-deformed rocks have protogranular to granular textures. They are relatively coarse-grained (minerals often 5–10 mm in size), with isometric grains of olivine and orthopyroxene often forming triple junctions. Visible evidence of deformations is rare, but is manifested in "cloudy" extinction of grains, deformation of cleavage surfaces and exsolution lamellae in pyroxenes, appearance of kink-bands in olivines and, more rarely, in orthopyroxenes. Orthopyroxene is often associated with clinopyroxene and Cr-spinel in systems of linear aggregates. Chrome spinels typically exhibit complicated wormlike forms in the interstices of larger silicate grains.

Porphyroclastic ultramafic rocks are characterized by greater deformation, forming two distinct types of grains: porphyroclasts, and small (<1 mm), polygonal grains without evidence of deformation evidence (neoblasts) and meeting at 120° triple junctions. Porphyroclasts are commonly elongated and neoblasts are grouped along their grain boundaries. Due to the differing competencies of the minerals, olivine most commonly occurs as neoblasts and orthopyroxene as porphyroclasts.

Increasing deformation at relatively low temperature (<800–900 °C) leads to the formation of cataclasites and blastomylonites with rare porphyroclasts of orthopyroxene and fine-grained olivine. At higher temperatures, extensive recrystallization of the rocks leads to the formation of granoblastic textures.

Current thought is that ultramafic rocks in ophiolitic associations represent the residual mantle after remelting to form basaltic magmas (e.g., Coleman 1977; Nicolas 1989; Sharma et al 1995).

Oceanic ultramafic rocks also occur along mid-oceanic ridges (MOR), oceanic plates, and active margins of oceans and continents. MOR and oceanic-plate peridotites are mainly found in the steep slopes of rift valleys and transform faults, in boreholes of deep-sea wells, and occasionally in outcrops on the surface (e.g., St. Paul's Rocks in the Atlantic). They exhibit a mineralogy similar to ophiolites; however, the proportion of pyroxene is higher in them on average – 30% orthopyroxene and 10% clinopyroxene (Laz'ko and Sharkov 1988). As a rule, lherzolites predominate over harzburgites and textures are protogranular to pophyroclastic. At St. Paul's Rocks, hornblende peridotites and hornblendites occur above the ultramafic rocks and are linked to an uplifted block of metasomatized, oceanic mantle (Roden et al. 1984). Another typical feature of this occurrence is the presence of plagioclase-bearing peridotites, mainly lherzolites. As a whole, oceanic peridotites lie close to xenoliths in alkaline basalts.

In contrast, peridotites which occur in slopes of deep-sea trenches of island arcs and back-arc basins are generally represented by depleted harzburgites and are similar to ophiolitic peridotites in both composition and texture (Laz'ko and Sharkov 1988). Garnet lherzolites are known in only one occurrence, an alnoite dyke from Malaita Island, Solomon Arc, Pacific (Neal and Nixon 1986; Neal 1988). These garnet lherzolites were-most likely derived from the mantle beneath the underwater Ontong-Java plateau (Bielski-Zyskind et al. 1984).

Mantle Xenoliths and Xenocrysts in Kimberlites and Lamproites

Our understanding of the deep continental upper mantle hinges on studies of xenoliths in kimberlites and lamproites. Mantle samples in kimberlites are less altered and have been more extensively studied than those in lamproites, where often only the diamond is retained. Xenoliths in kimberlites are of a great variety, including peridotites and pyroxenites (both garnet and spinel, and transitional spinel-garnet varieties), dunites, eclogites ("true" eclogites along with those containing corundum, kyanite, coesite, and sanidine), and specific titaniferous rocks: ultramafic rocks containing picroilmenite, rutile, phologopite, and hornblende (Cr-bearing pargasite or K-richterite), MARIDs (those containing the specific metasomatic assemblage mica, amphibole, rutile, picroilmenite and diopside; Dawson and Smith 1977), alkremites (composed of coarse-grained, pyropic garnet and dark green, hercynitic spinel; Ponomarenko 1975), and ilmenite-pyroxene intergrowths. Garnet peridotites predominate among xenoliths in kimberlites; however, there are pipes where eclogites predominate (Zagadochnaya in Yakutia, Sobolev 1977; Roberts-Victor and Orapa in South Africa; MacGregor and Carter 1970).

The origins of the titaniferous rocks are often best explained by processes involving mantle metasomatism. The designation "green" and "black" series is not accepted for xenoliths in kimberlites; however, many titaniferous rocks could be related to the "black" series of xenoliths in basalts. As with the former mantle-derived rocks, there is a spectrum of transitional, metasomatized varieties of peridotites with disseminated amphibole and phlogopite.

Xenocrysts in kimberlites are subdivided into two series: chromian and titaniferous. Minerals of the chromian association [pyrope garnet, chrome-spinel, chrome-diopside, and orthopyroxene (En_{90-94})] are identical in composition to the same phases in xenoliths of garnet- and spinel-peridotites, and their origin most likely was linked to disintegration of such subcrustal rocks. The same is true for megacrysts of garnet and clinopyroxene, which are similar in composition to these minerals in eclogites. The compositions of titaniferous megacrysts (orange, high-Fe- and -Ti, and low-Cr garnet, picroilmenite, subcalcic clinopyroxene, orthopyroxene [En_{85-90}], and phlogopite) are also similar in composition to minerals in the titaniferous rocks. The origin of the vast majority of olivine megacrysts is not yet clear; they vary widely in composition (Fo_{73-93}) and could be either phenocrysts of kimberlitic melts or fragments of differentiated subcrustal peridotites.

Diamond is an important xenocryst in kimberlites. Although diamonds rarely exceed 0.1 to 0.2 g, occasionally crystals can weigh 600 g or more. Sometimes, diamonds contain small inclusions of other minerals: garnet, clinopyroxene, olivine, orthopyroxene, chrome-spinel, sulfides, and even metal (Sobolev 1977). In some cases, diamonds are found in garnet peridotite and eclogite xenoliths. Inclusions in diamonds are divided into two distinct groups ("peridotitic" and "eclogitic") which correspond in composition to minerals from these two types of xenoliths (Sobolev 1977).

Garnet Peridotites

This is the most numerous and varied group of mantle nodules. Most are composed of the standard four mineral association of magnesian garnet (i. e., pyrope), orthopyroxene, clinopyroxene, and olivine, although the content of clinopyroxene is typically small (<5 modal%). True harzburgites are rare and form a continuum with dunites. Varieties with only one pyroxene (ortho- or clinopyroxene) are also rare. Magnesian spinel plays a subordinate role with a Cr/Al ratio higher in lherzolites than in harzburgites. Rare accessory minerals are represented by diamond, graphite, moissonite, sulfides (pentlandite, makinavite, chalcopyrite, pyrrhotite), rutile, ilmenite, and zircon.

Garnet peridotites are subdivided into two groups based on texture, mineral chemistry, and equilibration temperatures: 1) granular (undeformed and coarse-grained) peridotites containing xenomorphic garnet, high-Mg olivine ($>Fo_{91.5}$), low TiO_2 in garnets and diopsides (<0.2 wt.%), and that exhibit relatively low equilibration temperatures (<1100 °C), and 2) sheared (cataclasized or deformed) peridotites containing relatively low-Mg olivine ($<Fo_{91.5}$) high TiO_2 in garnet (up to 2 wt.%) and equilibration temperatures >1100 °C (Boyd 1987). In general, the undeformed garnet peridotites correlate with the protogranular and porphyroclastic textures in xenoliths from alkali basalts. The sheared variety often exhibits a combination of granoblastic and porphyroclastic textures.

High-temperature, sheared garnet peridotites are thought to come from relatively great depths (>150 km), and are consistent with an origin in the asthenosphere, often beneath the subcontinental mantle. Low-temperature, granular garnet peridotites are consistent with derivation from the subcontinental mantle which has been previously depleted by removal of komatiitic liquids (Boyd 1987).

Spinel and Garnet-Spinel Peridotites

There are four varieties of spinel and garnet-spinel peridotites: (1) garnet-spinel peridotites with discrete grains of spinel in equilibrium with the other minerals; (2) similar to (1), but without garnet; (3) peridotites without garnet which contain individual grains of Cr-spinel in symplectitic intergrowths with pyroxenes, amphiboles, and phlogopites; and (4) peridotites with a reaction texture of garnet replacing spinel. All of these peridotites have protogranular textures. Compositionally, harzburgites predominate in the first and third groups, harzburgites and lherzolites in the second, and in lherzolites in the fourth group.

Magnesiochromites are typical in rocks of the first group where they are associated with Cr-rich garnet, olivine of Fo_{92-93} composition, Cr-diopside, and low-Al (high-pressure) orthopyroxene. Spinels of this composition are also common in many rocks of the second and third groups. In the fourth group, Mg-rich garnet replaces Cr-spinel or Al-Cr-spinel, suggesting a transition form the spinel to the garnet facies.

Pyroxenites

Pyroxenites are subordinate to other mantle xenoliths in abundance in kimberlites. Garnet websterites predominate, although garnet-orthopyroxenites also exist. Spinel and garnet-spinel websterites and orthopyroxenites also occur, and may contain variable amounts of olivine, grading into lherzolites and harzburgites. In contrast to peridotites, pyroxenites contain a plethora of accessory phases: picroilmenite, amphibole, phlogopite, rutile, and apatite. Spinel- and garnet-bearing clinopyroxenites are rare.

The chemical compositions of minerals in many of the pyroxenites are similar to those in the peridotites. However, they differ from the peridotites by having higher Fe and lower Cr contents. Many of the Fe-rich garnet-websterites could be related to the "black" series xenoliths in basalts.

Dunites

Dunites are rare among mantle xenoliths in kimberlites. They are Al-poor and consist mainly of magnesian olivine (Fo_{92-94}) with small amounts of high-Cr, subcalcic garnet, magnesiochromite, and enstatite (En_{92-94}), allowing Sobolev (1977) to assign them to the dunite-harzburgite paragenesis. These rocks are very important in gaining and understanding of the occurrence of diamonds in kimberlites as their mineral compositions are identical to the olivines, chromites, garnets, and enstatites which have been found as inclusions in diamonds. Furthermore, the dunites may also host diamonds (Sobolev 1977; Dawson 1980).

Glimmerites and Rocks of the MARID Series

The essential identifying phase in these rocks is phlogopite; however, the rocks are extremely variable ranging form pyroxene and amphibole-bearing phlogopitite-rich rocks, ilmenite, rutile-ilmenite, and apatite-bearing rocks, and rocks of the olivine-phlogopite paragenesis. The term MARID is formed from the names of the main rock-forming minerals: mica, amphibole, rutile, picroilmenite and diopside (Dawson 1980). The proportions of minerals in MARIDs vary widely: phlogopite (15–95 modal%), amphibole (5–80 modal%), picroilmenite (1–10 modal%), diopside (5–40 modal%), and Cr- and Fe-enriched rutile (up to 3 modal%). Minor phases include titanite, zircon, apatite, and garnet. The textures of these rocks sometimes resemble cumulates, although evidence of deformation is prevalent in all minerals.

All silicates in MARIDs have high Fe and Ti contents and resemble phlogopite and amphibole-phlogopite veins in basaltic xenoliths from high-pressure complexes (see below). Oxides (including picroilmenite) are enriched in incompatible elements (Zr, Nb, Ta) (Dawson 1980).

Metasomatized Ultramafic Rocks

Some garnet-peridotite xenoliths contain an atypical assemblage of minerals such as ilmenite, phlogopite, K-richterite, amphibole, and accessory rutile, zircon, apatite, Ba- and Zn-titanites, and other minerals with Ti, K, P, Pb, Sr, Ba, Zr, Nb, and LREE (Sobolev 1974; Kornprobst et al. 1981; Ryabchikov 1988).

It has been suggested that such minerals were formed as a result of mantle metasomatism. Textures of such ultramafic rocks are protogranular or granoblastic. Sometimes, these secondary minerals occur as spots or veins, but in many cases they form single grains.

Other evidence of mantle metasomatism is manifested in the presence of variations in the composition of minerals or within individual grains (zoning), and unusual garnet-clinopyroxene intergrowths (finger, and/or selvedges of clinopyroxene and garnet) Dawson 1980; Laz'ko and Sharkov 1988; Sobolev et al. 1995).

Eclogites

Eclogites are essentially bimineralic rocks consisting of garnet and omphacitic clinopyroxene. Accessory minerals include kyanite, corundum, coesite, sanidine, orthopyroxene, amphibole, corundum, phlogopite, rutile, ilmenite, Fe-Ni-Cu sulfides, quartz, graphite, and diamond. Amphiboles and other hydrous phases are thought to be secondary and due to infiltration of the kimberlite host (Snyder et al. 1996). Due to the presence of diamonds in many of the xenoliths, their economic and petrologic importance far exceeds their abundance in most kimberlite pipes. Xenoliths of eclogites are typically classified by two different schemes. The first uses the designation group I and group II and is based on the Na_2O content of garnets and the K_2O contents of clinopyroxenes (Sobolev 1977; McCandless and Gurney 1989). Group I eclogite xenoliths have garnets with >0.09 wt.% Na_2O and clinopyroxenes with >0.08 wt.% K_2O. Subhedral to anhedral, rounded garnets set in a finer-grained matrix of altered clinopyroxenes with occasional large, clinopyroxenes poikilitically enclosing garnets is a common feature of group I eclogites (MacGregor and Carter 1970). Group II eclogites have garnets with <0.09 wt.% Na_2O and clinopyroxenes with <0.08 wt.% K_2O. Textural relations within group II eclogites are more diverse, but often include an interlocking network of equigranular clinopyroxene and garnet (MacGregor and Carter 1970). McCandless and Gurney (1989) have used this classification as a diamond exploration tool as nearly all diamondiferous eclogites fall within group I.

The second classification scheme uses the designation of groups A, B, and C (Shervais et al. 1988). This scheme was developed in an attempt to reconcile the classification of massif eclogites (Coleman et al. 1965) with that of eclogite xenoliths. Group A eclogite xenoliths are characterized by high whole-rock Mg's, occasional primary orthopyroxene or olivine as an accessory phase, low-jadeite, and Cr-rich clinopyroxenes, and Mg- and Cr-rich garnets. Group B eclogites have moderate jadeite moles in clinopyroxene, and Fe-rich garnets. Group C eclogites are characterized by high-jadeite content clinopyroxenes, and CaO-rich garnets. Each group also exhibits certain well-defined chemical and isotopic characteristics (e.g., Neal et al. 1990) that are outlined in Snyder et al. (1996). Group A eclogites are thought to be true mantle cumulates, whereas Groups B and C are considered fragments of subducted oceanic crust

(Shervais et al. 1988; Taylor and Neal 1989; Neal et al. 1990; Jerde et al. 1993; Jacob et al. 1994).

Helmstaedt and Doig (1975) were the first to present convincing mineralogic and petrographic evidence for a subducted oceanic crustal origin for eclogite xenoliths. Jagoutz et al. (1984), MacGregor and Manton (1986), and Shervais et al. (1988) noted that the chemical compositions of some eclogite xenoliths from southern Africa (type B and C of Taylor and Neal, 1989) favor an oceanic crustal protolith. A vast array of geochemical evidence has been promulgated favoring an origin form oceanic crust, but will not be detailed here (cf. Neal et al. 1990; Snyder et al. 1996). Mineralogic evidence in favor of such an origin for some eclogites includes the occurrence of coesite and sanidine, which are considered to be "evolved" phases in igneous crystallization models, throughout the entire compositional range of eclogites (Schulze and Helmstaedt 1988), and high-Al whole-rock compositions which point to a low-pressure anorthositic precursor (Taylor and Neal 1989). The high Na_2O contents of some eclogitic clinopyroxenes, as well as high-pressure experimental data on the composition of liquidus and subsolidus clinopyroxenes in basaltic systems (Thompson 1974; Rapp 1995), has also been used as evidence against a mantle origin and in favor of a crustal one (Ireland et al. 1994).

Abundant evidence exists that at least some eclogites are of mantle derivation with little or no involvement of ocean-floor basalt. Mineralogic characteristics of group A eclogites which are consistent with a mantle origin include high Mg/Fe, low Na in clinopyroxene, and pyropic garnet (Shervais et al. 1988). In addition, Shervais et al. (1988) and Taylor and Neal (1989) pointed out the higher Cr contents in garnets and clinopyroxenes, versus <0.1 wt.% Cr_2O_3 in crustal-derived groups B and C. Other mineralogic evidence which has been used to suggest that eclogite xenoliths represent cumulates form an evolving basaltic magma at high pressure (Smyth et al. 1989; Caporuscio and Smyth 1990) include (1) the compositional continuity of groups B and C; (2) the observation of banded eclogites that contain two or more chemically distinct eclogites; (3) garnet and kyanite exsolution form clinopyroxene indicative of cooling from near-solidus temperatures above 3 GPa pressure; and (4) reconstructed exsolved clinopyroxene compositions similar to the total kyanite-eclogite host.

High-aluminum eclogites are represented mainly by kyanite-bearing (1 to 25 modal%) varieties, where kyanite occurs either as laths locally included in garnet, as intergrowths with clinopyroxene, or as large (1-3-cm) poikiloblastic crystals including other minerals. Quartz inclusions in omphacite and garnet are often surrounded by radial cracks, suggesting the former existence of coesite. Rarely, eclogites contain large poikiloblastic zoisite with pyroxene and kyanite inclusions. Occasionally, sanidine and apatite are present. Clinopyroxenes in these rocks have high jadeite contents (37 to 48 mol.%). Garnets contain 20–29 mol.% grossular component (Elfadini et al. 1995). A peculiar type of high-Al eclogites are the grospydites (grossular garnet, clinopyroxene, and disthene=kyanite), which also include accessory coesite and sanidine (Bobrievich et al. 1960). The difference between grospydites and more common kyanite eclogites is the high grossular content (>50 mol%) of garnets in grospydites.

Certain rare xenoliths of similar composition to kyanite eclogites are the so-called alkremites (Ponomarenko 1975). The term alkremite is an acronym form the Russian words aluminy (aluminnum) + kremnezem (silicon). These unique high-alumina rocks are dominated by garnet (40–90 modal%) and spinel (10–60 modal%), and occasional kyanite. Garnets in alkremites vary form extremely low CaO varieties to those similar to the typical grossular-pyrope garnets of eclogites and kyanite eclogites.

Ultradeep (>300-km) Xenoliths

Ultradeep xenoliths are defined by the presence of majoritic garnet inclusions in diamond and in ultramafic xenoliths in kimberlites (Haggerty and Sautter 1990). Majoritic inclusions in diamonds have retained pyroxene (nominally $R_4Si_4O_{12}$) in garnet solid-solution, whereas the latter have exsolved pyroxene along {111} planes in garnet. Modal and chemical reconstructions of garnet + pyroxene yield majoritic solid solutions with ^{IV}Si and ^{VI}Si. Based on natural and pure experimental systems, such coordinated four- and six-coordinated Si requires stabilization at $P > 100$ kbar (300 km). The deepest xenoliths identified are suggested to be from the transition zone at ~400 km, although they have obviously been re-equilibrated in shallower holding stations, possibly in the asthenosphere, but certainly in the lithosphere at ~40 kbar and 1200 °C (Sautter et al. 1991).

High-Pressure Mantle Rocks Tectonically Emplaced in the Crust

Ultramafic rocks of high-pressure origin are known to occur not only as xenoliths in basalts and kimberlites, but also as isolated bodies and massifs among highly metamorphosed domains in Phanerozoic fold belts. These bodies and massifs are found in the so-called Gibraltar Arc in the Alpine Belt of southern Europe (the Ronda massif in Spain) and northern Africa (Beni Bousera massif in Morocco), in zones of deep-seated thrusts in the Western and Central Alps (Ivrea zone, Alpe Arami; e.g., Dobrzhinetskaya et al. 1996), and the Bohemian massif (Moldanubian). Less commonly, high-pressure ultramafic massifs are emplaced during opening of oceans; likely examples of such occurrences are Zabargad Island in the Red Sea, and St. Paul's Rocks in the Atlantic. A specific group of high-pressure massifs is represented by tectonic blocks of ultra high-pressure diamond-bearing metamorphic rocks in the Dabie Mountains in central China (Wang et al. 1995), in the Kokchetav massif, northern Kazakhstan (Shatsky et al. 1995), and in the Dora-Maira massive, western Alps (Compagnoni et al. 1995).

As a rule, high-pressure ultramafic rocks form small lenses, boudins, and tectonic blocks (from meters to tens of meters thick).Their inner structures are rather complex. Lherzolites and harzburgites predominate, whereas dunites are much less abundant. Garnet-, spinel-, and plagioclase-peridotites and various transitional varieties occur routinely. Ca-amphibole (tremolite), magnesian hornblende, and/or phlogopite also may be present.

Mineral compositions within these high-pressure rocks are similar to those in mantle xenoliths in basalts and in the ultramafic portions of ophiolite complcxes. Pyroxenes, especially in garnet and garnet-spinel peridotites, are enriched in Al_2O_3 (up to 8 wt.% in clinopyroxene). The Cr/(Cr+Al) ratio in spinels does not exceed 25, and is usually much less. Magnesian garnet (usually more than 70 mol% pyrope) contains 0.5–2.0 wt.% Cr_2O_3.

Various pyroxenites (websterites, ortho- and clinopyroxenites, hornblende pyroxenites), melanocratic gabbros, and pyroxene hornblendites also occur. These pyroxenites typically form veins and dikes from a few centimeters to several meters in width and exhibit sharp contacts with the surrounding rocks.

The Ronda Massif

This is the largest known lherzolite body (~300 km^2) and is approximately 1.5 km thick (Obata 1980). It is surrounded by garnet-sillimanite gneisses. An impressive feature of this massif is that it contains rocks from all three upper mantle facies – plagioclase, spinel and garnet – representing a wide range in depths. From NW to SE, garnet (garnet-spinel-) peridotites gradually give way to spinel and spinel-plagioclase peridotites, and finally to plagioclase-peridotites. Calculations show that the pressures of formation of garnet- and plagioclase-peridotites differ by at least 6 kbar (about 18–20 km) or more, yet the thickness of the spinel-peridotites is not more than 3 km. Therefore, such a large difference in pressure cannot be accounted for by lithostatic pressure alone (Obata 1980). In other words, these rocks could not have been in equilibrium at depth. Obata (1980) suggested a model of "dynamic cooling" whereby this apparent discrepancy is in response to the movement of a uniform block of garnet-spinel peridotites to the surface. In the process, the SW part of the massif cooled faster, preserving the primary paragenesis. The inner parts of the body were heated longer and were recrystallized to form spinel- and plagioclase-peridotites. An alternative explanation for this phenomenon could be linked to deep-seated thrusting in the mantle long before the tectonic emplacement of this block in the Earth's crust.

The Beni Bousera Massif

This occurs on the opposite side of the Mediterranean from the Ronda massif, in northern Morocco. Like the Ronda massif, it is composed mainly of lherzolites, but also contains subordinate pyroxenites (<10%) of varying lithology. The pyroxenites occur as sheetlike bodies or layers, occasionally with apophyses at their margins. As in all mantle occurrences, two groups of pyroxenites have been defined on the basis of their pyroxene chemistry: an Al-augite group, containing pink/gray Al-augite to sodic clinopyroxene with or without garnet, and a Cr-pyroxenite group, containing Cr-diopside (Pearson et al. 1989). Al-augite pyroxenites are the predominant type, representing over 90% of the total population. Two garnet-clinopyroxenite layers belonging to the Al-augite pyroxenite group have been found to contain graphite. This graphite occurs as octahedral multicrystalline aggregates form 0.5 to 7 mm in size which pseudomorph the cubic diamond symmetry, and has been interpreted as graphitized diamond (Slodkevich 1980; Pearson et al. 1989). Several of the graphite octahedra also contain faceted inclusions of garnet and clinopyroxene. Textures of these pyroxenites are similar to typical cumulates. Graphite makes up from 2 to 15 modal% of the pyroxenite, i.e., sometimes becoming an essential rock-forming mineral; the abundance and size of the octahedra decrease upwards in the layers (Slodkevich 1980). Geochemical (HREE enrichment and the presence of both positive and negative Eu anomalies) and oxygen isotope evidence (Pearson et al. 1991) suggests that these pyroxenites were crystallized from melts of subducted, hydrothermally altered oceanic lithosphere in the diamond stability field (>45 kbar and T of ~1100 °C).

Massifs of the Moldanubian Zone

Jewelry containing garnetiferous ultramafic rocks has been known from the Czech Republic since the Middle Ages. These ultramafic bodies occur mainly in the peripheral parts of the Moldanubian Zone – a large tectonic block of strongly metamorphosed Precambrian rocks (granulites, eclogites, marbles, gneisses, etc.). Garnetiferous ultramafic rocks form small (50 × 70 cm) inclusions in metamorphic rocks as well as larger massifs (0.7 × 1.1 km). In some cases, they have layered structures with alternating garnet-bearing dunites, harzburgites, lherzolites, and eclogites (Fiala and Padera 1977; Beard et al. 1992; Medaris et al. 1995).

Mineral proportions in the garnet-peridotites vary over a wide range (in modal%): garnet, 0.3–10% (rarely up to 15%); clinopyroxene, 0.2–9%, orthopyroxene, 0.3–9%; serpentine, 30–100%. Chrome-spinel, picroilmenite, Ni-Cu-Fe sulfides, rutile, apatite, and occasionally phlogopite and moissonite are found as accessory phases. The main feature of the garnets is the variable content of Cr_2O_3 (form 0.4 to 11 wt.%). The concentration of Al_2O_3 in orthopyroxenes is sometimes as low as 0.9–1.5 wt.%, corresponding to pressures of 30–40 kbar.

The Peridotite Massif of Zabargad Island (Northern Red Sea)

This is widely known by the unique occurrence of gem-quality chrysolite olivine, which has been mined since ancient times. This massif is a block of slightly depleted upper mantle material, similar to xenoliths in basalts of the region, which was tectonically emplaced in the Earth's crust in the Cenozoic (Bonatti et al. 1986). There are three groups of rocks here: spinel-, plagioclase-, and hornblende-peridotites and all transitional varieties. Spinel-lherzolites predominate, and plagioclase and hornblende peridotites occur as bodies and layers among them. The main feature of these rocks is their freshness, allowing their use in jewelry.

Ultrahigh-Pressure Diamond-Bearing Metamorphic Rocks

One of the major geologic discoveries in recent years was the occurrence of ultrahigh-pressure minerals such as diamond and coesite in crustal rocks, which yields evidence of burial of continental crust to depths of over 120 km. The best-studied of such occurrences were those in northern Kazakhstan (Kokchetav massif: Rosen et al. 1972; Sobolev and Shatsky 1990; Shatsky et al. 1995), in the western Alps (Dora Maira massif: Chopin 1984; Compagnoni et al. 1995), and in central China (Dabie Shan massif: Xu et al. 1992; Wang et al. 1995).

The Kokchetav massif is a large tectonic block of metamorphic rocks in the Kazakhstan Caledonides. Microdiamond-bearing rocks include biotite gneiss, garnet mica schist, pyroxene marble, and garnet pyroxenite (Sobolev and Shatsky 1990).The diamond-bearing carbonates contain abundant dolomite (>70 modal%), garnet, diopside, and minor phlogopite and magnesite. Inclusions of diamond, dolomite, graphite, biotite, and clinopyroxene occur in garnet. However, no coesite or its pseudomorphs was found in the marbles. Hence, the assemblage diamond + diopside + dolomite + magnesite + garnet formed at pressures >40 kbar and at temperatures >900–1000 °C (Liou et al. 1995). The associated diamondiferous biotite-garnet gneiss, on the other hand, contains coesite pseudomorphs in garnet and coesite in zircon. Diamond has not been found in the associated metamorphosed eclogites, although inclusions of coesite pseudomorphs were identified.

The Dabie Shan massif is another example of such an occurrence (Wang et al. 1995). It constitutes the eastern end of the Qinling orogen, central China, which was produced during Triassic continental collision, and consists of several welded gneiss terranes of variable metamorphic grade. One of these terranes, the eclogite zone, is sandwiched between two amphibolite-facies terranes, and consists of a >25-km-thick sequence of granoblastic gneiss with eclogite, marble, and minor ultramafic lenses. The eclogites in the gneiss have a common mineral assemblage of garnet + omphacite + zoisite + kyanite + phengite ± Ca-amphibole ± quartz + rutile and very rare coesite inclusions in garnet. Eclogites also occur as blocks within marble layers; they are characterized by absence of kyanite, the presence of diopside-rich pyroxene and grossular-rich garnet, with coesite, K-feldspar, and diamond inclusions (Okay 1993). Microdiamonds were also found in garnet pyroxenites and jadeitic pyroxenites (Xu et al. 1992).

The composition of minerals in both types of eclogites are different: eclogites in gneisses contain garnet with a higher pyrope content than eclogites in marbles. Jadeite contents in omphacite overlap, but are higher in clinopyroxenes from the marbles (Okay 1993).

References

Beard BL, Medaris LG, Johnson CM, Brueckner HK, Misar Z (1992) Petrogenesis of Variscan high-temperature Group A eclogites from the Moldanubian Zone of the Bohemian Massif, Czechoslovakia. Contrib Mineral Petrol 111:468–483

Beard BL, Snyder GA, Taylor LA, Fraracci KN, Sobolev NV (1995) Eclogites from the Mir Kimberlite, Russia: evidence of an Archean ophiolithe protolith. 6th Int Kimberlite Conf Novosibirsk. Extended abstracts, pp 41–43

Bielski-Zyskind H, Wasserburg GJ, Nixon PH (1984) Sm-Nd systematics in volcanics and ultramafic xenoliths from Malaita, Solomon Islands, and the nature of the Ontong-Java Plateau. J Geophys Res 89, 4:2415–2424

Bobrievich AP, Smirnov, GI, Sobolev VS (1960) The mineralogy of xenoliths of a grossular-pyroxene-kyanite rock (grospydite) form the Yakutian kimberlites. Geol Geofiz 3:18–24

Bonatti E, Ottonello G, Hamlyn PR (1986) Peridotites from the island Zabargad (St. John), Red Sea: petrology and geochemistry. J Geophys Res 91:N1

Boyd FR (1987) High- and low-temperature garnet peridotite xenoliths and their possible relation to the lithosphere-astenosphere boundary beneath southern Africa. In: Nixon PH (ed) Mantle xenoliths. John Wiley, New York, pp 403–412

Brey GP, Kohler T (1990) Geothermometry in four-phase lherzolites. II. New thermobarometers and practical assessment of exiting thermobarometers. J Petrol 31:1353–1378

Brey GP, Kohler T, Nickel KG (1990) Geothermometry in four-phase lherzolites. 1. Experimental results from 10 to 60 kb. J Petrol 31:1313–1352

Bulanova GP, Milledge HJ (1995) Origin and history of growth of microdiamonds form Yakutian kimberlites. 6th Int Kimberlite Conf Novosibirsk 1995. Extended abstracts, pp 77–79

Caporuscio FA, Smyth JR (1990) Trace element crystal chemistry of mantle eclogites. Contrib Mineral Petrol 105:550–561

Chopin C (1984) Coesite and pure pyrope in high-grade blueschists of the western Alps: a first record and some consequences. Contrib Mineral Petrol 86:107–118

Coleman RG (1977) Ophiolites. Berlin, Springer, Berlin Heidelberg New York, 229 pp

Coleman RG, Lee ED, Beatty LB, Brannock WW (1965) Eclogites and eclogites: their differences and similarities. Geol Soc Am Bull 76:483–508

Compagnoni R, Hirajima T, Chopin C (1995) Ultra-high-pressure metamorphic rocks in the western Alps. In: Coleman RG, Wang X (eds) Ultrahigh pressure metamorphism. Cambridge University Press, Cambridge, pp 206–243

Dawson JB (1980) Kimberlites and their xenoliths. Springer, Berlin Heidelberg New York

Dawson JB, Smith JV (1977) The MARID (mica-amphibole-rutile-ilmenite-diopside) suite of kimberlite xenoliths. Geochim Cosmochim Acta 41:309–323

Dawson JB, Smith JV (1982) Upper mantle amphiboles: a review. Miner Mag 45:35–46

Dobrzhinetskaya L, Green HW, Wang S (1996) Alpe Arami: a peridotite massif from depths of more than 300 kilometers. Science 271:1841–1845

Elfadini S, Demaiffe D, Andre L (1995) Origin of eclogite nodules form the Mbuji Mayi kimberlites (Kasai, Zaire): subducted ancient oceanic crust? 6th Int Kimberlite Conf, Novosibirsk, 1995. Extended abstracts, pp 146–148

Fiala J, Padera K (1977) The chemistry of the minerals of the pyrope dunite from borehole T-7 near Stare (Bohemia). Tschermacks Mineral Petrogr Mitt 24, 4:205–219

Haggerty SE, Sautter V (1990) Ultradeep (>300 km) ultramafic, upper mantle xenoliths. Science 248:993–996

Helmsteadt H, Doig R (1975) Eclogite nodules form kimberlite pipes of the Colorado Plateau samples of subducted Franciscan-type oceanic lithosphere. Phys Chem Earth 9:95–111

Helmstaedt H, Schulze DJ (1989) Eclogite-facies ultramafic xenoliths from Colorado Plateau diatrem breccias: comparison with eclogites in crustal environments, evaluation of the subduction hypothesis, and implications for eclogite xenoliths from diamondiferous kimberlites. In: Smith D (ed) Eclogite-facies rocks. Elsevier, Amsterdam

Ionov DA, Hofmann AW (1995) Nb-Ta-rich mantle amphiboles and micas: implications for subduction-related matasomatic trace element fractionation. Earth Planet Sci Lett 131: 341–356

Ireland TR, Rudnick RL, Spetsius Z (1994) Trace elements in diamond inclusions from eclogites reveal link to Archean granites. Earth Planet Sci Lett 128:199–213

Irving AJ, Frey FA (1984) Trace element abundances in megarcysts and their host basalts: constraints on partition coefficients and megacryst genesis. Geochim Cosmochim Acta 48:1201–1221

Jacob D, Jagoutz E, Lowry D, Mattey D, Kudrjavtseva G (1994) Diamondiferous eclogites form Siberia: remnants of Archean oceanic crust. Geochim Cosmochim Acta 58:5191–5207

Jagoutz E, Dawson JB, Hoernes S, Spettel B, Wanke H (1984) Anorthositic oceanic crust in the Archean Earth. Lunar Planet Sci Conf XV.:395–396 (Extended Abstr)

Jerde EA, Taylor LA, Crozaz G, Sobolev NV (1993a) Exsolution of garnet within clinopyroxene of mantle eclogites: major- and trace-element chemistry. Contrib Mineral Petrol 114:148–159

Jerde EA, Taylor LA, Crozaz G, Sobolev NV (1993b) Diamondiferous eclogites form Yakutia, Siberia: evidence for a diversity of protoliths. Contrib Mineral Petrol 114:189–202

Kornprobst J, Ohnenstetter D, Ohnenstetter M (1981) Na and Cr contents in clinopyroxenes from peridotite: a possible discrimination between "subcontinental" and "suboceanic" mantle. Earth Planet Sci Lett 53, 2:241–254

Laz'ko EE, Sharkov EV (1988) Magmatic rocks. Ultramafic rocks. Nauka, Moscow, 508 pp (in Russian)

Liou JG, Ogasawara Y, Zhang RY (1995) Occurrence and stability of diamond-bearing assemblages in ultrahigh-pressure metamorphic rocks. 6th Int Kimberlite Conf, Novosibirsk, 1995. Extended abstracts, pp 328–330

MacGregor ID, Carter JL (1970) The chemistry of clinopyroxenes and garnets of eclogite and peridotite xenoliths from the Roberts Victor Mine, South Africa. Phys Earth Planet Int 3:391–397

MacGregor ID, Manton WI (1986) Roberts Victor eclogites: ancient oceanic crust. J Geophys Res 91:14063–14079

McCandless TE, Gurney JJ (1989) Sodium in garnet and potassium in clinopyroxene: criteria for classifying mantle eclogites. In: Kimberlites and related rocks. Vol 2. Geol Soc Australia Spec Publ 14. Blackwell, Carlton, pp 827–832

Medaris LG, Beard BL, Johnson CM, Valley JW, Spicuzza MJ, Jelinek E, Misar Z (1995) Garnet pyroxenite and eclogite in the Bohemian Massif: geochemical evidence for Variscan recycling of subducted lithosphere. Geol Rundsch 84:489–505

Neal CR (1988) The origin and composition of metasomatic fluids and amphiboles beneath Malaita, Solomon Islands. J Petrol 29:149–179

Neal CR, Nixon PH (1986) Spinel-garnet relationships in mantle xenoliths form the Malaita alnoite, Solomon Islands, southwestern Pacific. Trans Geol Soc S Afr 88:347–354

Neal CR, Taylor LA, Davidson JP, Holden P, Halliday AN, Nixon PH , Paces JB, Clayton RN, Mayeda TK (1990) Eclogites with oceanic crustal and mantle signatures form the Bellsbank kimberlite, South Africa, Part 2: Sr, Nd, and O isotope geochemistry. Earth Planet Sci Lett 99:362–l379

Nicolas A (1989) Structures of ophiolites and dynamics of oceanic lithosphere. Kluwer, Dordrecht Academic Press, Boston, 367 pp

Obata M (1980) The Ronda peridotite: garnet-, spinel- and plagioclase lherzolite facies and P-T trajectories of a high-pressure mantle intrusion. J Petrol 21, 3:533–572

Okay AI (1993) Petrology of a diamond and coesite-bearing metamorphic terrain: Dab: Shan, China. Eur J Miner 4:659–676

Orlov YuL (1973) Mineralogy of the diamond. Nauka, Moscow, 224 pp (in Russian)

Pearson DG, Davies GR, Nixon PH, Milledge HJ (1989) Graphitized diamonds form a peridotite massif in Morocco and implications for anomalous diamond occurrence. Nature 338:60–62

Pearson DG, Davies GR, Nixon PH, Greenwood PB, Mattey DP (1991) Oxygen isotope evidence for the origin of pyroxenites in the Beni Bousera peridotite massif, North Morocco: derivation from subducted oceanic lithosphere. Earth Planet Sci Lett 102: 289–301

Pearson DG, Shirey SB, Carlson RW, Boyd FR, Pokhilenko NP, Shimizu N (1995a) Re-Os, Sm-Nd, and Rb-Sr isotope evidence for thick Archean lithospheric mantle beneath the Siberian craton modified by multistage metasomatism. Geochim Cosmochim Acta 59:959–977

Pearson DG, Snyder GA, Shirey SB, Taylor LA, Carlson RW, Sobolev NV (1995b) Re-Os isotope evidence for a mid-Archean age of diamondiferous eclogite xenoliths from Siberia and constraints on Archean tectonics. Nature 374:711–713

Ponomarenko AI (1975) Alkremite – a new variety of aluminous hyperbasite xenoith from the kimberlites of the Udachnaya pipe. Dokl Akad Nauk SSSR 225:928–931 (in Russian).

Rapp RP (1995) Is eclogite in the sub-continental lithosphere the residue from melting of subducted oceanic crust? Experimental constraints and implications for the origin of the Archean continents (extended abstract). Abstr 6th Int Kimberlite Conf, Novosibirsk, Russia, pp 457–459

Roden MF, Hart SR, Frey FA, Melson WG (1984) Sr, Nd, and Pb isotopic and REE geochemistry of St. Paul's Rocks; the metamorphic and metasomatic development of an alkali basalt mantle source. Contrib Mineral Petrol 85, 4:376–390

Rosen OM, Zorin YM, Zayachkovsky AA (1972) Discovery of diamond in connection with eclogites in the Precambrian of the Kokchetav massif. Dokl Akad Nauk SSSR, 203:674–676

Ryabchikov ID (1988) Geochemical evolution of the Earth's mantle. Nauka, Moscow, 36 pp (in Russian)

Sautter V, Haggerty SE, Field S (1991) Ultradeep (>300) ultramafic xenoliths: petrological evidence from the transition zone. Science 252:827–830

Schulze DJ, Helmstaedt H (1988) Coesite-sanidine eclogites form kimberlite: Products of mantle fractionation or subduction? J Geol 96:435–443

Sharkov EV, Laz'ko EE, Fedosova SP, Hanna S (1990) Xenoliths at the Quaternary Tell Danyn volcano (Southern Syria) and intraplate magmatism. Int Geochem 27, 6:82–94

Sharma M, Wasserburg GJ, Papanastassiou DA, Quick JE, Sharkov EV, Laz'ko EE (1995) High 143Nd/144Nd in extremely depleted mantle rocks. Earth Planet Sci Lett 135:101–114

Shatsky VS, Sobolev NV, Vavilov MA (1995) Diamond-bearing metamorphic rocks of the Kokchetav massif (northern Kazakhstan). In: Coleman RG, Wang X (eds) Ultrahigh pressure metamorphism. Cambridge University Press, Cambridge, pp 427–455

Shervais JW, Taylor LA, Lugmair GW, Glayton RN, Mayeda TK, Korotev RL (1988) Early Proterozoic oceanic crust and the evolution of subcontinental mantle: eclogites and related rocks from southern Africa. Geol Soc Am Bull 100:411–423

Simkin T, Smith JV (1970) Minor-element distribution of olivine. J Geol 78, 3:304–325

Slodkevich VV (1980) Polycrystalline aggregates of octahedral graphite. Dokl Acad Nauk SSSR 253, 3:697–700

Smyth JR, Caporuscio FA, McCormick TC (1989) Mantle eclogites: evidence of igneous fractionation in the mantle. Earth Planet Sci Lett 93:133–141

Snyder GA, Taylor LA, Jerde EA, Sharkov Ye, Laz'ko Ye (1993a) Petrogenesis of garnet pyroxenites and spinel peridotite xenoliths of the Tell-Danun alkali basalt volcano, Harrat as Shamah, Syria. Int Geol Rev 35:1104–1120

Snyder GA, Jerde EA, Taylor LA, Halliday AN, Sobolev VN, Sobolev NV (1993b) Nd and Sr isotopes form diamondiferous eclogites, Udachnaya kimberlite pipe, Yakutia, Siberia: evidence of primary differentiation in the early Earth? Earth Planet Sci Lett 118:91–100

Snyder GA, Taylor LA, Jerde EA, Deines P, Clayton RN, Mayeda T, Rossman GR, Sobolev NV (1995) Archean mantle heterogeneity and the origin of diamondiferous eclogites, Siberia: evidence from stable isotopes and hydroxyl in garnet. Am Mineral 80:799–809

Sobolev NV (1974) Deep-seated inclusions in kimberlites and the problem of the composition of the upper mantle. American Geophysical Union, Washington, DC, 279 pp

Sobolev NV, Shatsky VS (1990) Diamond inclusions in garnets from metamorphic rocks: a new environment for diamond formation. Nature 373:742–746

Sobolev VN, Taylor LA, Snyder GA, Sobolev NV (1994) Diamondiferous eclogites from the Udachnaya kimberlite pipe, Yakutia, Siberia. Int Geol Rev 36:42–64

Sobolev VN, Taylor LA, Snyder GA, Sobolev NV, Pokhilenko NP, Kharkiv AD (1995) A unique metasomatised peridotite xenolith from the Mir kimberlite, Siberian Platform. 6th Int Kimberlite Conf, Novosibirsk, 1995. Extended abstracts pp 555–557

Taylor LA, Neal CR (1989) Eclogites with oceanic crustal and mantle signatures form the Bellsbank kimberlite, South Africa, Part I: Mineralogy, petrography, and whole rock chemistry. J Geol 97:551–567

Thompson RN (1974) Some high-pressure pyroxenes. Mineral Mag 39:768–787

van Heerden LA, Boyd SR, Pillinger CT (1995) Carbon and nitrogen isotope characteristics of Argyle and Ellendale diamonds. 6th Int Kimberlite Conf, Novosibirsk, 1995. Extended Abstr, pp 225–227

Wang X, Zhang R, Liou JG (1995) UHPM terrane in east central China. In: Coleman RG, Wang X (eds) Ultrahigh pressure metamorphism. Cambridge University Press, Cambridge, pp 356–390

Wilshire HG, Shervais JW (1975) Al-augite and Cr-diopside xenoliths in basalts host rocks form Western United States. Phys Chem Earth, 9:257–272

Xu S, Okay AL, Sengor AMC, Su W, Liu Y, Jiang L (1992) Diamond from the Dabie Shan metamorphic rocks and its implication for tectonic setting. Science, 256:80–82

2.2.3 Composition and Evolution of the Upper Mantle

I. D. Ryabchikov

Origin, Heterogeneity, and Evolution of the Upper Mantle

Primitive, Depleted, and Enriched Mantle. Magmatic Differentiation and Mantle Metasomatism

As mentioned above (Chap. 2.1) primitive mantle or bulk silicate earth is the reconstructed average composition of the Earth without the metallic core. Its estimation is based on the postulated chondritic proportions of nonvolatile lithophile elements in undifferentiated cosmic material participating in the accretion of our planet. Analysis of covariations of diverse components in various classes of chondritic meteorites permitted the overall composition of the Earth including core to be estimated (Allegre et al. 1995; McDonough and Sun 1995).

The prevailing part of the upper mantle consists of plastic sublithospheric material constituting part of a global convective system. Decompression partial melting of mantle material in upwelling parts of this convective system is the leading mechanism of magma generation. Upward migration of magmas escaping from partially molten matrix is the principal cause of global geochemical differentiation, resulting in the depletion of the upper mantle in components concentrating in the melt (incompatible elements). This depletion is reflected in trace element concentrations and the isotopic composition of MORBs. The main complementary enriched reservoir is the continental crust. Mass balance calculations based on radiogenic isotope ratios in the primitive mantle, MORB-source material, and continental crust show that the depleted mantle constitutes from 30 to 50% of the total mass of the mantle (Galer et al. 1989). Allowing for uncertainty, this result does not contradict the assumption that the depleted mantle coincides with the upper mantle, including the transition zone.

Knowledge of the proportion of depleted mantle permits estimation of its composition by subtracting the continental crust from the respective mass of the primitive mantle (McKenzie and O'Nions 1991). This inferred composition is identical to the primitive mantle with respect to major components, compatible and moderately incompatible trace elements, but impoverished in highly incompatible elements. Figure 35 shows that the extent of depletion is inversely related to the bulk partition coefficients (ratios of concentrations in the mixture of crystalline phases and melt in equilibrium with them) for the majority of elements. The continental crust is variably enriched in these elements parallel to decreasing bulk partition coefficients. This suggests that mantle magmatism is the principal cause of the enrichment of the continental crust and the depletion of the upper mantle in highly incompatible elements. Notable exceptions to this regular pattern are elements like Nb and Ta, which are less depleted in the MORB source and less enriched in the crust by comparison with the elements of comparable incompatibility (e.g., Nb compared to U), or elements like Pb, B, As and Sb, which, on the contrary, are more depleted in the MORB source and more enriched in the continental crust by comparison with reference elements B/La, Pb/Ce, Sb/Ce, and As/Ce ratios are higher in the continental crust than in the depleted upper mantle, and their values differ from those in the primitive mantle. This peculiar behavior is also manifested in the composition of magmas confined to subduction zones. In the case of Nb and Ta, this is explained by intense coprecipitation with Ti-rich minerals during generation and fractional crystallization of calc-alkaline magmas which later are added to the continental crust (Hofmann 1988). On the other hand, Pb, B, As, Sb, and possibly Cs are removed from the subducting material by aqueous fluids arising during its dehydration. In this way, these fluid-mobile components are transferred to the source region of calc-alkaline magmas (Chauvel et al. 1995).

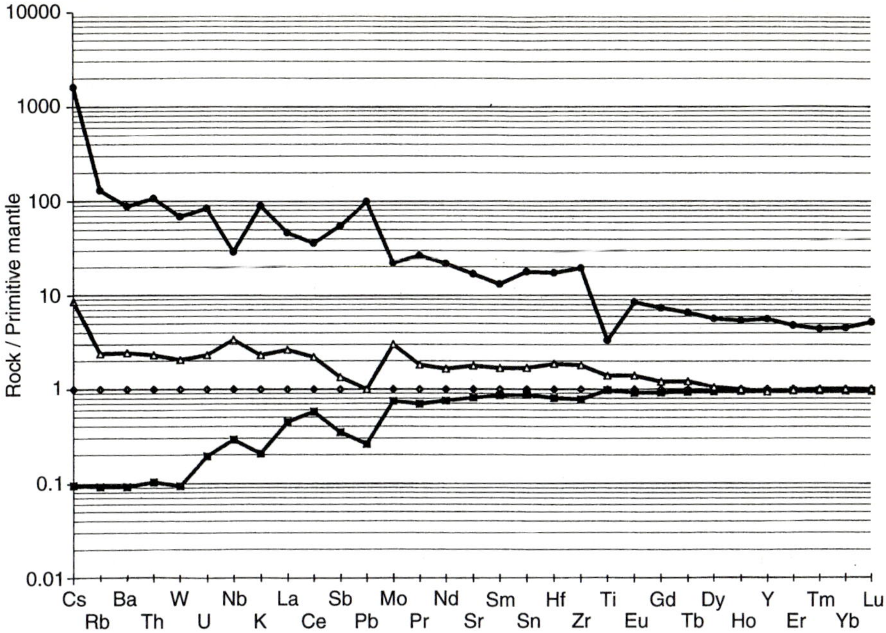

Fig. 35. Incompatible trace elements in N-MORB source mantle (*solid squares*), OIB source mantle (*triangles*) and continental crust (*circles*) anormalized by primitive mantle. (Data from Ryabchikov 1997)

Mantle peridotites with geochemical characteristics similar to the source of MORBs are represented by some massifs of alpine-type peridotites (e.g., Ronda, Spain – Frey et al. 1985; Reisberg et al. 1989). Most likely they are blocks of suboceanic lithosphere tectonically emplaced into the continental crust. Various amounts of basaltic melts were extracted from them in the past, and their average composition differs from the estimated composition of the convecting upper mantle, because they are depleted not only with respect to highly incompatible components but also with respect to mildly incompatible and magmaphile major elements as well. The most fertile members of these peridotites, with respect to major magmaphile elements (Ca, Al, Ti), are possibly the closest analogs of the prevailing part of the depleted upper mantle (Fig. 36).

Incompatible element-depleted peridotites serving as a source of modern MORBs are not the only type of material residing in the sublithospheric mantle. Isotopic investigations of oceanic island basalts reveal that their sources are less depleted and in some cases even enriched in highly incompatible elements by comparison with the primitive mantle.

There are also enriched in incompatible components reservoirs, as can be judged from the isotopic characteristics of magmas from intraplate oceanic islands (Zindler and Hart 1986). There are several types of such enriched

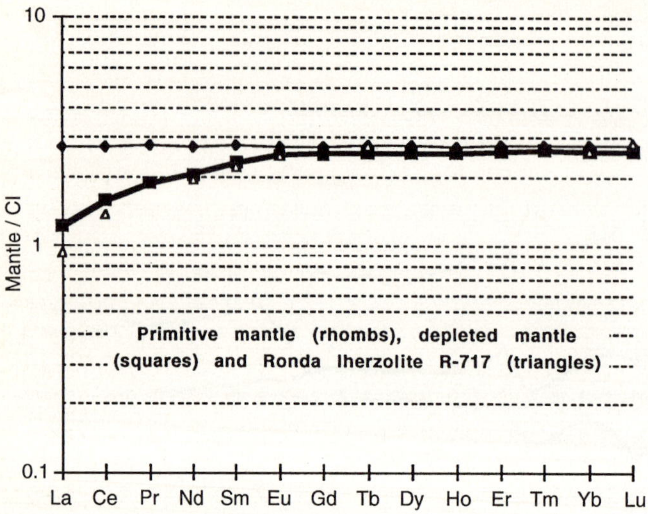

Fig. 36. CI-normalized REE abundances in bulk silicate Earth (McDonough and Sun 1995 *rhombs*), N-MORB-source mantle (Ryabchikov 1997 *filled squares*) and in Ronda lherzolite R-717 (Frey et al. 1985 *triangles*)

mantle: EMI (increased Nd/Sm but not Rb/Sr ratios; the explanation is contamination by pelagic sediments, or metasomatism by carbonate-rich melts), EMII (both Nd/Sm and Rb/Sr are elevated, possibly due to contamination of subducted terrigenous sediments or metasomatism by water-rich silicate melts or fluids) and HIMU (high U/Pb, most likely a contribution from the subducted oceanic crust).

The lithosphere is geochemically heterogeneous, comprising frozen blocks of depleted material, blocks enriched in incompatible components due to the processes of the migration of melts and fluids in the mantle, and also containing a certain amount of material similar to PM. It also includes some former crustal material (eclogites and, in rare cases, possibly event he relics of continental crust). Overall enrichment in highly incompatible elements and depletion in moderately incompatible elements has been suggested by McDonough (1990). The lithosphere contains a large proportion of harzburgites – buoyant restites floating on the denser lherzolitic asthenosphere. Many investigators maintain that the lower mantle consists at least partly of more primitive material. This is in part confirmed by geochemical balance calculations and by the elevated proportion of ^{3}He isotopes in some materials of deep origin (implying that they are less degassed during the entire Earth's history somewhere at great depth). These are also arguments in favor of multilayered global convection.

Magmatism is the main process responsible for the differentiation of mantle material. It causes the overall depletion of the upper mantle in magmaphile incompatible elements due to their transfer into the crust, and also to the enrichment of some parts of the mantle in these elements due to the migration of small fraction melts concentrating strongly in these components. When mantle-derived magmas are transported along macroscopic channels and solidify within the mantle as intrusive bodies, we have sensu stricto magmatic differentiation, whereas migration of melts through microscopic intergranular space and their reaction with previously existing mantle minerals is usually called mantle metasomatism. Mantle metasomatism may also be caused in some cases by dense aqueous fluids which at high pressures may exhibit gradual transition to water-rich melts. When metasomatizing agents are introduced into mantle rocks in small proportions, the mineral composition of the metasomatized rocks may remain intact, and the only result of this process may be the increase in the concentrations of highly incompatible elements. This case is described by the term cryptic metasomatism. When the proportion of metasomatizing melts or fluids is much higher, they may cause the formation of new minerals, in which case we speak of modal metasomatism. Newly formed minerals may be represented by amphiboles or phlogopites in the case of water-rich melts or fluids, whereas carbonate-rich melts may cause neoformations of clinopyroxene and the appearance of apatite in mantle rocks. In some cases, mantle metasomatism brings about enrichment in Ti, which may be manifested in the appearance of ilmenite and, in some cases, of rare titanates of Ba and REE.

References

Allegre CJ, Poirier J-P, Humler E, Hofmann AW (1995) The chemical composition of the Earth. Earth Planet Sci Lett 134:515–526

Chauvel C, Goldstein SL, Hofmann AW (1995) Hydration and dehydration of oceanic crust controls Pb evolution in the mantle. Chem Geol, 126:65–75

Frey FA, Suen CJ, Stockman HW (1985) The Ronda high temperature peridotite: geochemistry and petrogenesis. Geochim Cosmochim Acta, 29:2469–2491

Galer SJG, Goldstein SL, O'Nions RK (1989) Limits on chemical and convective isolation in the Earth's interior. Chem Geol, 75:257–290

Hofmann AW (1988) Chemical differentiation of the Earth: the relationship between mantle, continental crust, and oceanic crust. Earth Planet Sci Lett, 90:297–314

McDonough WF (1990) Constraints on the composition of the continental lithospheric mantle. Earth Planet Sci Lett 101:1–18

McDonough WF, Sun S-S (1995) The composition of the Earth. Chem Geol 120:223–253

McKenzie D, O'Nions RK (1991) Partial melt distribution from inversion of rare earth element concentrations. J Petrol 32:1021–1091

Reisberg L, Zindler A, Jagoutz E (1989) Further Sr and Nd isotopic results from peridotites of the Ronda Ultramafic Complex. Earth Planet Sci Lett 96:161–180

Ryabchikov ID (1997) Composition of Earth's upper mantle. Geochem Int 5:1–11

Zindler A, Hart SR (1986) Chemical geodynamics. Annu Rev Earth Planet Sci 14:493–571

Geothermobarometry of the Upper Mantle by Coexisting Minerals

Sources of information concerning the temperature in the mantle: correlation of seismic boundaries within mantle with phase transitions in $(Mg,Fe)_2SiO_4$, heat flux measurements, geothermobarometry based on phase equilibrium among mineral solid solutions, estimates of magma temperatures, and interpretation of seismic properties.

Thermobarometry, based on the distribution of major components between coexisting minerals of variable composition, is mainly confined for mantle rocks to Cpx+Opx+Ga assemblages (garnet lherzolites).

The two-pyroxene thermometer was first invented by Boyd (1973) and was initially based on the mutual miscibility of Cpx and Opx (Boyd and Schairer 1964) in the simple system $Mg_2Si_2O_6 - CaMgSi_2O_6$ (temperature dependence of En mole fraction in Cpx). Later, Wood and Banno (1973) suggested using the equilibrium constant of reaction

$$Mg_2Si_2O_6(Opx) = Mg_2Si_2O_6(Cpx)$$

(ratios of En activities). A simple ideal two-site model of pyroxene solid solutions was used with the assumption that Mg/Fe ratios are the same in M1 and M2 octahedral sites. It was also assumed that the pressure effect is negligible. Wells' (1977) formula for the Wood-Banno model is

$$T = \frac{7341}{3.355 + 2.44\, X_{Fe}\, Opx + \ln K_{En}}.$$

Recently proposed formulas are based on experimental data for complex systems in the range of compositions close to mantle peridotites (Brey and Koehler 1990).

The most reliable barometer for mantle rocks uses Opx-Ga equilibrium approximately expressed by the reaction

$$MgAl_2SiO_6 + 2\, MgSiO_3 = Mg_3Al_2Si_3O_{12}.$$

With increasing P, the Al content in Opx coexisting with Ga decreases, whereas T exerts the opposite effect. Thus, with T estimated from a two-pyroxene thermometer, P may be calculated from Opx and Ga compositions.

Other geothermometers are based on Mg/Fe partitioning for Ol-Ga, Opx-Ga, Cpx-Ga, and Ol-Sp pairs. The Ca contents in Opx coexisting with Cpx may also be used for this purpose. Approaches based on minor element distribution have also been proposed. Ni distribution between coexisting Ol and Ga seems to be quite promising (one-mineral thermometer, because Ni in Ga is much more variable than in Ol – Griffin and Ryan 1995).

Geobarometers are less abundant than geothermometers. Additional barometers use Ga-Sp equilibrium (yields satisfactory agreement with Ga-Opx barometer for transitional Ga-Sp lherzolites) and Ca solubility in Ol coexisting with Cpx+Opx. The latter is based on the reaction:

$$Mg_2SiO_4 + 2\ CaMgSi_2O_6 = Ca_2SiO_4 + 2\ Mg_2Si_2O_6.$$

This is applicable to peridotites in both spinel and garnet lherzolite facies, but it is not very sensitive at lower T (Koehler and Brey 1990).

The application of geothermobarometers is based on the assumption of complete equilibrium in natural rocks. As T and P change, the interphase distribution of components starts to accommodate itself to new conditions, but diffusion rates are different for various minerals and elements. So complete equilibrium may not be reached under new conditions, and there would be discrepancies for estimates from various thermometers and barometers. These differences may tell us about the direction and rate of T and P changes (geospeedometers).

Application of the Cpx+Opx+Ga geothermobarometer to garnet lherzolite xenoliths from kimberlites yields points on the P-T plane which lie near a common line with a slope close to 15/kbar (about 5 °C/% km) and passing near 1000 °C at 40 kbar. This, within unavoidable uncertainties, is consistent with heat flux measurements for stable continental cratons (about 40 mW/m^2). Approximately half of this flux comes from the mantle (half from radioactive decay in the crust).

Given the formula for heat conduction:

$$dQ/dt = k(dT/dz),$$

with $k \sim 4$ W/°C/m, we obtain from 0.02 W/m^2 gradient of 5 °C/km. This geotherm corresponds to the rigid lithosperhic part of the mantle where heat is transported by conduction. Below the lithosphere we have the convecting mantle, where the geothermal gradient should be close to adiabatic. For adiabatic isoenthropic) conditions:

$$(dT/dP)S = (dV/dT)PT/CP \text{ or } (dT/dz)S = \alpha \cdot gT/CP,$$

and substituting into these equations α and CP for peridotitic material results in (dT/dz)S of a fraction f degree per km. There are direct experimental measurements of adiabatic gradient for olivine and some other minerals at high pressures and temperatures. Perhaps the best estimate of this value for the solid mantle is close to 0.3 °C km.

The intersection of a conductive geotherm approximately corresponding to the surface heat flux close to 40 mW/m^2, with the adiabatic gradient passing through 1400 °C at 400 km, lies at about 200 km and 1200–1300 °C. These temperatures are close to the lower limit above which olivine-rich rocks are able to flow plastically. Therefore, the described temperature distribution is, in general, consistent with the structure of mantle. However, the seismic data do not distinctly define the lower boundary of continental lithosphere, and some investigators (e.g., Jordan) place it at much lower depths (about 400 km – continental keel). It is possible that in future the thermal structure of the subcontinental mantle will be reconsidered.

The interpolation of the adiabatic gradient from 1400 °C at 400 km to CMB gives 2300–2400 °C. Taking into account Boehler's estimate of T at the core side of CMB (about 3700 °C), we would have a temperature jump (thermal boundary layer) in excess of 1000 °C. This large, fast change corresponds to the one-layer model of mantle convection. For multilayer models, this value may be distributed between several thermal boundary layers. These resources of thermal energy stored in the inner parts of the Earth are essential for large-scale magma generation processes in the mantle. The growth of plumes, or, more generally, an increase in the convective activity of the mantle, is linked to an increment in the buoyancy of the material in D″ (above CMB) softened by heat flux coming from the core. When plumes ascend from D", this layer is thinned and causes an increment of the heat flux and consequent growth in the vigor of outer core convection. During this vigorous convection its pattern remains constant, and polarity is not reversed during long periods of time. Relaxation of superadiabatic thermal gradient at CMB by superplumes is considered to be the cause of correlations between superchrones and periods of intense magmatic activity (also high sea level and a number of other geological phenomena).

References

Boyd FR (1973) A pyroxene geotherm. Geochim Cosmochim Acta 37:2533–2546

Boyd FR, Schairer JF (1964) The system $MgSiO_3$-$CaMgSi_2O_6$. J Petrol 5:275–309

Brey GP, Koehler T (1990) Geothermobarometry in four-phase lherzolites II. New thermobarometers, and practical assessment of existing thermobarometers. J Petrol 31:1353–1358

Griffin WL, Ryan CG (1995) Trace-elements in indicator minerals: area selection and target evaluation in diamond exploration. 6th Int Kimberlite Conf Extended Abstr, Novosibirsk, pp 201–202

Koehler TP, Brey GP (1990) Calcium exchange between olivine and clinopyroxene calibrated as geothermobarometer for natural peridotites from 2 to 60 kb with applications. Geochim Cosmochim Acta 54:2375–2388

Wells PRA (1977) Pyroxene thermometry in simple and complex systems. Contrib Mineral Petrol 62:129–139

Wood BJ, Banno S (1973) Garnet-orthopyroxene and orthopyroxene-clinopyroxene relationships in simple and complex systems. Contrib Mineral Petrol 42:109–124

Oxygen Fugacities and Redox Conditions

As shown below, H_2O and CO_2 are prevailing volatiles in the mantle. It would be more accurate to say that the dominant volatiles are compounds in the C-O-H system, because, depending on f_{O_2}, they may be represented not by water and carbon dioxide, but by elemental carbon, methane, and H_2. Oxygen fugacities may be estimated for the mantle by a number of methods, the most reliable at the moment being thermodynamic analysis of Sp+Opx+Ol equilibria in mantle rocks (reaction $6\,Fe_2SiO_4 + O_2 = 2\,Fe_3O_4 + 6\,FeSiO_3$ with the estimation of the activities of respective components – Ryabchikov et al. 1986; Wood et al. 1990) and the utilization of f_{O_2}-Fe^{3+}/Fe^{2+} experimentally calibrated dependence in comparison with the Fe^{3+}/Fe^{2+} ratios in basaltic glasses (Christie et al. 1986). It has also been suggested using redox reactions with chromium (olivine contains predominantly Cr^{2+} while in spinel we have Cr^{+3}) to estimate f_O2. In spite of the very wide span in estimated values (–3 to +2 log units relative to QFM buffer), they all lie in the stability fields of $H_2O + CO_2$, H_2O + carbonates or H_2O + elemental carbon, but not CH_4 or H_2. The relatively oxidized nature of the mantle, at least down to the base of the continental lithosphere, is corroborated by the presence of dense CO_2 fluid inclusions in the rocks of the uppermost mantle, the occurrence of carbonates in mantle peridotites, and by the discovery of volatile-rich microinclusions in coated crystals of natural diamonds in which water and carbonates strongly predominate over the reduced volatile species (Navon et al. 1988). It cannot be excluded, however, that a P above 100 kbar the equilibria between mantle minerals and fluids is shifted in such a way that CH_4 may become important volatile compound. This may follow from the stabilization at very high pressure of the components containing ferric iron. In particular, this may be brought abbot by the stabilization of the Fe_3O_4 component in silicate spinel and distorted spinel (beta- and gamma-orthosilicates) phases (O'Neil et al. 1993).

The present redox state of the mantle may be the result of the global degassing of our planet. At lower oxygen fugacities the evolving gases (particularly at low pressures) are represented by H_2- and CO-rich gases, whereas in melts and crystalline minerals H_2O, OH^-, CO_2, and CO_3^{2-} predominate. Separation of reduced gases results in the oxidation of magma or minerals, and this process continues until the system is sufficiently oxidized, so that the equilibrium gas contains mainly oxidized species (CO_2 and H_2O).

References

Ballhaus C (1993) Redox states of lithospheric and asthenospheric upper mantle. Contrib Mineral Petrol 114:331–348

Christie DM, Carmichael ISE, Langmuir CH (1986) Oxidation states of mid-ocean ridge basalt glasses. Earth Planet Sci Lett 79:397–411

Navon O, Mutcheon ID, Rossman G, Wasserburg GI (1988) Mantle-derived fluids in diamond microinclusions. Nature 335:784–789

O'Neil HStC, Rubie DC, Canil D, Geiger CA, Ross CR, Seifert F, Woodland B (1993) Ferric iron in the upper mantle and transition zone assemblages: implications for relative oxygen fugacities in the mantle. In: Evolution of Earth and Planets, geophysical monograph 74; IUGG vol 14, pp 73–88

Ryabchikov ID, Ukhanov AV, Ishii T (1986) Redox equilibria in upper mantle ultrabasites in the Yakutia kimberlite province. Geochem Int 23, 2:38–50

Wood BJ, Bryndzya LT, Johnson KE (1990) Mantle oxidation state and its relationship to tectonic environment and fluid speciation. Science 248:337–345

Mantle Volatiles and Water; Fluid Inclusions in the Mantle Minerals

The presence of volatile components in the mantle is proved by the occasional occurrence of volatile-bearing minerals in mantle xenoliths (amphiboles, phlogopite, apatite, carbonates, elemental carbon), by their presence in mantle-derived magmas, and by findings of volatile-rich microinclusions in the minerals of mantle rocks (Navon et al. 1988) The simple approach to the estimation of the average volatile abundances in the mantle using the analysis of mantle xenoliths is rather doubtful because of the very inhomogeneous distribution of volatiles in mantle rocks (sporadic occurrence of volatile-bearing minerals) and because their contents are very much affected by secondary processes. The most promising source of information are fresh volcanic glasses (particularly basaltic glasses quenched by subaqueous eruption) and melt inclusions in magmatic rocks of mantle origin.

Water is present in MORBs in small amounts, and, down to rather small pressures, it is not degassed from the melt. Therefore, the water content of MORB glasses is taken as the concentrations in magmas. On the other hand, water does not appreciably enter nominally anhydrous minerals of mantle rocks or solid phases crystallizing from magma and, therefore, it may be considered as an incompatible minor component. Indeed, it has been noted that water contents in MORB glasses exhibit good correlation with other incompatible elements like potassium and LREE (Moore 1970; Dixon et al. 1988; Michael 1988; Jambon and Zimmermann 1990). These correlations lead to the conclusion that the primitive mantle contains around 1000 ppm H_2O (1300 ppm – Jambon and Zimmermann 1990) and the depleted mantle about 100–250 ppm (Michael 1988). Infrared spectra of mantle peridotite anhydrous minerals reveal certain amounts of OH in them. Small amounts of water estimated as average abundances for the mantle (0.1 wt.% or less) can, in principle, be accommodated by the lattices of these “anhydrous” minerals. The current estimates of mineral/melt partition coefficients for water in these solid phases are: 0.001–0.007 (Ga), 0.01 (Ol), 0.02 (Opx), and 0.04 (Cpx) (Bell and Rossman 1993). These values characterize water as a highly incompatible element which would easily pass from peridotitic minerals into interstitial melt or dense fluid.

Bell and Rossman (1992) investigated by IR spectroscopy hydroxyl contents in nominally anhydrous mantle minerals. They found that pyroxenes are the most hydrous (200–500 ppm H_2O). They concluded that nominally anhydrous minerals constitute a significant reservoir for mantle hydrogen, possibly accommodating all water in the depleted mantle and providing a possible mechanism to recycle water from the Earth’s surface into the deep mantle. Solubility of water in olivine was found to reach ca. 0.1 wt.% near 130 kbar and 1100 °C, and much higher water contents were detected under fluid-saturated conditions in spinel-like modifications of magnesium-iron orthosilicate at higher pressures (Kohlstedt et al. 1996).

Water interacts with peridotites, producing hydrous minerals in the subsolidus region. At low temperatures we have serpentine, which at about 500 °C gives way to talc, and later to amphibole. The latter reaches water-saturated solidus but has an upper stability limit with respect to pressure (about 30 kbar). At pressures above 40 kbar we may have DHMS [dense hydrous magnesian silicates, e.g., $4Mg_2SiO_4$ $Mg(OH)_2$], although the importance of these phases has been questioned (Luth 1995). This may mean that the prevailing form of water at great depths is OH^- groups in nominally anhydrous minerals.

Water enters the silicate melt in the form of OH^- groups at low concentrations and as H_2O molecules at higher contents. The structure of the silicate melt changes significantly with the addition of water, and water-saturated solidus temperatures decrease with increasing pressure to about 20 kbar, when garnet appears among the near-solidus phases. Dissolved water shifts cotectics among major silicate minerals in such a way that, in the system with low alkali content, melts become more silica-rich by comparison with the dry system (boninite, high-Mg andesites), while for more alkali-rich sources (containing omphacitic clinopyroxene) we have alkaline melts (basanites).

Phlogopite, as well as amphibole, may be present near the solidus, but it is stable up to much higher pressures. It is possible that at 50–70 kbar in lherzolite with ordinary K_2O content phlogopite disappears because all the potassium may be dissolved in Cpx. Both Amph and Phl are resorbed during melting not far from the solidus of lherzolite.

At high pressures, fluids which may be present in the Lhz-H_2O system are represented not by pure water, but by rather concentrated aqueous solutions of silicate components. The most abundant nonvolatile components in such fluids are silica, buffered by the equilibrium with Ol and Opx, and silicates and aluminosilicates of K and Na, the concentrations of which are controlled by the presence in the mantle of omphacitic Cpx and small amounts of Phl. The total solubility of silicate components in fluids interacting with peridotite at 100 km depth may exceed 50 wt.%, i.e., it may reach values characteristic of readily soluble salts (Fig. 37). This proves the efficiency of fluid mass transfer in the upper mantle and the possibility of supercritical (gradual) transition between aqueous solutions and H_2O-rich silicate melts (Ryabchikov 1993).

Carbon dioxide solubility in basaltic magmas is much lower as compared to water. Therefore, analyses of basaltic glasses do not provide very reliable information because of the danger of partial degassing in the intermediate magma chamber and on the way to the ocean floor. Taking maximum CO_2/H_2O values for basaltic glasses, we may estimate the CO_2 content for the depleted mantle as 30–100 ppm and for the primitive mantle as 100–200 ppm. If, on the other hand, we assume that part of the CO_2 was lost in the intermediate magma chamber situated close to oceanic crust-mantle boundary, then we are forced to accept that the initial CO_2 content in magma was higher than CO_2 solubility at 2–2.5 kbar (1000 ppm or more). In this case, estimated abundances for mantle would be too high.

Another approach is based on the comparison of CO_2 and ^3He juvenile fluxes. The investigation of basaltic glasses and CO_2-rich waters from the region of young mantle-derived volcanism provides $C/^3He$ ratios close to

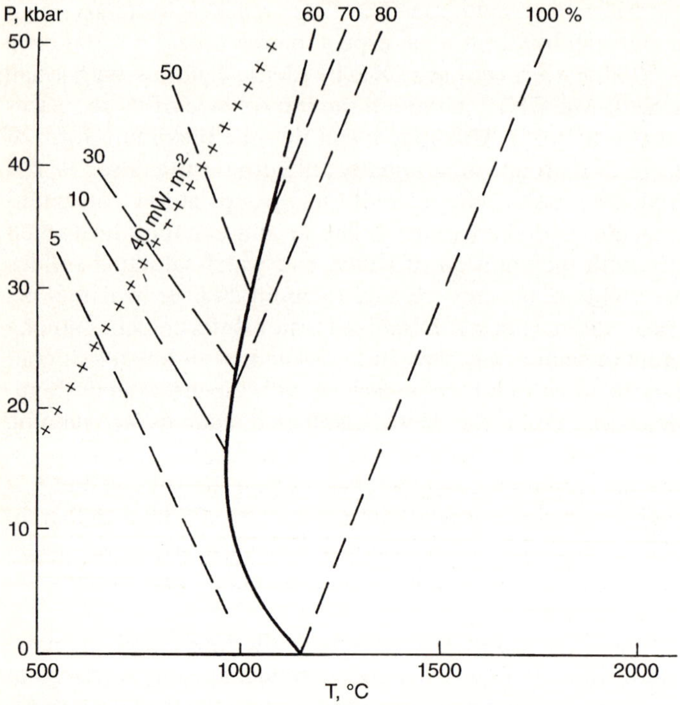

Fig. 37. P-T diagram of the system lherzolite-H_2O (Ryabchikov 1993), with isopleths showing total concentrations of nonvolatile components in fluid and in melt (*dashed lines*). The critical end point which terminates the curve of fluid-saturated solidus (*solid line*) is provisionally shown at 35 kbar. The crosses refer to the geotherm of stable continental regions with the surface heat flow of 40 mW/m^2

10^9 (O'Nions 1987). Taking into account the ^3He flux from the Earth's surface, and assuming that this isotope was mainly transported from the mantle by MORB-like magmas, we may estimate ca. 100 ppm CO_2 for the convecting upper mantle, which is not far from the above values. Trull et al. (1993) concluded on the basis of a similar approach that the upper mantle carbon content is probably in the range of 50–250 ppm C (200–1000 ppm CO_2).

Sulfur abundance in the mantle may also be based on its content in MORBs. McDonough and Sun's (1995) estimate is 250 ppm. A similar value was given by Lorand (1993), who mainly used the analyses of orogenic peridotites. Peridotitic nodules, however, in many cases contain much lower and very variable amounts of S (Ionov et al. 1992). Desulphurization processes operating in the mantle with the participation of fluid phase are not excluded.

Cl, F, and P contents may also be estimated from the data on basaltic glasses. For primitive mantle they are 20–50, 20–30, and 90 ppm respectively.

References

Bell DR, Rossman GR (1992) Water in Earth's mantle: the role of nominally anhydrous minerals. Science 255:1391–1397

Bell DR, Rossman GR (1993) The trace element partitioning of H in a high-pressure igneous system – megacrysts from the Monastery kimberlites. EOS 74, 16:340

Dixon JE, Stolper E, Delaney JR (1988) Infrared spectroscopic measurements of CO_2 and H_2O in Juan de Fuca Ridge basaltic glasses. Earth Planet Sci Lett 90:87–104

Ionov DA, Hoefs J, Wedepohl KH, Wiechert U (1992) Content and isotopic composition of sulphur in ultramafic xenoliths from central Asia. Earth Planet Sci Lett 111:269–286

Jambon A, Zimmermann JL (1990) Water in oceanic basalts: evidence for dehydration of recycled crust. Earth Planet Sci Lett 101:323–331

Kohlstedt DL, Keppler H, Rubie DC (1996) Solubility of water in the α, β and γ phases of (Mg,Fe)2SiO_4. Contrib Mineral Petrol 123:345–357

Lorand JP (1993) Comment on "Content and isotopic composition of sulphur in ultramafic xenoliths from central Asia." by Ionov DA, Hoefs J, Wedepohl KH and Wichert U. Earth Planet Sci Lett 119:627–634

Luth RW (1995) Is phase A relevant to the Earth's mantle? Geochim Cosmochim Acta 59, 4:679–682

McDonough WF, Sun S-S (1995) The composition of the Earth. Chem Geol 120:223–253

Michael PJ (1988) The concentration, behavior and storage of H_2O in suboceanic upper mantle: implications for mantle metasomatism. Geochim Cosmochim Acta 52:555–566

Moore JG (1970) Water content of basalts erupted on the ocean floor. Contrib Mineral Petrol 28:272–279

Navon O, Hutcheon ID, Rossman G, Wasserburg GJ (1988) Mantle-derived fluids in diamond micro-inclusions. Nature 335:784–789

O'Nions RK (1987) Relationships between chemical and convective layering in the Earth. J Geol Soc London 144:259–274

Ryabchikov ID (1992) Fluid transport of ore metals in ultramafic mantle rocks. Proc 8th Quadr IAGOD Symp, pp 425–433

Trull T, Nadeau S, Pineau F, Polve M, Javoy M (1993) C-He systematics in hotspot xenoliths: implications for mantle carbon contents and carbon recycling. Earth Planet Sci Lett 118:43–64

Zhang Y, Zindler A (1993) Distribution and evolution of carbon and nitrogen in Earth. Earth Planet Sci Lett 117:331–345

Upper Mantle Source of Metals

The differentiation of mantle-derived magmas intruded into the Earth's crust results in many cases in the formation of economic concentrations of various metals. Under favorable circumstances, this may lead to the formation of very large mineral deposits.

Depending upon the partition of a given metal between silicate melt and residual phases, its deposits are linked to different types of mantle magmatism: compatible elements are concentrated in magmas produced at high degrees of partial melting (komatiites, tholeiitic basalts), whereas the contents of incompatible metals are much higher in near-solidus magmas generated at low extents of partial melting.

Compatible ore elements include Cr, Ni, and highly chalcophile elements such as platinum group elements (PGE). The compatible behavior of Cr is due to its intense concentration by such mantle minerals as clinopyroxene and spinel. During the progressive partial melting of mantle peridotites, Cr-rich spinels appear due to the incongruent melting of clinopyroxene. Thus, primary magmas of tholeiitic basalt series are saturated with respect to chromite from the very beginning, and, on cooling, they start to precipitate chromite together with olivine. However, during this process, chromite is formed in very small proportions, and the main problem of the formation of chromitite lodes is separation of this mineral from olivine and other silicate phases. Irvine (1977) suggested that this may be achieved, during the mixing in the magma chamber, of melt partially differentiated by fractional crystallization with the newly injected portion of more primitive magma. This mixed melt must have chromite alone on the liquidus, and this mineral crystallizes without silicate phases for some time. The sinking of chromite to the bottom of the magma chamber results in the formation of chromitite seams. Chromite deposits are widespread in intrusive rocks of ophiolitic complexes produced by the crystallization of tholeiite basaltic magmas.

The compatible behavior of PGEs is mainly governed by their very intense extraction by sulfide melt immiscible with silicate magma. During the partial melting of mantle rocks, this heavy liquid remains in crystalline restite. One source of PGEs is chromitite lodes associated with ophiolite complexes. These ores are characterized by relative enrichment in metals of the Os-Ir group. The primitive mantle contains PGEs in nearly chondritic proportions, whereas tholeiitic basalts exhibit relative enrichment in Pd and Pt. This implies that Os and Ir are coprecipitated with early crystallizing chromite and residual melt is depleted in these metals. The exact mechanism of this process is a matter of debate. Some investigators suggest that PGEs may be coprecipitated with chromite in the form of microscopic grains of metallic phase (micronuggets), others maintain that Os and Ir may be present in chrome-spinelids in solid solution, or that these metals in isomorphous form, may enter, into early crystallizing olivine.

Very large reserves of PGEs are confined to layered intrusions of basaltic composition (Bushveld, Stillwater). These magmatic complexes were formed by the injection of large masses of mantle-derived magmas into the continental crust. The assimilation of crustal material is likely to be important for the formation of PGE deposits of this type (Irvine et al. 1983). Very probably, this is also true for sulfide lodes associated with komatiites and mineral deposits of the Norilsk type.

Mineral deposits of P, Nb, and many other incompatible components are associated with the products of differentiation of mantle-derived melts generated at low degrees of partial melting, such as alkali basalts and related magmas. These magmas are initially noticeably enriched in volatiles including CO_2 as compared to more columinous tholeiitic melts. It is possible that during the fractional crystallization the field of immiscibility of silicate and carbonate liquids is intersected. Liquid immiscibility is usually thought to be confined to compositions sufficiently rich in alkalies. Phonolitic and nephelinitic melts equilibrated with carbonate-rich liquids are strongly peralkaline (Hamilton et al. 1989; Wyllie et al. 1990). Carbonate-silicate liquid immiscibility was documented in melt microinclusions in silicate magmatic rocks from carbonatitic complexes. The crystallization of carbonate-rich immiscible melts separated from alkaline magmas results in the formation of crustal carbonatites.

A popular point of view holds that carbonatititc melts may extract some economically important components, which explains why carbonatites are an important source of some metals and minerals. This may be the case for REE (for LREE $C^{sil}/C^{carb}<1$ at $P>2-3$ kbar – Hamilton et al. 1989), and there are important carbonatitic deposits of REE (Mountain Pass, California, with ores containing bastnesite). Carbonatites are an important source of Nb (pyrochlore-$NaCaNb_2O_6F$). This seems to be strange, in view of data on the low solubility of HFSE in primary carbonatitic melts. This is corroborated by high C^{sil}/C^{carb} values reported by Hamilton et al. (1989) and not very high saturation Nb concentration in F- and Na_2CO_3-bearing carbonate melts (ca. 0.5 wt.%). It seems possible that at 0.n% level of Nb in carbonatitic melt pyrochlor-rich cumulates may be formed. Carbonatites are also a source of apatite, phlogopite, Ti-magnetite, perovskite, baddeleite (ZrO_2), and fluorite.

Primary magmas of alkaline series are relatively rich in many incompatible elements due to their generation at low degrees of partial melting. This enrichment is further augmented during prolonged fractional crystallization. As a result, residual magmas represented by nepheline syenites are capable of producing certain types of mineral deposits. One spectacular example of such deposits is huge bodies of apatite ore confined to the Khibina pluton of nepheline syenites. Alkaline intrusions are also the source of Nb (loparite = perovskite with $NaNbO_3$ in solid solution, the Lovozero massif), Zr (eudialite-rich bodies in many alkaline massifs), and REE (REE-rich apatites).

The genesis of apatite deposits associated with alkaline massifs was the subject of heated debate. Many investigators believe that they are the result of liquid immiscibility: magma splits into silicate and phosphate liquids. The experimental investigation of the Ne-Di-Ap system indicated that the two-liquid field is very narrow and is situated at very high temperatures far from compositions of natural magmas (Kogarko 1990). Furthermore, melt inclusions in Khibina apatite are represented by silicate melts with only 1% of P_2O_5 (Kogarko 1990). This is also an over-all average content of P_2O_5 in apatite-bearing intrusion, and it is in agreement with experimental data on apatite saturation in nepheline syenite melt. A logical conclusion is that apatite ores are cumulates formed during the crystallization of nepheline syenitic magma. The problem is how to accumulate apatite, which constitutes only a small part of the precipitating silicate minerals (nepheline, K-feldspar, aegirin). Kogarko suggested that silicate minerals settled to the bottom of the magma chamber, whereas very small apatite crystals remained suspended in the melt stirred by convection. In this way, apatite stays in a suspended state until late stages, when the increase in effective viscosity results in the termination of convection. This mechanism is an agreement with the very good sorting of apatite grains revealed by granulometric analysis.

A similar mechanism was probably operative during the formation of eudialite-rich and loparite bodies (Kogarko 1996).

References

Hamilton DL, Bedson P, Esson J (1989) The behaviour of trace elements in the evolution of carbonatites. Ibid 405–427

Irvine TN (1977) Origin of chromitite layers in the Muskox intrusion and other stratiform intrusions: a new interpretation. Geology 5: 273–277

Irvine TN, Keith DW, Todd SG (1983) The J-M platinum-palladium reef of the Stillwater complex, Montana: II. Origin by double-diffusive convective magma mixing and implications for the Bushveld complex. Econ Geol 78: 1287–1334

Keays RR (1995) The role of komatiitic and picritic magmatism and S-saturation in the formation of ore deposits. Lithos 34: 1–18

Kjarsgaard BA, Hamilton DL (1989) The genesis of carbonatites by immiscibility. In: Bell K (ed) Carbonatites. Genesis and evolution. Unwin Hyman, London, pp 388–404

Kogarko LN (1990) Ore-forming potential of alkaline magmas. Lithos 26: 167–175

Kogarko LN (1996) Geochemical models of supergiant apatite and rare metal deposits related to alkaline magmatism. Geochem Int 33(4): 129–149

Kogarko LN, Khapaev VV (1987) Formation of apatite deposits of the Khibina massif (Kola Peninsula). In: Parsons I (ed) Origins of igneous layering. NATO ASI Series, Reidel, Dordrecht, pp 589–611

Wyllie PJ, Baker MB, White BS (1990) Experimental boundaries for the origin and evolution of carbonatites. Lithos 26: 3–19

2.3 The Lower Mantle

L. S. Dubrovinsky, N. A. Dubrovinskaia, and S. K. Saxena

The lower mantle starts just below the major mantle discontinuity near 650 km (Fig. 38). The depth of this discontinuity varies, perhaps by as much as 100 km, and is variously referred to as the 650-km discontinuity or 670-km discontinuity. In recent Earth models there is a region of high sound velocity gradient for another 50–100 km below the discontinuity. This is probably due to phase changes, but it could also represent a chemical gradient. The lower mantle proper therefore does not start until a depth of about 750 or 800 km. Below this depth, the lower mantle is relatively homogeneous according to seismic data until about 300 km above the core-mantle boundary. If there is a chemical difference between the upper and lower mantle, then in a convecting dynamic mantle, the boundary will not be at a fixed depth.

The mineralogy of the lower mantle is hard to determine. According to arguments based on cosmic abundance, the major components of the lower mantle are magnesiowustite (Mg,Fe)O (Mw) and stishovite SiO_2 (St) (F. Birch's

Advanced Mineralogy, Volume 3
Mineral Matter in Space, Mantle, Ocean Floor, Biosphere, Environmental Management, and Jewelry (ed. by A. S. Marfunin)

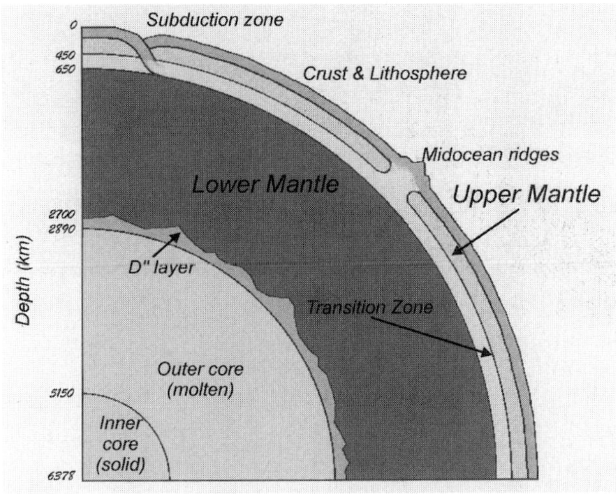

Fig. 38. Major elements of the Earth's interior according to geophysical data

model). According to A.E. Ringwood's hypothesis, the Earth's lower mantle consists primarily of $MgSiO_3$-perovskite ($MgSiO_3Pv$).

CaO and Al_2O_3 are likely to be the next most abundant components, but their concentrations are expected to be low, particularly if the material in the lower mantle has experienced low-pressure melting and removal of the basaltic components.

Several methods can be used to estimate the composition of the lower mantle from seismic data; perhaps the most direct is to compare shock-wave densities at high pressure of various silicates and oxides with seismically determined densities. However, minerals of quite different composition, say $(Mg,Fe)SiO_3$ Pv and mixture (Mg,Fe)O (Mw) + SiO_2 (St), can have identical densities, and mixtures involving different proportions of Mw and St can satisfy the density constraints. The structures of these minerals are shown in Fig. 39. Silicon atoms both in $(Mg,Fe)SiO_3$ Pv and SiO_2 (St) occupy octahedra. Mg and Fe atoms in magnesiowustite have coordination number 6. Orthorhombic distortions of cuboctahedra in $(Mg,Fe)SiO_3$ Pv decrease effective coordination number of (Mg,Fe) in this mineral to 8. That explains why $(Mg,Fe)SiO_3$ Pv and mixture (Mg,Fe)O (Mw) + SiO_2 (St) have close densities.

It has been shown that a chondritic and pyrolite composition for the lower mantle gives satisfactory agreement between density and seismic (PREM) data (Fig. 40).

Another approach for estimation of the mineralogical composition of the lower mantle is to extrapolate seismic data to zero pressure with the assumption that the lower mantle is homogeneous and adiabatic. A variety of equations of state are available that can be used to fit density, ρ, share moduli, G, isobaric bulk moduli, K_S, sound velocities, V_P and V_S, in the lower mantle, and the zero-pressure parameters can be compared with values inferred or measured for

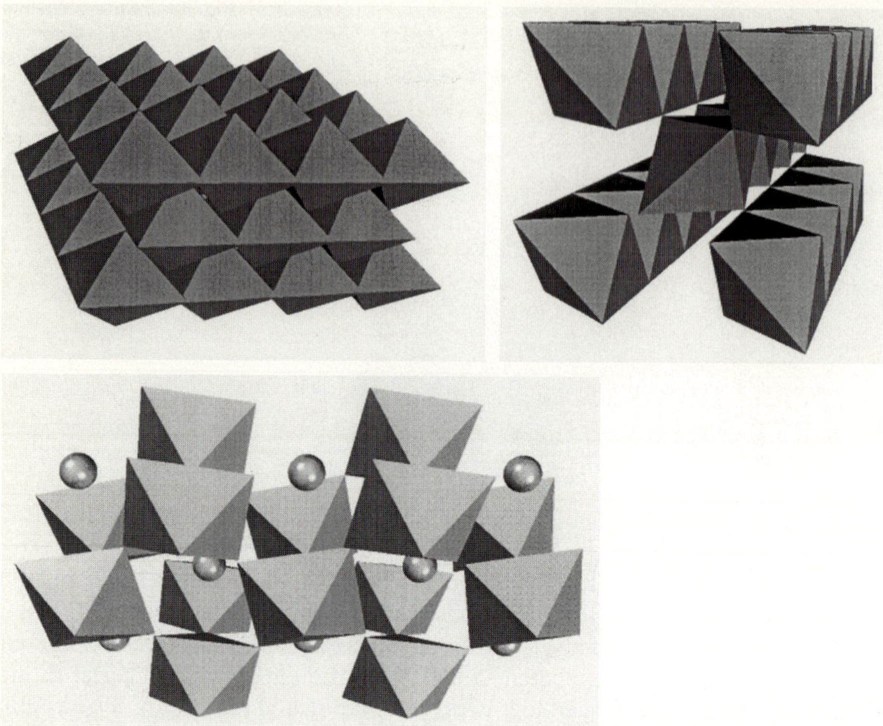

a b

c

Fig. 39 a–c. Structures of periclase (**a**) (octahedra MgO_6 are shown), stishovite (**b**) (SiO_6 octahedra), and $MgSiO_3$ perovskite (**c**) (SiO_6 octahedra; Mg-atoms (balls) occupy free space between octahedra)

various candidate minerals and compositions. The temperature corrections to be applied to the extrapolated lower-mantle values, are, of course, uncertain. Taking into account all uncertainties of this approach, $(Mg,Fe)SiO_3$ Pv or mixture Mw + St are good candidates for lower-mantle minerals.

It is clear that the most important „mineralogical" property of mantle minerals is their density. In order to derive the equation of state (EOS) from shock compression data, or from acoustic data obtained by ultrasonic or Brillouin scattering method, it is necessary to use high-temperature thermal properties (e.g., thermal expansivity) at ambient pressure. In the case of high-pressure minerals, such as perovskite, abnormal behavior is often observed when these minerals are heated to high temperature at ambient pressure. For such minerals, the ideal method for determining the equation of state is to measure the unit cell volume very accurately over a wide range of pressure and temperature conditions by in situ X-ray diffraction. Recently, P-V-T measurements of $MgSiO_3$ perovskite at pressures of 21–29 GPa and temperatures of 300–2000 K were carried out using MA-8-type apparatus (Fig. 41) which employed sintered diamond anvils, combined with synchrotron radiation. The thermal expansion at

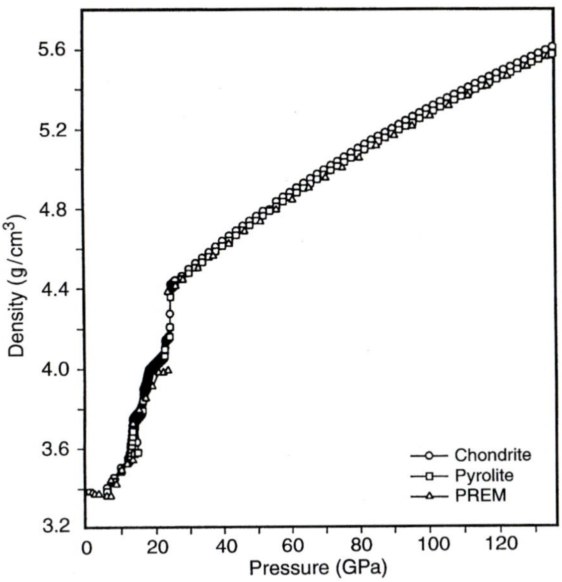

Fig. 40. Comparison of densities calculated and obtained from seismic data (PREM) densities. Concerning the Earth's mantle, both the pyrolite and the chondrite models for the $MgO-FeO-SiO_2$ system give a satisfactory approximation of the mantle's chemical composition

25 GPa up to 2000 K was determined from interpolation of the new P-V-T data. It is clear from Fig. 42 that the most reliable EOS obtained still has large uncertainty when large extrapolations in pressure and temperature are made. The lack of thermal expansion data to significant temperature at ambient pressure because of the instability of $MgSiO_3$ perovskite makes it difficult to determine the equation of state for perovskite. On the other hand, if accurate data can be

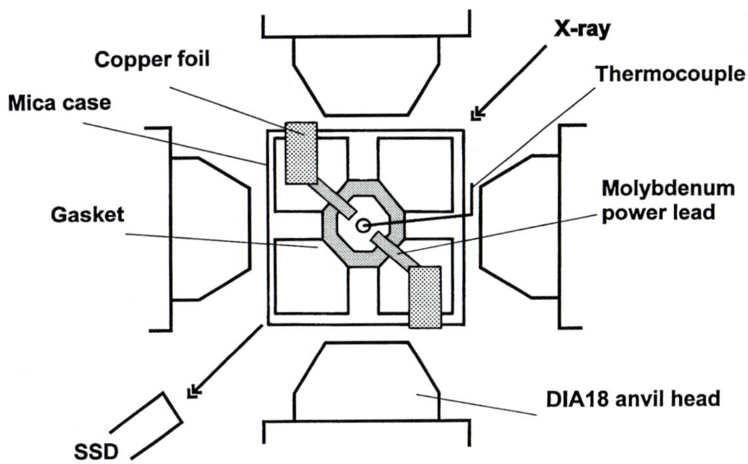

Fig. 41. Schematic diagram of MA-8 apparatus which have been used for in situ X-ray study of $MgSiO_3$ perovskite up to 30 GPa and 2000 K

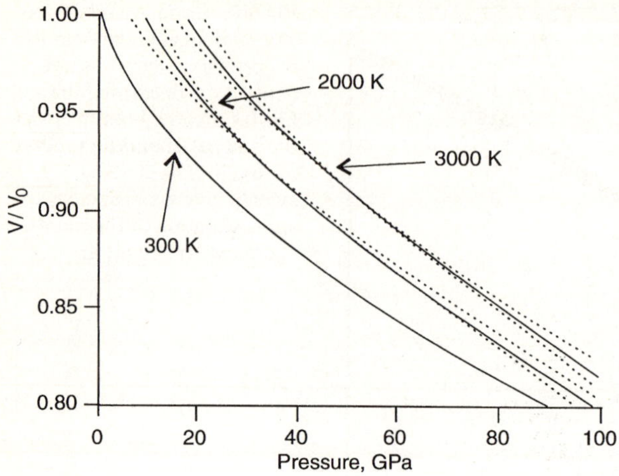

Fig. 42. P-V-T relations for $MgSiO_3$ perovskite. *Dashed lines* show uncertainty of extrapolations at high pressures and temperatures

obtained above 50 GPa, the accuracy of the determination will improve dramatically.

The lowermost 200 km of the mantle (Fig. 38), region D″, has long been known to be a region of generally low seismic gradient and increased scatter in travel times and amplitudes. There is a shear-velocity jump of 2.8% in this region that may vary in depth by up to 40 km. A large shear-velocity discontinuity exists at about 280 km above the core, in a region of otherwise low velocity gradient. The basic feature of a 2.75 – 0.25% velocity discontinuity is present for each of several distinct paths.

D″ may represent a chemically distinct region of the mantle. If so, it may vary laterally, and discontinuity in D″ would vary considerably in radius, the hot regions being elevated with respect to the cold regions. A chemically distinct layer at the base of the mantle that is only marginally denser than the overlying mantle would be able to rise into the lower mantle when it is hot, and sink back when it cools. The mantle-core boundary, being a chemical interface, is a region of high thermal gradient, at least in the colder parts of the lower mantle.

In the inhomogeneous accretion model the deep interior of the Earth would be initially rich in Fe and CaO-Al_2O_3-rich silicates. D″ may therefore be more calcium- and aluminum-rich than the bulk of the mantle. At D″ pressures this may be denser than the "normal" mantle.

Quite possibly, within D″, the old hypothesis of Birch (1952) that silicates at great depth decompose to oxides is valid; in fact, this agrees well with geophysical data. The D″ layer is often not singled out in the lower mantle, and it may therefore be said – at any rate at the base of the lowest mantle – that the Ringwood hypothesis about its perovskite composition is valid. The pressure dependence of melting of MgO has not been measured previously because of its high initial melting temperature of 3063 K, which prevented the use of thermocouples and conventional high-pressure techniques. However, recently,

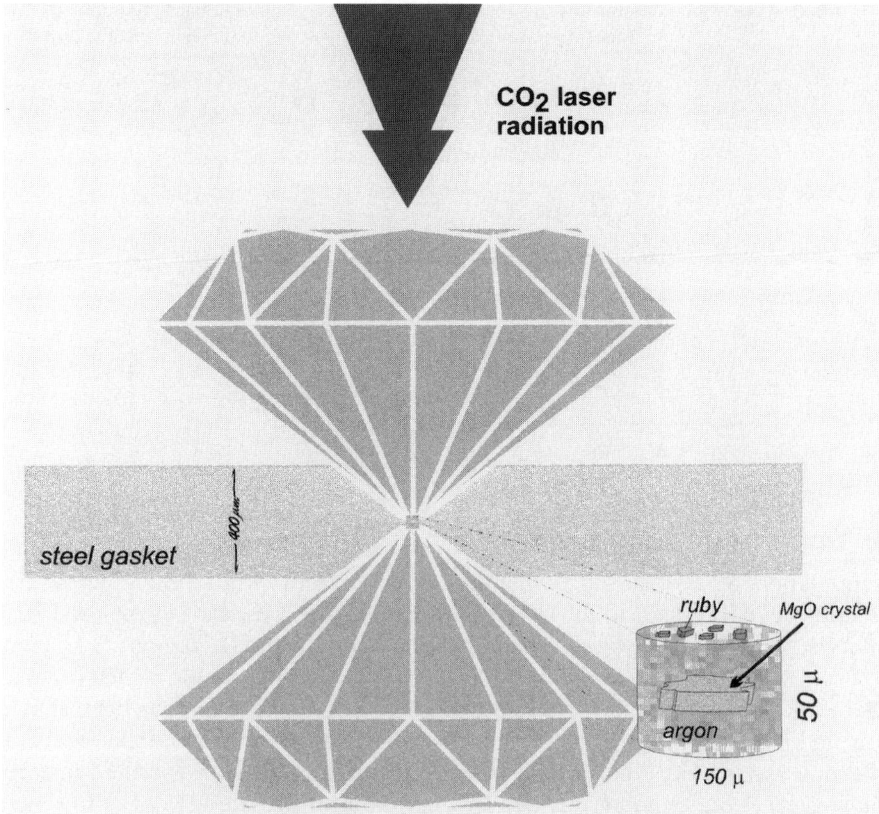

Fig. 43. Sample environment in the experiment on melting Mg in CO_2-laser-heated diamond anvil cell. A piece of MgO crystal is surrounded by argon. Ruby is used for pressure determination

CO_2-laser heating techniques and thermally insulated samples (Fig. 43) were used, and it was shown that the MgO melting curve crosses that of $(Mg,Fe)SiO_3$ perovskite at ~50 GPa. This implies that the MgO-SiO_2 phase diagram changes substantially at high pressures and that perovskite is not the low-melting phase in this system (Fig. 44).

The average temperature of the lower part of the lower mantle has been estimated to lie between 2550 and 2750 K. The thermal gradient across the core-mantle boundary is at least 1300 K, estimated from melting measurements of pure iron and on iron-oxygen mixtures. Therefore, the temperature at the very bottom of the lower mantle, in contact with the core, is on average ~4000 K, or still ~1000 K below the melting temperature of magnesiowustite. Melting of parts of the lower mantle cannot therefore be ruled out in the early history of the Earth, but would now require a substantially lower solidus temperature of the system, or additional heat sources such as frictional heating. Gravitational

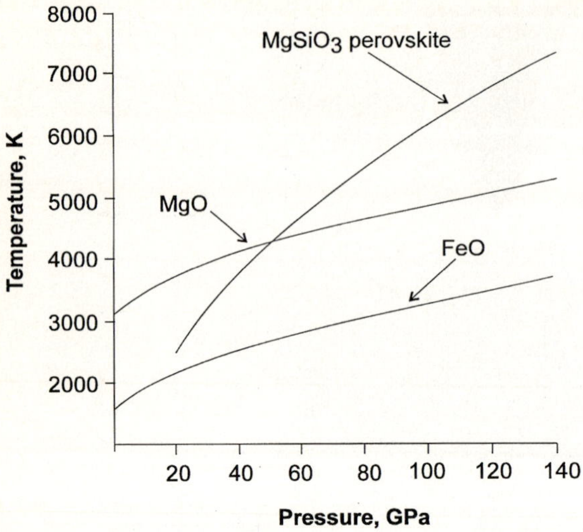

Fig. 44. Melting curves for MgO, FeO, and $MgSiO_3$ perovskite

segregation of iron-enriched magnesiowustite from $(Mg,Fe)SiO_3$ perovskite would inevitably lead to an iron enrichment at the very bottom of the lower mantle, thus providing an alternative explanation for the D″ region.

At lower mantle pressures (100–130 GPa), the major factor, which regulates relative stability of the different phases is their densities. It has been shown (using known compressibilities and systematics of the pressure dependence of thermal expansivity) that for the lowermost mantle, the iron-enriched magnesiowustite (Mg,Fe)O phase is denser than the iron-depleted $(Mg,Fe)SiO_3$ pervoskite phase by 2%; but the density of stishovite is still 2–5% lower than is necessary to stabilize the mixture periclase (MgO) + stishovite (SiO_2) with respect to $MgSiO_3$ perovskite. Such an appropriate high-pressure silica phase is not known experimentally, and theoretical methods must be applied to check if such phase (or phases) can exist.

Using pair potentials to model the interatomic interactions within the system, thermoelastic properties, phase relations, melting, and other related high-temperature phenomena may be simulated in a number of ways. Major theoretical techniques are molecular (MD) and lattice dynamics (LD). These theoretical methods are especially useful for modeling possible phases at physical conditions of the lower mantle. A new silica phase can be one example of successful predictions.

Various experimental and theoretical studies have suggested the possible existence of another, so-called post-stishovite phase – crystal chemical analog of AX_2-type oxides [such as α-PbO_2-, fluorite (CaF_2)-, cottunite (α-$PbCl_2$)-like phases, and baddeleyite (ZrO_2)-like phase] observed experimentally and the phases proposed for explaining the results of high-pressure experiments on SiO_2 (α-PbO_2, Fe_2N, disordered nykeline NiAs, and $CaCl_2$ structures). Despite all attempts, none of the possible post-stishovite phases with energy lower than

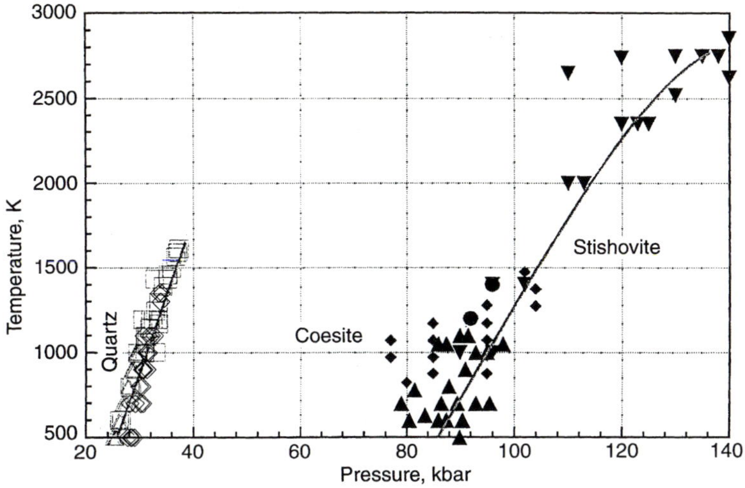

Fig. 45. Silica phase equilibria calculated by LD method (*lines*) compared to experimental data (◇ Akaogi and Navrotsky 1984; △ Bohlen and Boettcher 1982; □ Mirwald and Massonne 1980; ● Pacalo and Gasparik 1990; ◆ Suito 1977; ▲ Yagi and Akimoto 1976; ▼ Zhang 1992)

that of stishovite was simulated at pressures up to 150 GPa, except for the transition to the $CaCl_2$-like structure.

The following results were obtained by LD and MD methods with ab initio and empirical interatomic potentials starting form all the above described structural types considered to be reasonable high-pressure modifications of SiO_2:

1. The quasiharmonic lattice dynamic (QLD) allows us to reproduce the known part of the silica diagram in good agreement with experimental data (Fig. 45).
2. Fluorite-, α-PbO_2-, modified α-PbO_2- (with space group I2/a), and α-$PbCl_2$-like phases of SiO_2 are unstable with respect to stishovite up to pressures 150 GPa and temperature up to 2500 K. The phase closest to stishovite in stability is modified α-PbO_2 (with space group I2/a).
3. At pressures higher than 80 GPa and temperature 300 K according to QLD (100 GPa according to MD), orthorhombic distortion of stishovite unit cell gradually increases, which may be interpreted as a transition from stishovite to a $CaCl_2$-like structure. The value of distortion increases smoothly with pressure. This possible phase transition stishovite-$CaCl_2$-like modification of SiO_2 is probably of the second order.
4. Calculated pressure dependence of Raman frequencies under hydrostatic and nonhydrostatic conditions are shown in Fig. 46. Experimental results can be reproduced up to a pressure of about 40 GPa. Agreement with the experiment above that pressure can be achieved by assuming the presence of a nonhydrostatic stress about 1.5 GPa at 43 GPa and 2.5 GPa at 60 GPa

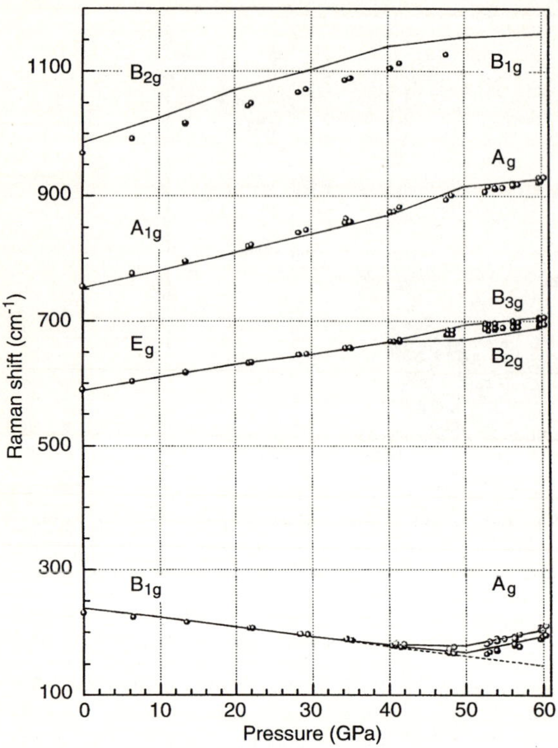

Fig. 46. Pressure dependence of calculated (*lines*) Raman frequencies of stishovite compared to experimental (*spheres*) data. Raman frequencies at pressures above 40 GPa are calculated assuming deformed stishovite ($CaCl_2$) structure under stress. The value of stress varied from 1.5 to 2.5 GPa at pressures from 40 to 60 GPa. The appearance of higher frequency branch of $B_{1g} \rightarrow A_g$ mode was calculated assuming the same stress causing c/a deformation of stishovite structure. Pressure dependence of $B_{1g} \rightarrow A_g$ without stress is shown by *dashed line*

with close agreement with estimation of possible nonhydrostatic stress in diamond anvil cell (DAC).

5. Starting form the baddeleyite structure, the structure with the Pnc2 space group was obtained (Fig. 47). This structure can be considered as an intermediate between α-PbO_2 (Pbcn) and baddeleyite ($P2_1/c$). Indeed, octahedra of this structure are gradually distorted and transformed to seven-apex polyhedra (characteristic of baddeleyite) during pressure decrease to 10 GPa. At pressures higher than 70 GPa, the density of the Pnc2 structure becomes greater than that of stishovite, and at 150 GPa and 300 K the difference amounts to 0.8%, at 150 GPa and 2000 K 1.6%, according to QLD. The Pnc2 structure becomes more stable than stishovite (or rather $CaCl_2$-like silica) at pressure above 90 GPa according to QLD. The transition to Pnc2 structure has weak dependence on temperature up to 3000 K. There-

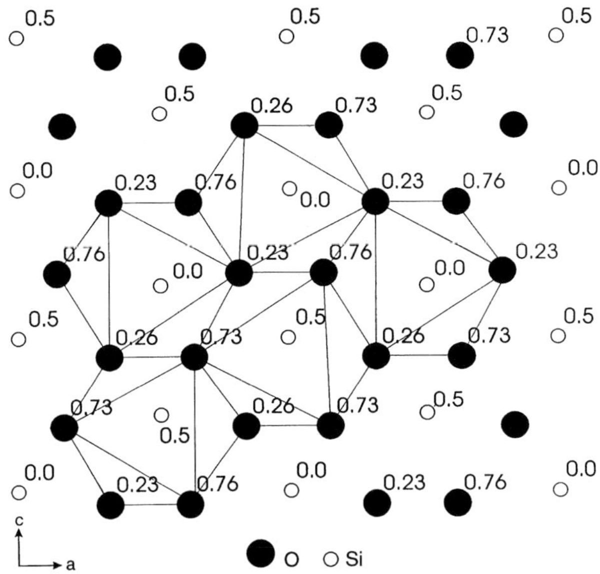

Fig. 47. Projection along the b-axis of the SiO_2 structure (space group C_{2v}^6-Pnc2) simulated with LD at 100 GPa and 300 K [a = 4.2275 A, b = 3.8900 A, c = 4.7225 A, Z = 4, Si(1) (0., 0., 0.145), Si(2) (0.5, 0., 0.8271), O(1) (0.8322, 0.2636, 0.3680), O(2) (0.6672, 0.7586, 0.1050)]. *Numbers* indicate positions of atoms along the vertical axis. This structure can be considered as intermediate between α-PbO2 (Pbcn) and baddeleyite (P21/c). *Thin lines* show octahedra with an arrangement similar to α-PbO_2 structure

fore, the new structure (space group Pnc2) is a good candidate to represent the silica phase in the Earth's lower mantle.

From the above considerations of lower mantle mineralogy, one may conclude that the major part of the lower mantle has (Mg,Fe)SiO_3-perovskite composition, but at the bottom a mixture of oxides may be more stable.

References

Belonoshko AB, Dubrovinsky LS (1995) Molecular dynamics of stishovite melting. Geochim Cosmochim Acta 59:1883–1889

Birch F (1952) Elasticity and constitution of the Earth's interior. J Geophys Res 57:227–286

Bohler R (1993) Temperature in the Earth's core from melting-point measurements of iron at high pressures. Nature 363:534–536

Dubrovinsky LS, Belonoshko AB, Dubrovinsky NA, Saxena SK (1996) New high-pressure silica phase obtained by computer simulation. In: High pressure science and technology. World Scientific Publ Co, Singapore pp 921–923

Duffy TS, Hemley RJ, Mao HK (1995) Equation of state and shear strength at multimegabar pressures: magnesium oxide to 227 GPa. Phys Rev Lett 74:1371–1374

Funamori N, Yagi T, Utsumi T, Kondo T, Uchida T (1996) Thermoelastic properties of $MgSiO_3$ perovskite determined by in situ X-ray observations up to 30 GPa and 2000 K. J Geophys Res 101:8257–8269

Jeanloz R, Hazen RM, MacCammon C (1993) Composition of limits of Fe_xO and the Earth's lower mantle. Science 261:923–925

Kingma KJ, Cohen RE, Hemley RJ, Mao HK (1995) Transformation of stishovite to a denser phase at lower-mantle pressures. Nature 374:243–245

Meade C, Mao HK, Hu J (1995) High-temperature phase transition and dissociation of $(Mg,Fe)SiO_3$ perovskite at lower mantle pressures. Science 268:1743–1745

Navrotsky A (1994) Physics and chemistry of Earth materials. Cambridge University Press, Cambridge, 417 pp

Ross NL, Hazen RM (1990) High-pressure crystal chemistry of MgSiO3 perovskite. Phys Chem Minerals 17:228–237

Saxena SK (1996) Earth mineralogical model: Gibbs free energy minimization computation in the system MgO-FeO-SiO_2. Geochim Cosmochim Acta 60:2379–2395

Stacey FD (1992) Physics of the Earth. Brookfield Press, Brisbane, Australia 513 pp

Stixrude L, Bukowinski MST (1993) Thermodynamic analysis of the system MgO-FeO-SiO_2 at high pressure and the structure of the lowermost mantle. In: High-pressure research in mineral physics. Terra Scientific Publ Co, Tokyo, American Geophysical Union, Washington, DC Geophysical Monograph 74, IUGG vol 14, pp 131–141

Stixrude L, Hemley RJ, Fei Y, Mao HK (1992) Thermoelasticity of silicate perovskite and magnesiowustite and stratification of the Earth's mantle. Science 257:1099–1101

Wang Y, Weidner D (1994) Thermoelasticity of $CaSiO_3$ perovskite and implications for the lower mantle. Geophys Res Lett 21:895–898

Wang Y, Weidner DJ, Lieberman RC, Zhao Y (1994) P-V-T equation of state of $(Mg,Fe)SiO_3$ perovskite: constraints on composition of the lower mantle. Phys Earth Planet Inter 83:13–40

Weidner DJ (1986) Mantle model based on measured physical properties of minerals. In: Saxena SK (ed) Advances in physical geochemistry. Springer, Berlin Heidelberg New York, pp 251–274

Yagi T (1994) Recent advances in high pressure and high temperature in-situ X-ray studies using sintered diamond and synchrotron radiation. Proc 14th AIRAPT Conf, Colorado, pp 1621–1624

Zerr A, Reichmann H, Euler H, Boehler R (1993) Hydrostatic compression of γ-$(Mg_{0.6},Fe_{0.4})_2SiO_4$ to 50 GPa. Phys Chem Minerals 19:507–509

2.4 The Earth's Core

S. K. Saxena, and L. S. Dubrovinsky

Seismic data show that the Earth's core consists of an outer part, which has all the properties of a liquid, and the inner core, which is solid. From cosmochemical, geophysical, and geochemcial considerations, the dominant species in the core has to be iron (over 90%). Although not experimentally shown, but based on the available data on the compressibility of iron, it appears that iron itself has a density higher than the density of the core. Shock-wave data demonstrate that the density of the outer core is 8–11% below that of Fe-Ni under

Advanced Mineralogy, Volume 3
Mineral Matter in Space, Mantle, Ocean Floor, Biosphere, Environmental Management, and Jewelry (ed. by A. S. Marfunin)

similar pressure and temperature conditions, implying the presence of 5–16% of a light element. Sulfur (S) was suggested as a single light element of the core; but sulfur is a moderately volatile element and its abundance in the bulk Earth is related to the abundance of other similarly volatile elements. Arguments are presented that the content of sulfur in the bulk Earth is similar or even lower than that of zinc (Zn). This leads to an upper limit for the bulk Earth S-content of 0.56% and S-content for the core of 1.7%. This is not enough to decrease the density of pure iron at the pressures and temperatures of the Earth's core. Therefore, the Earth's core may contain several light elements. However, dilution of iron by addition of sulfur or another light element is not essential. The outer core is liquid because of the phase transformations in iron and its melting behavior. Thus, the problem of the Earth's core condition is a problem of relationships of iron phases at high temperature and pressure.

Figure 48 shows a summary of the currently available experimental data, until the year 1993, on phase equilibrium relations obtained from static devices, i.e., techniques with in-situ heating of samples under pressure sustained over a period of time. The phase relation boundaries α-γ, γ-ε and α-ε meet at one point (triple point). Recent theoretical study on iron properties at high-P-T has shown that an experimental determination of the data on ε-γ phase transformation and the ε-γ-melt triple point are particularly crucial in extending our currently available results to understand the state of the Earth's core. The argument is that the pressure-temperature of the triple point determines the behavior of the melting curve at high pressures. Thermodynamically, the pressure-tem-

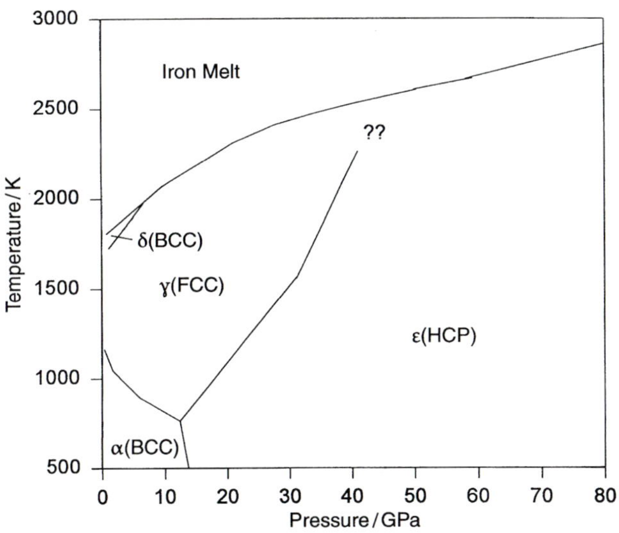

Fig. 48. Phase diagram of iron form the data available up to 1993. The various curves drawn do not include the new experimental data. The stability regions are shown for the four iron polymorphs: α (bcc), δ (bcc), γ (fcc), ε (hcp)

perature slope of the ε-γ phase equilibrium curve constrains the estimated enthalpy, entropy, and other physical properties of the ε phase which are largely unknown.

A triple point, similar to the one shown in Fig. 48, should exist for the coexisting γ, ε, and melt. All the available experimental data on the ε-γ transition appear to indicate that such a triple point would lie around 2700 K between 60 and 70 GPa. The problem is that shock-wave data by Brown and McQueen had located a solid-solid transition at 4400 ± 500 K at a pressure of 200 ± 2 GPa. As can be seen in Fig. 48, this phase transition could not be the ε-γ phase transformation. The triple point is located at too low pressure for this to be possible. This apparent inconsistency between the sets of data led Boehler to conclude that there should exist a fifth iron phase. The existence of this phase is supposed to solve two problems. First, it would explain the phase transformation recorded by Brown and McQueen in their shock-wave studies, and second, by reacting a third triple point, it would let the melting curve slope appropriately fit the existing geophysical data. This led to the start of a search for a new phase of iron and determination of the iron melting at pressures higher than that at the core/mantle boundary (130 GPa). Achieving such extreme pressures in the laboratory requires special techniques, for example, the laser-heated diamond-anvil cell.

The diamond-anvil cell experimental setup for generating high pressure is discussed in several articles. Here a brief description of experimental setup specific to Uppsala Laboratory is given (Fig. 49). The use of a high-power

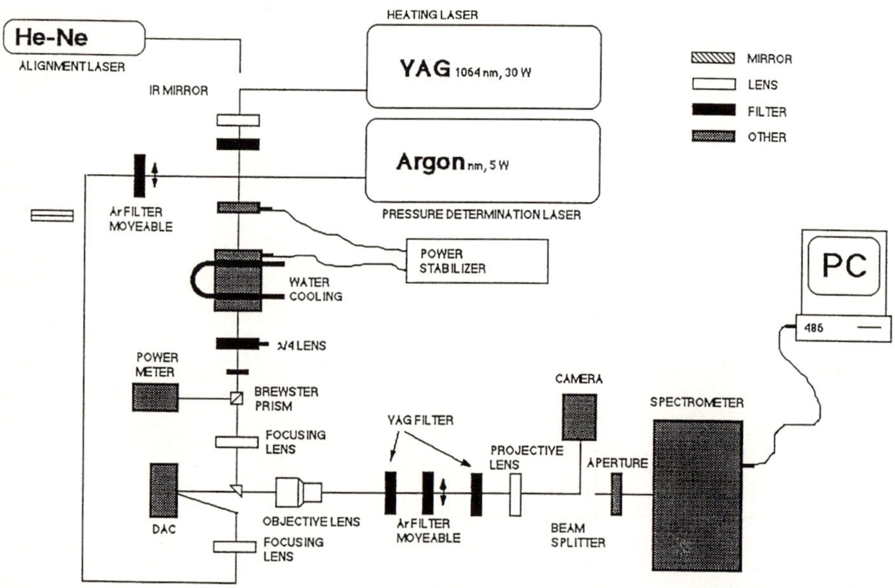

Fig. 49. Experimental setup for the study of iron in a laser-heated diamond-anvil cell

Nd:YAG laser operating in CW TEM_{00} mode at a wavelength of 1064 nm is required. The laser provides maximally 35 W of vertically polarized light with a feedback-controlled stability better than 0.5% peak to peak. To achieve real high pressure beyond 100 GPa, one needs to use the standard Drukker type IA beveled diamond anvils with 0.3-mm diameter culet faces. For pressure determination, the calibrated pressure shift of ruby R1 fluorescent line is used. Melting is determined by visual observation when possible and by plotting the laser power against the temperature. At the melting temperature, there is a distinct change in slope. The temperature is determined by spectroradiometry. The 50 μ in diameter slit of the spectrograph collects the radiation from an area of 3 μ in width of the sample. A two-dimensional CCD detector is used to obtain temperatures measured over the entire length (30 μ) of the slit area. The temperature gradient over the central area (3 μ) is only a few degrees. The laser power fluctuations are minimized to a root mean square value $<0.5\%$ by using a laser power stabilizer. Iron is surrounded by thermal insulators such as oxides and silicates. An area of 3 μ^2 is sampled by the spectrograph to measure pressure; this reduces errors in pressure determination resulting form the presence of pressure gradient in the chamber. Prior to measurement, a chosen spot is heated by laser to relax the sample mechanically. The pressure usually drops to 15% of the initial unrelaxed value. The pressure drop is almost linear and can be stabilized by repeated heating. The following problems are cited with the diamond-anvil cell experiments: (1) reaction with the pressure medium and (2) temperature and pressure gradients. In experiments with iron, the cells are usually dried carefully. The oxide pressure media (Al_2O_3, MgO, BaO etc.) show no visible reaction with iron. At low pressure, melting can be registered by visual observation, but at high pressures (above 40–50 GPa) visual observations become difficult and then one must use the laser-power/temperature function which changes slope during the melting.

In Fig. 50, results of some recent experiments are displayed. The figure shows the stability of a new phase (β-Fe) and melting of iron to a pressure of 200 GPa. The issue of the new phase is not fully resolved. It is likely that we have many additional structural modifications of iron. In-situ phase transformation of α-β transformation of Fe confirm that β-Fe is a thermodynamically stable phase and it can from the Earth's core if no other more stable from of iron is be found; but it is likely that a higher density phase of iron exists.

The new experimental data can be used to obtain an internally consistent thermodynamic database to model the iron phase diagram and the core energetic. Figure 51 shows the calculated phase transformations and melting in iron at pressure from 0 to 350 GPa. Computational models permit decrease in melt compressibility at a higher rate at high temperatures by including the lower limits of the shock-wave data from Brown and McQueen. On the simulated melting curve there are two triple points; the first (melt-β-γ) is located at 2746(100) K and 56.5(4) GPa; the second (ε-θ-melt) is located at 4080 K and 196 GPa, ensuring that the shock-wave constraint on melting is satisfied.

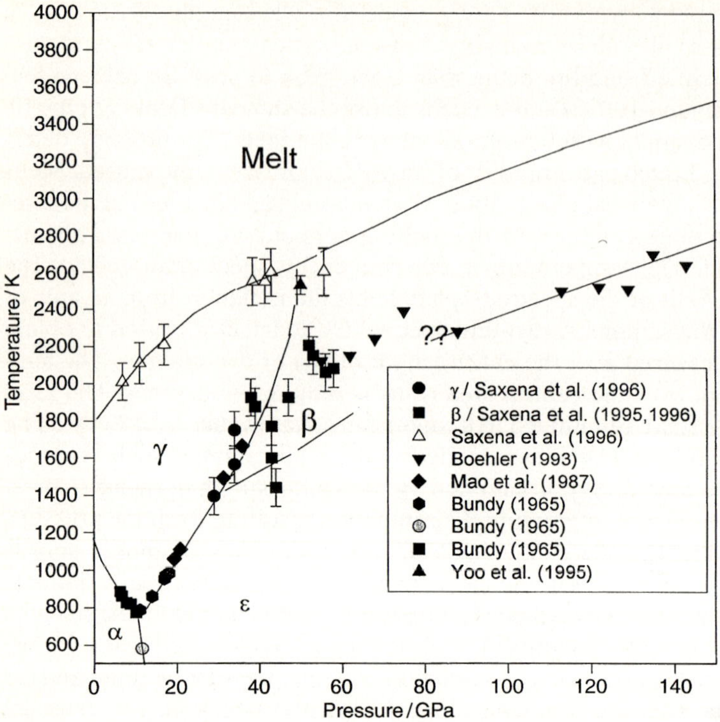

Fig. 50. Phase diagram of iron from experimental data available up to 1996. Only some important experimental data are shown. Although recent studies indicate that a new iron phase with a dhcp structure may be stable, we need many more data to establish its field of stability. The melting curve now extends to nearly 200 GPa

A model of the Earth's core should (1) explain why the inner core is solid and the outer core liquid, (2) use a chemical composition which is consistent with cosmochemical data, and (3) be similar in intensity to the seismic density. We note that the answer to the first problem lies in the special melting characteristics of iron associated with phase transformation and the location of the triple point ε-β-melt. The calculated melting curves in Fig. 51 show that we may expect iron melting to occur at central core pressures (363.85 GPa, PREM model) between 6000 and 6700 K. Then, if we assume that the temperature in the inner core is largely adiabatic, we may start the calculation of adiabat, starting where the pure iron melting curve intersects the inner-core/outer-core seismic discontinuity. The calculated temperature at the Earth's center is 6150 K.

The experimental iron phase diagram and the thermodynamic data result in a rather simple model for the core. The relationship between the Earth and meteorites has been discussed thoroughly in the literature. The class of meteorites called irons and stony-irons is considered to be debris from differentiated broken-up planetary bodies. What the present results tell us is that this picture

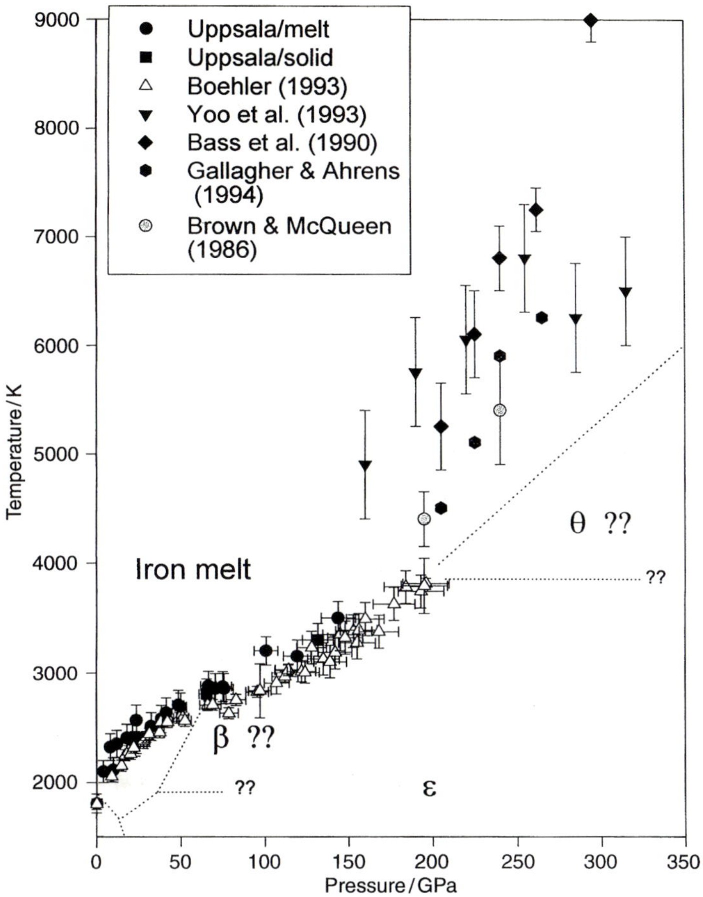

Fig. 51. Iron phase diagram (*lines*) calculated form the thermodynamic data. Hugoniot data (*diamonds*, *hexagons*, and *circles*) are shown for comparison with the static pressure melting data (*triangles* and *inverted triangles*) extrapolated with modeled thermodynamic data. The two points of Brown and McQueen (1996) signify a solid-solid and solid-melt transition. The high entropy-enthalpy hypothetical θ phase can be stable with a triple point θ-β-melt between 190 – 200 GPa, as suggested by Anderson (1993). Such a phase could have a significantly high melting temperature

fits the model of Earth's core well. Thus, the core has simply the same composition as the irons and stony-irons. The density from such compositions fits the seismic density well. Dilution of iron addition of sulfur or another light element may not be essential. The outer core is liquid because of the phase transformations in iron and its melting behavior. Current research on fluids opens the possibility that it may be liquid because of the dissolution of fluid species in molten iron.

References

Anderson OL (1993) Physics of iron. Conf Proc for the Association for the International Advancement of High Pressure, Colorado Springs, Colorado 28 June – 2 July 1993, Lawrence Livermore National Laboratory, pp 123 – 127

Anderson OL (1995) Mineral physics of iron and of the core. Rev Geophys Suppl: 429 – 441

Boehler R (1993) Temperatures in the Earth's core form melting-point measurements of iron at high static pressures. Nature 363 : 534 – 536

Brown JM, McQueen RG (1986) Phase transitions, Grüneisen parameters and elasticity for shocked iron between 77 GPa and 400 GPa. J Geophys Res 91 : 7485 – 7494

Chen GQ, Ahrens TJ (1995) High pressure melting of iron – new experiments and calculations, Lindhurst Laboratory of Experimental Geophysics, Seismological Laboratory, California Institute of Technology, Pasadena, California, pp 1 – 28

Goarant F, Guyot F, Peyronneau J, Poirier J-P (1992) High-pressure and high-temperature reactions between silicates and liquid iron alloys, in the diamond anvil cell, studied by analytical electron microscopy. J Geophys Res 97 : 4477 – 4488

Karato S (1993) Inner core anisotropy due to the magnetic field-induced preferred orientation of iron. Science 262 : 1708 – 1711

Poirier JP, Shrankland TJ (1993) Dislocation of iron and the temperature of the inner core, revisited. Geophys J Int 115 : 147 – 151

Saxena SK, Dubrovinsky LS (1996a) Iron phases at high pressures and temperatures: phase transition, melting and thermodynamics, US-Japan Seminar on High Pressure-Temperature research: properties of Earth and Planetary Materials, Jan 22 – 26, 1996, AGU monograph. American Geophysical Union, Washington, DC

Saxena SK, Dubrovinsky LS (1996b) Nd-YAG laser heating temperature measurements in a diamond anvil cell. In: Advanced materials '96. Proc 3rd NIRIM ISAM '96, Tsukuba, Japan, March 4 – 8, 1996, pp 137 – 142

Saxena SK, Shen G, Lazor P (1993) Experimental evidence for a new iron phase and implications for Earth's core. Science 260 : 1312 – 1314

Saxena SK, Shen G, Lazor P (1994) Temperatures in Earth's core based on melting and phase transformation experiments on iron. Science: 254 : 405 – 407

Saxena SK, Dubrovinsky LS, Haggkvist P, Cerenius Y, Shen G, Mao HK (1995) Synchrotron X-ray study of iron at high pressure and temperature. Science 269 : 1703 – 1704

Saxena SK, Dubrovinsky LS, Häggkvist P (1996) X-ray evidence for the new phase β-iron at high temperature and high pressure. Geophys Res Lett 23 : 2441 – 2444

Shen G, Lazor P, Saxena SK (1993) Melting of wüstite and iron up to pressures of 600 kbar. Phys Chem Minerals 20 : 91 – 96

Yoo CS, Holmes NC, Ross M (1993) Shock temperatures and melting of iron at earth core conditions, Phys Rev Lett 70 : 3931 – 3934

Yoo CS, Akella J, Campbell AJ, Mao HK, Hemley RJ (1995) Phase diagram of iron by in situ X-ray diffraction: implications for Earth's core. Science 270 : 1473 – 1475

CHAPTER 3

Mineralogy and Mineral Resources of the Ocean Floor

3.1 Stages of the Great Discoveries in the Ocean's Geology, Metallogeny, and Mineralogy

G. N. Baturin

The pioneering HMS *Challenger* expedition of 1872–73 opened the era of great geologic discoveries on the sea floor. Unusual mineral deposits were among the remarkable findings (Murray and Renard 1891). On 18 February 1873, 160 miles southwest of the island of Ferro in the Canary Island Group, the Challenger recovered deep-sea manganese nodules enriched in metals such as copper, nickel, and cobalt. Similar nodules were subsequently recovered in many other places throughout the Atlantic, Indian, and especially Pacific oceans. At shallower depths along the continental margins the Challenger also recovered numerous samples of phosphorite, a sedimentary deposit that was known from marine sedimentary strata on land. Just at the time of the Challenger Expedition, the first such rock deposits, from South Carolina, began to be sold internationally as an alternative to then-predominant guano deposits as phosphorus fertilizer.

In ensuing decades, many other oceanographic expeditions expanded knowledge of the ocean floor, each usually focusing on special geographic areas or ocean phenomena. None has duplicated the monumental scope and breadth of the *Challenger's* investigations and the treatises that documented them.

A new period of geologic investigations in the Ocean began after World War II, along with general efforts to increase knowledge of the oceans for scientific and economic reasons, and because of military rivalries. During the next 40 years, large scientific fleets were equipped with sophisticated devices capable of carrying out detailed geological and geophysical investigations in the remotest and deepest parts of the Ocean. They used acoustic bottom mapping (bathymetry), magnetic and seismic profiling, and sample recovery by cores, grab samplers, and manned submersibles.

Probably the most important achievements in investigation of the geologic history of the oceans have been made thanks to the deep ocean drilling programs, the first of which was launched as the Deep-Sea Drilling Project in 1968, sponsored by the US National Science Foundation. In 1984, the successor, Ocean Drilling Program (ODP), became intentional in sponsorship. By the end of 1995, more than 1000 holes had been drilled, penetrating up to 1000 m below the sea floor. The cores and numerous geophysical measurements performed in the boreholes allowed unprecedented possibilities to decipher the history of tectonic movements, evolution of sedimentation, supply of metals, fluctuations in biological productivity, and global climate.

These enormous research activities led to a series of fundamental scientific discoveries. Among the most important ones were sea-floor spreading, involv-

The author wishes to thank Prof. F. T. Manheim for helpful discussions and advice.

ing the evolution of young ocean crust at spreading centers, and propagation of rifting movements of oceanic and continental plates. Others included the development of seismic stratigraphy, which permitted development of a world-wide delineation of both sedimentary sequences, as well as the effect of ocean level fluctuations that created major transgressive and regressive cycles. Micro-faunal observations showed that fossil organisms mutated and evolved at rates that permitted the dating of ocean sediments to within a million years. Formation of massive stagnation periods, and evaporative sequences, forming brine and salt deposits could be studied on a regional and historic basis.

Ferromanganese Oxide Deposits

In the area of mineral deposits, the most important development was the discovery of marine manganese nodules and crusts, which formed sea-floor resources rivaling base metal deposits on the continents for a number of metals. Following the enthusiastic advocacy of John Mero, in a series of articles dating from 1959 (Mero 1965) a number of national and international consortia formed to develop the special technologies needed to recover nodules at 5000 m water depth, and to process these for manganese, copper, cobalt, and nickel. National claims were registered with the United Nations for future nodule exploitation in the Clarion-Clipperton area of the Eastern Equatorial Pacific. The extensive early studies of Soviet scientists, under the leadership of P.L. Bezrukov, during the period 1960–1975, were likewise stimulated by the expected economic potential of nodules.

In the late 1960s another type of ferromanganese deposit was reported. This was peculiar manganese "nodules" that formed in the upper parts of seamounts (subsea mountains). They were especially noted by the English marine geologist, D.S. Cronan (1977), in the Pacific Ocean. It was to solve the riddle of the seamount "nodules" that the German Ministry of Technology, together with the German National Science Foundation, supported a special cruise to the Central Pacific seamount areas by the research vessel, *Sonne* in 1981. Although nodule-like oxide concretions occured on the seamounts, this investigation discovered that, unlike the deep-ocean nodules that formed on or embedded in soft sediment, the seamount deposits were actually blanketlike crusts formed over hard substrates (Halbach et al. 1982). By this time, political disagreements about how exploitation of nodules in international waters should be managed had helped discourage mining in international waters. Unlike the deep-ocean nodules, most of the valuable ferromanganese crust deposits occurred within national economic jurisdiction (200 nautical miles). Their high cobalt content (up to 2%) therefore led to increased interest in them in ensuing years. Investigations by detailed dredging, bathymetric mapping, and observation by manned deep-sea submersibles identified the largest cobalt concentrations in seamounts of the Line Islands trend, SE of Hawaii, and the Marshall Islands (Hein et al. 1987; Manheim and Lane-Bostwick 1989; Bogdanov et al. 1990).

In recent years, the Asian nations, Japan, Korea, and, most recently, China, have taken active interest in crust deposits.

Hydrothermal Discharge and Ore Deposits

The shattering of traditional postulates of classical geology, brought about by the discovery of the global rift systems, led to other startling findings. The new global tectonics made it possible to reconstruct former positions of the continents and oceans in the course of planetary history. It stimulated new understanding of the nature of hydrothermal activity of modern and ancient oceans. The essence of this new idea was first formulated by J. Corliss (1971), though Skornyakova (1964) and Bostroem and Peterson (1966) already deduced the action of hydrothermal exhalation from anomalous sediment compositions on the East Pacific Rise. Corliss postulated that seawater penetrates into porous and brecciated rift basalts as deep as several kilometers. It is then transformed into high-temperature acidic hydrothermal metalliferous solutions that extract metals from surrounding rocks. Its subsequent upward migration leads to spasmodic ejections and formation of hydrothermal plumes, metalliferous sediments, and massive sulfides. Mapping of metal distribution of Pacific Ocean sediments and crusts permits indirect delineation of areas where iron and manganese are enriched due to hydrothermal discharge. In contrast to the nodules and cobalt-rich crust areas, the more oxidized, hydrothermally formed sediments and crusts are depleted in valuable metals. Metals such as zinc, copper, and silver have, however, been found enriched to significant degrees in the sulfide deposits.

Before the phenomenon of hydrothermal discharge from the sea floor was actually observed, the role of volcanism in the ocean had already had a long history. John Murray and his coworkers in the late 1800s had already understood the powerful influence of volcanism on the deep ocean floor. Pumice and ash from volcanic eruptions were widely distributed, and remains of basalt flows were dredged from many ocean areas. A special episode of this history is associated with the Russian admiral, Makarov, in an expedition of the *Vityaz* in the Red Sea in the 1880s. While sampling the water in the central part of this semiisolated water body, Makarov found unusually warm and saline water in the deepest water sampled by bottles, at 600 m water depth. In the next 80 years, the phenomenon was confirmed but not explained.

In 1964 the English oceanographic vessel, *Discovery* found not only hot brines, but evidence of hydrothermal deposits in the Red Sea deeps (Charnock 1964). In 1965 an American expedition on board the Atlantis II (Woods Hole Oceanographic Institution) revisited the area. This time, extensive investigations found a sequence of brilliantly multicolored sediments, several meters thick, beneath a large brine pool. The results were described in a major monograph, edited by E. T. Degens and D. A. Ross (1969) and later probed in even greater detail by German investigations (Baecker and Schoell 1972) that later evolved into a mining venture together with Saudi Arabia.

These sediments consisted of alternating layers of hydrous iron oxides, sulfides, along with other phases like anhydrite. The bottom of the rift zone was floored by basaltic rocks, believed to provide the heat for the hot brine pool. The hypersaline salt concentrations were the result of dissolution of evaporites, and the metals were believed to be leached from the thick sediment deposits on the floor of the Red Sea basin by the brines.

In the same year as the *Discovery* finding K. Zelenov published the results of this observations on the subsea slope of the Indonesian volcano, Banu Wuhu, where he found "live hydrothermal solutions enriched in metals" (Zelenov 1964). He interpreted these as indication of the possible worldwide occurrence of such phenomena.

Early interpretations of the Red Sea deposits tended to regard the hydrothermal sulfides as unique, protected from oxidation by the layer of hot brine. Such brine layers would not be expected in the open ocean, it was thought. Given these assumptions, some disseminated sulfide occurrences on the bottom of the open ocean seemed anomalous. One such finding had been made in 1967 during the cruise of the Russian research vessel *Akademik Kurchatov* in the NW Indian Ocean, in a triple-junction rift area: Hydrothermally altered rocks with sulfide stockwork mineralization were recovered. Thin veins permeating this rock consisted of pyrite, chalcopyrite, ilmenite, covellite, and hematite. The character of the chalcopyrite (copper-iron ore) crystal twinning suggested a temperature as high as 500 °C. Later, in 1972, another Russian expedition, on board the R/V *Dmitri Medeleev* recovered 1 ton of hydrothermally altered rocks from the bottom of the Hess Deep in the Equatorial Pacific. These included brecciated gabbrodolerites with sulfide cement and calcareous sands, consisting of Recent foraminifera shells filled with pyrrhotite, troilite, and marcasite. These mineral assemblages, too, were interpreted as being Recent, and formed at temperatures close to 450 °C.

After traces of hydrothermally activity in seawater were found in 1975 in the Galapagos rift area, a series of investigations revealed huge plumes of hydrothermal suspensions enriched in Mn, Fe, and Cu, as well as ^{3}He, interpreted as being outgassed from the igneous mantle. The interest raised by these findings led to a detailed investigation by the American submersible, *Alvin*, in 1977. A monograph on hydrothermal processes on the sea floor describes the early findings in detail (Rona et al. 1983). *Alvin* dives found mounds up to 30 m high, made of ferruginous mud covered with black manganese hydroxides. Interstitial waters recovered from these mounds were warm, proving the presence of hydrothermal activity. The unexpected phenomena that electrified the scientific world, however, were an unprecedented profusion of crabs and gigantic mollusks, supported directly or indirectly on abundant bacterial populations that used hydrogen sulfide solutions rather than conventional organic matter as their nutrient source.

Subsequently, hydrothermal sediments consisting of talc with sulfide inclusions were found in the Baha California rift valley. The next year, 1978, marked the beginning of another long series of discoveries, the first being sulfide chimneys up to 5 m in diameter and up to 10 m high, observed in a Franco-American-Mexican expedition near the southern end of Baha California. In this expedition the French submersible, *Cyana*, observed a veritable forest of such chimneys, many of which discharged metalliferous fluid at a temperature of 375 ± 25 °C. The earlier-described, anomalous "vent" fauna was likewise observed. Since then, massive sulfides of various types have been found in many areas throughout the World Ocean, including the Juan de Fuca Ridge (1982), back-arc spreading centers in the Western Pacific, the southwest and northern Mid-Pacific Ridge, and the Mid-Atlantic Ridge (Rona et al. 1993).

A series of experimental studies on pressure-volume-temperature relations in seawater in simulated magma chamber conditions (see Bischoff 1980) revealed fundamental physicochemical constraints on the hydrothermal vent phenomena. Metal solubilization at 21°N, East Pacific Rise, was interpreted to take place only above 400 °C, with effective water/rock ratios in the geothermal system being between 7 and 16. Under these conditions, effective metal transport is ineffective at temperatures below 350 °C. Later studies have shown that considerable metal is transported in vapor phase at temperatures above 400 °C. Thus, the largest deposits of sulfides will be near to or on the sea floor in cavities, lava tubes, and evacuated lava lakes in the top few tens of meters below the sea floor. Anhydrite and barite form as the hydrothermal fluid mixes with sulfate-rich bottom waters. Fe greatly exceeds the heavier metals, Zn and Cu, in the original fluids. Therefore, where sphalerite or copper mineralization is found, either pyrite or pyrrhotite must predominate in the subsurface. The hydrothermal emanations channeled through flanking sedimentary formations, or formed late in the rift zone hydrothermal cycle tend to be low in trace element concentrations. Black "smokers" tend to have a pyrite-chalcopyrite mineral composition, whereas white smokers tend more to sphalerite-marcasite paragenesis. Concentrations of gold and silver as high as 12 ppm Au and 200 ppm Ag have been observed in the Mid-Atlantic Ridge TAG hydrothermal field. Spreading rates in lateral directions have been estimated at about 1.2 cm/year, with about 55 km between spreading segments (Rona et al. 1993).

Among the major scientific conclusions from the ocean investigations of hydrothermal deposits is that these close resemble sulfide deposits or their oxidation products (gossans) in Cyprus, Oman, Turkey, Japan, the Ural Mountains, and in Newfoundland. The genesis of these deposits is therefore interpreted as resulting from earlier sea-floor hydrothermal processes. Although the economic potential of the oceanic polymetallic sulfide deposits has been a subject of much interest, and deposits as large as 25 million tons have been estimated, so far no serious industrial interest in the deposits has emerged.

Phosphorites

For nearly a century after their discovery, it was widely assumed that oceanic phosphorites lying exposed on the sea floor were Recent in age. However, radiometric dating of samples from most known occurrences in a widely cited study from 1969 showed that all exceeded 1 million years in age (Kolodny 1969). This posed a dilemma, since there appeared to be no contemporaneous model for the formation of widely occurring and economically important phosphorite deposits on land. The problem was resolved by the concurrent discovery of phosphorites in the process of formation on the Namibian shelf (SW Africa) by Baturin (1969). Initial reactions of some scientists were to dismiss the new findings as unique and exotic phenomena, having little relationship to the big commercial deposits consisting of thick layers of rich phosphorite. However, new studies along the Peru-Chile continental margin found similar young phosphorites (Baturin et al. 1972). In this case, the evidence indicated that similar conditions hat produced volumes of phosphate deposits comparable to some of the largest ancient deposits. In both cases, the phosphorite enrichment processes took place in diatomaceous sediments initially rich in organic matter.

It is now widely recognized that such processes preferentially take place in highly productive zones formed by upwelling of deep, nutrient-rich waters, predominately caused by the impingement of eastern boundary currents on open oceanic shelf-slope regions. The concentration of phosphatic material is not formed by a single process, but results from a succession of processes, beginning with bioassimilation and sedimentation of phosphate in planktonic organisms. Primary phosphate minerals then form by diagenesis near the sediment-water interface. Subsequently, phosphate-enriched material is enriched by repeated sediment reworking and winnowing that removes fine clayey fractions by bottom currents or wave action, especially during periods of sea level change. Where the complete cycle occurs, significant phosphate may accumulate. A partial cycle (omitting the enrichment by reworking) is more common, resulting in only disseminated phosphatic particles within nonphosphatic sediments.

A different, and as yet controversial type of phosphorite is found on isolated subsea mountains in the deeper ocean. Here, phosphorites are frequently covered or replaced by ferromanganese crusts. An extensive review of marine phosphatic deposits can be found in Baturin (1978, 1982).

The Cost of Marine Geologic Discoveries

How much did the above mentioned discoveries cost? Based on recent prices, the cost of one large modern oceanographic research vessel is between $20 and $40 million, and a day of work at sea costs between $10000 and $15000. The scientific fleet of the former Soviet Union included nearly 100 vessels of different sizes, of which only a part were engaged in geological and geophysical marine research.

The Institute of Oceanology of the Soviet Union, which, along with the All-Union Ministry of Geology, conducted the majority of marine geological investigations during the period, 1949–1992, organized about 100 cruises lasting from 30 to 130 days each. The aggregate cost of all these investigations was therefore about $100 million. The same cost can be roughly ascribed to seven large and medium-sized oceanographic vessels operated along with the manned submersibles, *Mir* and *Pisces*. A further cost for repairs and refitting was $50 million. All in all, the total cost of one of the world's largest oceanographic institutions for marine geologic research during nearly 40 years amounted to about $300 million. The expenses of other organizations, specially the Ministry of Geology of the USSR, which sent expeditions for manganese nodule and crust prospecting to the Pacific, Indian, and Atlantic oceans, may have equaled these costs. If we assume that the combined costs of similar investigations carried out by the USA, European countries, Japan, India, and China are more than three times those of the former USSR, we arrive at a grand total expenditure of $3 to $4 billion – a relatively modest cost for scientific progress in one of the major branches of earth science.

3.2 Minerals of the Sea Floor: Manganese Nodules, Crusts, and Phosphorites

3.2.1 Manganese Nodules

General Features of Deep-Sea Mineral Formation

G. N. Baturin and N. F. Chelishchev

The mineralogical composition of manganese nodules is complex and variegated owing to the weak crystallization and intergrowth of major minerals as well as to the poligenic nature of many mineral phases.

The major components of manganese nodules are manganese minerals, which host the greater part of valuable base metals. During the early stage of their study, which begin in the 1950s with X-ray method, three manganese minerals with characteristic diffraction patterns were found: δ-MnO_2 (2.4 and 1.4 A), 7 Å manganite (7.2, 3.6, k2.4, and 1.4 Å) and 10 Å manganite (9.8, 4.8, 2.43, and 1.42 Å). Later, the 7 Å manganite was identified as birnessite and the 10 Å manganite as todorikite, whereas δ-MnO_2 was long considered as a disordered variety of birnessite.

The new stage of mineral studies of Mn nodules began with the advent of the special variants in investigation methods for fine disordered particles of deep-sea mineralization: transmission and analytical electron microscopy and microdiffraction, energy dispersive microprobe analysis, special variants of X-ray diffraction, as well as infrared and Mösbauer spectroscopy, EXAFS, XANES, locating mineralization by natural radioactivity measurements, radioisotope energy dispersive X-ray analysis, high-sensitivity sensors to analyze gases, metals, and isotopes.

Owing to this new approach, it was established that δ-MnO_2 is an independent mineral (vernadite), whereas 10 Å manganite represents a series of minerals. Among less abundant manganese minerals found in nodules are pyrolusite, ramsdellite, nsutite, romanechite (psilomelane), chalcophanite, and ransieite.

The principal minerals of manganese nodules and crusts belong to a single and extremely specific group: autigenic oxyhydroxides of tetravalent manganese, fine and ultrafine or poorly crystallized aggregates (which need special methods for their identification), disordered metastable tunnel and layer structures with ion exchange cations and molecular water, postdepositional stagger transformations, redox processes, metal adsorption by exchange processes, participation of microorganisms in mineral formation, and slow growth rates (mm in a million years).

The main manganese minerals of the nodules and crusts can be divided into two structural groups: layer phyllo-manganates and tunnel manganese oxides.

In both types, the basic structural unit is MnO_6 octahedra linked together to form layers in phyllo-manganates or frameworks with different tunnel cross-sectional dimensions (one-one, one-two, two-three, and so on in MnO_6 octahedra) in tunnel oxides. Thus, the basic part of the formulas of these manganese oxides is MnO_2.

All other components (1) large cation Ka, K, Ca, Ba, (2) water molecules, and (3) ore elements Co, Cu, Ni, Ti, Pt, Ir, REE, Tl, Bi, Te, Mo occupy interlayer spaces in layer phyllo-manganates or channels in tunnel oxides, and are exchangeable. To compensate the charge of the Na^+, Ca^{2+}, Ba^{2+} and other cations in channels or interlayer space, tetravalent manganese can be changed partially into Mn^{3+} and Mn^{2+}.

Tunnel Mn oxides of the nodules and crusts:

Pyrolusite [1x1] MnO_2
Ramsdellite [1x2] MnO_2
Hollandite [2x2] $MnO_2 \cdot$ Ba, K, Pb
Cryptomelane [2x2] $Mn_{8-z}(O, OH)_{16} \cdot K_{2-y}$
Coronadite [2x2] $Mn_{8-z}(O, OH)_{16} \cdot Pb_{2-y}$
Romanechite [2x3] $Mn_5O_{10} \cdot$ Ba, K, Co $\cdot\, nH_2O$
(psilomelane)
Todorokite [3xn] $Mn_5O_{12} \cdot (Ca, Mg, Ba)_{12} \cdot nH_2O$

Mixed channel:

Nsutite [1x1][1x2] (pyrolusite + ramsdellite) $(Mn^{4+}, Mn^{3+}, Mn^{2+})(OH)_2$

Layer phyllo-manganates of the nodules and crusts:

Birnessite $(Mn_6^{4+}O_{14})$ Mn^{2+}, Mg, Na, K, Ca $\cdot\, nH_2O$
Buserite $(Mn^{4+}O_2)$ Mn^{2+}, Na, K, Ca $(OH, H_2O)_n$
(buserite-like with additional interlayer water molecules and larger c parameter)

Asbolane ($Mn^{4+}O_2$) Ni, Co, Ca · nH_2O
Vernadite ($Mn^{4+}O_2$) Ni, Co, Fe · nH_2O

All these layer oxides are hexagonal, with a = 0.282–0.287 nm, c = 0.708–0.731 (birnessite), 0.960–0.970 (buserite, asbolane), 0.470 (vernadite).

Mixed layer:
Asbolane-buserite, asbolane-vernadite, asbolane-birnessite (can be considered as a microepitaxy accretion of the minerals).

The general sequence of the diagenetic transformation of Mn minerals in the pseudomorph-alteration model of ocean manganese nodule formation is: protomineral Fe-rhodochrosite-vernadite-buserite-birnessite-asbolane.

Ion exchange reactions play an important role in the deep-sea mineral formation. Manganese minerals of nodules show the selectivity to the heavy metals increased in a row: Ni < Cu < Co < Pb. The mobility of atoms in the manganese minerals of nodules differs for various structural positions: disordered interlayer cations (Na, K, Ca, Sr, etc.) > ordered interlayer cations in a halcophanite position (Cu, Cd, Zn, Pb, Mo, etc.) > atoms of asbolane layers (Ni, Co) > atoms of anion matrix (Mn^{4+}, Co^{3+}).

The transition of cobalt sorbed by ion exchange mechanism in a non exchangeable position by redox processes in a disordered highly oxidized matrix of vernadite determines the high concentration of cobalt in vernaditic nodules and crusts. The concentration of copper and zinc in buseretic nodules is determined by an availability in buserite of the structural positions of halcophanite type. The heavy metals can replace manganese and the alkaline earth metals in this position. Formation of asbolane can be connected with the hydrolyzed precipitation of the heavy metals in an interlayer space of buserite. Transformation of buserite to asbolan occurs in two stages in hydrolysis of metals sorbed by ion exchange mechanism. Therefore mixed-layer asbolane-buserite can be considered as an intermediate product of transformation from buserite to asbolane.

The process of solution-crystallization on the redox barrier of the sea floor can occur with the participation of microorganisms, playing the role of catalyst. The contradiction between the speed of nodule growth and their absolute age can be connected with the permanent process of solution-crystallization.

There are two types of ion exchange equilibrium: the oceanic type (ion exchange equilibrium of minerals with the near-bottom ocean water) and the diagenetic type (ion exchange equilibrium between minerals and interstitial water of compact deposits). The fact that oceanic nodules and phillipsite occur at the surface of the bottom sediments means that they are in ion exchange equilibrium with the bottom water, whose composition is close to that of ocean water. Disruption of mass exchange with ocean water as a result of compaction of the sediments causes a considerable change in the compositions of interstitial solutions, which is reflected in the cation composition of nodules. Their extreme selectivity for heavy metals permits oceanic nodules to be considered as the geochemical barrier of sorption type at which nonferrous and rare metals are concentrated from seawater (Co, Ni, Cu, Zn, Cd, Hg, Sb, Bi, Mo, Sn, TR, Y). Data on the partition coefficient of nonferrous metals indicate that in relation to Cu, Ni, and Zn the nodules are close to equilibrium with ocean water. The partition coefficients of Co and Pb are nearly 50 times higher than the equilibrium coefficient. As a result, Co and Pb concentration is considerably higher than the equilibrium value.

There are two mineral types of nodules: vernaditic (with iron oxide feroxyhyte δ-FeOOH) located on the tops of the subsea mountains and slopes of the ocean hills, and buseritic (with goethite α-FeOOH) forming a large field at the floor of the oceanic depressions. Contents of metals Cu, Ni, Co, Zn, Pb vary from 0.8–1.5 in vernaditic nodules to 0.6–1.3 in vernadite-buseritic and up to 3 in buseritic nodules.

The change in paragenetic associations correlates with genetic type, microcrystal sizes and forms, and main metals:

- Vernadite-feroxyhyte paragenesis, hydrogenetic, 0.1 mkm, twist sheets, Co.
- Buserite early diagenetic, 1.0 mkm, thick plates, Cu.
- Buserite-asbolane diagenetic, 10 mkm, plates, Cu, Ni.
- Birnessite-asbolane late diagenetic, 100 mkm, thick plates, Co, Cu, Ni.
- Todorokite early metamorphic and hydrothermal metasomatic, > 100 mkm thick plates, Cu, Ni, Co.

The disordered metastable vernadite transforms in more ordered layer phases buserite-asbolane-buserite-birnessite.

Typical textures of the nodules are spheroidal and dendritic. Analysis of the texture relation and experimental data suggest that the nodules are fine-grained pseudomorphs of Mn and Fe oxides after a protomineral Fe-rhodochrosite metastable under oxidation conditions of the ocean floor.

Along with major minerals, manganese nodules contain a large spectrum of accessory minerals of authigenic, detrital, and probably volcanogenic origin, including the following:

Rare manganese oxides and hydroxides: manganite, manganosite, cryptomelane, hausmannite, braunite, woodruffite.
Iron oxides and hydroxides: hydrogoethite, hematite, hydrohematite, magnetite, ilmenite, maghemite, wüstite; along with iron oxides, some rare ultramicroscopic inclusions of bunsenite (NiO) have been reported.
Clay minerals: montmorillonite, kaolinite, illite, chlorite, and nontronite.
Sulfides: pyrite, markasite, pyrrhotite, greigite, chalcopyrite, covellite, violarite, bornite, and nickeline.
Native metals and intermetallic minerals: iron, aluminum, chromium, gold, silver, platinum, β-cobalt, and taenite.
Other minerals (detrital, authigenic, biogenic, and volcanogenic): quartz, feldspars, fuchsite, phillipsite, boehmite, diopside, amphibole, prenite, anatase, rutil, sphene, calcite, apatite, barite, and gypsum.

The most amazing feature of manganese nodule mineralogy was the discovery of sulfide minerals ranging in size from ultramicroscopic particles to well-developed crystalline pyrite inclusions up to several millimeters in cross-section. The formation of sulfides in highly oxidizing deep-sea environment has been interpreted in two ways. On the one hand, it may be bound with activity of microorganisms including sulfate-reducing species which develop when organic detritus such as remains of bottom fauna attached to nodules are available. On the other hand, sulfides might be provided by hydrothermal or volcanogenic action whose more or less distinct trances have been found in some parts of large nodule fields.

However, the findings of native metals such as iron and aluminum remain enigmatic, since these are considered to be formed at very high temperatures and pressures available only in the course of magmatic processes. In any case, this problem relates as well to oceanic sedimentation in general because the particles of native metals and intermetallic minerals have been equally found in sediments in the vicinity of active or extinct volcanic zones.

The chemical composition of nodules throughout the ocean is highly variable owing to different environmental factors, but average values are rather representative, especially for higher grade nodules since most part of the analyzed material has been provided from the largest nodule fields of assumed economic value.

Among these three types of nodules, the micronodules are seemingly enriched in manganese, nickel, copper, vanadium, arsenic, and rare earth elements which might be interpreted as more efficient mobilization and concentration of these elements at the earliest stage of diagenetic nodule formation owing possibly to the comparatively larger volume of the surrounding sediments.

On the contrary, marine nodules are low in manganese and all essential base metals but enriched in iron, phosphorus, and carbon, which may reflect the appropriate difference in chemical composition of oceanic and marine sediments as a major source of material for nearly all types of nodules.

The comparison of average contents of elements in deep-sea manganese nodules and deep-sea sediments allows the concentration coefficients of elements in nodules to be calculated relative to sediments, which is essential for understanding the behavior of various groups of elements in oceanic ore-forming processes. Among the elements which prove to be concentrated in nodules at higher grade (more than ten times compared with sediments) are manganese, thallium, bismuth, tellurium, molybdenum, cadmium, antimony, lead, platinum, and platinum group metals except palladium, whereas the large group of elements are either not concentrated (e.g., berillium, boron, strontium, barium, and gold) or are even depleted (carbon, nitrogen, scandium, rubidium, mercury, and halogens). The arranging of these data on the Periodic Table reveals that all elements concentrated in nodules belong to periods 4 to 6 and that group VIII is the most abundant in this respect.

Origin of Manganese Nodules and Crusts

The origin of manganese nodules an crusts raises a series of controversial problems such as the sources of metals, their ways of migration, and their mechanisms of deposition and concentration.

The major part of sedimentary material enters the ocean through riverine input, supplying annually 35000–44000 km^3 of water, 13–22 billion t of suspensions, and 3.3–3.8 billion t of dissolved matter. Most of the metals in riverine runoff are bound to suspensions. For example, the share of dissolved Mn is only 0.4 Mt, as compared to 16–20 Mt of suspended Mn.

Another source of metals, especially Mn and Fe, in the Ocean is submarine hydrothermal activity. According to different estimates, this source supplies from 0.5 to 10 Mt of Mn per year, but calculation of the depositional balance of Mn in the Ocean has shown than the input of hydrothermal Mn is modest compared to the upper limit of the above estimates.

To explain how the dissolved metals fall out from the sea-water, a series of mechanisms have been suggested, including sorption, coagulation, coprecipitation, and oxidation.

Owing to the very low content of Mn, Fe, and most base metals in seawater (less than 1 nM kg^{-1}), their chemical precipitation from the water column is highly improbable, as confirmed by natural observations coupled with a number of experiments and thermodynamic calculations. According to the now prevailing opinion, the major mechanism of metal deposition is their bio-assimilation and bioprecipitation on being incorporated in biogenic detritus, and especially by means of pelletal transport, whose intensity is apparently adequate to account for the total downward flux of dissolved manganese.

The formation of Mn nodules and crusts begins at or near the bottom surface owing to the hydrogenetic and diagenetic factors which are conventionally divided according to the Mn/Fe ratio, being of the order ~1 ÷ 2.5 in hydrogenetic and 4 ÷ ≥5 in diagenetic formations, whereas the intermediate values are thought to be evidence in favor of the influence of both factors. This gradation is supported by the apparent difference in chemical composition of nodules half-buried in sediments whose lower part is enriched in Mn contrast to the upper part, which is enriched in Fe.

The most developed theoretical background to explain the processes leading to the formation of Mn nodules and crusts is based on the interplay of a series of partial chemical and physicochemical equilibria in line with the activity of several groups of microorganisms fueled by the consumption of organic matter, according to the general scheme:

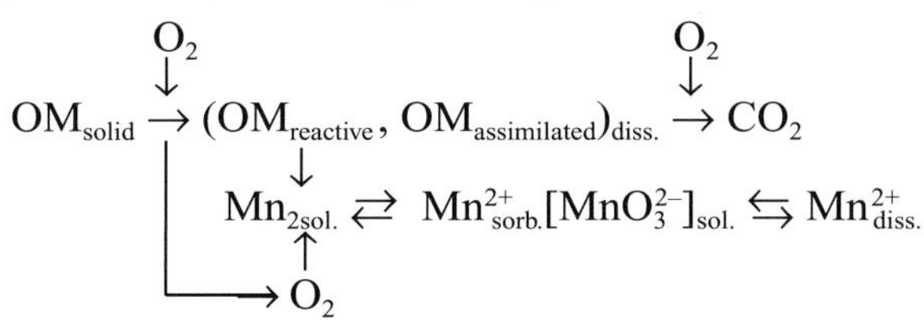

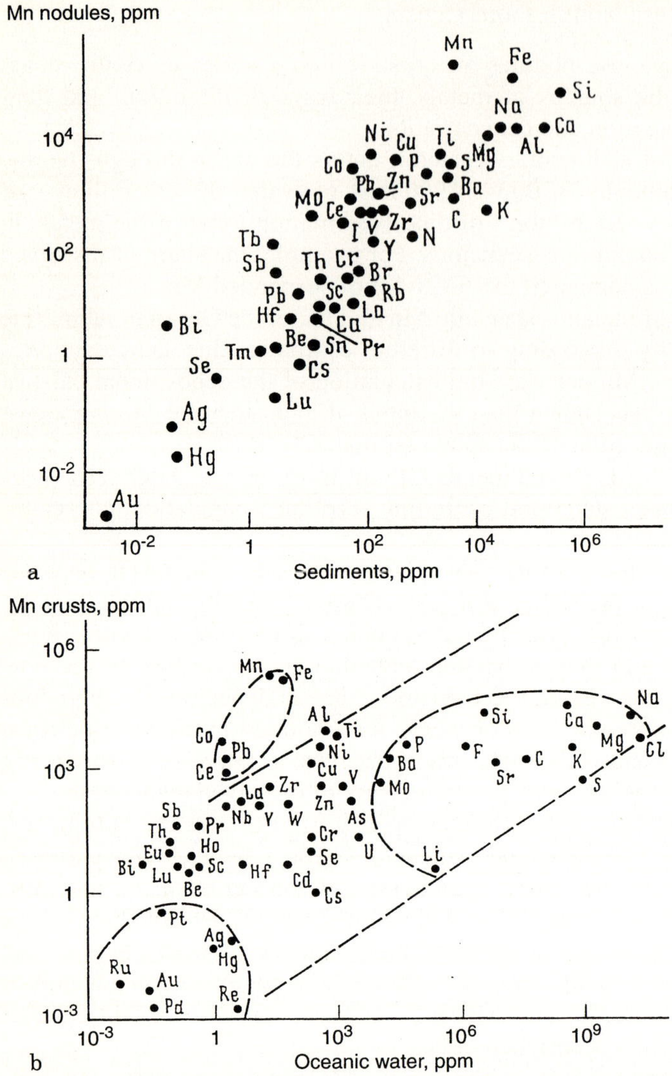

Fig. 52. Correlation of average contents of chemical elements. **a** In manganese nodules versus deep-sea sediments. **b** In manganese crusts versus oceanic water

It is, however, hardly possible to describe the intermittent cycling regime of oxidizing-reducing interactions of such a system by a simple steady-state model.

The behavior of Mn on solid surfaces in the marine environment is determined by two processes: coalescence and fractionation of dispersed manganiferous particles; but the chemical properties of MnO_2 induce the predominance of coalescence of particles such as Mn^{2+}_{sorbed} $[MnO_3^{2-}(MnO_2)_n]_{solid}$, which is helped by coalescence of MnO_2 with Fe hydroxides leading to irreversible

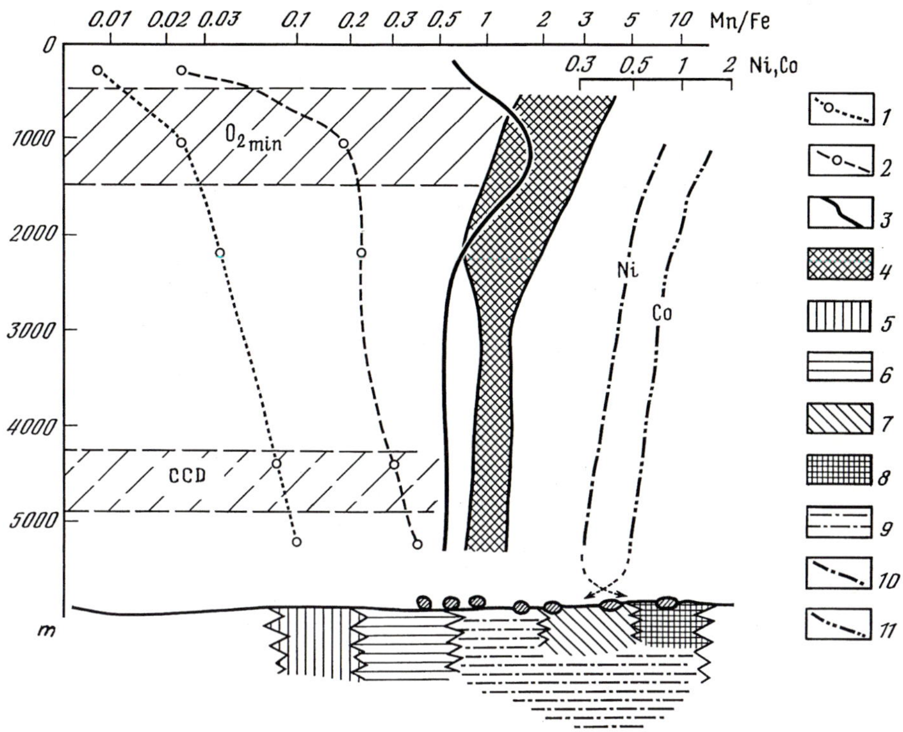

Fig. 53. Mn/Fe ratios in manganese nodules and crusts versus various components of marine environment. *1* Suspended matter from Northern Pacific Ocean; *2* chemically mobile part of suspended matter; *3* oceanic water; *4* Mn crusts from subsea mountains; *5* sediments of the tropical radiolarian belt of the Pacific Ocean with most abundant nodules; *6* integrated mobile fractions of sediments; *7*, *8* fractions of sediments extracted by hydroxylamine and acetic acid; *9* interstitial waters from the same sediments; *10*, *11* Ni and Co contents in crusts; *12* Mn nodules. O_{2min} – oxigen minimum zone; *CCD* carbonate compensation depth

reaction, which may be termed colloidal-chemocoalescent binding of solid manganese phases.

According to the above general scheme, there is no fundamental difference in the formational mechanism of diagenetic Mn nodules and hydrogenetic Mn crusts, since both are fed by dissolved metal fluxes supplied by sediments via interstitial waters (nodules) or directly by near-bottom seawater (crusts), with compositional differences owing to these two sources.

These differences are demonstrated by general correlation patterns of elements in nodules versus deep-sea sediments and in crusts versus seawater (Fig. 52).

The attempt to interpret the origin of Mn nodules and crusts in the light of Mn/Fe ratios in various components of the marine environment is presented in Fig. 53, which shows that nodules extract these metals from the geochemically most mobile fractions of sediments and their interstitial waters, whereas the crust extracts metals predominantly from the ambient oceanic water.

Distribution, Resources, and Prospects of Exploitation

G. N. Baturin

Manganese nodules of various form and size are common on the bottom of all oceans, especially in the Pacific Ocean (Fig. 54).

The greater proportion of nodules is lying on the bottom surface, the rest being buried in sediments as separate nodules or their layers. The occurrence and composition of nodules depends on the depth of the ocean, bottom relief, and the nature of underlying or enclosing sediments. Nodules with the highest content of base metals (Ni + Cu + Co) are most abundant on abyssal plains with low deep-sea hills at depths of 4 to 5 km, and lie on very slowly accumulating clayey-siliceous oozes or deep-sea red clays.

The most detailed studies have been carried out in the great nodule field between the Clarion and Clipperton fractures stretching in sublatitude direction from 116 °W to Line Rise in the east. The distribution of nodules inside the fields is not uniform, but in its western, central, and eastern parts there are arcas where more than 50% of the bottom surface is covered by nodules whose coverage reaches 22 kg m^{-2}. The lower coverage is typical for the abyssal plain and hill foots, whereas the highest coverage has been found in hollows between hills as well as on their slopes and upper parts. The extent of nodule coverage

Fig. 54. Distribution of manganese nodules over the world ocean. Major nodule fields: *1* Clarion-Clipperton; *2* South Pacific; *3* Wake-Necker; *4* Central Pacific; *5* Hawaii; *6* California; *7* Menard; *8* Drake Passage-Scotia Sea; *9* Diamantina; *10* Central Indian Ocean; *11* West Australia; *12* Somali; *13* North America-Guyana; *14* Cape-Agulhas; *15* Blake Plateau

and the metal content in nodules are as a rule inversely correlated. The typical contents of metals in nodules of this field are:

$$Mn = 27\text{–}31\,\%,\ Ni + Cu + Co = 2.7\text{–}3.5\,\%.$$

The total resources of manganese nodules on the sea floor have been calculated and recalculated many times. The first estimate made in the early 1960s amounted to 1.66×10^{12} t, but seems now to be too high. According to later data, the territory of the World Ocean where nodule coverage exceeds 2.5 kg m^{-2} is nearly 41 M km^2 or about 11% of the World Ocean area. Less than half of this territory (18.5 km^2) is characterized by nodules containing more than 0.88% of Ni + Cu, and even less (6.5 M km^2) is considered as promising for potential exploitation. The greater part of this promising territory (4.2 M km^2) is situated in the North Pacific, whereas the Clarion-Clipperton field occupies 2.2 M km^2.

The total mass of nodules in areas where their coverage exceeds 2.5 kg m^{-2} is evaluated at 500 billion tons, of which half contains more than 0.88% Ni + Cu.

The mass of richest nodules which could be considered as reserves amounts to 33 billion tons, containing about 6 billion tons of manganese, 290 million tons of nickel, 240 million tons of copper, and 60 million tons of cobalt.

Such an impressive tonnage of metals contained in oceanic nodules, especially nickel and cobalt, exceeding their on-land reserves, stimulated large-scale investigations to evaluate the technological and economic prospects of their potential exploitation, on the basis of an assumed average recovery of 3 Mt nodules per year, which would cover an essential part of world demand for these metals (Table 11).

Among the systems offered for nodule recovery, three seemed most promising, namely continuous line bucket system (CLB), air lift, and hydrolift.

The CLB system uses buckets attached to a long, continuous, buoyant rope loop which is driven by shipboard traction motors. The problem to be solved

Table 11. Tentative evaluation of resources and annual production of metals in case of recovery of 3 Mt of manganese nodules and 1 Mt of manganese crusts per year

World resources, t	Mn	Ni	Cu	Co
On continents	8×10^9	100×10^6	100×10^6	3×10^6
In the Ocean:				
Nodules	6×10^6	290×10^6	240×10^6	60×10^6
Crusts	2.5×10^6	40×10^6	7×10^6	80×10^6
World production, t y^{-1}				
On continents	15×10^6	700×10^6	14×10^6	50×10^6
In the Ocean:				
Nodules	0.75×10^6	36×10^6	30×10^6	7.2×10^6
Crusts	0.25×10^6	5×10^6	0.7×10^6	8×10^6

when applying this system is the efficient design of buckets and the strength of rope and pulleys, as well as the apparent necessity of employing two ships for the operation, to avoid the rope tangling.

When applying airlift or hydrolift systems, the operation consists of two steps: collecting nodules on the seafloor and hoisting them on board ship.

The production levels (up to 10 000 t day^{-1}) require a wide, up to 30 m, unwieldy collector operating on the sea bottom which can vary in physical characteristics from sticky grease to a very firm surface, exhibiting brittle features. Numerous patents have been issued for nodule-gathering devices which are analogous to conventional agricultural harvesting or fishing devices, and to adaptation of wet-terrane vehicles.

The airlift and hydrolift systems consist of a combination of pipes, pumps, and air compressors allowing nodules and their fragments to be hoisted on board ship. The airlift system was tested for this purpose beginning from early the 1970s at water depths from 900 to 5000 m, but its efficiency was evaluated at only 30 to 50%.

The hydrolift has been considered the more promising technique, offering predictable, efficient results. The technological developments required for this system are the placement of the various pumps along the pipe length, the engineering of 5000-m-long, wide-diameter pipe strings, and pipe-handling equipment. The main disadvantage of early hydrolift systems was the placement of moving mechanical parts at depth, whereas for the airlift the air compressors can be mounted on the deck of the surface vessel.

The latest development in these systems by Russian specialists led to new combination of both methods supported by the construction of more efficient suction units based on utilizing the kinetic energy of swirled water jets, which allows the energy consumption to be reduced by two to three times.

Among other projects offered for the recovery of nodules are some seemingly less realistic, such as the use of robot devices or cargo submersibles. In particular, a group of chemists from Vladivostok offered a quite new principle of nodule exploitation by isolating a certain area of the bottom from the ambient seawater by a special cover, dissolving of the nodules by pumping acid under this cover, and hoisting the metal-bearing solution on board ship for subsequent recovery of pure metals.

The processing of manganese nodules (as well as manganese crusts) can be effected by either pyrometallurgical or hydrometallurgical process, which can be summarized as follows:

1. Pyrometallurgical: smelting with a flux and sulfur source to produce a high-manganese slag and metal concentrate.
2. Hydrometallurgical: cuprion process using an ammonicia leach in a reducing atmosphere, and recovery of nickel and copper by solvent extraction-electrowinning with separate cobalt-molybdenum recovery.

Projects to build plants for nodule processing have been designed in the USA in Hawaii and in the former USSR in Vladivostok.

According to economic estimates made before 1980, the capital investment for a venture able to recover and process 3 Mt of nodules per year may amount to $1.5–2 billion. Thus, the appropriate capital and exploitation costs per ton of nodules are 230–270 and $70.

The full price of the final product or metals recovered from 1 ton of nodules (Mn + Ni + Cu + Co) amounted in 1989 to $220–250 in the case of iron-manganese concentrate (78% Mn) or to $505–550 in the case of manganese concentrate (99% Mn).

The assumed profit of the whole venture has been calculated as being not more than 8.5–9.5%, whereas the net income left after taxes might be as low as 3 to 4.5%, which probably means postponing the beginning of nodule exploitation to the next century.

Environmental Impact of Deep-Sea Nodule Mining

The large-scale recovery of manganese nodules at the rate of the order of 10 000 t day^{-1} might lead to unprecedented disturbance in the near-bottom environment and to pollution of the whole water column.

It is assumed that the dredge head of the mining device must be able to scrape the ocean floor within a close tolerance over a width of 12 to 30 m, and the whole device would weight about 90 t. The area plowed per day by such a device would reach 1 to 2.5 km^2. Along with nodules, this device will remove a surface layer of sediment of up to 650 000 m^3 of sediments per day.

Most of these sediments in major nodule fields are clayey-siliceous oozes or red clays whose greater part consists of fine particles: colloidal (less than 0.001 mm) up to 40% and pelitic (up to 0.01 mm) up to 70%. If such sediments were disturbed, the fine fractions would be held suspended and prevented from settling even by weak bottom currents whose velocity is only 2–4 cm s^{-1}, which is the case in many deep-sea areas. This effect would be aggravated if collected nodules were washed from adhering sediments on the spot, i.e., before hoisting on board ship.

To imitate the bottom disturbance by nodule recovery, a huge plow was dragged across the limited area covered by loose sediments in the Central Pacific. Observations made in the same area several months later showed no appreciable changes in bottom environment including the bottom fauna; but it must be kept in mind that such a short and limited experiment has nothing in common with large-scale recovery processes of long duration and over vast areas, where a sporadic increase in bottom currents might produce the so-called bottom storms able to provoke unpredictable changes in sediment redistribution.

The impact of nodule recovery on surface water and underlying water layers is also inevitable owing at least to the long stay of large recovery vessels, whose displacement would reach 300 000 t or more in one and the same place. In the case of nodules being cleaned of sediment on board ship, the amount of suspensions thrown overboard might reach 5000 t day^{-1}, since material hoisted by hydraulic systems contains at least one third sediment along with nodules. It is assumed that the plumes of fine suspensions might spread over an area of about 400 km wide before they settled to deeper water, thus decreasing the transparency of the upper water layer, and having negative effects on phytoplankton production and fish larvae growth. To avoid such an effect, it has been proposed to discharge all production wastes into deeper water through pipes, but the chance to perform such tests has not yet been realized.

It is in any case quite evident that future nodule recovery must be preceded by more thorough theoretical and experimental study of this possible impact on both near-bottom and surface-water environments.

3.2.2 Manganese Crusts

G. N. Baturin

The mineral composition of manganese crusts has been investigated by the same methods as manganese nodules.

Similarly to nodules, crusts consist of manganese and iron oxides and hydroxides in association with several groups of minor minerals. In contrast to nodules, however, the dominant manganese mineral in hydrogenic crusts is ferrugeneous vernadite, whose content in the bulk material may reach 60 to nearly 100%.

Iron minerals are less abundant and are represented by X-ray amorphous hydrogoethite $FeOOH \cdot nH_2O$ epitaxially intergrown with manganese hydroxides, forming fine-grained mineral associations with the admixture of the alumosilicate fraction.

On the other hand, the deep-sea crusts which have occasionally been found on the surface of unconsolidated sediments may consist of asbolane-buserite as the major mineral phase, owing to diagenetic processes in the underlying sediments.

Among the accessory minerals of Mn and Fe in hydrogenetic crusts of the Pacific Ocean are todorokite, birnessite, asbolane-buserite, Na-cryptomelane, manganosite, natrojarosite, goethite, magnetite, and maghemite. The non-metalliferous minerals are represented mostly by apatite (up to 22%), feldspar (up to 5%), and quartz (up to 2.6%). The apatite is usually well crystallized and forms thin layers or veins, or is dispersed throughout the Mn-Fe mass in the form of separate tabular crystals. Grains of calcite of biogenic origin are common, especially in the upper parts of crusts. Among the less frequent accessory minerals are barite, phillipsite, clinoptilolite, smectite, chlorite, caolinite, and orthoclase.

It has been found that quartz and feldspar in crusts of the Equatorial Pacific are less abundant as compared to the Northern Pacific, which may reflect the role of aeolian input from Asia. On the other hand, apatite is relatively more abundant in the Northern Equatorial Pacific owing to its higher biological productivity.

The manganese crusts from the Atlantic Ocean consist of an equal association of highly oxydized manganese and iron minerals, as demonstrated by typical samples recovered from subsea mountains in the Cape Verdo Deep (Fig. 55).

The major minerals of this crust are ferruginous vernadite and manganiferous feroxyhite, with minor admixtures of geothite, asbolane-buserite, and birnessite.

The manganese crusts, as well as nodules formed under the influence of hydrothermal processes, are in contrast to hydrogenetic crusts and nodules, composed mainly of todorokite with some birnessite, minerals which are less oxidized as compared to vernadite.

The manganese crusts recovered near the Kuril Islands contain three types of todorokite: disordered, 9.75 Å and 19.5 Å todorokite, as well as 14 Å birnessite and minor vernadite without any traces of iron, which has not been found anywhere else.

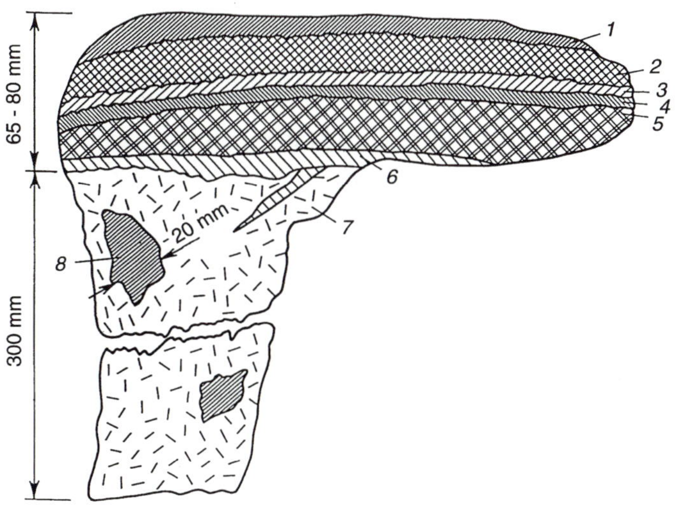

Fig. 55. Manganese crust on the hydrothermally altered hyaloclastite from the Krylov Seamount in Cape Verde Basin. *1* Upper layer of black hydroxides (5–7 mm): Fe-vernadite, Mn-feroxyhite and rarer birnessite with admixture of asbolane-buserite and goethite; *2* brownish-black hydroxides (23–30 mm): goethite, Fe-vernadite, Mn-feroxyhite, rarer birnessite with admixture of asbolane-buserite, and smectite with palygoskite; *3* black hydroxides (3–7 mm): Mn-feroxyhite, Fe-vernadite, and goethite; *4* black Mn and Fe hydroxides (2–7 mm); *5* brownish-black material (5 mm) impregnated by Mn-Fe hydroxides, goethite, lesser vernadite, and apatite; *6* basal layer of black hydroxides (2–10 mm): goethite, Fe-vernadite, with lesser asbolane-buserite, and traces of todorokite; *7* hydrothermally altered hyaloclastite (Fe-palygorskite); *8* inclusions and veins of Mn and Fe hydroxides

The 14 Å birnessite has been also met in hydrothermal crusts from the Tadjura Rift, whereas todorokite-birnessite association has been found in manganese crusts recovered from subsea heights and volcanoes of the Sea of Japan. The rarer varieties of hydrothermal crusts such as found on a subsea volcano near the Commander Islands consist of ferrugenous nontronite, hisingerite, 14 Å birnessite, and iron hydroxides.

The chemical composition of manganese crusts throughout the Ocean is as variable as that of manganese nodules, with major Mn and Fe associated in various proportions with base metals and nonmetallic and trace elements.

The hydrogenetic crusts enriched in Cu and Ni are concentrated mainly in the Pacific Ocean. Some crusts, especially in the Atlantic and Indian Oceans, are rich in cerium, but its practical value is up to now questionable, as well as that of platinum, whose concentrations are, as a rule, below 1 ppm.

Manganese crusts, which cover various kinds of rocks and sediments exposed on the sea floor in a more or less continuous layer from less than 1 to more than 100 mm thick, are, in general, similar to manganese nodules in their chemical composition and are spread at various depths in pelagic as well as in marginal oceanic environments, but, in contrast to nodules, they tend to occur predominantly on subsea mountains and heights. Since the Pacific Ocean is the most abundant in subsea mountains (more then 10 000 summits), the greatest part of oceanic manganese crusts is concentrated in this ocean, in particular on Mid-Pacific subsea mountains, on the Necker Ridge, on subsea slopes of the Hawaiian Ridge, Tuamotu Islands, Marshall Islands, and in numerous other places.

On the Mid-Pacific mountains, which are one of the areas of greatest occurrence, crusts cover crystalline rocks or consolidated sedimentary and volcanogenic deposits including basalts, gravelstones, limestones, marls, and flints. In most cases, they occur on relatively steep slopes of subsea mountains and abraded surfaces of volcanic cones at water depths from about 1000 to 3400 m.

There is some relation between the thickness of the crusts and the water depth. Crusts which are more than 6 cm thick occur predominantly at depths ranging from 1500 to 2100 m, whereas at shallower as well as a greater depths, their thickness is reduced to 0.5–3.5 cm. The thickest crusts (13–15 cm) occur on basalts and breccias, the medium ones on flints, and the thinnest, including films or stainings, are found on limestones and unconsolidated sediments. Some of these features, with local variations, have been found in other areas of large-scale crust occurrence on subsea mountains.

The crusts considered most promising for future exploitation contain as much as 14–25% Mn, 0.6–1.2% Co, 0.4–0.6% Ni, and 0.05–0.11% Cu; their growth rates are very slow owing to the slow accumulation of metals from the overlying seawater.

Along with such hydrogenetic crusts, another type is found in areas of subsea hydrothermal activity; these are characterized by very large variations in Mn to Fe ratio and extremely low base metal contents, since their formation is related to rapid deposition of metals from hydrothermal plumes rich in Mn or

Fe, but poor in base metals. The latter crusts are considered as having no value as a mineral resource.

The total resources of hydrogenetic crusts can be evaluated very approximately by analogy with the few well-studied areas, such as Magellan Sea mounts in the Western Pacific, where crust occupy 9700 km^2 of the bottom surface with an average abundance of 105.2 kg m^{-2} and resources amounting to 1021.4 Mt wet or 665.6 dry ore. One of the isolated guyots in the same region with 738 km^2 of crust-covered area and 115.6 kg m^{-2} abundance contains 85.3 t of crusts, of which 56.8 Mt are enriched in cobalt above 0.6%.

On the Horizon Guyot in the Northern Pacific, the proportion of bottom surface covered by crusts is 52%, with total resources of 7.5 Mt, whereas for the S. P. Lee Guyot, calculations showed 24% and about 20 Mt of ore material.

The majority of subsea mountains in the Pacific Ocean are situated inside the exclusive economic zone of the USA. The area considered as promising for manganese crust exploration on these mountains and subsea slopes of Hawaii, Guam, Mariana, and Marshall islands reaches 500 000 km^2. The total area of oceanic bottom with manganese crust occurences in the Pacific, Indian, and Atlantic Oceans may be about 4 M km^2. If the average proportion of this area covered by crusts with 50 kg m^{-2} abundance is only 25%, the total crust resources would amount to 50 billion t or at least 1 order of magnitude less than manganese nodules.

According to recent estimates, crusts of potential economic value must be more than 4 cm thick and contain no less than 0.8% Co.

Among the areas where such crusts might be abundant outside exclusive economic zones are the Ogasawara Plateau, Wake Rise, the Mid-Pacific mountains, Necker Ridge, the Magellan and Emperor seamounts in the Pacific, the Corner Rise, some mountains in the North American Basin, Sierra Leone Rise, Walvis Ridge and several parts of the Mid-Atlantic Ridge in the Atlantic Ocean, Equator Mountain, Cocos Rise, and isolated mountains in the Madagascar Basin in the Indian Ocean. The assumed resources of mineable crusts throughout all these areas may reach about 10 billion t containing up to 2.5 billion t cobalt, 40 Mt nickel, 7 Mt copper, and 6 Mt molybdenum.

The techniques offered for crust recovery include the continuous line bucket system, large volume grabs, and autonomous scrapers weighing up to 100 t and moving at a rate of 0.2 m s^{-1}, combined with a hydraulic pipeline hoisting system. The processing of crusts might be similar to that of nodules.

If cobalt-rich manganese crusts were recovered at the rate of 1 Mt per year, it would cover about 20% of world demand in cobalt and 2% in manganese (see Table 11). At present, however, the economic considerations for crust exploitation are too speculative and risky, forecasting from 20% loss to 30% profit per year for capital investment.

3.2.3 Phosphorites

G. N. Baturin

According to conventional practice, the term phosphorite is applied to sedimentary rock containing no less than 15% P_2O_5, whereas its maximum values may exceed 32%. The empirical coefficient to recalculate the P_2O_5 content into phosphatic mineral lies between 3 and 3.15, which means that up to half of the whole rock may consist of other minerals originating from the surrounding sediments. This circumstance is most essential when regarding the mineral composition of phosphorites which contain various amounts of terrigeneous, biogenic, volcanogenic, and chemogenic components along with phosphatic material.

Morphologically, oceanic phosphorites are represented by phosphatic grains, nodules, slabs, breccia, conglomerates, and more or less phosphatized rocks, predominantly limestones.

Phosphatic grains whose average diameter lies within the limits 0.1–0.5 mm, irrespective of geographic location, may be subdivided into oolitic, nucleated, and structureless varieties. Along with phosphate, the grains usually contain some minor amount of amorphous silica, organic matter, and clastic material.

The mineral composition of phosphatic nodules, as well as slabs, breccia, and conglomerates, is much more complex, as shown in Fig. 56, which represents a typical nodule from the Catham Rise east of New Zealand, where phosphate is associated with more or less phosphatized chalk, goethite, glauconite, and faunal debris.

Phosphate in these morphological types of phosphorites may be represented by several generations with various degrees of crystallinity, including phosphatic grains, debris of earlier nodules, and phosphatic cementations.

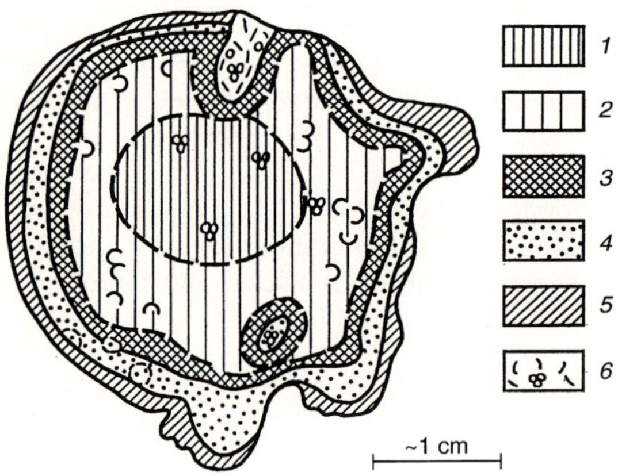

Fig. 56. Schematic representation of phosphatic nodule from the Chatham Rise. *1* Slightly phosphatized chalk; *2* phosphatized chalk; *3* goethite zone; *4* phosphorite zone; *5* outer glauconitic shell; *6* fauna debris

The biogenic components include four types: phosphatic, calcareous, opaline, and carbonaceous. Biogenic phosphate consists of fish bones, teeth, and scales, made of hydroxyapatite $Ca_5(PO_4)_3OH$.

The calcareous components, whose content may reach 40–50% of the whole rock, are represented mostly by remains of planktonic foraminifera and nannoplankton, but in some places by debris of shallow-water bottom fauna.

The biogenic opaline silica include remains of diatoms, radiolaria, and sponge spicules, whereas remains of organic matter are usually represented by black microscopic flocks, films, and stainings.

The most common authigenic components associated with phosphate are glauconite and pyrite. The former may be represented by grains, including detrital grains from eroded sediments, as well as by infillings in foraminifera shells and films or stainings. Pyrite may form globular or irregular aggregates and impregnate various components in phosphorite.

The detrital material may be represented by a large spectrum of minerals including quartz, feldspar, pyroxene, mica, heavy minerals, as well as debris of various sedimentary, volcanogenic, and metamorphic rocks.

This variability in the mineral composition of phosphorites is reflected by their chemical composition, characterized by substantial fluctuation in concentration of both major and minor elements.

The composition of pure phosphatic mineral is determined only approximately owing to the close intergrowth of phosphate with other components on the ultramicroscopic level as well as to the changeable composition of phosphate itself, whose most general formula is accepted to be

$$Ca_{10-a-b}Na_aMg_b\,(PO_4)_{6-x}(CO_3)_{x-y-z}(CO_3\cdot F)_y(SO_4)_zF_2,$$

where Ca might as well be substituted by a number of other cations such as K^+, Sr^{2+}, Ba^{2+}, Cd^{2+}, Zn^{2+}, Mn^{2+}, Y^{3+}, REE^{3+}, Sc^{3+}, Bi^{3+}, and U^{4+}, whereas groups of anions may include also CrO_4^{2-}, $CO_3\cdot OH^{3-}$, AsO_4^{3-}, VO_3^{3-}, and SiO_4^{4-}, and fluorine may be associated with OH^-, Cl^-, Br^-, and O^{2-}.

Based on the bulk chemical analyses, the recalculated formula of phosphatic mineral from nodules of the Spanish continental margin proved to be:

$$Ca_{9.71}(PO_4)_{4.97}(CO_3)_{1.36}F_{2.2},$$

whereas the phosphate predominating in grains of the large phospharite deposit on the outer Namibian shelf in the Southern Atlantic might be represented by the more extended formula:

$$Ca_{9.42}Mg_{0.36}Sr_{0.03}Na_{0.32}(PO_4)_{4.46}(CO_3)_{1.02}(SO_4)_{0.5}F_{1.95}(OH)_{1.0}.$$

Owing to its complex composition, there is no general agreement about the name of this mineral. A number of terms have been applied for its denomination (dahllite, podolite, collophane, francolite, staffelite, kurskite, carbonate-apatite, carbonae-fluorapatite), of which the latter seems to be the most acceptable as reflecting its chemical composition, which is typical for most oceanic phosphorites.

The phosphorites on the present sea floor are mostly concentrated along continental margins in the eastern and northwestern Atlantic and eastern Pacific, as well as on subsea mountains predominantly in the Pacific Ocean, where phosphorites are usually overlain by manganese crusts. Some less voluminous phosphorite occurrences have been found in many other areas throughout the ocean, in particular in the See of Japan, along the eastern Australian continental margin, and in the northwestern Indian Ocean (Fig. 57).

Morphologically, phosphorites are represented by phosphatic sands, nodules, slabs, conglomerates, and various more or less phosphatized rocks, predominantly limestones. Among the largest deposits are the layers of relatively pure phosphatic sands of Miocene age situated on the shelf in Southeastern USA, with total resources of about 1.3 billion t of P_2O_5, on the outer Namibian

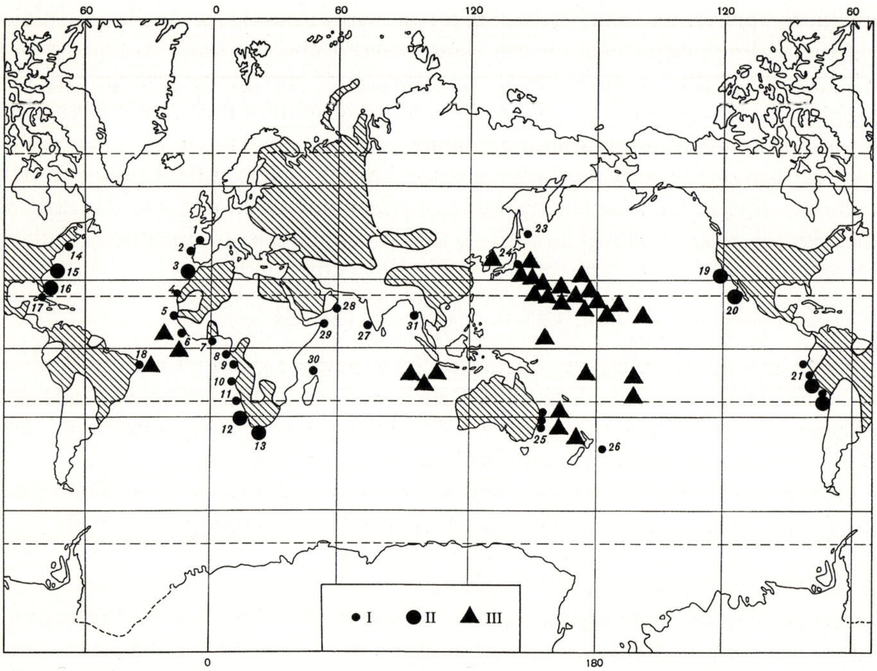

Fig. 57. Distribution of phosphorites over the World Ocean. *I* Isolated samples and small occurrences of phosphorites; *II* significant resources of phosphorites; *III* Phosphorites on subsea mountains. Phosphorites on continental margins, shelves and slopes: *1* Spain; *2* Portugal; *3* Morocco; *4* Sahara; *5* Senegal; *6* Guinea; *7* Ghana; *8* Gabone; *9* Congo; *10* Angola; *11* Namibia; *12* South Africa; *13* Agulhas Bank; *14* Newfoundland Bank; *15* Georgia and North Caroline; *16* Blake Plateau; *17* Pourtales Terrace; *18* Pernambuku Plateau; *19* California Basin; *20* Mexican California; *21* Peru; *22* Chile; *23* Sakhalin Island; *24* Subsea heights in the Sea of Japan; *25* Eastern Australia; *26* Chatham Rise; *27* South-Western India; *28* South-Eastern Arabian Peninsula; *29* Sokotra Island; *30* Cosmoledo Atolls; *31* Amirant Islands. *Hatched areas* may have considerable phosphorite resources on land

shelf (more than 4 billion t P_2O_5) and on the Mexican part of the California shelf (1.5 to 4 billion t P_2O_5).

Many more lower-grade phosphorites are spread along other parts of continental margins, their total resources amounting to 20–30 billion t of P_2O_5.

The resources of phosphorites on subsea mountains have been evaluated for the Sea of Japan, where several heights are covered by sediments intermixed with phosphatic nodules and slabs, and their fragments occasionally concentrated in lenses, their total resources reaching, according to incomplete data, about 200 Mt P_2O_5. On this basis, it might be concluded that total phosphorite resources on subsea mountains throughout the whole ocean, might reach no less then 10 to 20 billion t of P_2O_5, which have been accumulated during several phosphogenic episodes from the Cretaceous to the Miocene.

The quality and agricultural uses of oceanic phosphorites have been tested by several fertilizer companies with a view to their potential practical use in South Africa, New Zealand, the USA, and the former USSR.

All these phosphorites of various age and morphology recovered from the Chatham Rise, Agulhas Bank, Namibian shelf, the heights of the Sea of Japan, and Mid-Pacific mountains proved to be efficient fertilizers even in their natural state but powdered to phosphorite flour without any chemical treatment. The percentage of citric-soluble fraction in oceanic phosphorites, which is the highest compared to on-land phosphorites, is essential for plant nutrition.

Additional experiments performed by Russian fertilizer company on phosphorites recovered from the Sea of Japan have shown that they are equally good for the production of simple and double superphosphate, ammonium phosphate, phosphoric acid, and fodder for cattle and poultry.

Phosphatic sands from all oceanic shelves are easily beneficated by ordinary sieving, and there is no problem in their recovery owing to the shallow depth, short distance to the shore, and loose consistency of the phosphatic sediment.

According to preliminary considerations, the recovery of phosphatic sands like those occurring on the Namibian shelf would cost much less compared to many on-land deposits, but since even high-grade phosphorites are much cheaper than manganese nodules (about \$30–35 per ton), the most serious problem is marketing and transportation. However, the need for phosphatic fertilizers is growing in many regions of the world with no appreciable phosphorite resources in the land, such as the Russian Far East, Japan, Korea, China, and some of the African and South American countries.

It appears that the exploitation of oceanic phosphorites will depend in any specific case primarily on transport expenses, or the distance between the local deposit and the place of destination.

References

Baecker H, Schoell M (1972) New deeps with brines and metalliferous sediments in the Red Sea. Nature 240:153–158

Baturin GN (1969) Authigenic phosphate concretions in recent sediments of the southwest African shelf. Doklady Akad Nauk, Earth Science Sections (English translation) 189: 227–230

Baturin GN (1971) Stages of phosphorite formation on the ocean floor. Nature, Phys Sci 232:61–62

Baturin GN (1978) Fosfority na dne okeanov (Phosphorites on the ocean floor). Nauka, Moscow, 231 pp (in Russian)

Baturin GN (1982) Phosphorites on the sea floor. Elsevier, Amsterdam

Baturin GN (1988) The geochemistry of manganese nodules in the Ocean. Reidel, Dordrecht

Baturin GN, Dubinchuk VT (1989) Microstructures of ferromanganese nodules. Nauka, Moscow (in Russian)

Baturin GN, Rozanova TV (1972) Evidence of ore mineralization in the rift zone of the Indian Ocean. In: Udintsev GB (ed) Investigations of the rift zones of the World Ocean. vol 2 Nauka, Moscow, pp 192–202 (in Russian)

Baturin GN, Merkulova KI, Chalov PI (1972) Radiometric evidence for recent formation of phosphatic nodules in marine shelf sediments. Mar Geol 13, 3:M37–M41

Bischoff JL (1980) Geothermal system at 21°N, East Pacific Rise: physical limits on geothermal fluid and role of adiabatic expansion. Science 207:1465–1469

Bischoff JL, Piper DZ (eds) (1979) Geology and Oceanography of the Pacific manganese nodule province. Plenum Press, New York

Bogdanov YuA, Sorokhtin OG, Zonenshain LP, Kuptsov, VM, Lisitsyna NA, Podrazhanskii AM (1990) Zhelezo-margantseviye Korkii Konkretsii Podvodnykh Gor Tikhogo Okeana (Iron-manganese crusts and concretions on submarine mountains of the Pacific Ocean. Nauka, Moscow, 229 pp

Bostroem K, Peterson MNA (1966) Precipitates from hydrothermal exhalations on the East Pacific Rise. Econ Geol 61:1258–1265

Burns RG, Burns VM (1979) Manganese oxides. In: Burns RG (ed) Marine minerals. Reviews of mineralogy. vol 6. Washington DC, pp 1–46

Burns RG, Burns VM, Stockman HW (1985) The todorokite-buserite problem: further consideration. Am Mineral 70, 1–2:972–980

Charnock H (1964) Anomalous bottom water in the Red Sea. Nature 203:591

Chelishchev NF, Gribanova NK (1983) Sorption properties of oceanic iron – manganese nodules. Geochem 20, 3:73–80

Chelishchev NF, Malikov AV (1988) Pseudomorphous origin of subaqueous manganese nodules and crusts. Doklady Acad Nauk. Earth science sections vol 298, no 1, pp 100–103 (in Russian)

Chukhrov FV, Goshkov AI, Drits VA (1989) Hypergene manganese oxides. Nauka, Moscow (in Russian)

Corliss JB (1971) The origin of metal-bearing submarine hydrothermal solutions. J Geophys Res 76, 33:8128–8138

Cronan DS (1977) Deep-sea nodules: distribution and geochemistry. In: Glasby JP (ed) Marine manganese deposits. Elsevier, Amsterdam, pp 11–44

Cronan DS (1980) Underwater minerals. Academic Press, London

Degens ET, Ross DA (eds) (1969) Hot brines and recent heavy metal deposits in the Red Sea. Springer, Berlin Heidelberg New York

Giovanoli R (1985) Layer and tunnel structures in manganates. Chem Erde 44, (3):227–244

Glasby GP (ed) (1977) Marine manganese deposits. Elsevier, Amsterdam

Govorov IN, Baturin GN (eds) (1995) Guyots of the Western Pacific and their mineralization. Nauka, Moscow (in Russian)

Halbach P, Manheim FT, Otten P (1982) Co-rich ferromanganese deposits in marginal seamount regions of the Central Pacific Basin – results of the Midpac ’81, Erzmetall 35, 9:447–453

Halbach P, Friedrich G, von Stackelberg U (eds) (1988) The manganese nodule belt of the Pacific Ocean. Ferdinand Enke Verlag, Stuttgart
Hein JR, Schwab WC, Davis AS (1988) Cobalt- and platinum-rich ferromanganese crusts from the Marshall Islands. Mar Geol 78:255–283
Hein JR, Morgenson LA, Clague DA, Koski RA (1987) Cobalt-rich ferromanganese crusts from the exclusive economic zone of the United states and nodules from the oceanic Pacific. In: Scholl DW, Grantz A, Vedder JG (eds) Geology and resource potential of the Continental Margin of western North American and adjacent ocean basins – Beaufort Sea to Baja California. Circum-Pacific Council for Energy and Mineral Resources, Earth Science Series, Houston, Texas, vol 6, pp 753–771
Kolodny Y (1969) Are marine phosphorites forming today? Nature 224:1017–1019
Manganese nodules: dimensions and perspectives (1979) Natural Resources Forum Library. UN Ocean Economics and Technology Office. Reidel, Dordrecht
Manheim FT, Lane-Bostwick CM (1989) Chemical composition of ferromanganese crusts in the world ocean: a review and comprehensive data base. Open-File Rep 80–020. USGS, Woods Hole
Mero J (1965) The mineral resources of the sea. Elsevier, Amsterdam
Morozov AA (1995) Manganese in early diagenesis, 2. About the mechanism of manganese nodule formation. Lithol Miner Res 390–401 (in Russian)
Murdmaa IO, Skornyakova NS (eds) (1986) Manganese nodules of the Central Pacific. Nauka, Moscow (in Russian)
Murray J, Renard AF (1891) Report on deep sea deposits, vol 1 In: Report on the scientific results of the voyage of HMS Challenger, Eyre and Spottiswoode, London, pp 1–525
Nicholson K, Hein JR, Buhn B, Dasgupta S (1997) Manganese mineralization: geochemistry and mineralogy of terrestrial and marine deposits. Geol Soc Lond Spec Publ 119
Rona PA, Bostroem K, Laubier L, Smith KL Jr (1983) Hydrothermal processes at seafloor. spreading centers, NATO Conferences Series, IV, Marine Sciences, vol 12, 796 pp
Rona PA, Hannington MD, Raman CV, Thompson G, Tivey MK, Humphris SE, Lalou C, Petersen S (1993) Active and relict sea-floor hydrothermal mineralization at the TAG hydrothermal field Mid-Atlantic Ridge. Econ Geol 85:1989–2017
Skornyakova NS, Murdmaa IO (1995) Ferromanganese nodules and crusts of the Atlantic Ocean. Lithol Miner Res 4:339–361 (in Russian)
Sorem RK, Fewkes RH (1979) Manganese nodules: research data and method of investigation. Plenum Press, New York
Uspenskaya TYu, Skornyakova NS (1991) Textures and structures of oceanic ferromanganese nodules and crusts. Nauka, Moscow (in Russian)
Zelenov KK (1964) Iron and manganese in exhalations of subsea volcano Banu Wuhu, Indonesia. Dokl Akad Nauk 155, 6:1317–1320 (in Russian)

3.3 Hydrothermal Mineralization in the Rift Zones of Mid-Ocean Ridges

Yu. A. Bogdanov

The discovery in 1977–1978 at the East Pacific Rise of high-temperature submarine hot springs with sulfide deposits and unusual fauna was a real sensation. Since that time, scientists have been searching for an investigating hydrothermal sites on the ocean floor. To date, more than 100 active and inactive (relict) hydrothermal fields have been discovered (Fig. 58), mostly in the

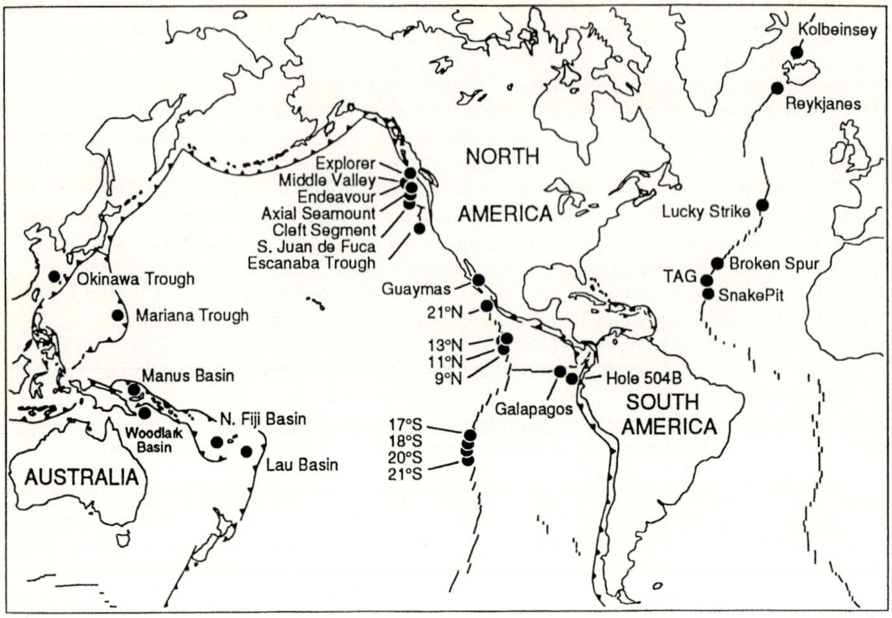

Fig. 58. Hydrothermal sites at ocean spreading centers (After Mannington et al. 1994)

Pacific Ocean, but sites in the Atlantic, Indian Ocean, Red Sea, and the Mediterranean are known, leading to the conclusion that these geological formations are not unique and are widely distributed on the sea floor.

Hydrothermal deposits are enriched in many chemical elements, some of which, e.g., Cu, Zn, Pb, Ag and Au, are of economic interest: deposits have been discovered which contain millions and tens of millions of tons of ore.

Polymetallic sulfide deposits on the ocean floor are of interest not only as a future mineral resource. These deposits are analogs of ancient hydrothermal sulfide deposits exploited on land. Scientists now have a natural laboratory where they can study the formation of hydrothermal ores in detail and create genetic models of this process.

Tectonic Control of Ore Deposition

Mid-ocean ridges, where the majority of hydrothermal sulfide deposits have been discovered, encircle the earth in all the oceans and form a united rift system 55 000 km long. Similar geological structures (back-arc spreading centers) have also been found in the marginal seas. In these structures lithospheric plates diverge, mantle material rises from depth, and new ocean crust is formed.

Although the mid-ocean ridges have common structural elements, their morphologies depend to a large extent on spreading rate (Fig. 59). The central

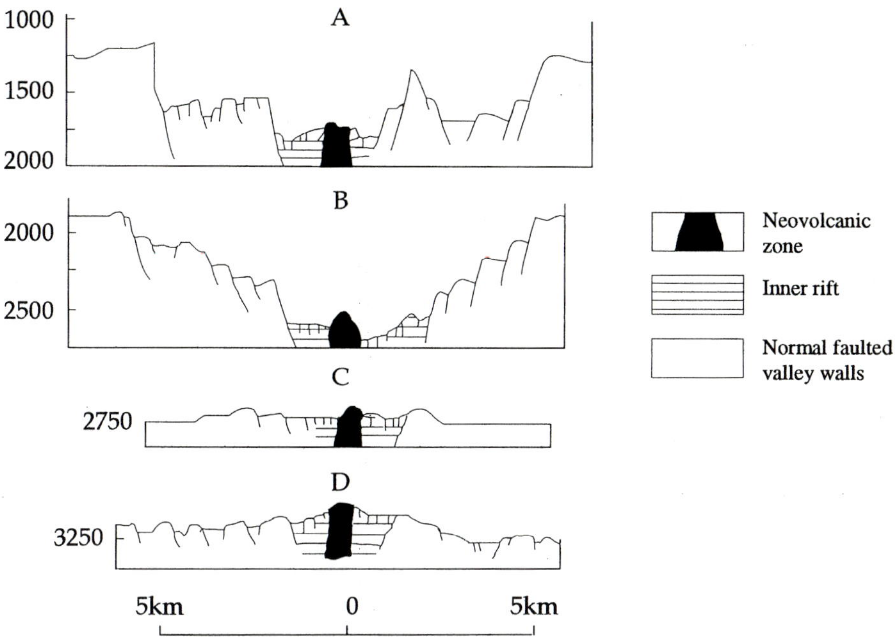

Fig. 59 A–D. Cross-sections of the rift zones, with different spreading rates. **A** The Red Sea rift at 18°N, spreading rate 1.5 cm a^{-1}. **B** The Mid-Atlantic ridge at 37°N, spreading rate 2.5 cm a^{-1}. **C** The East Pacific Rise at 21°N, spreading rate 6 cm a^{-1}. **D** The East Pacific Rise at 3.5°S, spreading rate 15 cm a^{-1}

part of the rift is occupied by a 10–20-km-wide tectonically and volcanically active zone. The tectonic zone includes the inner rift (or rift valley) and the block-faulted marginal scarps within a clearly tensional regime. As a rule, in the central part of the inner rift there is a neovolcanic (or extrusive) zone several hundred meters wide.

Ocean rifts are divided by transform faults into individual segments several tens to hundreds of kilometers in length. Within the neovolcanic zone of each segment there is one or more accumulative volcanic structure (dome) which may be thermally active in comparison with the surrounding rift elements. They represent the surface expression of subsurface magma chambers. Formation and growth of a magma chamber is a cyclic process. After a period of magmatism, a volcanic phase is replaced by a nonvolcanic one with intensive vertical faulting, collapse of the chamber roof, and the formation of axial grabens and calderas. The duration of such a cycle is 5000–10 000 years.

Hydrothermal sites are predominately found within neovolcanic zones, on the surface of the domes, and amongst young volcanics, the integrity of which has been disturbed by tectonic processes. It is important to emphasize that this indicates that the main part of hydrothermal deposits are formed during the nonvolcanic cycle after the extinction of volcanic activity and collapse of the magma chamber roof.

Recently a new type of hydrothermal deposits was discovered on abnormal walls of an inner rift of the mid Atlantic Ridge.

Two types (normal and abnormal) of faulted scarps bounding the central rift are known. Normal walls of the central rift include fault terraces forming a staircase structure. Fault scarps facing the inner floor dip steeply with an angel of ~60° with terraces corresponding to back-tilted blocks rotated 5–10°. Fault scarps may coincide with listric faults.

Abnormal walls of the central rift are formed either by blocks of ocean crust uplifted considerably higher than normal walls, with gabbro and sometimes ultramafic outcrops in the lower parts of the scarps, or by protrusions of serpentinized ultramafics and serpentinites. Abnormal rift valley walls may be the surface manifestation of serpentinization of mantle ultramafics resulting from reaction with seawater.

Classification of Mid-Ocean Ridge Hydrothermal Sulfide Deposits

Two main groups of hydrothermal sulfide deposits have been discovered in ocean rifts (1) deposits of axial parts of inner rifts associated with the interaction of seawater with mafic rocks, and (2) deposits of abnormal fault scarps of rifts (Table 12) associated with the interaction of seawater with ultramafic rocks.

Morphology, composition, and properties of axial hydrothermal deposits depend considerably on the conditions of hydrothermal discharge. The three main types of discharge that occur at ocean rifts are: (1) at the surface of a volcanic basement (bare rock or sediment-starved systems), (2) within or at the surface of soft sediments overlying volcanic basement (sediment hosted), and (3) at the surface of a volcanic basement where there is a high-salinity bottom-water brine.

Sources of Ore Minerals Composing Hydrothermal Deposits of Ocean Ridges

Hydrothermal ore-bearing fluids are formed due to the interaction of seawater with rocks of the ocean crust. Within ocean rifts, seawater penetrates into the ocean crust through a system of open fractures continuously formed due to the divergence of lithospheric plates. Circulating seawater is heated to the critical temperature, acquires buoyancy, and rises to the sea floor. Hydrothermal circulation systems are formed (1) in the central rift where hydrothermal cells form above shallow magma chambers (axial circulation systems), and (2) at slow-spreading ridges with narrow individual magma chambers where seawater

Table 12. Classification of hydrothermal deposits from ocean rifts

Main rock types	Types of hydrothermal deposits	Location of major hydrothermal deposits
A. Deposits associated with the alteration of mafic rocks	I. Deposits of sediment-starved ridges	
	1. Mature rifts	Galapagos rift, East Pacific Rise, Juad de Fuca Ridge (southern, Axial Seamount, Endeavour Segment), Mid-Atlantic Ridge (TAG, Snakepit, Broken Spur, Rainbow, Menez Gwen), Reykjanes Ridge
	2. Rifts in the early stage of evolution	Franklin Seamount (Woodlark Basin of the SW Pacific)
	3. Rifts of back-arc-spreading zones	Lau Basin, Manus Basin, North Fiji Basin, Mariana Trough
	II. Deposits of sediment-covered ridges	
	1. Mature rifts	Middle Valley (N. Juan de Fuca), Escanaba Trough (S. Gorda Ridge)
	2. Rifts in the early stage of evolution	Guaymas Basin (Gulf of California)
	III. Deposits of rifts with high salinity bottom brines	
B. Deposits associated with the alteration of ultramafic rocks		14°54′ and 14°45′N, Mid-Atlantic Ridge

penetrates through fractures into the deeper crust and the uppermost mantle (deep circulation systems).

Axial Circulation. In the axial circulation system, seawater heated to ≈400 °C reacts with extrusive basalts or sheeted dolerite dykes, which undergo greenschist facies metamorphism. During such interactions there is chemical exchange between the fluid and rocks. The degree of seawater transformation to a high-temperature ore-bearing fluid can be shown on a plat of chemical element concentrations of fluids compared to seawater (Fig. 60), Mg, SO_2^{2-}, and U enter the solid phase during the water-rock interaction, though some seawater sulfate is replaced by sulfide, reducing the pH. Concentrations of Sr, B, Na, and Cl are not changed during this interaction although there may be chemical exchange. For example, the $^{87}Sr/^{86}Sr$ isotopic ratio of hydrothermal fluids

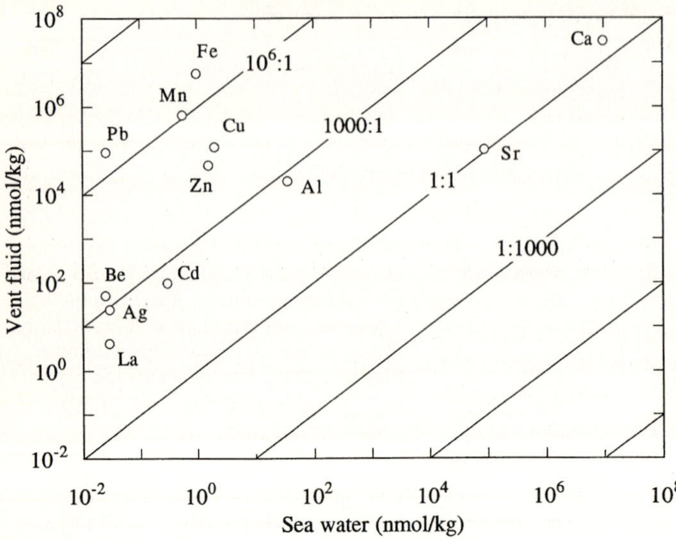

Fig. 60. Comparison of the chemical composition of hydrothermal fluid compared with seawater

reflects a mantle source although concentrations are similar or that of seawater, whilst in the case of B, the $^{11}B/^{10}B$ ratio is altered through low-temperature interactions during downwelling. Most other chemical elements are enriched in hydrothermal fluid relative to seawater. Concentrations of the main hydrothermal ore metals (Cu, Zn, Fe, and Mn) in fluids are 10^3–10^7 times higher than seawater.

The relative similarity of the compositions and properties of submarine hydrothermal fluids and hydrothermal deposits from different ocean rifts results from the similarity of P-T conditions during the transformation of seawater to high-temperature ore-bearing fluids in the reaction zone, and from participation of the same components (seawater and MORB) in the reaction. Some dissimilarities in hydrothermal fluids and chemical composition of hydrothermal deposits from back-arc spreading centers and ones from mid-ocean rifts are due to the participation of nonbasaltic rocks (e.g., andesites, rhyolites) in the seawater transformation.

Most of the metals present in hydrothermal fluids form chloride complexes which are stable in a wide range of physicochemical conditions. These complexes are destroyed when the high-temperature reduced fluid mixes with strongly oxidizing seawater. As a result, released metal ions interact with sulfide ions present in the hydrothermal fluid and form sulfide minerals.

Deep Circulation. In the deep hydrothermal circulation system, the reaction zone is located deeper than in the axial circulation system under conditions of higher temperature and pressure. The rocks of the deeper layers of the ocean crust and of the uppermost mantle, including ultramafics, interact with seawater. At 400–500 °C and 1 kbar, serpentinization takes place. Serpentinization reactions occur before, during, and may persist after mantle uplift by downwelling of seawater along crust-penetrating faults. Primary ore-bearing fluids resulting from this reaction should differ from ones of the axial circulation system by higher temperature and salinity as well as by chemical composition. For example, such fluids have been found to be extremely enriched in methane compared to fluids of the axial circulation systems. Considerable changes in hydrostatic pressure during migration of the fluids through the whole ocean crust towards the sea floor cause their repeated boiling, during which the physical and chemical characteristics of the fluids are changed. Boiling results in oversaturation and precipitation of some chemical elements carried by the fluids, as well as the formation of disseminated mineralization within the ocean crust. Only the residual ore-bearing solution reaches the sea floor.

Deposition of Hydrothermal Matter from Axial Hydrothermal Circulation

Deposits of Sediment-Starved Rifts. Deposits of this type are the widest-spread in ocean rifts. At sediment-starved rifts, primary high-temperature ore-bearing fluids leave the volcanic basement at the sea floor. Probably <5% of hydrothermal elements supplied by the ore-bearing fluids are deposited near vents to form massive sulfide deposits. The rest dispersed as particulates or in dissolved form by bottom currents.

Hydrothermal deposits accumulating on bare rock surfaces usually form mineralized chimneys 2–30 m in height. A great variety of chimney morphologies have been found: columns, spires, bulbous "onion domes", with single or multiple orifices, with or without side flanges. Over time, collapsed and consolidated chimney debris may form mounds meters to tens of meters across and several meters high. The mean reserve of ore in these mounds is about 1000 t but several gigantic mounds are known. For example, the active mound in the TAG hydrothermal field (26°N, Mid-Atlantic Ridge) contains about 5 Mt of hydrothermal matter, and a relict hydrothermal mound containing about 10 Mt of hydrothermal matter was discovered in 1991. In addition, there are several mounds containing 3–5 Mt at the Explorer Ridge in the Pacific Ocean and at 13°N, East Pacific Rise.

As a rule, there occurs a well-pronounced lateral zonation in the distribution of minerals within hydrothermal deposits (Fig. 61). Higher-temperature minerals are typical in the central active parts of the mound, and are gradually replaced by lower-temperature minerals towards the margins. The lateral succession of the composition of hydrothermal deposits reflects the change of temperature in ore-forming fluids. The temperature, composition, and properties of ore-forming fluids depend on the degree of mixing of primary hydrothermal fluids with seawater. This general pattern is complicated due to shifting of the

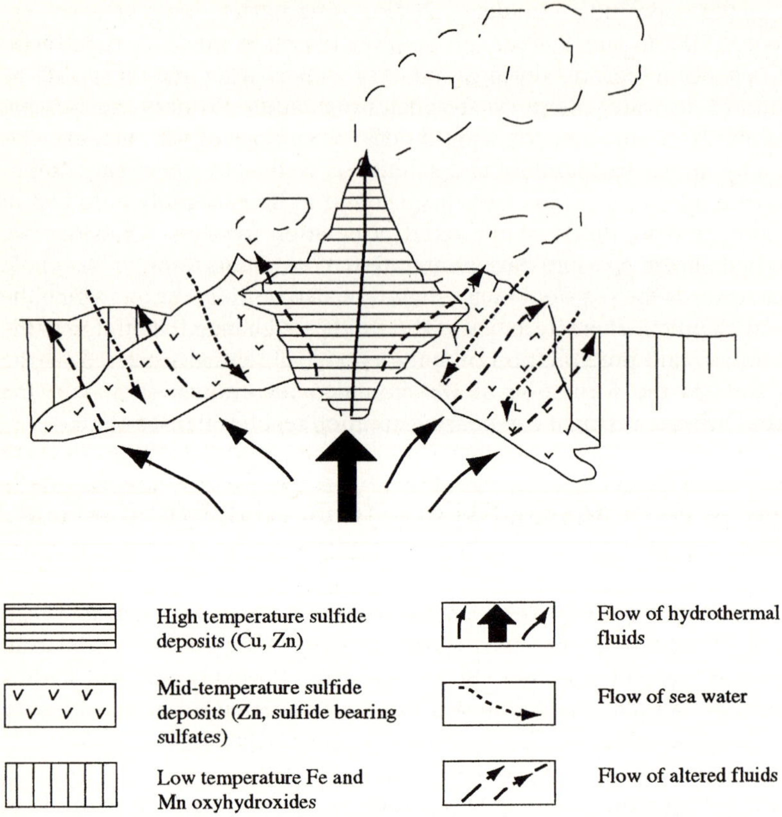

Fig. 61. Schematic cross-section of sulfide zonation in hydrothermal deposits of sediment-starved ridges

hydrothermal conduits and metasomatism caused by repeated warming up of the mound.

Sulfides of Fe, Cu (pyrite, chalcopyrite), and, more rarely, of Zn occur in higher-temperature deposits in the central parts of large mounds. They are replaced predominantly by sulfides of Zn and Fe (sphalerite, wurtzite, pyrite, and marcasite), more rarely of Pb (galena). Nonore metals (sulfates, silica, calcite, etc.) become more important components of the marginal parts of the mounds where sulfides occur in disseminated form. Reduced facies of hydrothermal deposits are replaced by oxidized marginal facies. Transitional facies conditions are usually represented by nontronite deposits. Oxides and oxyhydroxides of Fe and Mn are the main ore minerals of the oxidized facies of hydrothermal deposits. The lateral succession of solid hydrothermal phases is often disturbed due to the instability and cyclicity of hydrothermal activity.

Sometimes, only the mid- and low-temperature parts of this succession occur on the sea floor. In these cases, the high-temperature members of the succession are absent due to their precipitation in subsurface layers of the ocean crust where primary hydrothermal fluids meet cold bottom water circulating through fissures. The interaction of rising hydrothermal fluids with subsurface high-temperature deposits can lead to the process of element remobilization and redeposition on the sea floor. High concentrations of Ag and Au in sphalerite and sulfide-barite deposits may be related to the process of remobilization.

Deposits of Sediment-Covered Ridges. Sediment-covered ridges occur in areas of very high fluxes of sedimentary material. For example, the recent sedimentation rate in the Gulf of California is 1000 times higher than at mid-ocean ridges. Here, spreading and volcanism forming the new ocean crust occurs under a thick layer of sediments. High-temperature ore-bearing fluids emanate from the volcanic basement below the sediment layer and the rising hydrothermal fluid migrates to the sea floor through this layer (Fig. 62).

The boundary between bottom sediments and the volcanic basement acts as the meeting zone between primary high-temperature, ore-bearing fluids and cold porewaters. In this zone ore deposition begins. We can expect deposition of high-temperature ores near this boundary and in the central part of the upwelling zone to be similar to deposits accumulated on sediment-starved rifts. However, deposits of the peripheral part of his zone and of hydrothermal

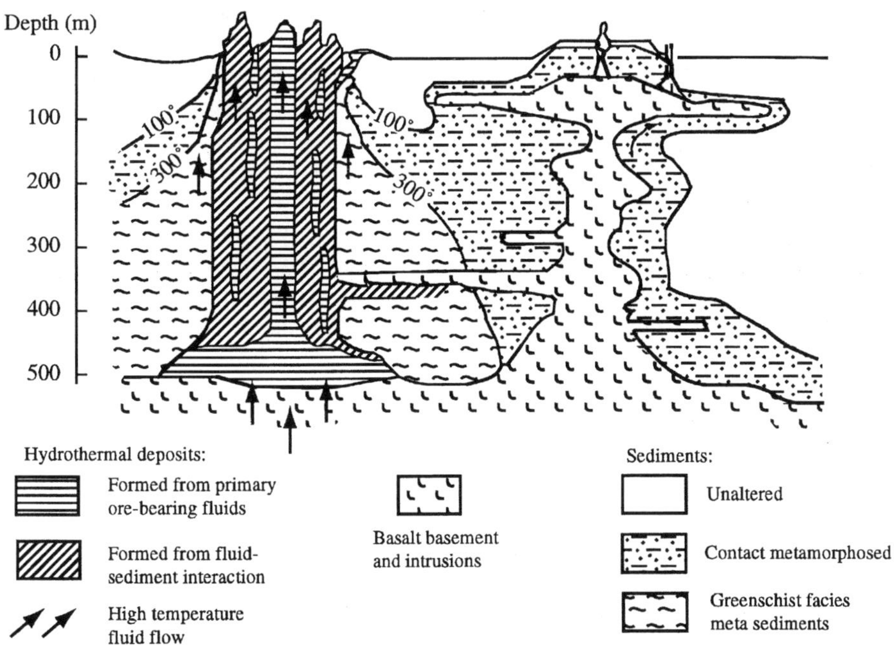

Fig. 62. Schematic cross-section of hydrothermal deposits and alteration at sediment-covered ridges, Guaymas Basin, Gulf of California

mounds on the sea floor will be compositionally distinct. During migration through the sediment layer, there is chemical exchange between the solutions and sediments and the primary high-temperature fluid is transformed. Hydrothermal solutions are enriched in K, Rb, Ca, Sr, and Ba due to leaching of biogenic and terrigenous matter and the pH of the fluid increases to 6. The sediments undergo greenschist facies metamorphism. Thermolysis of plantonogenic organic matter results in increased NH_4 and generation of hydrocarbons.

In sea-floor deposits, nonore minerals are common, e.g., silica, calcite, and sulfates of Ba and Ca. The main sulfide minerals are pyrrhotite, sphalerite, and chalcopyrite. It is important to emphasize that pyrite is absent in these deposits. High contents of hydrocarbons are found in the porewaters of massive hydrothermal deposits.

Hydrothermal systems of sediment-covered rifts are considerably more efficient at forming large deposits than those of sediment-starved rifts. Reserves of hydrothermal matter in these deposits reach several tens of millions of tonnes.

Deposits of Rifts with High-Salinity Bottom Brines. Such deposits are typified by the Red Sea rift. The Red Sea rift is very young: rifting of the thinned continental crust and beginning of spreading is dated as 3.4–4.0 Ma. Thick layers of Miocene salts were formed before the beginning of the spreading episode on the rift walls. Salts dissolved in seawater form high-salinity brines with reduced Eh, which fill some deeps in the rift zone. In these deeps, high-temperature fluids discharge into the brines. A small part of hydrothermal matter is deposited near hydrothermal orifices; most is precipitated and is dispersed within the bottom waters. The reducing Eh of the brines preserves the sulfides from oxidation and dissolution. In the Atlantis II deep, a layer of metalliferous sediments up to 10–30 m thick has accumulated around hydrothermal orifices (Fig. 63). It is predominantly composed of fine-grained muds with a high (up to 95%) content of high-salinity porewaters. Fe and Mn oxyhydroxides, Fe, Cu, and Zn sulfides, smectites, Ca and Ba sulfates, Cu and Fe chlorides, and native metals have been identified in the muds. Within the layer of metalliferous sediments the concentrations of these minerals are very viable. Reserves of hydrothermal matter in the Atlantis-II deep reach some hundred Mt.

Deposits Associated with the Alteration of Ultramafic Rocks

Hydrothermal deposits associated with the alteration of ultramafics have been found at 14°54′N and, more recently, at 14°45′N. The deposits at 14°15′N occur at the intersection of the Mid-Atlantic Ridge with the Fifteen-Twenty Fracture one, in the form of a series of dome-shaped features 1.5 km in diameter. Dredging has confirmed the presence of gabbroic and serpentinized ultramafics in the domes, which are interpreted to be either mantle diapirs or detachment fault blocks. The hydrothermal fluids have a high ratio of gases to metals, and are particularly enriched in methane.

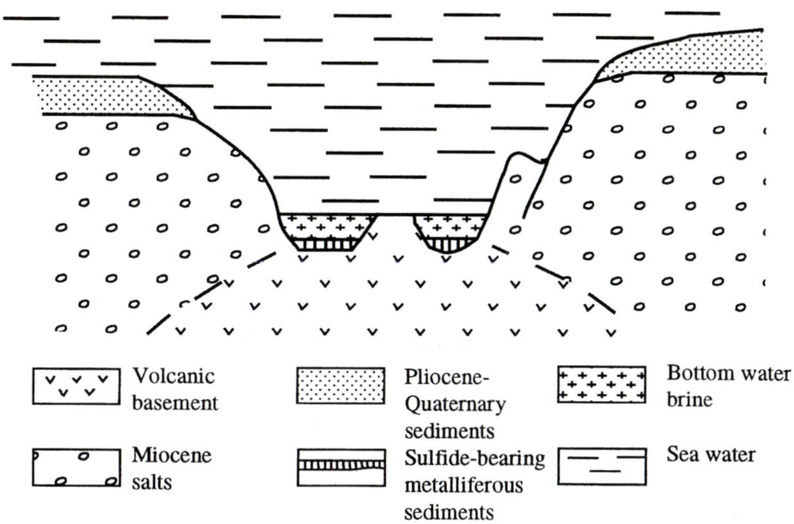

Fig. 63. Origin of hydrothermal deposits in rifts with near-bottom high-salinity brines (Atlantis II Deep, Red Sea)

At 14°45′N in the Mid-Atlantic Ridge rift zone, the "normal" western slope of the rift valley is formed by a system of normal faults (typical listric faults) composed of basalts. The eastern slope of the rift valley is much steeper than the western slope and is composed of ultramafic rocks with different degrees of serpentinization (Fig. 64).These rocks are exposed on the steep faults of the slope. Flat parts of the bottom are covered by a layer of soft carbonate and metalliferous sediments up to 2.5 m thick. The hydrothermal field is located in a zone of serpentinized ultramafic rocks.

The hydrothermal field has developed as a mound 200 m long, 100 m wide and 5–10 m high. This construction extends up the slope in the southeastern direction between the depths of 3010 and 2930 m. Three active hydrothermal areas up to 10 m in diameter have been discovered within the field at 3005, 2950, and 2940 m. Each active area includes three to four black smokers. The marginal part of the hydrothermal mound is composed of iron-silicate deposits with veins and inclusions of Fe-oxyhydroxides and nontronite. They are covered by a thin black crust formed of Fe and Mn oxyhydroxides. Sulfide deposits occur near the black smokers. These deposits are similar to deposits of other known hydrothermal fields of the Mid-Atlantic Ridge. Pyrite, chalcopyrite, bornite, and sphalerite are the main mineral phases, but the sulfides are relatively enriched in Co and As. Hydrothermal pentlandite (Ni sulfide) is common in surrounding serpentinized ultramafics and has been found in the sulfide deposits.

Above, we have described the main types of hydrothermal deposits formed in ocean rifts. Within each type it is possible to distinguish variations in the nature of the hydrothermal deposits connected with such factors as the maturity

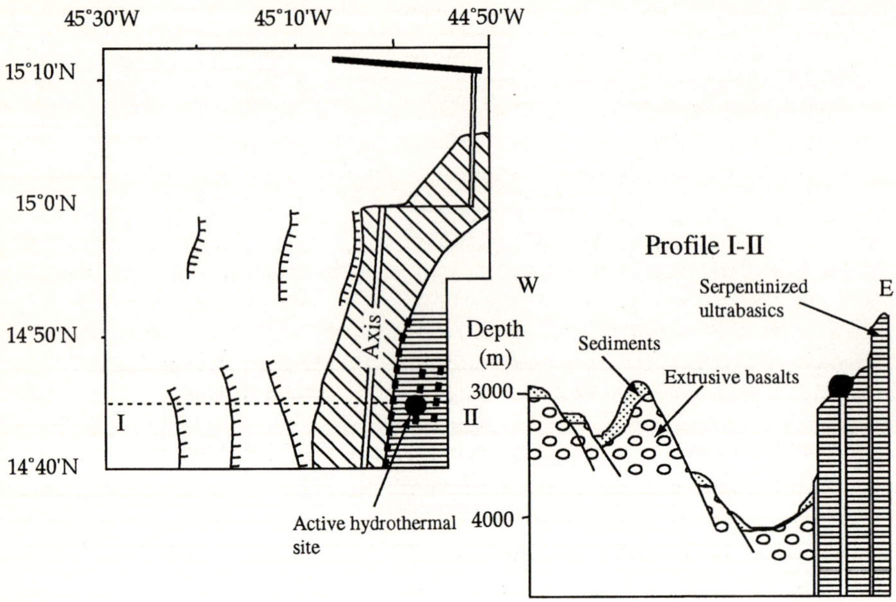

Fig. 64. Geological structure and cross-section of the 14°45′N hydrothermal field, Mid-Atlantic Ridges

of rifts, the spreading rate, the position of spreading centers (open ocean, back arc basin), etc., but these variations in the composition of deposits are considerably less than between deposits of different types.

The Dispersion of Hydrothermal Material and Formation of Metalliferous Sediments

Only a fraction of hydrothermal material supplied to the sea floor in ocean rifts accumulates in hydrothermal deposits. As hydrothermal fluids mix with seawater, some chemical elements are precipitated from solution and transferred to suspended matter. A part of this suspended matter sinks to the ocean floor and forms metalliferous sediments. A residual part of dissolved chemical elements and gases is dispersed in seawater.

The mineral and chemical composition of suspended matter depends on the physical and chemical characteristics of hydrothermal solutions. Black smokers occur in the highest temperature (250–400 °C) zones of discharge. Sulfides (pyrrhotite, pyrite, sphalerite, wurtzite, chalcopyrite, marcasite), sulfates (anhydrite, gypsum, barite), silicates of Fe and Ca, elemental sulfur, and graphite have been identified in black smoke. Amorphous silica, barite, native sulfur, and pyrite are formed from low and intermediate temperature hydrothermal solutions (white smoke). These minerals are precipitated directly near the orifices of hydrothermal vents and are dispersed at relatively short range (10s–100s of meters).

High-temperature hydrothermal fluids have positive buoyancy, and after entering the ocean, rise through the water column. The vertical flow of hydrothermal fluids entrains seawater and mixes with it. During this mixing, rising hydrothermal fluids are diluted by seawater several tens of thousand times, lose buoyancy, and cease to rise at the depth of neutral buoyancy. This new water mass (the "hydrothermal plume") moves laterally with water currents and is gradually dispersed. Hydrothermal plumes are enriched in some chemical elements supplied by hydrothermal fluids, but also in elements scavenged or coprecipitated from entrained seawater. Fe and Mn oxyhydroixides are the principal minerals of suspended matter formed near hydrothermal vents. This matter falls out of plumes, sinks to the sea floor, and takes part in the formation of metalliferous sediments.

Four types of metalliferous sediments have been found in hydrothermal fields of the ocean rifts. The first type of sediment is mainly composed of unsorted coarse-grained hydrothermal debris accumulated during periods of intensive hydrothermal discharge and destruction of hydrothermal mounds. They consist of Fe and Mn oxyhydroxides, nontronite, and sulfides. The hydrothermal debris is cemented by silty-clayey material mainly composed of Fe oxides, euhedral sulfide crystals (as well as rare sulfide druses), and sometimes of amorphous silica. The second type of sediment is mainly represented by variegated sandy, partly sorted material composed of the same components as the first type. Both types appear to have resulted from the horizontal transport of hydrothermal mound debris. These sediments occur at a distance of less than 100 m from the hydrothermal mounds.

The third and fourth type of sediments are typical metalliferous, formed during particle-by-particle sedimentation. The difference between these two types is that the third-type sediments have a significant admixture of euhedral sulfide crystals and represent proximal plume fallout deposits. Occasionally, the sulfides represent the major component of the sediments and form sulfide bands. Fe oxyhydroxides are the main hydrothermal components of the fourth-type sediments. Very often, pelagic material containing biogenic remains (foraminifera and coccoliths), terrigenous clay minerals and zeolites occur in metalliferous sediments. The third-type sediments cover the sea floor up to 500 m from the hydrothermal vents. The fourth-type sediments are the most widespread and sometimes occur at the distance up to 2000 km from hydrothermal sources.

Conclusions

Hydrothermal processes of the ocean rift occupy a particular place in the global system of formation, transportation, and deposition of sedimentary material. A volume of seawater equal to that of the World Oceans passes through the hydrothermal circulation system every ≈3 Ma, i.e., almost instantly from a geological point of view. During hydrothermal circulation, some chemical elements and ions (Mg, SO_4^{2-}, etc.) are extracted from seawater by the ocean crust. Hydrothermal fluids supply large amounts of some chemical elements to the oceans. For example, the hydrothermal input of Mn to the oceans is eight to ten times higher than the continents. Considerable parts of Si, Li, and other elements in ocean water are also of hydrothermal origin. Undoubtedly, in the near future, widely distributed massive hydrothermal deposits will be enormous interest for economic geology.

Considerable parts of hydrothermal solutions and gases are dispersed in ocean water and in the atmosphere, and influence (or maybe determine) their chemical composition. Moreover, hydrothermal processes produce a large amount of sorbents (Fe and Mn oxyhydroxides) that extract chemical elements, including biogenic ones (P, C, and N compounds) from seawater and transport them into sediments.

The significance of ocean floor hydrothermal systems exceeds the limits of the study of ore-forming processes. Only now is a true picture of the huge scale of hydrothermal matter input, its influence on the atmosphere, hydrosphere, and lithosphere (i.e., on the evolution of the exosphere) beginning to be drawn.

References

Barrett TJ, Fox JS (1988) Seafloor hydrothermal mineralization. Can Mineral 26(3):429–888

Bogdanov YuA, Zonenshain LP, Kuzmin MI, Lisitzin AP, Podrazhanskii AM, Sagelevich AM (1990) Mystery of the deeps. Progress, Moscow, 383 pp

Bogdanov YuA, Sagalevich AM, Chernyaev ES, Ashadze AI, Gurvich EG, Lukashin VN, Peresypkin VI (1995) 14°45′N hydrothermal field of the Mid-Atlantic Ridge. Dokl Rossi Akay Nauk 343(3):353–357 (in Russian)

Gramberg IS, Ainemer AI (eds) (1992) Hydrothermal sulfide ores and metalliferous sediments in the ocean. Nedra, St Petersburg, 278 pp (in Russian)

Humphris S, Zierenberg RA, Mulineaux LS, Thomson RE (eds) (1995) Seafloor hydrothermal systems. Geophysical monograph series, 91, American Geophysical Union, Washington, DC

Jones M (ed) (1985) The hydrothermal vents of the Eastern Pacific Ocean: an overview. Bull Biol Soc Wash, 6

Lisitzin AP, Bogdanov YuA (eds) (1986) metalliferous sediments of the Red Sea. Nauka, Moscow, 288 pp (in Russian)

Lisitzin AP, Bogdanov YuA, Gurvich EG (1990) Hydrothermal deposits of ocean rift zones. Nauka, Moscow, 256 pp (in Russian)

Morton JL et al. (eds) (1994) Geological, hydrothermal and biological studies at Escanaba Trough, Gorda Ridge, offshore northern California, US Geol Surv Bull, 2022:

Parson LM, Walker CL, Dixon DR (eds) (1995) Hydrothermal vents and processes. Geol Soc Lond Spec Publ 87:396 pp

Rona PA (1984) Hydrothermal mineralization at seafloor spreading centers. Earth Sci Rev 20:1–104

CHAPTER 4

Biomineralization

4.1 Tracers of Evolution. Impact on the Biosphere

I. S. Barskov

"Biomineralization is many different things to different organisms" (Chave 1984); however, there exists a general significance of the phenomenon of biomineralization as a whole. It belongs to fundamental aspects of the characteristics of nature and has to be considered in the philosophical systems of the world.

Biominerals are the result of interaction between two different types of the state of matter: mineral (inert) matter and the stuff of life. Moreover, the stuff of life itself, as well as the phenomenon of life, must include as their immanent constituents mineral matter (albeit as their inert part). This was the case in the initial, permanent separation and coexistence which already began in cosmic space with coexistent mineral dust and organic molecules in the interstellar clouds. This is the interaction of the living and inliving; the creation of the "secondary" inliving nature by the living nature by use of its own laws. It manifests their coevolution in geological history as an indispensable condition of biological evolution.

Life as one of the types of the states of matter in the Universe includes a biomineral framework formed by the organisms and interacting with their organic matrix.

The emergence of multicellular animals and diversification in the Cambrian explosion is closely related to biomineral skeleton formation.

Only the biomineralized components of extant organisms compose the paleontological chronicle of the evolution of life. Because of these biomineralized fossils, the description of life can be obtained not only for its contemporary state but as a whole from its beginning and through all its diversification.

Paleontology, having as its only empirical material fossil remains of organisms, owes its existence to biominerals.

- The preservation of biomineralized fossils in geological history is responsible for its biogeochronological scale. Stratigraphy has its basis in biomineralized fossils.
- Bacterial interaction with the inorganic lithosphere led to the formation of limestones, siliceous sedimentary rodes, phosphorites, iron quartzites, manganese nodules on the ocean floor, crusts of weathering, and zones of oxidation in ore deposits. Geochemical cycling of sulfur, carbon, oxygen, phosphorus, silicon, iron, and other elements is closely connected with bacterial activity.

4.2 Diversity and Phylum Distribution of Biominerals

I. S. BARSKOV and A. Y. ROZANOV

There are four types of interaction between minerals and organisms:

1. Biomineralization proper: minerals formed by organisms (including bacteria) and plants, and existing as skeletons, shells, carapaces, scales, teeth, bones.
2. Biomineralization processes mediated by microorganisms.
3. Permineralization (organomineralization) of extant organisms and plant tissues (wood petrification, soft-bodied animals, and plant permineralization).
4. Possible mineral-induced (clays, zeolites, abiogenic organic compounds) origin of life.

Selectivity of Biominerals

The composition of the stuff of life, in spite of the multitude of species, is determined principally by a few elements: oxygen (70%), carbon (18%), hydrogen (10%), and 2% all other elements. Similarly, this limited set of elements determines the composition of biominerals formed by organisms: Ca, which is most widespread, CO_3 and PO_4, and Si as silica.

Thus, only three principal groups of biominerals exist:

1. Calcium phosphates as apatite (two varieties: fluorcarbonate apatite francolite and hydroxycarbonate apatite dahllite; amorphous precursors of apatite, and rarely other calcium phosphates).
2. Calcium carbonates (calcite and aragonite; rare vaterite and protodolomite).
3. Silicon always and only as opal.

Iron minerals (magnetite and others, mostly in magnetotactic bacteria) have a special place in biomineralization.

About 80 biominerals have now been determined. Those occurring in only a few species are called exotic (celestine in radiolarian, barite fluorite, and others).

A list of the principal biominerals and their phylum distribution is presented in Table 13. Other biominerals are related to physiological, metabolic, and behavioral functions of organisms, element deposition (mainly calcium and phosphorus), detoxication of organisms, function of space orientation (otoconia and otolites in different mobile animals from medusae to Man), and geomagnetic field reception by organisms (from magnetotactic bacteria up to fishes, birds, and Man).

There are also several organic biominerals (mostly calcium oxalates) which are mainly products of pathological metabolism.

Table 13. Distribution of the main groups of biominerals in the highest taxa

Minerals	Taxa of the organic world																	
	Bacteria	Animals													Plants			
		Protozoa		Metazoa											"Algae"			High
	1	2	3	4	5	6	7	8	9	10	11	12	13	14	15	16	17	18
Calcium carbonates																		
Calcite	+	+		+	+	+	+	+	+	+	+	+		+	+			+
Aragonite		+				+	+	+		+				+				
Vaterite								+						+			+	
Protodolomite												+		+				
Monohydrocalcite	+							+						+				
Amor. hydrous carbonate	+							+	+					+				+
Phosphates																		
Hydroxylapatite	+						+						+	+				
Francolite	+							+			+			+				
Dallite	+							+						+				
Brushite								+										
Am. calcium phosphates						+	+	+	+	+		+		+				
Silica																		
Opal			+	+			+	+	+			+				+		+

Explanations: 1 – Bacteria as whole, 2 – Foraminiferida, 3 – Radiolazian, 4 – Porifera, 5 – Archaeocyata, 6 – Coelenterata, 7 – Annelida, 8 – Mollusca, 9 – Arthropoda, 10 – Bryozoa, 11 – Brachiopoda, 12 – Echinoderinata, 13 – Conodonta, 14 – Chordata, 15 – Red, brown, green algae, 16 – Diatomean, 17 – Charophyta, 18 – Vascular plants.

"Strange minerals": Special Features of Biominerals

- Occur in organisms as ultrafine grained crystallites, of nanometer or micron size, usually in well-ordered faceted particles: biomagnetite cubo-octahedra, needle-shaped fluorite crystals, calcitic and at agonic platelets, prisms, etc. The single crystals form different types of structures: globular, spherulitic, prismatic. The aggregate units are arranged in an orderly fashion with crystallographic axes aligned.
- Each of these microarchitectural units is enveloped by an organic matrix sheath and includes an intracrystalline organic matrix.
- Special state of the biominerals: their behavior as participants of the living processes continuously during the whole life time of an organism, with renewal of the mineral matter.
- Amorphous "preminerals" or "precursors" are characteristic as the initial stage of biomineral formation, and their transformation within several days

in crystalline biominerals; neutron and X-ray amorphography are used to study these stages (amorphous apatite precursor → apatite, amorphous calcite → calcite, etc.).

- The genetic ability of organisms to form different biominerals is related to structural genes that control matrix-mediated formation of different biomineral species (genes of apatite, genes of magnetite, etc.).

Other minerals occur more rare or are exotic.

Halides (fluorite and hieralite) occur in gastropod jaws as crushers of shelled prey, in some pelagic crustaceans as gravity receptor statocysts.

Within sulfate minerals, gypsum in the statoconia of medusae is a gravity receptor. The presence of celestine in some radiolarians is unusual. The mineral is a regulator of bouyancy in one of the ontogenetic stages, and of skeletal material in an other stage. The function of barite forming in the mantle of some gastropods is unknown.

Iron oxides – goethite, lepidocrocite – present in jaw structures of mollusks and vertebrates strengthen these structures. Widely distributed ferrihydrite as a temporary storage of iron in various organisms.

Manganese oxides (todorokite, birnessite) and a wide variety of sulfides (pyrite, hydrotroilite, sphalerite, wurtzite, galena, greigite, mackinawite, and others), jarosite and Fe-phosphates such as vivianite and others are products of bacterial activity.

The fundamental process of biological regulation of crystallization (Sikes 1993) has been studied by comparison between computer modeling of the interaction of the basal surface of calcite and protein molecules (this surface exhibits hexagonal morphology and is a homoplane of calcic or carbonate ions) with direct observation at the atomic level at atomic force micrographs of protein bound to calcite surfaces.

Trace element impurities, stable isotopes, crystal habit can be used to distinguish biogenic from inorganically formed minerals. Mechanisms, rates, and periodicity of biomineralization are different in individual groups of organisms.

Two Types of Biomineral-Forming Processes

1. Organic matrix-mediated or biologically controlled process (for example, in numerous animals): macromolecules of the organism can control; (1) sites of a biomineralization (cell vacuoles, intracellular organelles, intercellular space, specially formed membranes, extrapallial space, etc.); (2) mineral nucleation (transportation of the appropriate ions and initiation of primary nucleation and growth); (3) crystal growth (primary and secondary); (4) termination of crystal growth.

The biomineral species, crystallographic orientation, and microarchitectures are under genetic control.

2. Biologically induced process (for example, in some bacteria, green and brown algae; formation of sulfide minerals by sulfate-reducing bacteria; pro-

ducts of metabolism): macromolecules are responsible only for sites and nucleation of organic matrices; whole composition and habitus of crystals depend on the environment and are similar to those produced by precipitation from inorganic solution.

Calcification, Phosphatization, Silification: Biological Functions and Geological Evolution. The absolutely predominant biominerals (by number of minerals and volume of biomineralization products) are calcium minerals and phosphorous minerals. This predominance reflects the biological necessity and origin of biomineralization.

Calcium is one of the most biologically active elements, playing a major role in all vital functions and at all stages of the evolution: the appearance of mitosis, metazoan, nervous system are connected with fine regulation of the calcium content in the cytoplasma of a cell. Therefore, calcium balance regulation, including removal of its toxic surpluses from the cytoplasma, is one of the main problem in the life of organisms.

Only two carbonate minerals compose nearly all products of calcification: calcite and aragonite (and their amorphous precursors); vaterite and protodolomite are rare biominerals.

The most ancient biogenic formation of carbonates is connected with cyanobacterial activity (stromatolites of 3.7 billion years) and belongs to the type induced biomineralization. This type predominates also in modern prokaryotes.

Calcification takes place in marine and freshwater macroalgae, coccolithophorids, bacteria and fungi, corals, mollusks, arthropods, and vertebrate biomineralization. Calcite of biominerals contains up to 30% substitutional Mg, biogenic aragonite up to 0.012 Sr: Ca ratios. In plant biomineralization, Chlorophyta and Phaeophyta are formed by aragonite, Rhodophyta by Mg-calcite and aragonite. Mollusk shells are composed of low Mg-calcite and (or) aragonite (rare vaterite and amorphous $CaCO_3$).

Two skeleton-forming processes have been distinguished in coral biomineralization (Goreau 1977): calcification (intracellular deposition of aragonite) and skeletogenesis (expelled aragonite crystals to extracellular sites). The mechanism, rates and periodicity of calcification have been studied by measuring ^{45}Ca and ^{14}C uptake, ^{18}O paleotemperature equation, observations of periodic growth patterns, and annual rings.

Rates of calcification vary: 7–20 mm/year for macroalgae, 10^4–10^5 g $CaCO_3/m^2/a$ (Chave 1984).

There are differing trends in the evolution of carbonate biomineralization.

1. The most ancient occurrence of biogenic carbonates (aragonites) is known from the Ordovician. However, calcite or aragonite formation in the skeleton is strictly determined by the biological characteristics of organisms. Therefore, the presence of a taxon which normally secretes certain biominerals serves as indication for their existence even if their skeleton is diagenetically altered. Thus, occurrences of mollusks in the early Cambrian indicates the existence of biogenic aragonite in that period.

2. In the Phanerozoic, a total increase in biogenic carbonates occurs. The main amount of carbonates is connected with the reef-building organisms: archaeocyates in the Cambrian, tabulates, calcareous algae, rugose corals in the Ordovician, Silurian, Devonian, rugose corals and calcareous algae at the end of the Paleozoic, scleractian corals and coralline algae since the Triassic, stromatoporates in the Jurassic, and rudist bivalves in the Cretaceous. Maximum biogenic carbonate formation was reached in the Cretaceous.
3. There is a general predominance of aragonite over calcite biomineralization. However, in some groups of organisms, the tendency can vary. An evolutionary shift from one biocarbonate to another can be demonstrated by the example of the coral evolution: Paleozoic tabulates and rugosa corals have a calcitic skeleton, while their Mezozoic and Cenozoid descendents build an aragonitic skeleton (but Mezozoic coccolitophorids and planktonic foraminifers have a calcitic skeleton).
4. Phosphorus is a vital element in the metabolism of all organisms, participating in energy exchange in the cells and playing a structural role in DNA. The use of phosphorus in biomineralization is also evolutionary stipulated.

Some 20 phosphatic biominerals are recognized in crystalline and amorphous states. Only apatite, however, is predominant among phosphate biominerals (carbonate-hydroxy apatite or flour-carbonate apatite). Amorphous apatite precursors are a characteristic stage of phosphate biomineralization. Crystallographic and mineralogical features of biogene appetites were investigated by X-ray diffraction, energy disperse spectrometry, scanning electron microscopy, infrared spectroscopy, chemical analysis for calcium, sodium, magnesium, carbonate, fluoride, chloride, and rare earths.

The lattice energy of Mg and Ca phosphate biominerals has been calculated and used for the description of cation substitutions (Taylor and Simkiss 1994).

However, calcium-phosphate crystals of vertebrates occur which exhibit short-range arrangements of atoms and ions that are not found in crystalline apatite (nonapatite environments). Due to the enormous specific surface area and their reactivity, the labile environments of biogene apatites are sensitive indicators of the maturity of phosphates in the tissues.

1. Processes of phosphatization are almost immediate. The preservation of cyanobacterial filaments and coccoids and delicate epithelial cover within ammonite siphunele (Barskov 1996) suggests that this process lasted for a few days; experiments on modern cyanobacteria have shown that it takes only several hours. Phosphorus is first transferred from the valutine granules to the amorphous phosphate accumulated on the surfaces of trichosomes. Later, it transforms from the X-ray amorphous state into well-ordered crystals. The rates of phosphatization are similar to those of carbonatization and silification.
2. Phylum distribution of crystalline and amorphous phosphate biominerals is shown in Table 13. In Monera, struvite and hydrophosphate Fe (1Y) occur,

in Protoctista only rare dahllite and amorphous calcium phosphate, in Fungi very rare dahllite (only in Deuteromycota). In molluscs, arthropods and insects all amorphous and crystalline phosphate biominerals occur.

Phosphorus biominerals are unknown in plants. Amorphous calcium phosphate is used by cephalopods and arthropods as gravity receptors or for phosphorus storage.

The formation of all the earliest phosphorite deposits (except those composed of shells) is found to be related to the cyanobacterial mats (communities) (Rozanov and Zhegallo 1989).

3. Biogenic silification takes place invariably in the form of amorphous hydrated silica: opal. Among more than 1000 silicate and silica minerals, there are no biominerals except opal; because of strong covalent bonding Si–O and the tendency to glass formation, biogenic opal is nearly always in an amorphous state or weakly transformed into crystobalite-tridimite opal. Extremely regular sophisticated structures built up from well-ordered monodisperse silica particulates ranging from 2–15 nm form an amazingly intricate and architecturally elegant morphology such as fibrils, tubules, and sheets. Silica is always hydrated (water in porous interspaces) and associated with organic matter (particulates are not firmly attached to each other).

   Diatoms, silicoflagellates, radiolarian, and sponges utilize opal for forming their hard parts. Huge quantities of biogenic opal form in the ocean. Silica is a widespread biomineral in plants: "silica is as important for plant life as calcium for animal life" (Harrison and Lu 1993).

4. A correlation exists between the three principal processes of biomineralization (calcification, phosphatization, silification) in the evolution of the biosphere and geological history (Table 14). From the very beginning of the Cambrian explosion, organisms used almost all known types of biomineralization and formed all presently known principal biominerals.

   There is a connection between the type of skeletal biomineralization and the level of organization of the organic world. Siliceous skeletal biomineralization is characteristic for unicellular plants (diatomean) and for primitive

Table 14. Distribution of the main types of skeletal biomineralization

Types of biomineralization	Main plant and animal taxa													
	Dia	Coc	For	Rad	Arc	Por	Coe	Ver	Mol	Bry	Art	Bra	Con	Cho
Silicification	+		+?	+*		+								
Calcification		C	C		C	C	CA	C	CA	CA?	C	C		CA?
Phosphatization											+?	+	+	+

Explanations: Dia – Datomean, Coc – Coccolithophorida, For – Foraminiferida, Rad – Radiolarian, Arc – Archaeocyatha, Por – Porifera, Coe – Coelenterata, Mol – Mollusca, Bry – Bryozoa, Art – Arthropoda, Bra – Brachiopoda, Con – Conodonta, Cho – Chordata. C – calcite, A – aragonite, +* – mainly opal and celestite in one group (Acantaria).

animal organisms (radiolaria, sponges). Phosphatic biomineralization takes place in building the skeleton of highest invertebrates and vertebrates (see Table 13).

5. Unusual and unexpected cases of biomineralization are presented by some other biominerals in a few types of organisms:

   Fluorite in the gizzard plates of the gastropods *Scaphander* and *Meloscaphander* and in the statoliths of opossum shrimps (*Mysidacea*), amorphous fluorite precursor in the spicules of certain marine gastropods, celestine in the skeletons of some marine radiolaria.

   Citrates (earlandite) and oxalates (whewellite, weddelite, glushinskite, manganese, and copper oxalates) serve as a temporary storage of calcium, citric acid, or detoxication products. Many of these minerals are products of pathological mineralization in higher vertebrates, including man.

6. Plant biomineralization occurs essentially in all types of plants. There are three groups of biominerals. Calcification (carbonatization) takes place in macroalgae, coccolithophorids, silification in diatoms and some flowering plants, apatite in some algae; calcium oxalates are predominent in higher plants.

   In plants, in regions with ore deposits, exotic biominerals are recognized: gold, sheelite, pyrite, hydrotroilite, sphalerite, wurtzite, and others.

References

Addadi L, Weiner S (1986) Interaction between acidic macromolecules and structured crystal surfaces. Stereochemistry and biomineralization. Mol Cryst Liq Cryst 134:305–322

Allemond D, Curf J-P (eds) (1994) Biomineralization-93. 7th Int Symp on Biomineralization, Monaco, Bull de l'Institut Oceanographique Monaco, Numero special 14,1

Barskov IS (1975) Microstructural and biochemical methods in paleontology. Results of science and technics. Stratigraphy and Paleontology, vol 6. Viniti, Moscow, pp 5–59

Barskov IS (1996) Phosphatized blood vessels in the siphuncle of Jurassic ammonites. Bull Inst Oceanogr (Monaco) 14, 4:335–341

Berman A, Hanson J, Leiserowitz L, Koetzle TF, Weiner S, Addadi L (1993) Biological control of crystal texture: a widespread strategy for adapting crystal properties to function. Science 259:776–779

Borowitzka MA (1987) Calcification in algae: mechanisms and the role of metabolism. CRC Crit Rev Plant Sci 6:1–45

Borowitzka MA (1989) Carbonate calcification in algae initiation and control. In: Mann S, Webb J, Williams RJP (eds) Biomineralization: chemical and biochemical perspectives. Weinheim, FRG, pp 63–94

Chave KE (1984) Physics and chemistry of biomineralization. Annu Rev Earth Planet Sci 12:293–305

Crick RE (eds) (1988) Origin, evolution and modern aspects of biomineralization. Elsevier, Amsterdam

Deffeyes KS (1965) Carbonate equilibria: a graphic and algebraic approach. Limnol Oceanogr 10:412–426

Dodd JP, Crips EL (1982) Non-linear viariation with salinity of Sr/Ca and Mg/Ca ratios in water and aragonitic bivalve shells and implication for paleosalinity studies. Paleogeogr Palaeoclimatol Palaeoecol 38:45–56

Epstein S, Buchsbaum R, Lowenstam HA, Urey HC (1953) Revised carbonate-water isotopic temperature scale. Bull Geol Soc Am 64:1315–1326

Erez J (1978) Vital effects on stable-isotope composition seen in foraminifera and coral skeletons. Nature (Lond) 273: 199–202

Frankel RB, Blakenure (eds) (1990) Iron biominerals. Plenum Press, New York

Goreau TJ (1977) Coral skeletal chemistry physiological and environmental regulation of stable isotopes and trace metal in *Monastres annularis*. Proc R Soc Lond Ser B 196: 291–315

Gratz AJ, Hillner PE (1993) Poisoning of calcite growth viewed in the atomic force microscope (AFM). J Crystal Growth 129: 789–793

Gregoire C, Duchateau GH, Florkin M (1990) La trâme protidique des nacres et des perles. Ann Inst Oceanogr Monaco 3: 1–36

Hillner PE, Gratz AJ, Mann S, Hansma PK (1992) Atomic-scale imaging of calcite growth ad dissolution in real time. Geology 20: 359–362

Korago AA (1992) Introduction to biomineralogy. Nedra, S Petersburg, 238 pp (in Russian)

Leadbeater BSC, Riding R (eds) (1986) Biomineralization in lower plants and animals. Clarendon Press, Oxford

Legeros RZ (1981) Apatites in biological systems. Prog Crystal Growth Charact 4: 1–45

Lippmann F (1973) Sedimentary carbonate minerals. Springer, Berlin Heidelberg New York

Lowenstam K, Weiner S (1989) On biomineralization. Oxford University Press, Oxford, 324 pp

Mann S (1983) Mineralization in biological systems. Struct Bonding 54: 125–174

Mann S (1993) Biomineralization: the hard part of bioinorganic chemistry. J Chem Soc Dalton Trans 1–9.

Mann S, Webb J, Williams RJ (1989) Biomineralization. In: Meyer JL, Eanes ED (1978) A thermodynamic analysis of the amorphous to crystalline calcium phosphate transformation. Calcif Tissue Res 25: 54–68

Nakahara H (1983) Calcification of gastropod nacre. In: Westbroek P, de Jong EW (eds) Biomineralization and biological metal accumulation. Reidel Dordrecht, pp 225–230

Nancollas GH (ed) (1982) Biological Mineralisation and demineralisation. Springer, Berlin Heidelberg New York

Nylen MV, Eanes ED, Termin JD (1972) Molecular and ultrastructural studies of non-crystalline calcium phosphates. Calcif Tissue Res 9: 95–108

Omory M, Watabe N (eds) (1980) The mechanisms of biomineralisation in animals and plants. Nkai University Press, Tokio

Perry CC (1989) Biogenic silica. In: Mann S, Webb J, Williams RJP (eds) Biomineralisation, chemical and biologial perspectives. VCH, Weinheim, 233 pp

Popov LE, Bassett MG, Holmer LE, Laurie J (1993) Phylogenetic analysis of higher taxa of Brachiopoda. Lethaia 26, 1: 1–5

Rozanov AY (1992) The Cambrian radiation of shelly fossils. TREE 7, 3: 84–87

Rozanov AY, Zhegallo EA (1989) To the problem of the genesis of the ancient phophorites from Asia. Lithol Mineral Resour 3: 67–82

Schraer H (ed) (1970) Biological calcification: cellular and molecular aspects. Appleton Century Croft, New York

Sikes CS, Fabry V (1994) Photosynthesis, $CaCO_3$ deposition, and the global carbon cycle. In: Tolbert EN; Preiss J (eds) Photosynthetic carbon matabolism and regulation of atmospheric CO_2 and O_2. Oxford University Press, New York

Sikes CS, Wilbur KM (1980) Calcification by coccolithophorids: effects of pH and Sr. J Phycol 16: 433–436

Sikes CS, Wierzbicki A, Fabry VJ (1994) From atomic to global scales in biomineralization. In: Allemand D, Cuzif J-P (eds) Biomineralization-93. 7th Int Symp on Biomineralization, Monaco, Bull de l'Institut Oceanographique Monaco, vol 1, 1–49

Simkiss K (1994) Amorphous minerals in biology. In: Allemand D, Cuif J-P (eds) Biomineralization-93. 7th Int Symp on Biomineralization, Monaco, Bull de l'Institut Oceanographique Monaco, vol 1, pp 49–54

Simpson TL, Volcani BE (1981) Silicaon and siliceous structure in biological systems. Springer, Berlin Heidelberg New York

Smith SV, Buddenmeier RW, Redaile RC, Huck JE (1979) Strontium-calcium thermometry in coral skeletons. Science 204:404–407
Sognnaes RF (ed) (1960) Calsification in biological systems. AAAS, Washington
Sokal RR, Rohle FJ (1981) Biometry. Freeman, New York, 859 pp
Suharan S, Taylor MG, Simkiss K (1994) Computer modelling of phosphatic surfaces. In: Allemand D, Cuif J-P (eds) Biomineralization-93, vol 1, pp 81–84
Taylor MG, Simkiss K (1994) Cation substitutions in phosphate biominerals. In: Allemand D, Cuif J-P (eds) Biomineralization-93, vol 1: pp 81–84
Termine JD, Eanes ED, Conn KM (1980) Phosphoprotein modulation of apatite crystallization. Calcif Tissue Int 31:247–251
Trudinger PA, Swain DJ (eds) (1979) Biochemical cycling of mineral-forming elements. Elsevier, Amsterdam
Watabe N (1974) Crystal growth of calcum carbonate in biological systems. Crystal Crowth 24/25:116–122
Watabe N (1981) Crystal growth of calcum carbonate in the invertebrates. Prog Crystal Crowth Charact 4:99–147
Watabe N, Wilbur KM (eds) (1976) The mechanisms of mineralisation in the invertebrates and plants. University of South Carolina Press, South Carolina, Columbia
Weiss A, Herzoc A (1978) Isolation and characterisation of a silicon-organic complex from plants. In: Bendz G, Lindqvist I (eds) Biochemistry of silicon and related problems. Plenum Press, New York, pp 109–207
Westbroek P, de Jong EW (eds) (1983) Biomineralisation and metal accumulation. Reidel. Dordrecht
Wheeler AP, Sikes CS (1989) Matrix-crystal interactions in $CaCO_3$ biomineralization. In: Mann S, Weeb J, Williams RJP (eds) Biomineralization: chemical and biochemical perspectives. VCH, Weinheim, 95 pp
Wierzbicki A, Sikes CS, Drake B, Madura J (1994) Atomic force microscopy and molecular modeling of protein and peptide binding to calcite. Calcif Tissue Int 54:133–141
Wright AC (1988) Neutron and X-ray amorphography. Non-Cryst Solids 106:1–6
Zipkin Z (ed) (1973) Biological mineralisation. John Wiley, New York

4.3 Magnetotactic Bacteria. Microorganisms in Ore-Forming Processes

4.3.1 Magnetite Biomineralization, Magnetofossils, and Magnetoreception in Organisms

I. S. Barskov and A. Y. Rozanov

Since its initial discovery by Lowenstam in 1962 in the teeth of mollusks, studies of biomagnetite have evolved to a field covering biomagnetism, magnetoreception in organisms, bird and fish navigation and orientation in space, sensitivity of animals to small magnetic field variations, contribution of magnetofossils to the sediment paleomagnetism, magnetostratigraphy, natural remanent magnetization, and crystal growth of biomagnetite within cells of bacteria (Kirschvink et al. 1985).

Cells of *Aquaspirillum magnetotacticum* are ca. several micron long, but contain from 10–20 to several thousands of intracellular enveloped magnetite single crystals 40–50 nanometer in diameter. A membrane-bound structure called a magnetosome is incorporated in magnetoreceptive organelles.

High-resolution transmission electron microscopy (HRTEM), atomic force microscopy (AFM), 125-kV field-emission electron microscopy of highest resolution (FEEM), and selected-area electron diffraction (SAED) have revealed the crystal habit and structure of bacterial magnetosomes: cubo-octahedra (and elongated cubo-octahedra) of single-domane magnetite that generate permanent magnetic moments, aligned in linear chains, with the directions of magnetosome alignement coinciding with the primary axis [111] of magnetization in magnetite.

Superconducting S QUID magnetometers with sensitivities approaching 10^{-10} Gauss have been used for low-level magnetic measurements in magnetotactic bacteria.

Mössbauer spectroscopy studies indicated, in addition to magnetite, the presence of ferryhydrite ($5\,Fe_2O_3 \cdot 9\,H_2O$) (surrounded by the protein apoferritin). Iron deposited in specific areas in the form of ferryhydrite is converted within 1–2 days in well-ordered single-domane magnetite crystals and magnetosomes.

Biomagnetite is widespread in organisms, ranging from bacteria and algae through vertebrates and higher animals (fishes and birds) and man.

Biomineralization of pyrite, pyrrhotite, and ferrimagnetic greigite (Fe_3S_4) in magnetotactic bacteria also occurs.

The ability of magnetotactic bacteria to orientate and navigate along the geomagnetic field is due to intracellular crystal growth conditions controlled by the organic matrix.

References

Blakemore RP (1982) Magnetotactic bacteria. Annu Rev Microbiol 36:217–238

Diaz Ricci JC, Kirschvink JL (1992) Magnetic domain state and coercivity predictions for biogenic greigite (FE_3S_4): a comparison of theory with magnetosome observations. J Geophys Res 97, 17:309–317

Esquivel DMS, Lins de Barros HGP (1986) Motion of magnetotactic microorganisms. J Exp Biol 121:153–163

Frankel RB, Blakemore RP (1980) Navigational compass in magnetic bacteria. J Magnetism Magnetic Materials 15–18:1562–1564

Heywood BR, Bazylinksi DA, Garrant-Reed A, Mann S, Frankel RB (1990) Controlled biosynthesis of gregite (Fe_3S_4) in magnetotactic bacteria. Naturwissenschaften 77:536–538

Kirschvink JL (1980 a) South-seeking bacteria. J Exp Biol 86:345–347

Kirschvink JL (1980 b) Biogenic magnetite (Fe_3O_4): a ferromagnetic mineral in bacteria, animals, and man. Proc Int Conf on Ferrites, pp 135–138

Kirschvink JL (1992) On the magnetostatic control of crystal orientation and iron accumulation in magnetosomes. Automedica 14: 257–269

Kirschvink JL, Jones DS, MacFaddan BJ (1985) (eds) Magnetite biomineralization and magnetoreception in organisms. Plenum Press, New York

Mann S, Sparks NHC, Frankel RB, Bazylinski DA, Jannasch HW (1990) Biomineralization of ferrimagnetic gregite (Fe_3S_4) and iron pyrite (FeS_2) in a magnetotactic bacterium. Nature 343:258–261
Matsuda T, Endo J, Osakabe N, Tonomura A (1983) Morphology and structure of biogenic magnetite particles. Nature 302:411–412

4.3.2 Microorganisms in Supergenic Processes

L. K. YAKHONTOVA

Bacterial Processes in Sulfide Ore Deposits

Microorganisms, especially bacteria, are common in oxydation zones of ore deposits and take an active part in the formation of oxidized ores. As a rule, 1 g ore or 1 ml of mine waters contains from 10^4 to 10^9 bacterial cells.

Thionic oxidizing bacteria are the most common in sulfide ores and their analogs (arsenides, selenides, sulfosalts, and others). Their activity is provided by underoxidized forms of S, As, Se, Te, Sb, Bi, H_2S, and dissolved anions: AsO_3^{2-}, S^{2-}, SO_3^{2-}, SeO_2^{2-}, and others. They reduce the S of sulfates to S^{2-}, H_2S, and S^0.

The bacterial factor is very important in ore deposits and is 5 orders more intensive than the chemical (abiogenic) processes. A high rate of bacteria multiplication by cell division is a great significance (60–70 cell divisions daily). Bacterial activity is limited mostly by temperature conditions: optimal temperature is within the range of 20–30 °C. Most thionic bacteria are represented by aerobes and autotrophs, obtaining carbon from the atmosphere. They prefer acid environment with pH 1–6. Only some bacilla (*Th. thioparus, Th. "y"*) develop in almost neutral and even alkaline solutions (pH 4–9).

The major function of thionic bacilla is the destruction of primary minerals (sulfides and analogs) by oxidation of their structures. Electrons (oxidation energy) then pass into transport chains of bacterial cells and participate in the processes of energetic metabolism. The general pattern of sulfide oxidation by thionic bacteria is:

$$MeS + 1.5\ O_2 + H_2O = [MeSO_4]^+ + H^+ + 2e^-.$$

A universal oxidizer of sulfides and their analog (arsenides and others) is *Thiobacillus ferroxidans* – an aerobic autotrophic bacterium active in sulfuric acid solutions with $pH<3$.

Destroying the sulfide structure, bacteria enrich natural solutions with specific components – oxidation products, mainly sulfate complexes. They serve as a basis for the following induced biomineralization, leading to the formation of sulfates with varying water content. Some of them (melanterite, boothite, chalcantite, etc.) crystallize directly on sulfide bodies, others crystallize as arsenates on the arsenite surface (scorodite) or as basic sulfates (jarosite).

Experiments demonstrated that many trace elements, such as Ge, In, Cd, Ga, Re, and Au, were dissolved when sulfides were destroyed by thionic bacteria.

Thionic bacteria extract intensively selenium and tellurium from sulfide ores and involve them in supergenic migration as complex compounds such as $Fe_3(Se, Te)O_3(OH)_4$.

Biogenic degradation of antimonite and antimonic sulfosalts is quite specific. It proceeds in two stages: (1) with the participation of *Th. "y"*, Sb_2S_3 is transformed into Sb_2O_3 (senarmontite) and (2) $Sb_2^{3+}O_3$ is transformed into $Sb_2^{4+}O_4$ (stibioconite) by the culture of *Stibiobacter senarmontii*.

Microorganisms in Rock Weathering

Since the 19th century, microorganisms have been known to take part in rock weathering.

First data on the microbial weathering factor were related to anaerobic heterotrophic bacteria producing organic matter and acid metabolites, which appeared to destroy rock minerals. The fact that 1 g of rock material obtained from the depth of 1 – 1.5 m contained 10^6 living bacterial cells was impressive. The hypothesis developed, assigning a leading role to microorganisms as determining groundwater composition and geochemical cycling of elements.

In the past decades, there has been a series of works concerning the participation of aerobic autotrophic microorganisms (nitrifiers and silicate bacteria) in the weathering of silicate rock-forming minerals.

Studies of the interaction between silicate bacilla *Bacillus mucilaginosus* and major rock-forming minerals (quartz, feldspars, chlorites, smectites) demonstrated an active Si leaching, and its dependence on the degree of structural regularity of minerals, defects, and phase heterogeneity (quartz weathering depends on the degree of crystallinity, plagioclase weathering on anorthite content and structural regularity, the weathering of silicates of the disthene group on the type of Al coordination, the weathering of layer silicates on structural regularity, and substitutional impurities).

Recently, experimental data were obtained concerning the weathering of serpentinized ultrabasite samples from the Saranovskii and Kempisarai massifs (the Urals) by an accumulative culture of nitrifying bacteria. Bacteria leached out equivalent amounts of Mg and Si, decreased solution pH from 8 to 5.5 and enhanced Ni migration in organic complexes.

Life support of autotrophic nitrifying bacteria is connected with the oxidation of ammonia NH_4^+ (150 – 600 g/tn). In the first stage, the bacteria *Nitrosospira briens* oxidize NH_4^+ to nitrites. Then *Nitrobacter winogradskii* transforms nitrites into nitrates.

Up to 10^5 cells of thionic bacteria (*Th. ferrooxidans, Th. thiooxidans, Th. thioparus*) were found in 1 g of rocks in fissure zones of weathering crusts with sulfides. The bacteria oxydize pyrite, pyrrotine, and marcasite.

An anaerobic culture of *Pseudomonas chromatophila* develops due to the reduction of Cr^{6+} of chromates to Cr^{3+} of smectites of weathering crust (volkonskoit).

Sulfate-Reducing Bacteria in the Biogeochemical Cycling of Sulfur

Heterotrophic, anaerobic sulfate-reducing microorganisms occupy another place within the profile of oxidized ores. They are found in deep horizons with limited access to atmospheric O_2 and CO_2, or in the presence of organic matter (petrol waters, carbon-containing matter), and solution pH of 4–10, SRB are widespread and represented by two main genera – *Desulfovibrio* (sporeless vibrions) and *Desulfotomaculum* (spore bacilla), both of them being heterotrophic anaerobes. They were found in salt lakes, maritime beaches, oceanic silts, groundwaters of oil and gas occurrences, both in the tropics and in Antarctica at temperatures less than 5 °C. In ore deposits, SRB occur in poorly aerated domanes of a supergenesis zone with slowly moving or stagnant sulfuric solutions (pH 7–9) containing organic mater (glucose, oil, kerogen, coal particles). Usually, these domanes are concentrated near groundwater level in the profiles of supergenic zones. Microbiological analysis of sulfide-containing groundwaters of many regions demonstrates that 1 ml of water can contain up to 3×10^3 active SRB cells.

Within oxidized ore deposits, the horizons and areas with stagnant sulfate solutions containing sulfate reducing bacteria work as specific geochemical barriers.

The major function of SRB is the reduction of dissolved sulfates to sulfur and H_2S. This process is connected with electron transport to sulfate anion occurring in the cell transportation system during the process of cell energetic metabolism. Oxidizing organic matter and binding sulfate anions, the cell produces H_2S. As a result, sulfate reduction is accompanied by H_2S release and following processes of sulfidization and sulfur formation (fi green bacteria, oxidizing H_2S to sulfur) occur.

An example of the natural intensity of sulfate reduction in the mineral-forming sulfur cycle was observed in Somalia hollow, Indian Ocean. About 300 mg of sulfur are reduced there daily from 1 kg of silt. Annual accumulation of sulfate reduction products in the sediment of the ocean is approximately 10^8 t (recalculated to sulfur).

SRB activity results in the accumulation of biogenic sulfides, especially pyrite, marcasite, pyrrotine, sphalerite, and wurtzite, having a globular form of aggregates in this case. It is possible that, in many stratiform deposits, sulfide formation was biogenic and resulted from SRB activities. The highest temperature limit of some SRB development (ca. 150 °C) fits the conditions of "strata" sulfide formation.

Sulfate reduction is connected directly with the formation of sedimentary sulfur deposits, and processes of metal deposition as sulfides from different industrial sewages. When hydrogen sulfide binding is insufficient, poisoning of waters by hydrogen sulfide can occur, damaging water fauna. SRB activities result in the corrosion of metal constructions, underground tubes, mining machinery, and oil and gas boreholes.

SRB could also take part in the formation of ore deposits, in particular Pre-Cambrian iron quartzite ores. In this case, quartz, dissolved in sulfuric solutions, precipitates intensively due to decrease in pH induced by bacterial reactions, but does not affect the activity of SRB cells.

One of the criteria of biogenic mineral formation is isotopic analysis of sulfates, sulfides, hydrogen sulfides, and pure sulfur, since sulfate reduction is connected with the kinetics of isotope effect and with the fractioning of the isotopes S^{32} and S^{34}. Sulfates assimilated by microorganisms and microbial activity products are enriched by light sulfur. δS^{34} of SRB cells is close to -2.5%.

In sulfate deposits (gypsum, tenardite) and sulfate-containing waters with pH 6.3–6.8, sulfates are enriched by S^{34} (δS^{34} + 20% and +35%). At the same time, δS^{34} of biogenic hydrogen sulfide waters varied from −12 to +15%, and S^{32} content in solution was higher. The intensity of sulfate reduction and H_2S synthesis was close to 0.003–0.005 mg/l daily.

Bacteria in Iron-Manganese Mineralization of the Ocean Floor

Iron-manganese nodules and crusts are composed of Fe-Mn oxyhydroxides formed with participation of the bacteria *Metallogenium, Siderocapsa,* and *Siderococcus* able to oxidize Fe and Mn from sediments of oceanic waters.

The formation of tetravalent Mn oxides precedes only a very quick Mn oxidation which is not characteristic for a slow abiogenic process occurring in bottom oxygen-containing oceanic waters. Experimental studies have shown a possibility of biogenic formation of vernadite and Mn-Ferroxyhite in todorokite-containing sediments with the participation of *Metallogenium* culture after 3-day exposure. The growth of vernadite over todorokite was observed by electron microscopy.

The biogenic mechanism of Mn hydroxide formation (vernadite and Mn-ferroxyhite) includes: (1) the interaction of bacterial cells (strongest oxidizers) and the products of their metabolism (hydrogen peroxide) with complex and colloidal forms of Mn^{2+} and Fe^{2+} of oceanic waters and young sediments, and (2) direct precipitation of Mn^{4+} and Fe^{3+} hydroxides, and opal-like silica in the sphere of bacterial activities. Silica is always fixed in capsular material of bacteria and preserves biogenic relicts. Later, diagenetic solid-phase processes of aging take place in hydrogenous Fe-Mn crusts. They result in the formation of diagenetic phases – bouserite, bernessite, and partly asbolane-bouserite.

The morphological similarity between some iron-manganese hydroxides and relict ferrobacteria cells supports their biogenic origin. Biogenic ferrihydrite consists of cells of the stem Fe-bacterium, *Gallionella*.

References

Birnbaum SJ, Wireman JW (1985) Sulfate-reducing bacteria and silica solubility: a possible mechanism for evaporite genesis and silica precipitation in banded iron formation. Can J Earth Sci 22, 12: 1904–1909

Ivanov MB, Lein AYu, Belyaev SS (1980) Geochemical activities of sulfate-reducing bacteria in bottom deposits of the SW part of the Indian Ocean. Geokhimiya 8: 1238–1249 (in Russian)

Krouse HR, McCready RGJ (1979) Biogeochemical cycling of sulfur, Biogeochem Cycl Mineral Form Elem, Nostzandt, Amsterdam e.a., pp 401–430

Taylor BE, Weeler MC, Nodrstrom DK (1984) Isotope compositon of sulfate in acid mine drainage as measure of bacteria oxidation. Nature 308, 5959: 538–541

Trudinger PA, Chamers LA, Smith JW (1985) Low-temperature sulfate reduction: biological versus abiological. Can J Earth Sci 22, 12: 1910–1915

Vinogradov VI (1980) Role of sedimentation cycle in the geochemistry of sulfur isotopes. Nauka, Moscow, 192 pp (in Russian)

Vitovskaya IV, Bugel'skii YuYu (1982) Nickel-containing weathering crusts. Nauka, Moscow, 190 pp (in Russian)

Yakhontova LK, Nesterovich LG (1983) Supergenesis area of ore deposits as bio-abiotic systems. Moscow State University, Moscow, 57 pp (in Russian)

Yakhontova LK, Grudev AP, Zuev VV (1994) To the theory of bio-abiotic interactions in the ecosystems with a mineral substrate. Miner Zh 216, 2: 5–16

4.3.3 Biogeotechnology. Bacterial Mining Technology for Gold, Copper, and Uranium Ores

L. K. Yakhontova and E. V. Adamov

The biological method of processing mineral raw materials includes: (1) metal extraction (leaching) from ores, (2) changing the surface properties of mineral particles, and (3) metal sorption from suspensions and solutions. This is a technology with economical perspectives, having great advantages compared to traditional ones. Microorganisms, often bacteria and fungi, are used in biogeotechnology as technological reactants.

One of the most attractive aspects of biotechnology is the possibility of using raw materials either with quite a low metal content or of low quality, which is practically unprofitable in traditional processing. Biogeotechnology opens a realistic way to improve the structure of the raw material basis, leading to the conservation of agricultural and forested areas, the resolution of ecological problems, decreased volumes in industrial construction, and the possibility of providing complete automatization and improving labor conditions.

Industrial application of biogeotechnology started in the 1960s–1970s in the copper mining industry in the USA. In 1958, the first patent was obtained for bacterial leaching of copper from poor ores of the deposit Bingham Canyon, containing less than 0.4% Cu (Utah, Kennecott Copper Co.). Up to 80 000 t Cu is extracted there annually, practically from dumps. It costs five times less than traditional mining and processing.

Already in the early 1980s, the costs of copper and uranium produced in the USA by the bacterial method were $350 million and $300 million, respectively. Total production volume is planned to be equal to $5 billion by the year 2000.

Biotechnological processing of gold-containing (pyrrotine-pyrite-arsenopyrite) copper and uranium ores is being conducted successfully by several corporations and companies.

Up to 30% of the world copper is obtained biogeotechnologically. Forty of 70 biological plant sets, working all over the world, deal with copper ore processing. One ton of copper, obtained biotechnologically, costs from $10 to $30. It is five times cheaper than the copper obtained by traditional processing.

Thirty % of world Au resources consists of submicroscopic particles in fine-grained pyrite-arsenopyrite and arsenopyrite ores. Traditional processing technology comprises flotation, burning, cyanidation, and Au sorption by activated coal and extracts only up to 55% gold. The application of biotechnology with *Th. ferroxidans* or some thermophilic bacteria increases Au output to 95%.

Recently, an evident increase in world gold mining occurred due to the active application of biotechnological treatment of refractory arsenopyrite ores and concentrates.

Industrial bacterial leaching of Cu, Zn, U, Ni and Au from ores and concentrates proceeds in three ways – in situ, dump and tank leaching. In situ bioleaching, mainly of Cu and U, is conducted in natural ore deposits (virgin areas). The organization of in situ leaching is often problematic due to the complex relief and hydrogeologic conditions of the deposit area. Major expenses are always connected with special engineering works (construction of closed-loop underground pipelines, working underfloor adits, boring ring wells, setting off nuclear charges to crush ore mass, construction of settlers, installation of pumps and pump stations).

The first experience of in situ bioleaching of copper was obtained at the Degtyarskii deposit (Russia), the Miami deposit (Arizona, USA), and the Kosaki mine (Japan). Later, it was developed at the deposits Old Reliable and Beach Mike (Nevada, USA) and Avoka (Ireland). In addition, there are a few data about in situ leaching of uranium (the deposits Stenrock and Egnew Lake, Ontario, Canada, and Pitch, Colorado, USA).

Dump leaching is conducted by forming hills of crushed ore. They are 2–6 m high, up to 150 m long, and contain from dozens to 170 000 t of ore. The hills are created on flat plots with a sandy "bottom" covered by a plastic sheet. Hill irrigation by cultural solution is conducted with the help of extensions, fixed on plastic tubes, situated along the hill surface. The solution is collected in a tank with the help of drainage tubes, situated at the "bottom" of hills. The method can be applied to process ore heaps and industrial wastes.

Dump leaching is used primarily to treat oxidized and semioxidized copper ores (Nikolaevskii deposit, Russia; Kounrad deposit, Kazakhstan; Bluebird Mine and Bisby, Arizona, USA; Rio Tinto, Spain), to extract gold (Gold Ackress and Smoky Valley, Canada), and uranium (the deposit with phosphates and vanadates in Saint-Pierre, France, and Egnew Lake, Canada).

Tank leaching takes place in specially constructed tanks. Sometimes, there are three to five tanks connected to each other. Ore or concentrate is milled into the particles 40–50 μm in diameter. To provide equal wetting by a cultural solution, the pulp is mixed in the tank by forced air. The last stage of gold extraction by tank leaching is connected with cyanidation. Tank size and pulp weight do not affect process kinetics and metal output. The method is applied for leaching refractory gold-containing pyrite-arsenopyrite ores and concentrates, other types of sulfide ores (primarily copper-zinc, copper-nickel, and tin-arsenic), for coal desulfuration, bauxite desilication, and sewage purification.

Tank leaching is especially used for the processing of arsenic gold-containing ores and concentrates. The plants, processing daily up to 700–800 t of the concentrate and providing extraction of 97% gold and ecological security, work successfully in South Africa, Brazil, Ghana, and Australia. In Russia, half-industrial tests of these plants are being carried out with Au concentrates obtained from several Siberian deposits (Olimpiadinskoe, Taseevskoe, Darasunskoe). Up to 90% gold and 80% silver are extracted in test treatments.

Microorganisms used in biogeotechnology, include:

- Thionic bacteria, oxidizing sulfides and sulfur.
- Sulfate-reducing bacteria, used as metal precipitants in settlers in sewage purification.
- Microorganisms which accumulate and precipitate gold and other metals.
- Bacteria which oxidize and reduce Fe and Mn.
- Bacteria destroying silicates, carbonates, and oxides.

Table 15 contains a description of thionic bacteria, which are the most important microorganisms used in the biogeotechnology of sulfide and sulfuric ores. The culture of *Thiobacillus ferroxidans* is the most universal sulfide and sulfur destructor. It is able to destroy minerals completely or change their surface substantially. The interaction of a bacterial cell with a mineral particle proceeds as a mineral substrate oxidation. Oxidation products are dissolved as complex ions, their character depending on the Eh-pH of the solution. Mixed cultures of thionic bacteria are often used for ore and concentrate leaching, for example, *Th. ferroxidans* and *Th. thiooxidans.* Sulfate-reducing heterotrophic microorganisms are represented by two major genera: (1) *Desulfovibrio* (sigma species, spirella) and (2) *Desulfotomaculum* (bacillus). They produce sulfo-

Table 15. Thionic bacteria used in biogeotechnology

Genera and species	Optimal solution pH	Optimal temperature	Relation to oxygen	Carbon source
Sulfide- and sulfur-oxidizing				
Thiobacullus sp.				
Th. ferroxidans	2.5–2.8	28–35	Only aerobic	CO_2
Th. thiooxidans	2.0–3.5	28–30	Only aerobic	CO_2
Leptosririllum sp.				
L. ferroxidans	3.5–3.0	30	Aerobic	CO_2
Sulfobacillus sp.				
S. thermosulfidooxidans	1.7–2.4	50–55	Only aerobic	CO_2, glucose, saccharose, glutamine
Acidianus sp.				
A. brierleyi	1.5–2.0		Aerobic	CO_2, yeast extract, saccharose, lactase
Sulfur-oxidizing				
Thiobacillus sp.				
Th. acidophilus	3.0	25–30	Only aerobic	CO_2, glucose fructose, xylose
Sulfolobus sp.				
S. acidocalderius	2.0–3.0	70	Aerobic	CO_2, yeast extract, trip-tone, peptone, glutamine, xylose
S. solfataricus	3.0–5.0	75–90	Aerobic	CO_2, yeast extract, trip-tone, peptone, glutamine, xylose
Acidianus sp.				
A. infernus	2.0	0	Facult. aerobic	CO_2, lactose, saccharose

hydrides and dissolved metals as sulfides. The ability to reduce sulfates is used effectively for the purification of ponds, sewage, and settlers.

To accumulate and precipitate the gold extracted from sulfides to cultural media with the help of thionic bacteria, yeast (*Candida lipolitica* and *Candida utilos*), ascomycetes (*Aspergilla niger*), and microalgi (*Chlorella vulgaris* and others) are often used in biogeotechnology. In addition, these microorganisms actively absorb Ag and Ni, Co, Cu, Cd, Zn, U, Mo, and Sr, transformed preliminarily from ionic forms into metals with the help of amino acids produced by the bacillum *Pseudomonas*. The kinetics of metal biosorption is still rather low.

Table 16. The dependence of kaolinite bioleaching by *Bacillus mucilaginosus* on structural regularity (H), size of coherent dissipation blocks (N), fractal parameter (α), and intercalation time (τ). (Yakhontova et al. 1991 b)

Deposits	Characteristic of kaolinite pseudomorphoses after biotite	H	N, α, τ	Leaching result after 7 days, % of initial content		Final cell concentration in 1 ml
				SiO_2	Al_2O_3	
Glukhovetskoe (Ukraine)	Spiral growth, spiral dislocations, helioidal interlaying space	1.30	126 2.4 13	27	24	10^8
Dubrovskoe (Ukraine)	Lay-by-lay replacement, discrete interlaying space	0.67	47 2.7 19	12	16	10^6

To reduce and oxidize Fe and Mn, a large group of ferrobacteria is used. It includes heterotrophs, phototrophs, and some microalgi. *Bacillus polymyxa* is known to reduce Mn^{4+} to $Mn(OH)_2$ in freshwaters and sediments. In maritime waters and sediments, specific oxidation of Mn^{2+} to Mn^{4+} (pyrolusite and vernadite formation) involves bacteria of the genera *Arthrobacter* and *Metallogenium.*

Silicate bacteria (*Bacillus mucilaginosus*) are used for the destruction of silicate and quartz.

Different microorganisms are applied for decalcification, treatment of bauxites, and leaching Al from dawsonite.

The effectiveness of bio-abiotic interactions depends on the crystalline structure of the minerals (Table 16).

References

Brierley CL (1984) Microbiological mining technology status and commercial opportunities. Biotechnology-84 (Abstr), World Biotech Rep, Pinner, London, vol 1, pp 121–125

Brierley J, Luinctra L (1993) Heap concept for pretreatment of refractory gold ore. In: Torma AE et al. (eds) Biohydrometallurgical techniques, vol 1. The Minerals, Metals and Materials Society, Warrendale, Pennsylvania

Commercial-scale bioleaching of gold (1988) Bioprocess Technol 10, 8: 1–2

Dew DW, Miller DM, van Aswegen PC (1993) Genmin's commercialization of the bacterial oxidation process for the treatment of refractory gold concentrates. Beaver Creek'93 Randol, pp 229–237

Fridman ID (1995) Actual perspectives of processing technologies for Au-containing ores. Geoinfrommark, Moscow, 39 pp (in Russian)

Miller DM, Hansford GS (1992) Batch biooxidation of gold-bearing pyrite-arsenopyrite concentrate. Miner Eng 5, 6: 613–629

Moffat AS (1994) Microbial mining boosts the environment, bottom line. Science 264: 778–779

Nicholson H, Oti-Atakorah S, Lunt S (1993) Selection of refractory gold treatment process for the Sansu Project, Biomin93-Conference, Adereide, March 22, 1993, pp 20.1–20.11

Rossi G, Agate A (eds) (1988) Biogeotechnology of metals. Manual Center for Int Projects GKNT, Moscow, 350 pp (in Russian)

Torma AE (1988) A review of gold biohydrometallurgy. Proc 18th Int Biotechnol Symp, Paris, vol 2, pp 1158–1168

Yakhontova LK, Grudev AP, Krinari GA (1991) X-ray and intercalation characteristic of kaolinite as the parameters of its resistance in bio-abiotic interactions. Dokl USSR Acad Sci 320, 6:1459–1462

Yakhontova LK, Grudev AP, Zuev VV (1994) Problems of studying the system: mineral substrate-microorganism. Bull Moscow State Univ Ser Geol 5:80–92 (in Russian)

CHAPTER 5

Environmental Mineralogy. Radiation Mineralogy

H. U. BAMBAUER (ed.)

5.1 General Overview: The Global Problem of the Impact of the Production of Energy, Metals, Materials, Chemicals, and Radionuclides in the Modern Industrial Society on Air, Water, and Soil Pollution

U. Förstner

History of Environmental Pollution

Mining and smelting have been the cause of major pollution problems since the earliest days of these activities. However, significant increases in waste production occurred in the Middle Ages, when metallic mining industries in north and central Europe became capable of extracting ore from greater and greater depths. The extent of this evolution was summarized by Georgius Agricola in his *De re metallica*, published in 1556. Mercury, which readily forms amalgams with gold, silver, and many other metals, became intensively mined (Fig. 65). Although health hazards have long been known – inhaled mercury vapors affect the central nervous system and the peripheral nerves or cause behavioral disorders – primitive uses of mercury have persisted until recent years, for example during the present gold rush in Brazil (Lacerda et al. 1995).

The chronology of pollution can be seen from sediment cores, and Fig. 66 presents data from Lake Constance by Müller (1991). Heavy metals and polycyclic aromatic hydrocarbons show parallel evolution patterns, and it is concluded that both groups of pollutants could stem from a common source: combustion of coal and lignite as a consequence of increasing industrialization, chiefly in the Northern Hemisphere. The increasing substitution of coal and lignite by petroleum products during the past 30 years is documented by a decrease in heavy metals and PAH concentrations in the youngest sediment layers, whereas, on the other hand, petroleum-derived hydrocarbons show a sharp increase. The history of fecal pollution can be traced back in sediments even into preindustrial times with coprostanol, one of the principal sterols of excreta of higher animals and man. Since the Second World War, halogenated hydrocarbons have become significant toxic compounds in marine and other ecosystems. Polychlorinated biphenyls, which have been used as plasticizers in paints, plastics, resins, inks, copy paper, and adhesives (open use), first occur in sediments around 1935–1940, although their application had been legally restricted to closed systems in transformers and capacitors. The same holds true for DDT (totally banned in most western countries between 1970 and 1975) and its metabolites: concentrations begin to rise around 1945 and reach their maximum between 1960 and 1970. Lindane is a typical follow-up chlorinated pesticide after the ban of DDT. A group of chemicals not classified as hazardous are phthalate esters, which are widely used as plasticizers, and their application is closely related to the development of PVC production. Artificial radionu-

a

b

Fig. 65 a, b. a Mining and roasting of mercury ore in the Middle Ages (Georgius Agricola *De re metallica*, 1556). **b** Gold mining in central Brazil using mercury amalgamation. (Photograph Jorn Thöming)

clides (e.g., ^{137}Cs, ^{239,240}Pu, ^{55}Fe) were first released into the environment as a result of atmospheric weapons testing between 1952 and 1962. The Chernobyl catastrophe in May 1986 induced the so far last worldwide pulse of artificial radionuclides into the environment (Santschi et al. 1988).

Waste Production in the Modern Industrial Society

Contaminated solid materials, which may affect pollution of soil, air, surface water, groundwater, and associated biota include municipal solid wastes, sewage

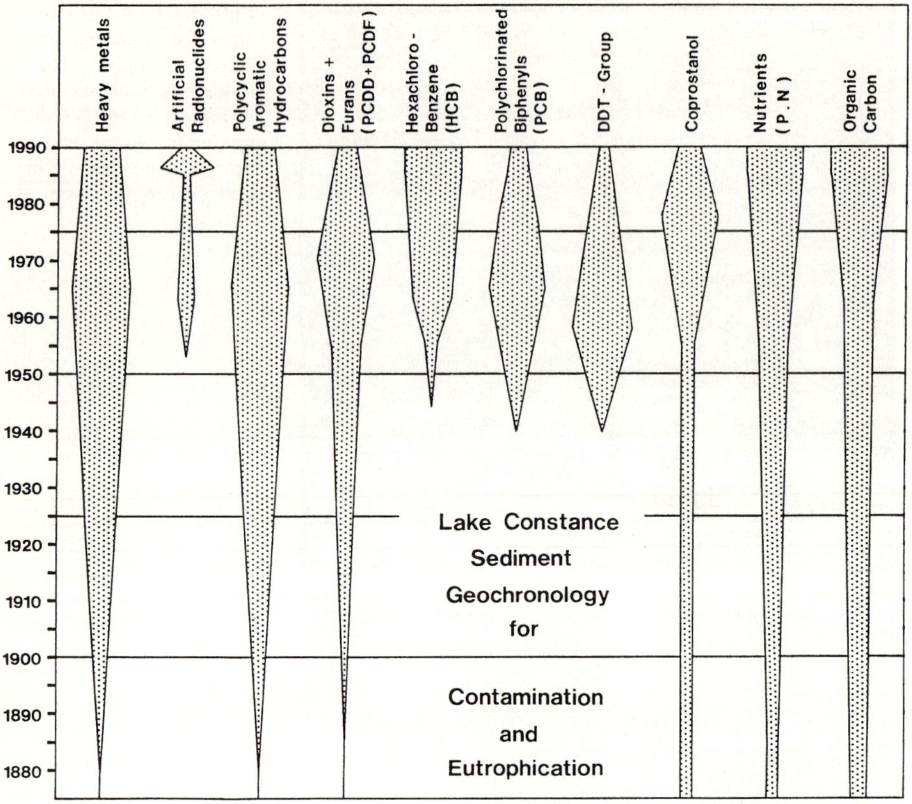

Fig. 66. Development and present situation of environmental pollution with specific contaminants as evidenced from dated sediment cores from Lake Constance (After Müller 1991)

sludge, dredged sediments, industrial by-products, wastes from mining and smelting operations, filter residues from wastewater treatment and atmospheric emission control, ashes and slags from burning of coal and oil, and from incineration of municipal refuse and sewage sludge. The problem of "contaminated land" was introduced with the accidental detection of large-scale pollution from industrial waste deposits which had been handled improperly.

Table 17 lists figures of the production of waste materials (estimated on the basis of global population according to data from Neumann-Malkau [1991]). Municipal solid wastes (see Sect. 5.3.6) and dredged materials amount to approximately 1 billion m^3/a, sewage sludge is even more. Mining residues are in the range of 20 billion m^3, the same order of magnitude as the actual sediment discharge to the ocean. It has been stressed that manufacture of an automobile generates waste about 20 times its actual weight. Since mining progressively proceeds to less concentrated ores, there is an increasing mass of

Table 17. Global waste balances and comparative data

Municipal solid waste	~ 1×10^9 m^3/a
Dredged material	~ 1×10^9 m^3/a
Sewage sludge (95% H_2O)	~ 3×10^9 m^3/a
Mining residues	17.8×10^9 m^3/a [a]
(Continental crust-archaikum	1.2×10^9 m^3/a [a])
(Sediment transport – prehuman	4.5×10^9 m^3/a [a])
(Erosion – present	26.7×10^9 m^3/a [a])

[a] Estimation after Neumann-Malkau (1991).

waste material per unit of extracted metal; it has been estimated that this mass is doubled each 20 years.

Dispersion of Metallic Pollutants in the Environment

Index of Relative Pollution Potential – Technophility Index. The degree of global pollution, for example of toxic metals, can be estimated by comparing the consumption of a certain element and its natural concentration in the different geospheres (lithosphere, pedosphere, hydrosphere, atmosphere). The ratio

$$I_{RPP} = \frac{\text{metal consumption (in t/a)}}{\text{average metal content in a specific sphere (in g/t)}}$$

can be used as a measure of the relative pollution potential of each element in a certain sphere (Förstner and Müller 1973). From Table 18 it becomes evident that the more "rare" (and more toxic) metals are enriched by an order of magnitude as compared with iron, manganese, nickel, and chromium. A similar approach has been proposed by Nikiforova and Smirnova (1975), estimating "technogenic migration" of a metal and the degree of utilization in the noosphere through its technophility index (TP), which represents a ratio of the annual output of a metal to its "Clarke", i.e., its mean concentration in the Earth's crust (Table 18). Of the technologically active elements, the greatest danger for living organisms is metals with maximum technophility, which are capable of forming highly contrasting anomalies in the environment.

Pathways of Pollutants – Global, Regional, and Local. There are many pathways or routes by which aquatic or terrestrial biota and, in particular, humans are exposed to metallic compounds, and these are changing as society uses more or less of the metal or changes the chemical form of the metal in the environment. To assess which elements may be of concern, four criteria have been proposed: (1) Has the geochemical cycle of the element been substantially perturbed by man, and on what scales? (2) Is the element mobile in geochemical processes because of either its volatility or its solubility in natural

Table 18. Index of relative pollution potential for the pedosphere (Förstner and Müller 1973) and technophility index (Nikiforova and Smirnova 1975)

	Consumption × 1000 t/a	Soil (mg/kg) (Bowen 1966)	Index of relative pollution potential	Technophility index (x 10^7)
Iron	400000	38000	1	5
Manganese	9200	850	1	5
Copper	6400	20	30	110
Zinc	4600	50	10	54
Lead	3500	10	35	160
Chromium	1700	100	2	20
Nickel	493	40	1	9
Tin	232	10	2	–
Cadmium	15	0.06	25	–
Mercury	10	0.03	30	150

waters, so that the effect of geochemical perturbations can propagate through the environment? (3) What is the degree of public health concern associated with the element? (4) What are the critical pathways by which the most toxic species of the element can reach the organ in man which is most sensitive to its effect?

Of the elements listed in Table 19 global perturbations are most dramatically seen for lead. Present-day levels of lead in Americans and Europeans are probably 2 to 3 orders of magnitude higher than those of pretechnological humans, as evidenced from studies on blood lead concentrations in remote populations. Changes on a regional scale are typically found for aluminum mobilization in soils and waters of low buffer capacity affected by acid precipitation; despite insignificant anthropogenic inputs of aluminum, increased solubility will induce toxic effects on both terrestrial and aquatic biota. Chromium usually represents examples of only local significance; here, elemental species exhibit characteristic differences, in that the hexavalent form is more toxic than the trivalent form. Other elements, such as lead and mercury in Table 19, may

Table 19. Perturbation of the geochemical cycles of selected metals by society. (After Andreae et al. 1984)

	Scale of perturbation			Diagnostic environments	Mobilizing mechanisms	Critical pathway
	Global	Regional	Local			
Lead	+	+	+	Ice, sediment	Volatilization	Air, food
Aluminum	–	+	–	Water, soil	Solubilization	Water
Chromium	–	–	+	Water, soil	Solubilization	Water
Mercury	(–)	+	+	Fish, sediment	Alkylation	Food (air)
Cadmium	(–)	+	+	Soil, sediment	Solubilization	Food

be mobilized by the biotic or abiotic formation of organometallic compounds. Accumulation of methyl-mercury in seafood, probably the most critical pathway of a metal to humans, has caused several thousand cases of poisoning in Japan. The first catastrophic event of cadmium pollution, causing the Itai-Itai disease in inhabitants in the Jintsu River area of Japan for 20 years after World War II, was due to effluents from zinc mine wastes, which flooded low-lying rice field areas.

Regional Metal Fluxes. Early recognition of limitations of resources and environmental impact requires a comprehensive understanding of man's activities in connection with nutrition, hygiene, habitation, and transport. From an operational point of view, "metabolic processes in the anthroposphere" must be controlled by the region, a geographically, economically, and politically defined entity (Baccini and Brunner 1991). The perceptions gained by material flux analysis complement those obtained in life sciences (e.g., human toxicology), environmental sciences (e.g., ecotoxicology), and social sciences (e.g., economics, sociology). Regional material management can serve as a political means for the "kybernetic control" of material fluxes, i.e., efficient measures to reduce potential impacts on essential resources in time and with appropriate priorities. In particular, however, this approach relates to the dispersion of pollutants in critical environmental compartments.

Figure 67 presents the example of zinc fluxes and reservoir in the test area of Bünztal/Switzerland (Baccini 1992). The anthropogenic zinc import is about 2 orders of magnitude higher than the inflow via water and air. The most important sink for Zn is the subsystem household/industry, which is increasing

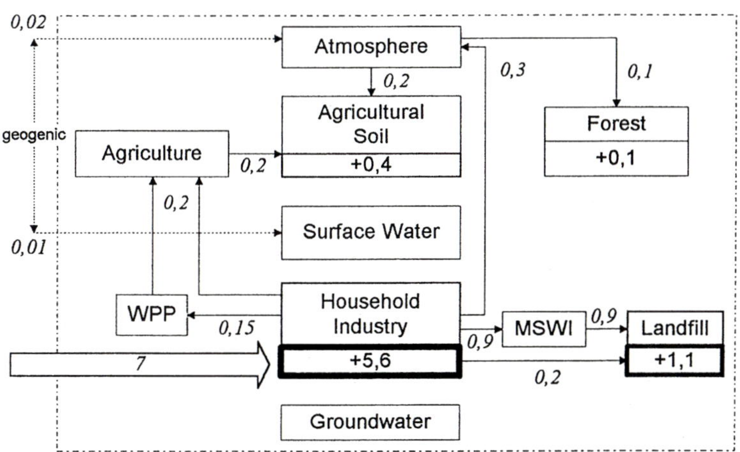

Fig. 67. Fluxes of zinc (kg per person and year) in the study area of Bünztal/Switzerland (Baccini, 1992). *WPP* Water purification plant; *MSWI* municipal solid waste incineration plant

by 5 to 6 kg per capita and year. The second important sink is landfills (+1.1 kg/cap/a, followed by agricultural soils (+0.4 kg/cap/a). The latter system, even at relatively low metal inputs, may exhibit adverse effects on sensitive plants, when metal mobility is enhanced due to simultaneous inputs of atmospheric acidity (Herms and Brümmer, 1980).

Final Storage Quality – Geochemical Engineering Approach

Storage of solid waste is one of the environmentally most critical processes in material management. Today's landfills require an additional material and energy flux for their operation, e.g., for leachate purification and gas treatment; in practice, they still produce leachates and gas fluxes to water and air (Baccini and Brunner 1991). Comparison of inorganic groundwater constituents upstream and downstream of 33 waste disposal sites in Germany (Arneth et al. 1989) indicates typical differences in pollutant mobilities, which may partly be related to releases during the initial phase of the landfill development. High *contamination factors* ("contaminated mean"/"uncontaminated mean"; Fig. 68) have been found for boron, ammonia, and arsenic; heavy metals such as cadmium, chromium, lead, copper, and nickel are significantly enriched in the leachates as well.

The final storage approach is one way to develop and control landfills on a conceptual basis: "landfills with solids of final storage quality need no further treatment of emissions into air and water". It has been defined by the Swiss

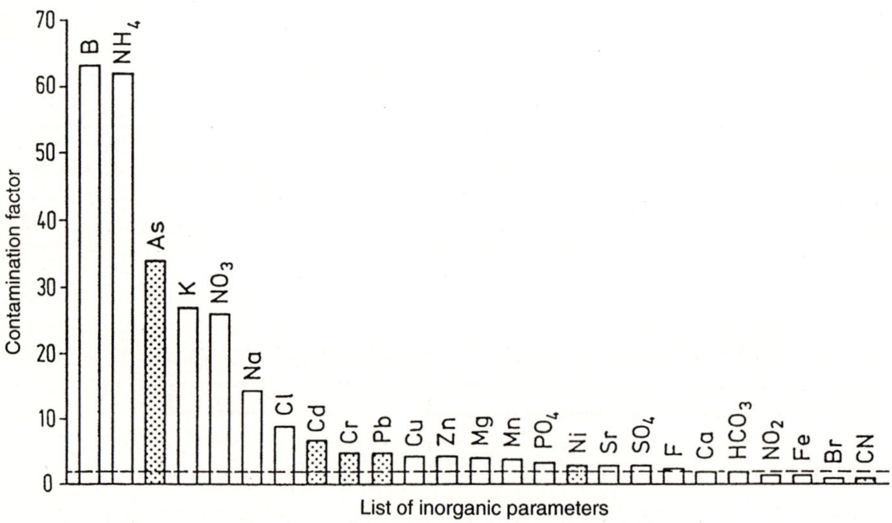

Fig. 68. Ranking of inorganic groundwater constituents based on their contamination factor (Arneth et al. 1989). *Dotted* Toxic elements, Group D, EC-Directive

Federal Government in 1986 and received wider attention through the book edited by Baccini (1989) on *Landfills – Reactor and Final Storage*.

Final storage properties can be achieved by using with typical geochemical engineering techniques such as (Förstner 1992): selection of favorable milieu conditions for the deposition of large-volume wastes such as dredged materials, selection of additives for the solidification and stabilization of hazardous waste materials, optimization of elemental distribution at high-temperature processes, e.g., incineration of solid waste materials, and chemical leaching procedures.

Geochemical Engineering – Engineering Geochemistry. The term geochemical engineering or engineering geochemistry, reminds of the earlier work of Russian geochemists, e.g., Vernadsky and Fersman in the 1930s, who introduced the expression geochemistry of technogenesis for defining the geochemical consequences of industrial and agricultural activities. According to Voronkevich (1994), engineering geochemistry is based on fundamental concepts of technogenetic migration of chemical elements and is concerned with:

1. Developing theoretical ideas concerning the interrelations between technogenetic alterations of geochemical parameters (pH, Eh, etc.) taking place in the geochemical environment and related changes in physicomechanical properties such as stress distribution, thermal and moisture conditions, and other physical quantities.
2. Studying analogous technogenetic geochemical processes such as hydrolysis, ion exchange, weathering, and leaching to forecast the technologically uncontrollable development of hazardous engineering phenomena.
3. Developing environmental technologies based on natural geochemical processes to remove contaminants from natural cycles.
4. Developing physicochemical methods to improve the conditions of the geological environment.

Topic (3), involving extractive procedures, is closely related to Schuiling's (1990) definition of geochemical engineering, whereas (4) widely corresponds to the concept of Salomons and Förstner (1988) for the same term, emphasizing "the increasing efforts of using natural resources available at the disposal site for reducing negative environmental effects of all types of solid waste materials".

Interlinkage of Controlling Parameters – Analytical Procedures. Analytical and experimental procedures as well as technologies related to final storage can be deduced from Salomons' (1993) concept of interlinkage between capacity controlling parameters and major biogeochemical cycles (Fig. 69): Gradients (driving forces) typically involve degradation of organic matter, which, in turn, will affect changes in redox and pH conditions as well as release of dissolved organic carbon and salt ions. Their interactions with solid matrices may induce nonlinear, delayed developments, so-called chemical time bomb processes (Stigliani 1992). This means that a waste deposit or contaminated hotspot

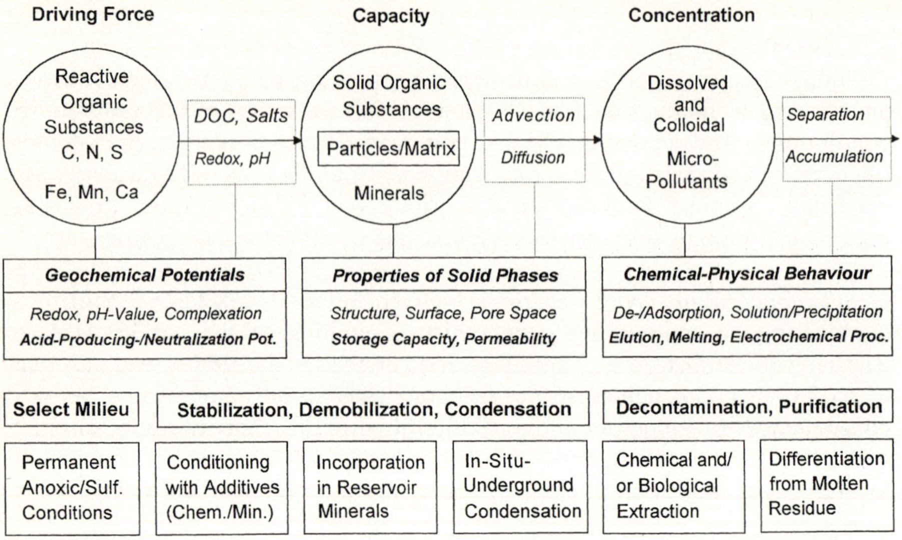

Fig. 69. Interlinkage of the various fields of environmental and engineering geochemistry. *Top* Factors for mobilization of micropollutants, *middle* analytical and experimental methods for long-term prognosis, *below* technologies for final storage quality

which initially appears to be relatively harmless can eventually have disastrous environmental effects as toxic contaminants are released.

Toxic metals can break through once the specific buffering capacity of a sediment or soil system as been surpassed either by direct oversaturation or through a fundamental change in a chemical property of the substrate that reduces its capacity to adsorb (or keep adsorbed) toxic materials.

System descriptions include chemical-physical, solid-phase, and geochemical characterizations. Most common of the former are determinations of elution and desorption rates of critical pollutants from solid substances (elutriate tests), while diffusion studies, e.g., on solidified waste residues, are predominantly based on the experience with radioactive material (see Sect. 5.4.2). Geochemical investigations are mostly aimed for long-term prediction of typical borderline conditions, such as redox, acid-producing, and complexation potential. Regarding the solid phase, major emphasis will be placed on the long-term capacities of minerals to carry and store critical pollutants, mainly metals (Bambauer 1991), under changing environmental conditions (see Sect. 5.2.1) and to form mineral barrier systems for solid waste disposal (see Sect. 5.2.2).

Outlook

Despite increased efforts in pollution control, a large number of environmental problems are related to the fact that many natural resources are underpriced and

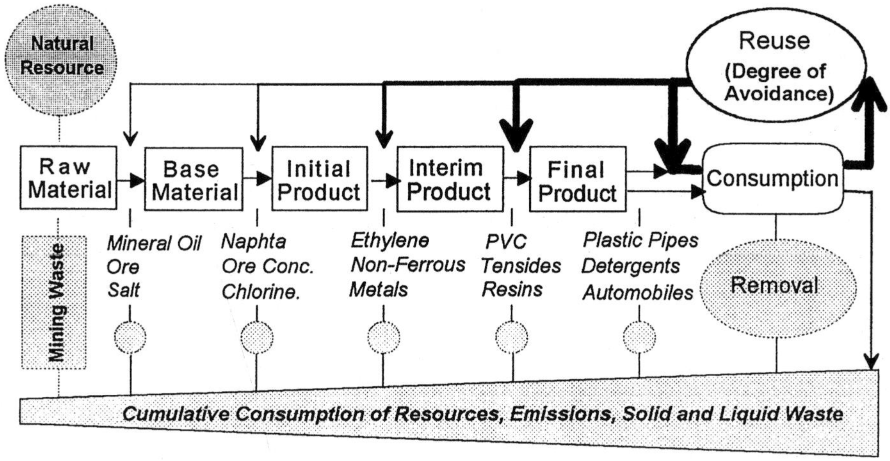

Fig. 70. Product cycles, emissions, and degree of avoidance. (After Sutter et al. 1994)

overused; it is also clear that many industrial processes involve multiple steps, resulting in a low level of system efficiency, especially in comparison with biological systems (Ayres 1989). In the technological sequence from natural resource to final product via raw and base material, initial and interim product, there is a general increase in its value. However, simultaneously, consumption of material and energy resources, emissions to water and air, as well as production of solid an liquid waste is enhanced (Fig. 69). This implies that recycling should always take place at the highest possible level of complexity to avoid a maximum of emissions and resource consumption. The sum of the argument suggests a long-run imperative favoring an industrial metabolism that results in reduced extraction of virgin materials, reduced loss of waste materials, and increased recycling of useful ones. At the same time, it is clear that ideal recycling is not possible, due to dissipative losses of material and energy during abrasion, corrosion, aging, etc.; in many instances, recovery of dispersed components would require an economically unjustified effort. On the other hand, the question has been raised, what could and should be the life design of everything we create? Such a life-cycle perspective might provide guidance for instances in which "dematerialization" rather than materialization should be the eventual objective (Herman et al. 1989).

References

Agricola G (1556) De re metallica libri XII. Froben, Basel

Andreae MO, Asami T, Bertine KK et al. (1984) Changing biogeochemical cycles – group report. In: Nriagu JO (ed) Changing metal cycles and human health. Dahlem-Konferenzen, Life Sciences Research Report 28. Springer, Berlin Heidelberg New York, pp 359–373

Arneth J-D, Milde G, Kerndorff H, Schleyer R (1989) Waste deposit influences on ground water quality as a tool for waste type and site selection for final storage quality. In: Baccini P (ed) The landfill – reactor and final storage. Lecture Notes in Earth Sciences, vol 20. Springer, Berlin Heidelberg New York, pp 399–415

Ayres RU (1989) Industrial metabolism. In: Ausubel JH, Sladovich HE (eds) Technology and environment. National Academy Press, Washington, DC, pp 23–49

Baccini P (ed) (1989) The landfill – reactor and final storage. Lecture Notes in Earth Sciences, vol 20. Springer, Berlin Heidelberg New York

Baccini P (1992) Vom Abfall zum Stein des Weisen. Introductory Lecture, May 20, 1992 at ETH Zürich, 18 pp

Baccini P, Brunner PH (1991) Metabolism of the anthroposphere. Springer, Berlin Heidelberg New York

Bambauer HU (1991) The application of mineralogy to environmental management. ICAM '91 Int Congr on Applied Mineralogy, September 2–4, 1991, Pretoria/RSA, vol I, C. 133

Bowen HJM (1966) Trace elements in biochemistry. Academic Press, London

Förstner U (1992) Contaminated aquatic sediment and waste sites as toxic chemical time bombs. 1992 RA Vollenweider Lectureship in Aquatic Sciences. Burlington/Ontario, October 7, 1992. National Water Research Institute, Burlington, Canada 42 pp

Förstner U, Müller G (1973) Heavy metal accumulation in river sediments: a response to environmental pollution. Geoforum 14:53–61

Herman R, Ardekani SA, Ausubel JHG (1989) Dematerialization. In: Ausubel JH, Sladovich HE (eds) Technology and environment. National Academy Press, Washington, DC, pp 50–69

Herms U, Brümmer G (1980) Einfluß der Bodenreaktion auf Löslichkeit und tolerierbare Gesamtgehalte an Nickel, Kupfer, Zink, Cadmium und Blei in Böden und kompostierbaren Siedlungsabfällen. Landwirtschaftl Forsch 33:408–423

Lacerda LD, Malm O, Guimaraes JRD, Salomons W, Wilken R-D (1995) Mercury in the New Gold Rush in the South. In: Salomons W, Stigliani WM (eds) Biogeodynamics of pollutants. Springer, Berlin Heidelberg New York, pp 213–245

Müller G (1991) Sustainable future for planet earth? Freedom from smoke. In: Geotechnica. International Trade Fair and Congress for Geo-Sciences and Technology. Cologne/Germany, September 18–21, 1991, pp 45–47

Neumann-Malkau P (1991) Anthropogenic mass movement – interfering with geologic cycles? In: Geotechnica Congress Cologne, September 1991, pp 153–154

Nikiforova EM, Smirnova RS (1975) Metal technophility and lead technogenic anomalies. Abstr Int Conf Heavy Metals in the Environment, Toronto, Part C, pp 94–96

Salomons W (1993) Non-linear and delayed responses of toxic chemicals in the environment. In: Arendt F, Annokée GJ, Bosman R, van den Brink WJ (eds) Contaminated soil '93. Kluwer Dordrecht, pp 225–238

Salomons W, Förstner U (eds) (1988) Environmental management of solid waste: dredged materials and mine tailings. Springer, Berlin Heidelberg New York

Santschi PH, Bollhalder S, Farrenkothen K, Lueck A, Zingg S, Sturm M (1988) Chernobyl radionuclides in the environment: tracers for the tight coupling of atmospheric, terrestrial, and aquatic geochemical processes. Environ Sci Technol 22:510–516

Schuiling RD (1990) Geochemical engineering, some thoughts on a new research field. Appl Geochem 5:251–262

Stigliani WM (1992) Chemical time bombs, predicting the unpredictable. In: Chemical time bombs. European State-of-the-Art Conf on Delayed Effects of Chemicals in Soils and Sediments. Veldhoven, The Netherlands, Sept 2–5, p 12

Sutter H, Mahrwald B, Grosse-Ophoff M (1994) Die Bedeutung der ökologischen Rangfolge im neuen Kreislaufwirtschafts- und Abfallgesetz. Umwelttechnol Aktuell 6/94:427–436

Voronkevich SD (1994) Engineering geochemistry: problems and applications. Appl Geochem 9:553–559

5.2 Concepts and Methods for Applications of Mineralogy to Environmental Management

The following flow sheet (Fig. 71; Bambauer 1991, lit. in Sect. 5.2.2) illustrates some main outlines of environmental management with respect to pollution. Besides important technical and chemical issues, environmental management involves a great deal of mineralogical (including crystallographic, petrologic, and geochemical) methods and ways of thinking. The first step towards the management of an environmental problem is a thorough analysis; based on this, an optimal and realistic approach to a desired absolute solution has to be found, but never will be reached for technical limits and an unavoidable variety of interrelated detail problems; each attempt may create new problems.

An important aspect of environmental management is the identification of emission sources. Examples of source identification by mineralogical and geochemical methods, including literature, are given by Bambauer (1991).

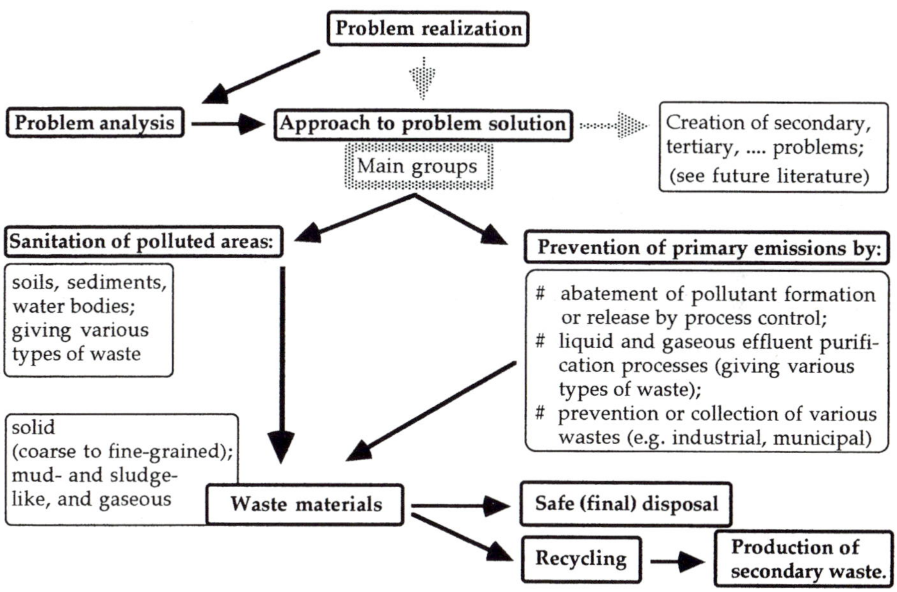

Fig. 71. Schematic representation of the management of environmental problems with reference to pollution. (Bambauer 1991)

5.2.1 Chemical-Mineralogical Speciation

H. U. BAMBAUER and H. PÖLLMANN

While pollution cases are mostly documented by chemical bulk analyses only, chemical speciation of the (inorganic as well as organic) pollutants in question is of equal importance. Speciation involves the form in which an element, e.g., a heavy metal ion, occurs in a given compound with respect to valence, bonding, and coordination; for instance, as a heavy metal ion of a complex of specific composition in aqueous solution (e.g., arsenic as arsenate complex), or as a specific crystal element in a certain crystalline compound, e.g., $[AsO_4]^{3-}$-group replacing $[SO_4]^{2-}$ in ettringite.

For practical purposes, a special case of "mineralogical" speciation can be called carrier mineral, e.g., a crystalline compound, frequently of airborne origin, carrying a pollutant as a main or trace element. For instance, the thallium carrier Tl(Cl,Br) (Fig. 72). Also glasses may play an important role as a sort of carrier minerals (Table 20).

Fig. 72. Example of a "carrier mineral": cube of Tl[Cl,Br], partly overgrown by sylvinite and surrounded by calcite and syngenite, in the stack fly ash of a cement works, Germany; STEM/BF. (Bambauer and Schäfer 1984)

Table 20. Examples of heavy metal speciation as „carrier minerals"; samples from various stack fly ashes

Name	Chemical formula	Additional trace elements
Franklinite	$ZnFe_2O_4$	Ti, Cr, Mn
Spinel	$CoFe_2O_4$	
Calcium phosphate	$Ca_3[PO_4]_2$	V, (Zn?)
–	$(K,Na)_2(Pb, Ca, Zn)[SO_4,Cl_2]$	
Thallium chloride	Tl[Cl,Br]	
Galenite	PbS	
Glassy spheres	Ca-alkali- -alumosilicate	Sr, Ba, Sc, Ti, V, Cr, Mn, Co, Ni, Cu, Nb

A carrier mineral may become a (storing-up) reservoir mineral, such as a mineral neoformed in a pollutant-containing milieu that is able to incorporate pollutants in its crystal structure during growth. The main properties of these mineral reservoirs after formation, e. g., in waste disposal, can be summarized as follows:

1. Highly stable or resistant and slightly soluble (low leachability) in the geochemical cycle of a given disposal and its geological environment.
2. Variable chemical composition (solid solution) to enable the incorporation of various hazardous chemicals.
3. Formation preferentially from the waste material itself (or by adding of minor admixtures).

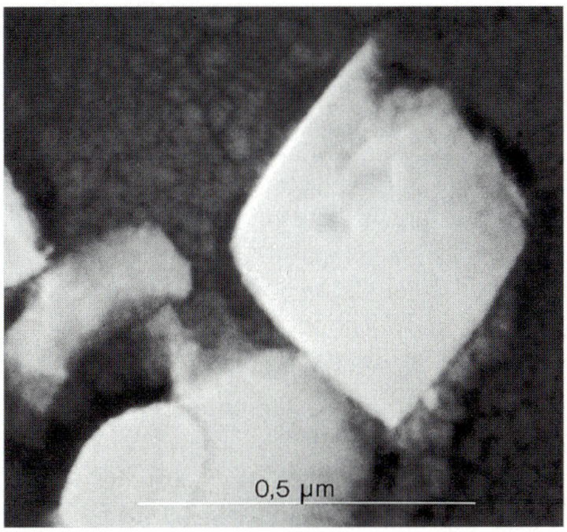

Fig. 73. Example of a "reservoir mineral": euhedral jarosite as an accessory of dumped pyrite calcines mainly consisting of hematite (microtome section, STEM/BF); the corresponding EDX-spectrum shows fixed heavy metals (Cu-peaks from copper grid). See formula in Section 5.2.2 (Steffes-Tun et al. 1989)

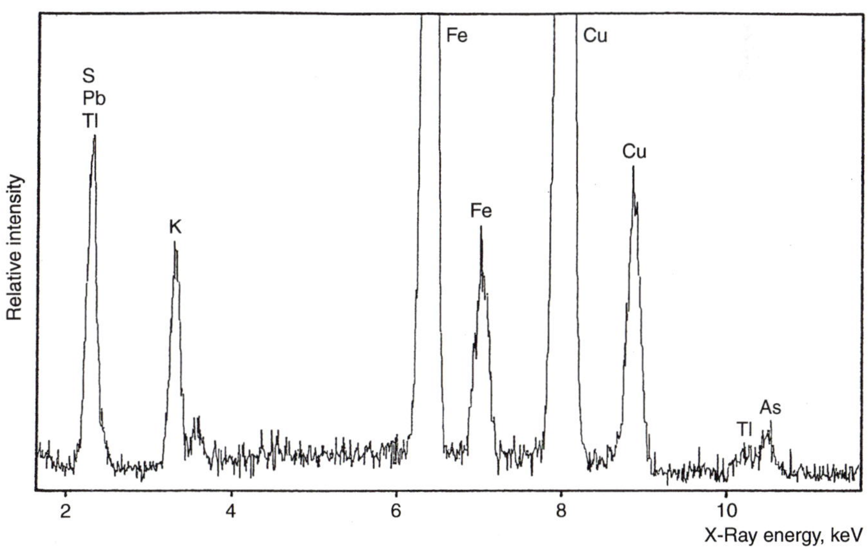

4. Giving rise to the formation of a dense microstructure in the disposed material.

Therefore, reservoir minerals are expected to have a selective immobilization capacity for certain pollutants or trace elements (e.g., a heavy metal or a radionuclide) in the long run (Fig. 73; Tables 21 and 22). Also, mineral phases with special surface sorption properties and amorphous materials may be considered here (Table 23). Reservoir minerals may form in a wide range from high-temperature processes to temperatures of diagenesis.

Speciation is often related to aggregation, as pointed out in Fig. 74.

The importance of localization of environmentally relevant elements or compounds is demonstrated in Fig. 75.

Table 21. Examples of inorganic reservoir minerals for hazardous elements occurring in various waste materials. I. High to medium temperature phases with isomorphic substitution

Compound	Hazardous elements	Occurrence
Chloroellestadite $Ca_{10}Cl_2[SO_4,SiO_4]_3$ apatite type	Zn, Cd, Pb, Sr, Ba, V, As, Se	Formation during thermal treatment of refuse incinerator flue gas purification residues
Alumosilicate glassy spheres	V, Cr, Mn, Co, Ni Cu, Sr, Ba, etc.	Formation in fly ashes of coal-fired power plants
Borosilicate glasses	Radionuclides	Synthetic
Hollandite type $BaAl_2Ti_60_{16}$	Radionuclides Cs, Sr, etc.	Synthetic: synroc B [a]
Fluorite type (U, Ce, Zr, etc.)O_{2+x}	U, Ce, Zr	Synthetic

[a] For more details about synroc see Section 5.4.2.

Table 22. Examples of inorganic reservoir minerals for hazardous elements from various waste materials. II. Low temperature phases with isomorphic substitution

Compound	Hazardous elements	Occurrence
Calcite $CaCO_3$	Mn, Co, Ni, Cu, Zn, Cd	Stabilized dredged sludge
Jarosite $KFe_3[(OH)_6SO_4]$	Tl, As, Pb	Dumped pyrite calcines, mainly composed of hematite
Brucite $Mg(OH)_2$	Zn; Mn (?)	Stabilized residues of lignite-fired power plants
Ca-monosulpho-aluminate hydrate type compounds	Various anions (Cl); heavy metals: Cd, Cr, etc.	Stabilized residues of lignite-fired power plants and various other sludge-like waste materials
Ettringite-type compounds	Various anions; heavy metals: Cr, Mn, Co, Ni, Zn, Sr, Pb, As	As above, see also Section 5.2.1

Table 23. Example of reservoir minerals by intracrystalline and surface sorption

Substance	Application or research field
Bentonite (smectite group), palygorskite	Additive for dump sealing materials and cements used for waste consolidation; hectorite with superficial heavy metal precipitation (as well as tetracalcium aluminate hydrate) is known as adsorber of highly poisonous organic compounds
Zeolites (natural, synth.); synthetic sodalite, analcime, leucite, kalsilite type compounds	Catalyst material for NO_x reduction in thermal power stations; pollution control, effluent clean up; pollutant adsorbing additive to cements used for consolidation of industrial waste and for barrier systems; fixation of radionuclides.
Calcium silicate hydrates	Known to adsorb chloride ions and heavy metals (see also Sect. 5.3.7); fixation of radionuclides

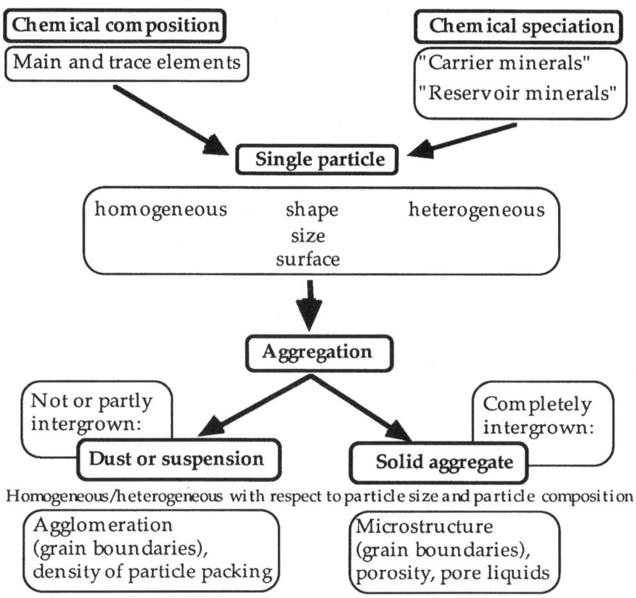

Fig. 74. Chemical composition vs. speciation and aggregation of solid geological and anthropogenic materials

Whereas the relation between major bulk chemistry and major constituents is obvious, major constituents, accessories, and grain boundaries may contribute to the bulk trace element content. In addition, according to the nature of grains and their position in the microstructure, a very different reactivity upon environmental processes is to be expected.

Some applications in practice and literature will be given in the following Section 5.2.2.

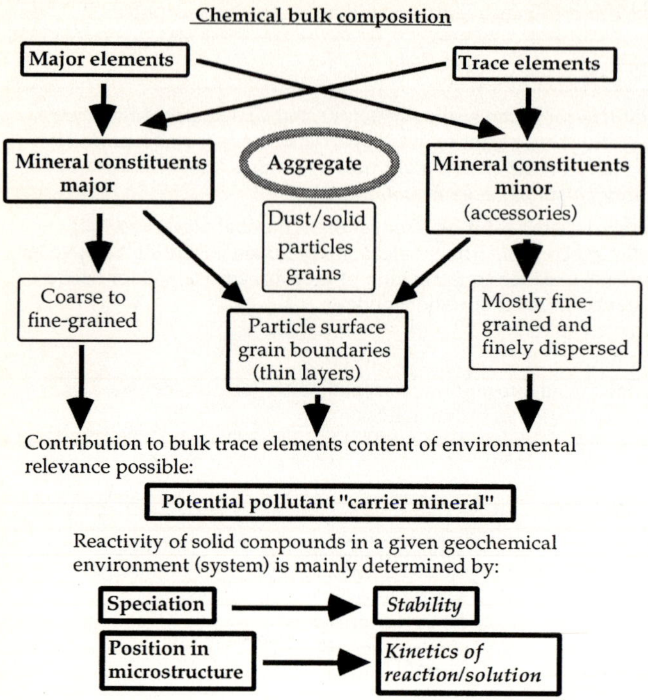

Fig. 75. Environmental behavior of particle/grain aggregates with respect to pollution

5.2.2 "Mineralogical" Barrier Systems

H. U. BAMBAUER and H. PÖLLMANN

According to the flow sheet (Fig. 71), large amounts of various types of waste have to be disposed of. Most of the inorganic waste materials are fully oxidized, e.g., residues of thermal processes, but also wastes of more "euxinic" composition, e.g., dredged sludges, are frequent (see Sect. 5.3.5). A special problem are radio-active waste materials (see Sect. 5.4.2). Most waste materials contain easily soluble or leachable components. In order to keep them apart from the geochemical cycle, a specific disposal strategy has to be developed for each waste material. In particular, waste material dumps have to be sealed from groundwater by *barrier systems.*

Usually, outer barrier systems are installed (see scheme below); but in certain cases, inner barrier systems could be realized either by mechanical encapsulation of waste particles into an added matrix material or by mineralogical pollutant immobilization as outlined in the following scheme and in Fig. 76.

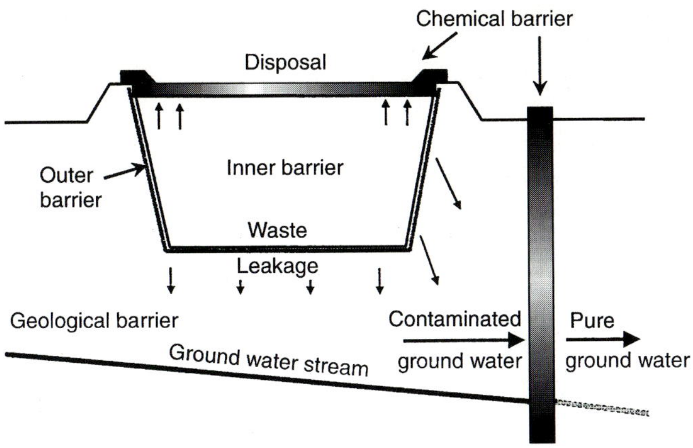

Fig. 76. Scheme of a disposal site showing the position of different barriers

Barrier Systems for Waste Disposals Including "Mineralogical" Barrier Systems

Outer Barrier Systems.
(a) Geological barriers: surrounding geology of the disposal and corresponding geochemical milieu.
(b) Conventional sealing, e.g., with top and basal clay linings, including additional special chemical barriers (resulting in the reduction of leakage contaminations).

Inner Barrier Systems.
Mineralogical pollutant immobilization inside the whole body of waste material of a given disposal.

Principle: A sequence of treatments or a single treatment of waste material to give various neoformed mineral phases which are expected to result in:

1. Formation of one or more generations of "reservoir minerals" under "magmatic-metamorphic" or "diagenetic" conditions which are able to immobilize certain pollutants (see Sect. 5.2.1).
2. Final consolidation and maximum condensation of micro-texture of the treated material (minimum porosity and permeability).

Expected result: the waste body develops its own, internal sealing property.

Also, a simple solidification resulting in mechanical encapsulation of pollutant particles into an added matrix material is in use.

Multiple Barrier Systems. A combination of outer an inner barrier systems (see also Sect. 5.4.2).

All barrier systems are based on minimizing the emission of hazardous elements or compounds into the environment. The expected properties of

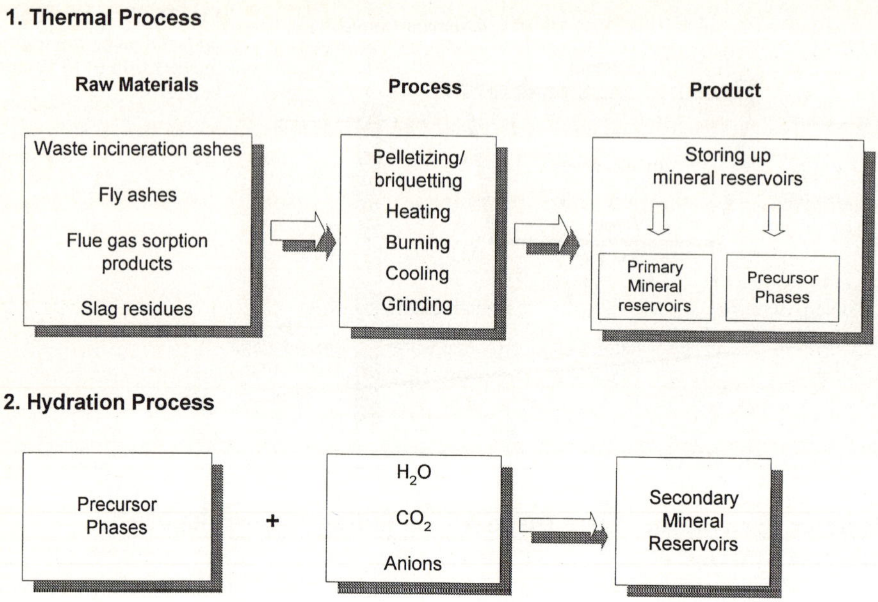

Fig. 77. Examples for the treatment of waste materials to give mineral reservoirs

the outer and geological barrier materials comprise very low water permeability ($Kf < 10^{-9} - 10^{-8}\ ms^{-1}$) and high chemical resistance (see Sect. 5.3.9). If practicable, the concept of the inner barrier system is based on the formation of a stabilized solid material.

In order to achieve an optimal immobilization effect in such a solid material, the attempt is made to form reservoir minerals synthetically by controlled thermal and hydraulic processes according to the given above general scheme (Fig. 77). It gives the proposed reaction sequence which is based on a two-step formation process of mineral reservoirs (according to Sect. 5.2.1). During the first (high-temperature) thermal step primary reservoir minerals and precursor phases are formed. These precursor phases are hydraulic minerals (reaction with water is possible). In the second low temperature step the precursor phases react with water and CO_2 to form secondary reservoir minerals.

In low-temperature environments, procedures similar to the setting of a cement may be practiced according to the general scheme:

$$\text{Primary mineral phases} + \underset{\text{(solution)}}{\text{mixing water}} \xrightarrow[\text{pH}]{\text{Temperature}} \text{Stabilized solid material}$$

$$\text{Paste} \longrightarrow \text{congealing} \longrightarrow \text{hardening} \longrightarrow \text{dense microstructure}$$

$$\text{Inert primary minerals} \longrightarrow \text{fillers}$$
$$\text{Reactive primary minerals} \longrightarrow \text{neoformed minerals}$$

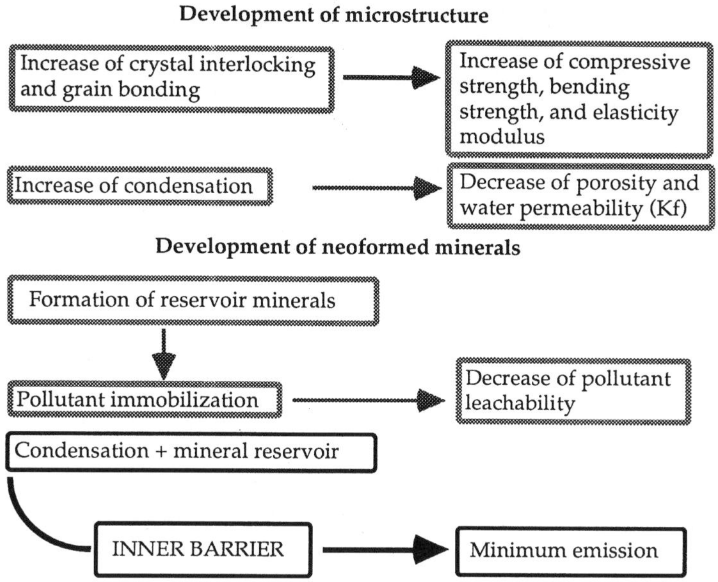

Fig. 78. The formation of an inner barrier by development of neoformed minerals and microstructure

In certain cases, the mixing water may also be a brine of known contribution to the running chemical reactions. The resulting technological and chemical properties of, for instance, a stabilized sludge, are controlled by the development of neomineralization in a given raw mix under given conditions. By that the waste body develops its own internal sealing barrier, and as a result the pollutant concentration of leachate of treated materials is found to decrease drastically according to the increasing content of mineral reservoirs. Some main relations are given in Fig. 78.

The formation of minerals in landfill areas by the reactions shown normally produce nature-identical products which are stable under surface conditions. Also, metastability is to be considered in some cases. Depending on the properties after the reaction and the bulk concentration of hazardous substances the resulting materials can mostly be used for landfill purposes. Also, a recycling utilization is to be expected in many cases, as has been practiced by the production of alinite cement from garbage combustion ashes. If leakages in the groundwater stream already occur, then the additional placement of reservoir minerals for clarifying purposes can be used as well.

Occasionally, reservoir minerals may form in a landfill area by nonpredicted reactions or at least unintentionally. As an example, the mean (somewhat simplified) formula of neoformed jarosite in an pyrite calcine dump is given as follows:

$$(K_{0.58\text{-}0.87}Tl_{0.07\text{-}0.18}Pb_{0.06\text{-}0.24})_{1\text{-}1.08}Fe_3[S_{0.92\text{-}1.00}As_{0\text{-}0.08})O_4]_2(OH)_6 .$$

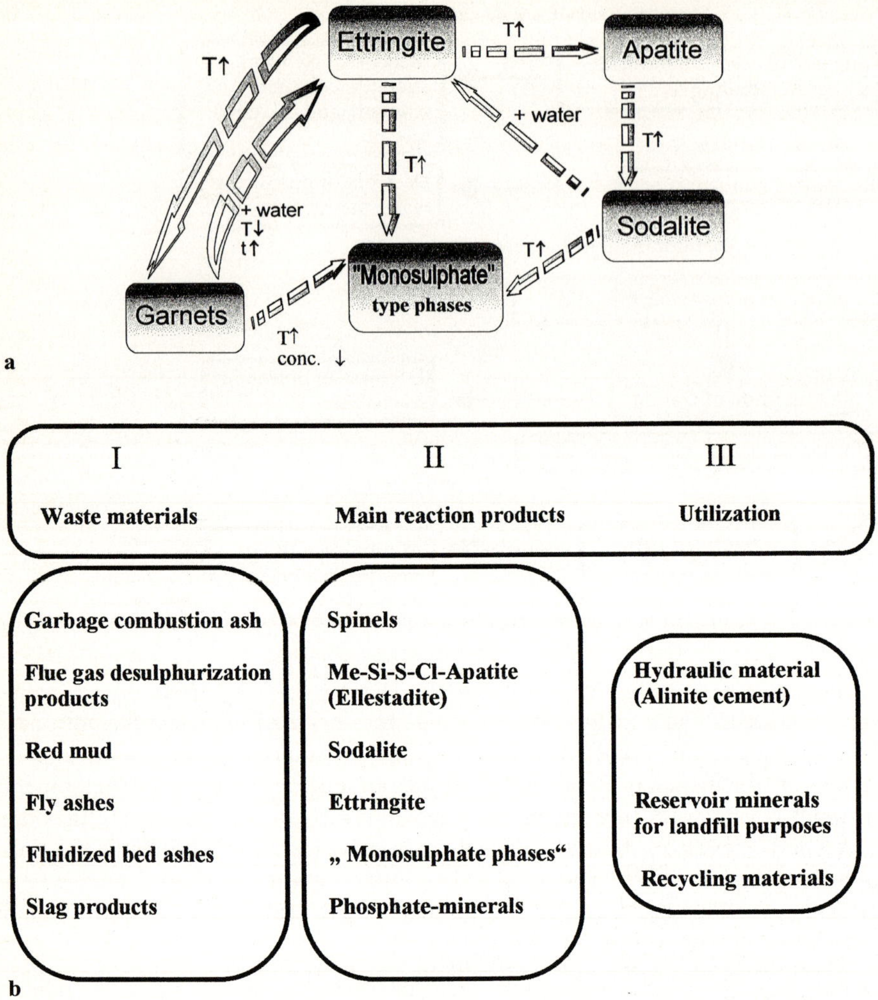

Fig. 79. a Examples for the relations between several reservoir mineral structure types (hydrates) as a function of temperature and water content. **b** A selection of waste materials, reaction products after treatment, and possible utilizations

Jarosite was found to be a reservoir mineral for Tl, Pb, As (see Fig. 73 and Table 22 in Sect. 5.2.1) in a low-pH environment.

In general, the investigation of waste materials and resulting products gives a broad range of chemical and mineralogical composition – even for products of a given industrial plant. The variables in composition, temperature, oxydation state, and techniques used influence drastically the resulting, coexisting phases. For this reason technologies for the optimization of reservoir mineral formation must consider these variables.

The reservoir minerals are stable under definite conditions. The scheme given above shows possible transformation reactions of selected reservoir minerals (represented by names of general structure types) as a function of temperature under normal pressure (Fig. 79a). These reactions are reversible under surrounding conditions. It can be shown that a fixing mechanism of certain elements is possible even under varying conditions by the replacement of one reservoir mineral by another. Even hydrogarnets can incorporate certain amounts of sulfate and chromate anions.

Table 24. Immobilization capacity of selected synthetic mineral reservoirs (wt.%)

Compound	Anion capacity		Compound	Cation capacity	
Ettringite type	Sulfate	>20	"Monosulphate" type[a]	Cobalt	>40
	Chromate	>20		Nickel	>40
	Selenate	>20		Copper	>40
				Zinc	>40
				Cadmium	>40
Apatite type	Sulfate	>25	Apatite type	Strontium	>50
	Chromate	>25		Lead	>50
	Phosphate	>25			
	Vanadate	>25			

[a] Calcium-monosulfoaluminate-hydrate type (lamellar Me-Me-hydroxisalt).

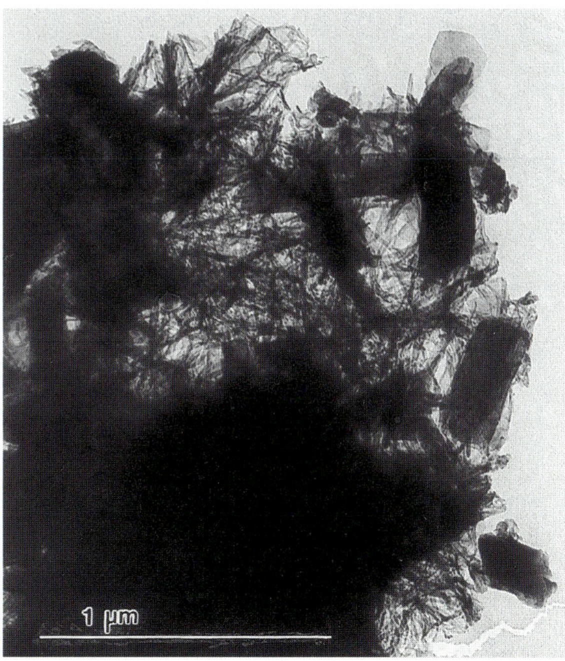

Fig. 80. Example of a low temperature paragenesis of reservoir minerals in a stabilized solid of lignite fly ash: columnar ettringite (first generation) in a web-like intergrowth of calcium silicate hydrates CSH (second generation). TEM/BF. (Bambauer 1992)

Figure 79b shows some selected examples which are already proven in practice: primary wastes, their reaction products, and final utilization.

Among the known reservoir minerals, various powerful materials were tested for their quantitative immobilization capacity, a selection of which is given in Table 24. Under optimal conditions these minerals incorporate certain cations and anions in their crystal structure up to the given limits.

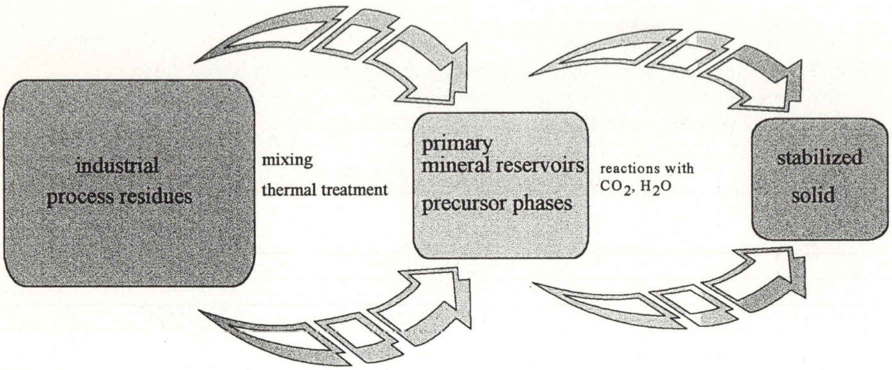

Fig. 81. Summarizing scheme: the formation of reservoir minerals from industrial process residues to give a stabilized solid

Fig. 82. A reservoir mineral to be used in the immobilization of Cl^- and CrO_4^{2-}: hexagonal-tabular synthetic crystals of the "monosulphate" type (lamellar Me-Me-hydroxisalt) $3CaO \cdot Al_2O_3 \cdot 1/2CaCrO_4 \cdot 1/2CaCl_2 \cdot nH_2O$. SEM/SE

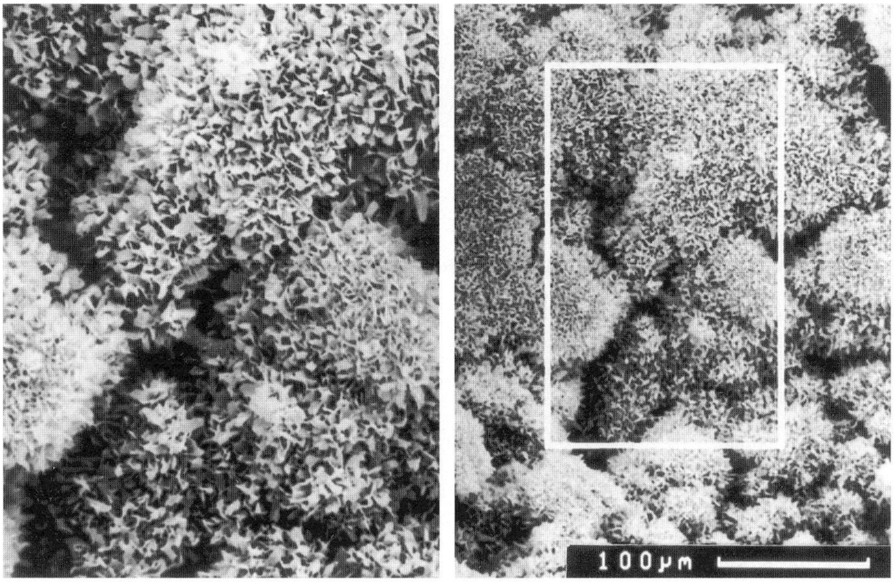

Fig. 83. In-situ crystallization of "monosulphate" type compounds (lamellar Me-Me-hydroxysalts) in a hardened cementitious system. SEM/SE

For the potential role of calcium silicate hydrates (CSH phases) as reservoir minerals (Fig. 80) see Section 5.3.7.

Figure 81 summarizes the mode of formation of the various reservoir minerals.

In addition to Section 5.2.1, selected examples of reservoir minerals are given in Figs. 82 and 83. For further examples see Bambauer (1991).

References

Auer St, Pöllmann H, Kuzel H-J (1990) Immobile Fixierung von CrO_4^{2-} in Zementmineralen. Ber Dtsch Mineral Ges 1 : 7

Bambauer HU (1991) The application of mineralogy to environmental management. An Overview. Proc ICAM 91, vol 1, Pretoria

Bambauer HU (1992) Brennst WK/TÜ Umw-Spec März

Bambauer HU, Schäfer H (1984) Der Mineralbestand eines thalliumhaltigen Reingasstaubs aus der Zementproduktion. Fortschr Min 62/1 : 33–50

Bambauer HU, Gebhard G, Holzapfel Th, Krause C, Willner G (1990) Blockage de pollutants dans des résidues stabilisés composés de cendres de lignite et de produits de désulfurisation de gaz de combustion. Industrie Minérale – Mines et Carrièrres, Jan-Fév 1990; (see also Bambauer et al. 1988 in Sect. 5.3.6 lit.)

Bambauer HU, Holzapfel Th, Schreiter P, Werner M (1995) Mineralogische Schadstoffimmobilisierung in Körpern aus Reststoffen von Braunkohlen-Kraftwerken. Handb Verwertung Braunkohlenfilteraschen in Deutschland. Hrsg: RWE AG Zentr Ber Forschung u. Entwicklung, Essen

Gebhard G, Pöllmann H (1993) The practical usage of mineral reservoirs for safety mechanisms of waste disposal sites. Int Conf Appl Min, Perth

Motzet H, Pöllmann H, Kuzel H-J (1990) Mögliche Einsatzgebiete für Calciumsulfit aus Rauchgasentschwefelungsanlagen. Ber Dtsch Mineral Ges 1:181

Neubauer J, Pöllmann H (1990) Einbau von Schwermetallen in Silikatapatite. Ber Dtsch Mineral Ges 1:191

Pöllmann H (1993) Capability of cementitious materials in the immobilization process of hazardous waste materials. Proc 15th Int Congr Cem Micr, Dallas

Pöllmann H (1995) Formation of reservoir minerals in waste disposals for immobilization purposes. Int Congr on Imm and Sol, Nancy

Pöllmann H, Kuzel H-J, Wenda R (1989) Compounds with ettringite structure. N Jahrb Mineral Abh 5; 133–158

Pöllmann H, Auer St, Neubauer J, Stemmermann P (1991) Bildung eines Innere-Barriere-Systems in Abfalldeponien – Immobile Fixierung von Chlorid und Schwermetallen in Speichermineralen. Umwelt-Forum GSF, pp 82–83

Steffes-Tun W, Bambauer HU, Krause C (1989) Jarosit als Thallium-Träger in einer Deponie von Schwefelkies-Abbränden. Tagungsber. ENVIROTECH Vienna, Hrsg Umwelt-Bundesamt, Wien

5.2.3 Analysis of Individual Airborne Mineral Particles

P. R. Buseck and J. R. Anderson

Introduction

Minerals in the micrometer to submicrometer size range are abundant in both aqueous environments and the atmosphere. Airborne mineral particles have significant effects on atmospheric visibility, climate, and, because they are continually inhaled, human health. Reactive sites increase and reactivity (including catalytic activity) is maximized as grain sizes decrease. Sizes and compositions reflect the sources as well as the health and environmental effects of particles. The challenge is to determine their identities through chemical and structural analysis, and to do this for individual particles. The same analytical procedures are useful for both natural and anthropogenic particles and so have been widely used for a range of environmental studies.

The goal of most procedures for analyzing aerosols is to strive for ever greater sensitivity in bulk samples. Such analyses produce useful data about average compositions, but conclusions about individual particles must be inferred. However, even when a given element is present at a very low concentration in a bulk sample, it may be highly concentrated within given particles. Common examples are Zr and Ba in desert-derived mineral dust; the great majority of these elements are, respectively, in individual particles of zircon and barite. Since many reactions impacting health are affected by individual compounds, i.e., individual particles, it can be important to know whether certain minerals or toxic compounds occur within a sample, even if only at

Table 25. List of Acronyms

Acronym	Instrument (technique)
EELS	Electron energy-loss spectrometer (spectrometry)
EMPA	Electron microprobe analyzer (analysis)
L^2MS	Two-step laser mass spectrometry
LAMMA	Laser microprobe mass analyzer (analysis)
REMPI	Resonance-enhanced, multiphoton ionization
SEM	Scanning electron microscope (microscopy)
STEM	Scanning transmission electron microscope (microscopy)
SXRFM	Synchrotron X-ray fluroescence microprobe
TEM	Transmission electron microscope (microscopy)
TOFMS	Time-of-flight mass spectrometer (spectrometry)
	Other terms
PAHs	Polyaromatic aromatic hydrocarbons
XANES	X-ray absorption near-edge structure
ZAF	Atomic number (Z), absorption (A), and fluorescence (F) corrections

low concentrations. In such cases, analysis of individual particles can prove exceedingly powerful.

The most common procedure for determining composition is to bombard particles with energizing radiation and monitor the resulting fluoresced X-radiation. Since many particles must generally be analyzed, such work can be tedious. However, with automated methods one can analyze many hundreds of particles in a day. On the other hand, crystallographic analysis of given individual particles may be necessary to distinguish them from among several structural forms. Such work is done almost exclusively with the transmission electron microscope (TEM). Electron energy-loss spectroscopy (EELS) is a TEM technique that can be useful for determining the degree of crystallinity or information about speciation in certain types of particles. Routine morphological analysis is best done with scanning electron microscopy. Acronyms are widely used; a brief list is given in Table 25.

Electron-Induced Excitation; Elemental and Structural Analysis

Electron beams can be focused into regions as small as 0.3 nm in diameter (far smaller than normal X-ray beams), and so they are ideal for the study of small regions of material. A portion of the energy of the incident electron beam is typically absorbed by the specimen through inelastic scattering, with the subsequent emission of X-rays that exhibit a range of energies (Fig. 84). These energies are characteristic of the elements in the sample and so can be used for chemical analysis by positioning a detector laterally above the specimen. Also, when the electrons in the incoming beam interact inelastically with the specimen atoms, there is an energy transfer whereby the beam electrons lose energy.

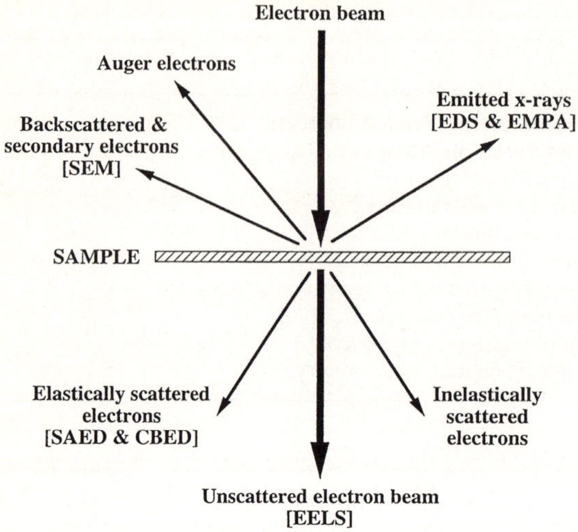

Fig. 84. Schematic diagram of the different types of signals produced when an electron beam strikes a sample. (Buseck 1992, Fig. 1–2)

This energy decrease can be measured by a detector that is placed below the sample, so that it records electrons after they have traversed the sample. This is the basis for EELS, which is useful for elements that are too light for normal X-ray spectroscopy. Finally, the incoming beam can also excite secondary electrons, such as are used in the conventional scanning electron microscope (SEM). These electrons produce information regarding surface morphology.

Several types of instruments are employed to make the measurements. The electron microprobe analyzer (EMPA) and SEM are the most widely used. The former is optimized for chemical analysis while the SEM is designed primarily to produce images with secondary or backscattered electrons. However, both instruments can and commonly do perform either function.

EMPA/SEM methods of analysis are extremely effective for most aerosol particles, but there are also particles where additional magnification or complementary techniques are required. Examples include aggregate particles, particles less than 0.1 μm in diameter, and compounds such as sulfates in which the proportions of light elements can assume various values. Magnification greater than is possible with the SEM or EMPA may be necessary for identifying the individual components of the aggregates. For the latter case, structural information is diagnostic. These requirements can be fulfilled by use of the TEM or the scanning TEM (STEM), which also are the instruments of choice for the smallest particles.

Although the EMPA and SEM can be operated in transmission mode, they are generally operated in reflection mode because the specimens are typically thick enough that the incident beam is completely absorbed within the sample; at normal accelerating voltages of 15 to 25 kV, the absorption limits resolutions in bulk samples to roughly 2 μm. Standard corrections then need to be applied for the effects of atomic number (Z), absorption (A), and fluorescence (F).

Most aerosol particles are smaller than the excited volumes (we routinely analyze particles as small as 0.1 μm in diameter). On irregularly shaped particles, the absorption effect can strongly deviate from that of ideal flat samples; this deviation, plus the problem of aggregation, can make the calculation of correction factors very difficult. With the TEM, samples must be electron-transparent and so are generally sufficiently thin that there is little spreading of the electron beam as it traverses the sample. ZAF corrections can then be avoided entirely. However, standardization is more difficult with the TEM than with the EMPA or SEM. In practice, analysis of typical natural and anthropogenic particles is complicated by problems of morphology, size, and aggregation so that only ratios of concentrations can be obtained. Generally, results of somewhat lower precision and sensitivity are produced with the TEM than with the EMPA.

Analysis such as are described above are commonly done by using solid-state, energy-dispersive detectors that are most sensitive for Na and heavier elements, although newer detectors can analyze light elements down to B if concentrations are sufficiently high. Large data sets that contain much information are obtained, and statistical methods are required to organize and then extract significant information. EELS is well suited for measurement of lighter elements, but such analyses are only done with the TEM or STEM.

TEM analyses must be done manually, much as we also used to do EMPA measurements. However, with the EMPA, images produced from backscattered electrons (BSE) can be used in automated runs to locate grains and distinguish them from their substrates (commonly polycarbonate Nuclepore filters). Using either a predetermined or dynamically determined gray-level threshold, the gray-level BSE image can be converted to a binary image that serves as a map of the particles in the field of view. This image can be used to determine size, shape, orientation, and particle position.

An X-ray spectrum can be acquired from each particle by either directing the beam to the center of mass of the particle or by rastering over a rectangle that just covers the particle. For rapid analysis using only qualitative chemical data, spectra may be accumulated for 5 to 10 s each. However, for more quantitative data, counting times of 60 s or more are necessary to improve the ratios of peaks to background. The spectra are processed by using a filtered least-squares fit of standard spectra. With a completely automated system, the location, analysis, and spectral processing of particles can be done unattended and therefore around the clock. The number of particles that can be analyzed in a day is determined by the time for spectral acquisition. For maximum accuracy, we favor a counting time of 100 s, in which case it takes about 36 h to analyze 1000 particles (the number we typically analyze for a given sample) and store the results on disk.

Crystallographic information is commonly necessary for particle identification, and then the TEM has unique capabilities. High-magnification images (Fig. 85) can provide detailed information on the structures and shapes of complex particles. Selected-area electron diffraction can be used to determine the

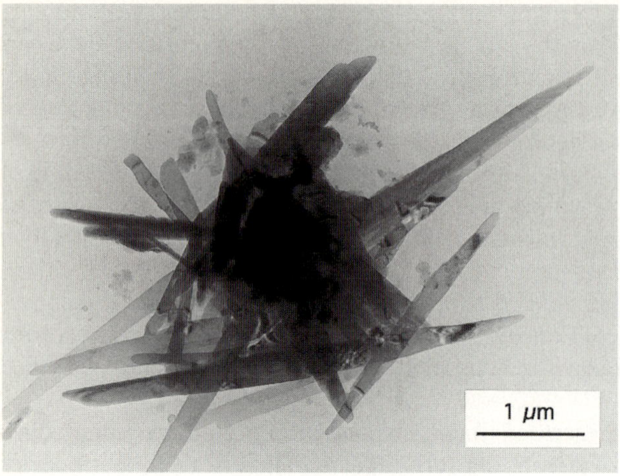

Fig. 85. Transmission electron microscope image of an aerosol particle collected over the ocean near the Azores after polluted European air arrived in the region. The central part consists of Na_2SO_4, and the thin spikes are mixed-cation sulfates. The particle was presumably converted to sulfate from sea salt by reaction with atmospheric SO_2 when ozone oxidized SO_2 to SO_4 in the aerosol water (Photo M. Posfai)

crystallinity and structure of portions of complex particles and to distinguish among compounds that contain different proportions of the same major elements, such as $PbSO_4$, $PbO.PbSO_4$, and PbS, as well as other metal sulfates and sulfides common in the atmosphere. Many of these species are especially difficult to distinguish because they occur primarily in the submicron size fraction; structural data offers the only hope of success. In some cases, specialized methods such as fractal analysis of TEM images can provide useful insights.

For examples of environmental applications, see Section 5.3.1.

Speciation; SXRFM, LAMMA, and L^2MS

Chemical measurements made with the instruments described above provide elemental data that must be interpreted to identify the specific mineral species providing the signal. While such interpretation is commonly straightforward, it can be complex when several minerals contain the same elements but in slightly different proportions, as commonly occurs for minerals of elements having multiple oxidation states (e.g., Fe, Mn, S, U). The problem is perhaps even more complex for organic compounds. For these purposes there are several specialized techniques that can be used. These include synchrotron X-ray fluorescence microprobe (SXRFM), laser microprobe mass analysis (LAMMA), and resonance-enhanced, two-laser mass spectrometry (L^2MS).

Most measurements that use X-rays as the radiation source, whether for diffraction to obtain structural information or fluorescence for chemical analyses, use broad beams and so are best applied to bulk samples. An exception is the bright beams that can be obtained using synchrotron radiation. Chemical analysis with the SXRFM is a relatively new procedure whereby a high-intensity X-ray beam can be collimated onto small areas of sample. As with the instruments mentioned above, the resulting emitted radiation is then measured. The advantage of SXRFM is that trace concentrations (0.1 to 1 ppm) of a wide range of elements (S or heavier) are accessible using either K- or L-line fluorescence. Spatial resolutions are in the range of 10 μm, although higher resolutions can be obtained through the use of narrower beams, but with a consequent loss of intensity. Oxidation states can be determined using X-ray absorption near-edge structure (XANES) spectroscopy.

The technique has great potential for studying small particles. For example, it has been used to measure the Se distribution in soil aggregates as well as variations in oxidation states in individual soil particles. The next step, should it be desirable, would be to disperse the soil or other sample into its constituent components and use SXRFM to measure these.

LAMMA has been applied to the study of aerosol particles as small as 0.5 μm. A particle is vaporized by a high-power laser pulse, producing atomic and molecular fragments, some of which are ionized. The ions are accelerated, collimated, and analyzed using a time-of-flight mass spectrometer (TOFMS) that separates the ions based on their mass-to-charge ratio. However, LAMMA is limited by its inability to distinguish between ions of the same nominal molecular mass. It has been used successfully in cases such as analyses for ammonia in marine aerosols; there is no significant interference for ammonia in such samples. LAMMA is also inherently more sensitive to trace metals than electron-beam techniques and is useful in cases where interferences are not severe. To date, the potential of LAMMA to analyze single particles has not been realized, but the method may become increasingly useful if its limitations are overcome.

Several laser-based techniques have been combined with a reflectron TOFMS to analyze individual particles. A single laser can be used for both vaporization and the ionization required for mass spectrometry. Recent work shows that particles between 0.1 and 10 μm diameter can be introduced directly from the ambient atmosphere into a differentially pumped particle beam. Depending on the laser fluence, the resulting mass spectra can contain fragmentation patterns, and these can be difficult to reconstruct (they are similar to those from a LAMMA microprobe). Particle sizes are determined indirectly by comparison of spectra to those of standard spheres.

Greater chemical selectivity can be achieved through the use of two lasers, where one either vaporizes the particle or desorbs its surface molecules and the other produces the subsequent ionization. Such two-step laser mass spectrometry (L^2MS) can utilize resonance-enhanced, multiphoton ionization (REMPI) to study the polyaromatic aromatic hydrocarbons (PAHs) desorbed from particle surfaces. These procedures have been used for on-line analysis of selected alkali salts and organic compounds, and real-time sampling is possible. Light scattering by the particles was used to determine their sizes prior to their entry into the TOFMS for chemical analysis. A similar procedure was used for a study of particles collected in an industrial area, the countryside, and in a highway tunnel. PAHs detected in this way, in contrast to LAMMA, are free of fragmentation and ionization matrix effects. The sensitivity is greater than can be achieved with gas chromatography combined with mass spectrometry, and nonvolatile, insoluble compounds can be studied directly on the collection filters.

The laser methods have great promise for studying individual particles and especially organic molecules on and within small particles. However, particle sizes are determined indirectly. Also, these methods result in complex spectra containing many peaks, so that considerable time and effort is required to interpret them.

Summary

A variety of methods are capable of being used to determine compositions and structures of small particles. The choice of technique depends on the information desired; it is commonly most productive to use a variety of methods in combination. Currently, the most widely used, general-purpose methods are electron microprobe analysis and transmission electron microscopy. When used in tandem, even the smallest inorganic particulate compounds in aerosols can be identified.

References

Electron-microprobe-based methods

Anderson JR, Aggett FJ, Buseck, PR, Germani MS, Shattuck TW (1988) Chemistry of individual aerosol particles from Chandler, Arizona, an arid urban environment. Environ Sci Technol 22:811–818

Anderson JR, Buseck PR, Saucy DA (1992) Characterization of individual fine-fraction particles from the Arctic aerosol at Spitsbergen, May-June, 1987. Atmos Environ 26A: 1747–1762

De Bock LA, Van Malderen H, Van Grieken RE (1994) Individual aerosol particle composition variations in air masses crossing the North Sea. Environ Sci Technol 28:1513–1520

Germani MS, Buseck PR (1991) Automated scanning electron microscopy for atmospheric particle analysis. Anal Chem 63:2232–2237

Goldstein JA, Newbury DE, Echlin P, Joy DC, Romig AD Jr, Lyman CE, Fiori C, Lifshin E (1992) Scanning electron microscopy and X-ray microanalysis. Plenum Press, New York

Katrinak KA, Anderson JR, Buseck PR (1995) Individual particle types in the aerosol of Phoenix, Arizona. Environ Sci Technol 29:321–329

Markowicz A, Raeymaekers B, Van Grieken R, Adams F (1986) Analytical electron microscopy of single particles. In: Spurny KR (ed) Physical and chemical characterization of individual airborne particles. Ellis Horwood, Chichester

Post JE, Buseck PR (1984) Characterization of individual particles in the Phoenix urban aerosol using electron-beam instruments. Environ Sci Technol 18:35–42

Post JE, Buseck PR (1985) Quantitative energy-dispersive analysis of lead halide particles from the Phoenix urban aerosol. Environ Sci Technol 19:682–685

Statistical treatment of data

Hopke PK, Gladney ES, Gordon GE, Zoller WH, Jones AG (1976) The use of multivariate analysis to identify sources of selected elements in the Boston urban aerosol. Atmos Environ 10:1015–1025

Kim D, Hopke PK (1988) Classification of individual particles based on computer-controlled scanning electron microscopy data. Aerosol Sci Technol 9:133–151

Saucy DA, Anderson JR, Buseck PR (1987) Cluster analysis applied to atmospheric aerosol samples from the Norwegian Arctic. Atmos Environ 21:1649–1657

Saucy DA, Anderson JR, Buseck PR (1991) Aerosol particle characteristics determined by combined cluster and principle component analysis. J Geophys Res 96/D4:7407–7414

Shattuck TW, Germani MS, Buseck PR (1985) Cluster analysis of chemical compositions of individual atmospheric particles data. Am Chem Soc Symp Ser 292:118–129

Shattuck TW, Germani MS, Buseck PR (1991) Multivariate statistics for large data sets: applications to individual aerosol particles. Anal Chem 63:2647–2656

Transmission electron microscopy-related data and methods

Buseck PR (ed) (1992) Minerals and reactions at the atomic scale: transmission electron microscopy. Mineral Society of America, Reviews in Mineral, vol 27

Buseck PR, Bradley JP (1982) Electron beam studies of individual natural and anthropogenic microparticles: Compositions, structures and surface reactions. In: Schryer DR (ed) Heterogeneous atmospheric chemistry. Geophys Monogr Ser 26:57–76, AGU, Washington, DC

Buseck PR, Self PG (1992) Electron energy-loss spectroscopy (EELS) and electron channeling (ALCHEMI). In: Buseck PR (ed) Minerals and reactions at the atomic scale: transmission electron microscopy. Mineral Society of America. Reviews in Mineralogy, vol 27: pp 141–180

Buseck PR, Cowley JM; Eyring L (eds) (1992) High-resolution transmission electron microscopy and associated techniques. Oxford University Press, New York

Katrinak KA, Rez P, Buseck PR (1992) Structural variations in individual carbonaceous particles from an urban aerosol. Environ Sci Technol 26:1967–1976

Katrinak KA, Rez P, Perkes PR, Buseck PR (1993) Fractal geometry of carbonaceous aggregates from an urban aerosol. Environ Sci Technol 27:539–547

Parungo F, Nagamoto C, Sheridan P, Schnell R (1990) Aerosol characteristics of Arctic haze sampled during AGASP-II. Atmos Environ 24A:937–949

Peacor DR (1992) Analytical electron microscopy: X-ray analysis. Rev Mineral 27:113–140

Pósfai M, Anderson JR, Buseck PR, Shattuck, TW, Tindale N (1994) Constituents of a remote Pacific marine aerosol: a TEM study. Atmos Environ 28:1747–1756

Sheridan P (1989) Characterization of size segregated particles collected over Alaska and the Canadian High Arctic, AGASP-II flights 204–206. Atmos Environ 28:2371–2386

Synchrotron X-ray fluorescence microprobe (SXRFM)

Bajt S, Sutton SR, Delaney JS (1994) X-ray microprobe analysis of iron oxidation states in silicates and oxides using X-ray absorption near edge structure (XANES). Geochim Cosmochim Acta 58:5209–5214

Bertsch PM, Hunter DB, Sutton SR, Bajt S, Rivers ML (1994) In situ chemical speciation of uranium in soil and sediments by micro X-ray absorption spectroscopy. Environ Sci Technol 28:980–984

Sutton SR, Rivers ML, Bajt S, Jones KW (1993) Synchrotron X-ray flurorescence microprobe analysis with bend magnets and insertion devices. Nucl Inst Meth B75:553–558

Tokunaga KT, Sutton SR, Bajt S (1994) Mapping of selenium concentrations in soil aggregates with synchrotron X-ray fluorescence microprobe. Soil Sci 158:421–434

Laser-based methods (LAMMA and L^2MS)

Dierck I, Michaud D, Wouters L, Van Grieken R (1992) Laser microprobe mass analysis of individual North Sea aerosol particles. Environ Sci Technol 26:802–808

Hintz KP, Kaufmann R, Spengler B (1994) Laser-induced mass analysis of single particles in the airborne state. Anal Chem 66:2071–2076

Otten Ph, Brynseels F, Van Grieken R (1987) Study of inorganic ammonium compounds in individual marine aerosol particles by laser microprobe mass spectrometry. Anal Chim Acta 195:117–124

Prather KA, Nordmeyer T, Salt K (1994) Real-time characterization of individual aerosol particles using time-of-flight mass spectrometry. Anal Chem 66:1403–1407

Xhoffer C, Wouters L, Van Grieken R (1992) Characterization of individual particles in the North Sea surface microlayer and underlying seawater: comparison with atmospheric particles. Environ Sci Technol 26:2151–2162

Zenobi R (1995) In situ analysis of surfaces and mixtures by laser desorption mass spectrometry. Int J Mass Spectrom Ion Processes 145:51–77

Zhan Q, Voumard P, Zenobi R (1995) Application of two-step laser mass spectrometry to the chemical analysis of aerosol particle surfaces. Rapid Commun Mass Spectrom 9:119–127

5.3 Special Examples of Applications of Mineralogy and Geochemistry to Environmental Problems

5.3.1 Atmospheric Dust

J.R. ANDERSON and P.R. BUSECK

Introduction

Suspended inorganic particles in the troposphere impact the environment on every scale, from global to local. Such particles constitute only a tiny fraction of the troposphere's total volume, but the uniformly high abundance of particles with diameters smaller than 10 μm dictates that they have a strong effect on the chemical and physical properties of the troposphere. Even in remote regions of the globe, concentrations in the environmentally significant size range of about 0.1 to 10 μm often exceed 10^6 particles/m^3 of air.

The tropospheric aerosol at a given location is a combination of multiple origins: primary particles from local and distant sources, products of gas-particle and liquid-particle interactions, and aggregates of particles from single or multiple sources. The chemistry and morphology of different particle types in an aerosol reflect their origins and histories. The greatest variability in chemical and physical properties, and the strongest resulting environmental impact, come from particles and particle-forming precursors of anthropogenic origin. Even much of the mineral dust in the troposphere results from man's activities that accelerate processes such as desertification.

On the largest scale, aerosol particles in the troposphere affect the solar radiation balance and the stability of global climate. Tropospheric aerosols present one of the greatest uncertainties in quantifying the climate forcing arising from anthropogenic alteration of the atmosphere's composition. On a regional scale, aerosol particles affect the pH of rain by (1) providing catalytic surfaces, (2) acting as reactants in processes that form acidic nuclei for rain droplets, and (3) acting as buffering agents. In many regions of the world, acid precipitation is a significant threat to both agriculture and biologic diversity. The incorporation of certain metal-bearing particles into acidic rain droplets can also have the secondary effect of solubilizing toxic metals.

Many local effects of aerosol particles in the lower part of the troposphere, the boundary layer, are obvious to anyone who lives in impacted areas. Visibility is impaired by the light-scattering from suspended particles, although much of such "haze" may result from organic rather than inorganic particles. Less obvious, but of greater impact, are the effects of inhaling toxic particles. The inorganic aerosol particles can also serve as important catalytic sites for the production of photochemical smog.

Natural Inorganic Aerosol Particles

Mineral dust from continental areas and sea salt from the oceans are two of the principal natural inorganic particle types. A third type results from chemical reactions in the atmosphere that ultimately result in the formation of particles; starting materials may be biogenically derived gases, but the final products are inorganic species such as ammonium sulfate and ammonium nitrate. Many key reactions are photochemically driven. Volcanic eruptions also produce significant quantities of dust but are highly periodic in nature.

On a global scale, the largest sources of mineral dust are the arid and semiarid areas of North Africa and central China. However, mineral dust is generated anywhere that soil or unconsolidated sediments are exposed. The suspension of mineral dust is greatly enhanced by human activities, for instance, farming and grazing on marginal, semiarid land and vehicular traffic on unpaved roads.

The minerals that occur as aerosol dust from continental areas are the same as those in the original soils and sediments. Quartz, feldspars, clays, micas, chlorite, calcite, dolomite, and minor minerals such as apatite, barite, rutile, and zircon are among those commonly observed in aerosols derived from North Africa and Central Asia. Gypsum is another common mineral observed, but it is difficult to determine if it was initially gypsum or was calcite that reacted in the atmosphere with sulfate of anthropogenic or biogenic origin. The size distributions of airborne particles are determined by many factors including the sizes at the sources, velocity of transporting winds, transport distance, and the structure of the troposphere over the transport path.

Over the marine regions that comprise 75% of Earth's surface, sea salt is the major component by mass of the aerosol in the tropospheric boundary layer. Seasalt aerosol is generated by the ejection of jet and film droplets produced by the bursting of air bubbles at the sea surface.

It is widely believed that the major dissolved components of seawater (Na, Mg, Ca, K, Cl, and SO_4) are partitioned into each aerosol particle in approximate proportion to their abundances in seawater, i.e., their chlorinity ratios are preserved in every particle. However, analyses of individual aerosol particles from the Arctic and other areas show the presence of several types of sea-salt particles. This compositional diversity is most pronounced in fine-fraction (< 1,5 μm) samples and is apparently the result of a fractionation process, such as mechanical fragmentation of primary sea-salt particles or differentiation by separation of crystalline products from residual brine in aerosol droplets.

The chemical alteration of mineral aerosol particles during transport, part of the process of aerosol "aging", may result in the formation of water-soluble species of elements otherwise locked in relatively insoluble silicate structures. Mineral particles may be altered by sulfuric, nitric, or other acids (e.g., methanesulfonic acid and HBr). Multiple cycles of cloud processing accelerate mineral-aerosol alteration.

The secondary generation of particles is also a major source of aerosol in the remote marine troposphere, and it is presumably responsible for the large fraction of the sulfate that is not associated with sea-salt (nonsea-salt or n.s.s. sulfate).

Important secondary processes include (1) the reaction among gaseous species and (2) condensation of gaseous species and subsequent reaction with various types of particles, particularly sea-salt particles.

The origins of n.s.s. sulfate in the remote marine atmosphere have been widely debated. For instance, is the oxidation of biogenic, reduced-sulfur gases such as dimethyl sulfide and hydrogen sulfide or the oxidation of anthropogenic sulfur dioxide the major source of n.s.s. sulfate in the remote marine atmosphere? Clearly, anthropogenic sulfate aerosol is abundant in large parts of the Northern Hemisphere and should have a major impact on the climate. However, n.s.s. sulfate aerosol occurs in significant concentrations in remote marine areas not impacted by anthropogenic emissions and is most likely biogenic in origin. It is now widely held that n.s.s. sulfate of both biogenic and anthropogenic origin is important to global climate forcing.

Examples of tropospheric aerosol particle types, both natural and anthropogenic, are given in Table 26. Each study shown used different analytical methods to identify the particles. X-ray diffraction, used in the Denver study, determines the mineral types present but provides no direct information about particle composition. Electron microprobe analysis, equivalent to scanning electron microscope analysis, was used in the Antwerp study and provides individual-particle compositions but no information about crystal structure. Transmission electron microscope (TEM) analysis, used in the New Mexico study, provides semi-quantitative compositional data and can provide structural data when electron diffraction is employed. The differences in methods are reflected in the types of data shown for each study in Table 26.

Anthropogenic Inorganic Aerosol Particles

The variety of inorganic aerosol particles of anthropogenic origin is almost limitless. Major categories include particles from (1) industrial sites (e.g., fac-

Table 26. Examples of particle types identified in several studies

A. Denver, USA urban aerosol. Method: X-ray diffraction (Davis 1984).
Types (excluding carbonaceous particles): quartz, muscovite, microcline, limonite, oligoclase, kaolinite, gypsum, anthophyllite, calcite, sodium chloride, fly ash, biotite, magnetite, ammonium sulfate, ammonium lead sulfate

B. Antwerp, Belgium urban aerosol. Method: electron microprobe (Van Borm et al 1989).
Types (excluding biological particles): (1) soil dust (chemically similar to quartz, kaolinite, montmorillonite, feldspars, and rutile) and coal fly ash, (2) auto exhaust (lead and ammonium-lead sulfates, lead halides), (3) sulfates (including gypsum and ammonium sulfate), (4) sea salt (sodium chloride and potassium chloride), and (5) miscellaneous anthropogenic types (including oil fly ash, iron oxides, oxides of Pb, Cu, Zn, Sn, and Cr, barium sulfate, and others)

C. New Mexico, USA desert aerosol. Method: analytical (transmission) electron microscope (Sheridan et al 1993)
Types (excluding carbonaceous particles): (1) sulfates (including probable sulfuric acid, ammonium sulfate, and sulfates with crustal particle core), (2) fourteen classes of crustal particles (not identified as to specific mineral), (3) calcium sulfate, (4) calcium carbonate, (5) eight Fe- and Ti-rich types, (6) sodium chloride, (7) sodium sulfate, (8) potassium chloride, (9) coal fly ash, and (10) Zn-rich

tories, foundries, and coal-fired power plants), (2) agriculture, (3) internal-combustion engines, and (4) mining and smelting. Large numbers of particles are also produced by fires, especially the seasonal (and deliberately set) biomass burning that occurs in southern Africa and areas of South America.

An example from an urban area impacted by smelting serves to illustrate some of the diverse anthropogenic particle types in the troposphere. Phoenix, Arizona, is within an arid region of desert. Several major copper smelters lie about 120 km to the southeast and contribute chemically distinct particle types when air-mass trajectories are from that direction. Particles whose sources are from within the urban area include those from foundries, manufacture of semiconductors, and motor vehicle exhaust. The samples discussed here were collected prior to the imposition of stricter controls on smelter emissions; particles are in the size range of 0.4 to 2 μm. For this example, let us consider observed particles that contain Pb, Zn, Cu, or As.

Most Pb-bearing particles from Phoenix fall into one of the three categories: (1) Pb plus S, commonly also with Zn; (2) Pb with halogens, primarily Cl, but also Br in some particles; (3) Pb without S or halogens. Wind data indicate that the Pb-S particles (assumed to be sulfate) are from the smelters. The halogenated Pb is from combustion of leaded gasoline, while the third type is most probably aged exhaust particles that have lost their halogens. Of the Cu-bearing particles, most are Cu-S (again, presumably as sulfate) and are of smelter origin. Another Cu-bearing type has no S, but has the association Cu-Si; wind data suggest these are from the urban area. Several Zn-bearing types are also present: a group of Zn-S types from the smelters, a Zn-P type probably from spray-painting of metal primer, and a group of Zn-Fe types probably from foundries in the urban area. As-bearing particles appear to have at least two sources: (1) the smelters, and (2) local semi-conductor plants. The As-bearing smelter particles always have S, whereas the latter type does not.

These compositionally distinct particle types appear to be typical components of the Phoenix aerosol. Another urban area is likely to have its own suite of particle types, some of which may be in common with the Phoenix aerosol and others of which are not. Regional differences in culture, climate, geographic location, level of economic development, and many other factors are responsible for the diversity of the anthropogenic aerosols from major urban centers.

Examples of tropospheric Aerosols in Remote Marine Regions

Tropospheric aerosols in the vast remote marine regions of Earth have a significant impact on global climate. In such regions, the aerosol consists of sea salt, particles generated by secondary processes of gas-to-particle conversion and coagulation/condensation, and anthropogenic and natural particles transported in the troposphere. Any of these particle types can be altered by cloud processing or processes involving gas-to-particle conversion to produce additional "aged" particle types. Two examples, one from Spitsbergen in the Arctic and the other from Bermuda in the North Atlantic, serve to illustrate the suites of aerosol particle types at remote marine sites.

At the Norwegian air-monitoring station near Ny Ålesund, Spitsbergen (78°55′N, 11°57′E), the influence on the High Arctic of polluted air from northern Eurasia is frequently observed. The strongest pollution events occur in winter, when the Arctic Front is to the south of many of the major industrial centers of northern Eurasia. However, even in the spring and summer, when the Arctic troposphere is more isolated from these pollution sources, anthropogenic particles can be found along with the natural marine and continental particle types present in the Arctic tropospheric aerosol. The samples described here were collected in May and June, a relatively "unpolluted" period.

The principal particle types in the size fraction from 0.1 to 2 μm are listed in Table 27. Additional particle types in lower abundance were found with one or more of As, Pb, Ni, Cr, Cu, and Zn. The compositions are given in atomic fractions and do not include H, C, N or O (which were not analyzed). The particle types are identified by arbitrary names (C3, C5, etc.) given to statistically determined clusters. The variation in total fine particle concentration and mean average diameter are shown in Fig. 86.

The first eight particle types are silicates, interpreted to be minerals because of the temporal association of types with the compositions of quartz and clays. However, some silicate particles are aggregated with sulfates and others are associated with Br, possibly from reaction with HBr or other gaseous Br-bearing species that have their seasonal maxima in the Arctic spring.

Many of the other major particle types are S-rich. Some are probably sulfates of biogenic origin formed as a result of the photo-oxidation of methane-

Table 27. Major clusters in fine-fraction samples from Ny Ålesund, Spitsbergen, May 13 through June 11, 1987. Compositions are in atomic fractions, diameters are in microns. (Anderson et al. 1992)

Cluster	Centroid composition (atom frac. ≥0.03)	Integrated concentration[a]		
		By count	By volume	Mean average diameter
A. Major particle types				
C3	Si 0.92 S 0.03	21.04	13.78	0.59
C5	Si 0.57 Al 0.16 Br 0.08 K 0.05 S 0.05 Fe 0.03	4.73	20.49	1.34
C6	Si 0.49 Al 0.26 Fe 0.08 Mg 0.05 K 0.03	7.19	11.41	1.12
C7	Si 0.54 Al 0.26 K 0.05 Na 0.05 S 0.05 Fe 0.03	14.55	38.05	1.05
C10	Si 0.66 S 0.24 Ca 0.06	3.43	1.18	0.54
C13	Si 0.58 Br 0.16 S 0.09 Na 0.09 K 0.04 Fe 0.03	8.73	3.14	0.68
C25	Si 0.48 S 0.24 Br 0.21 Ca 0.03	3.26	0.37	0.43
C28	S 0.53 Si 0.47	4.61	0.34	0.39
C31	Fe 0.61 Si 0.12 Na 0.12 S 0.06 Al 0.05	6.05	2.13	0.51
C33	S 0.43 Ca 0.42 Si 0.07 Na 0.05	11.63	4.25	0.53
C34	Na 0.44 S 0.33 Ca 0.16 Mg 0.03	2.79	0.75	0.43
C35	Cr 0.90 Si 0.10	5.78	9.22	0.45
C36	S 0.61 Ca 0.32 Si 0.03 Al 0.03	4.07	0.63	0.37
C37	Ca 0.87 S 0.05 Si 0.05	2.82	5.12	0.45
C38	Na 0.60 S 0.40	3.33	0.20	0.23
C39	S 0.47 Na 0.45 Mg 0.05 Al 0.03	38.61	1.76	0.24
C40	Na 0.64 S 0.30 Mg 0.05	21.76	1.19	0.26
C41	S 0.91 Mg 0.09	257.34	5.31	0.19
C42	Na 0.52 Cl 0.36 S 0.05 Mg 0.05	8.93	4.42	0.44
C44	Na 1.00	26.55	2.33	0.22
C48	S 0.46 Mg 0.38 Na 0.11 Al 0.03	2.37	0.06	0.21
C57	K 1.00	2.31	0.10	0.25
B. Minor particle types				
C19	Fe 0.28 Al 0.38 Mg 0.17 Si 0.10 S. 0.07	0.77	2.16	1.20
C26	Fe 0.27 Si 0.38 S 0.35	0.23	0.02	0.42
C32	Fe 0.42 S 0.42 Si 0.16	1.04	0.10	0.36
C30	Ti 0.82 Si 0.10 S. 0.06	0.70	0.40	0.61
C46	Na 0.41 Ca 0.14 Cl 0.23 S 0.19	0.33	0.08	0.48
C47	Mg 0.32 Na 0.24 Ca 0.10 S 0.31	0.53	0.33	0.45
C50	S 0.43 K 0.30 Na 0.25	1.79	0.04	0.27
C53	Pb 1.00	0.32	0.08	0.43
C52	As 0.36 Pb 0.23 S 0.22 Cl 0.10 P 0.04 Na 0.04	0.83	0.74	0.72
C56	As 1.00	0.84	0.03	0.17
C59	Ni 1.00	1.01	0.01	0.16

[a] Concentration units are (by count) number of particles $\times 10^4$ per m^3 of air and (by volume) μm^3 of a particle type per m^3 of air.

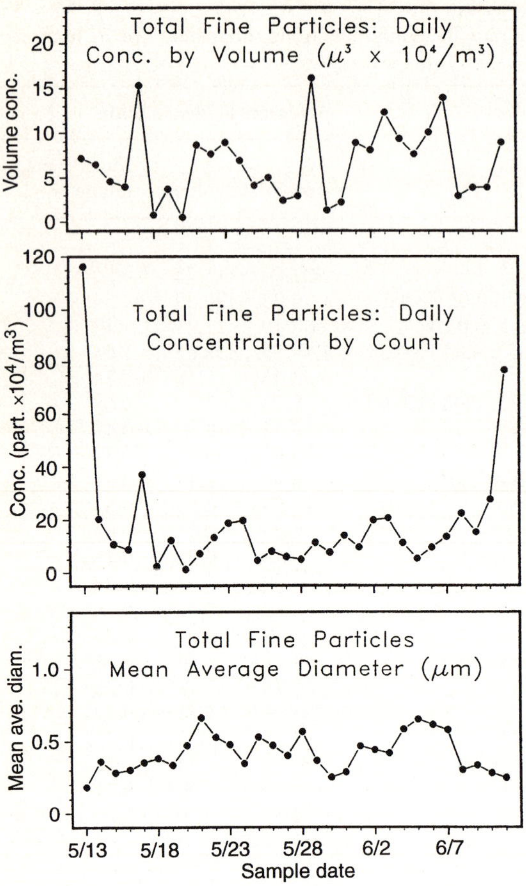

Fig. 86. Total fine-fraction particles; plots of time vs. concentration by volume (above) concentration by count (center) and mean diameter (below) for Ny Ålesund, Spitsbergen 1987 aerosol samples. (Anderson et al. 1992)

sulfonic acid, and others, particularly with $CaSO_4$ or $MgSO_4$, may be from sea salt. Few undisturbed sea-salt particles (type C42) remain in the aerosol, despite the surrounding marine environment. The NaCl from sea salt has reacted to form Na_2SO_4 (type C40), other Na-S species (type C39), or perhaps mixtures (type C38). Most of the S is probably from the surge of algal growth that accompanies Arctic spring.

The most abundant particle type clearly of anthropogenic origin is Cr-bearing C35; a smelter origin is possible. The Ni-bearing particles, not temporally associated, are probably from smelting in the northern part of the former Soviet Union.

The marine aerosol at Bermuda is in contrast to that at Spitsbergen. The data presented are from samples collected at the AEROCE site on Tudor Hill from May to September, 1989. The principal particle types in the fine fraction (0.1 to 2 μm) are shown in Table 28 (again, the particle types have arbitrary names). Sea-salt particles, both reacted and apparently unreacted, are in much greater relative abundance than at Spitsbergen. Particularly striking are the

Table 28. Compositions, concentration sums (particles x 10^4 per m^3 of air), and mean average diameters (microns) of individual-particle types of Bermuda, 1989 samples. Compositions are in atomic fractions and represent cluster averages. (Anderson et al. 1995)

A. Unreacted/uncoated silicates		Conc. sum ($\times 10^4$)	Mean Av. diam.
BC1	Si 0.93 S 0.03	650.0	0.68
BC2	Si 0.61 Al 0.28 Fe 0.07 K 0.03	507.9	1.48
BC3	Si 0.57 Al 0.25 Fe 0.12 S 0.04	217.7	1.27
BC4	Si 0.69 Al 0.17 Fe 0.06 S 0.06	141.5	1.21
BC5	Si 0.59 Al 0.20 Mg 0.11 Fe 0.07 K 0.04	125.5	1.55
BC6	Si 0.62 Al 0.19 K 0.14 S 0.04	90.5	1.04
BC7	Si 0.63 Al 0.21 Na 0.12 S 0.03	123.6	0.95
BC8	Si 0.58 Al 0.33 S 0.04 Fe 0.03 K 0.03	254.8	1.03
BC38	Si 0.47 Fe 0.25 Al 0.23 S 0.05	55.2	1.16
B. Silicates with Br			
BC9	Si 0.64 Al 0.24 Br 0.06 Fe 0.04	65.4	1.48
BC10	Si 0.65 Al 0.15 Fe 0.09 Br 0.08 K 0.03	112.9	1.47
BC11	Si 0.74 Br 0.09 K 0.07 Na 0.06 Al 0.04	52.0	0.93
BC12	Si 0.71 Br 0.16 Fe 0.06 S 0.04	92.5	0.92
C. Silicates with Br and S			
BC13	Si 0.61 S 0.18 Br 0.09 Al 0.07 Fe 0.06	60.9	0.97
BC42	Si 0.44 S 0.32 Na 0.11 Br 0.08 Fe 0.04	11.2	0.97
D. Silicates with S			
BC14	Si 0.47 S 0.26 Al 0.18 Fe 0.06 Na 0.04	109.7	1.06
BC15	Si 0.50 Al 0.19 S 0.17 Ca 0.08 Fe 0.06	80.4	1.15
BC16	Si 0.53 Al 0.24 S 0.15 Fe 0.05	182.7	1.10
BC17	Si 0.72 S 0.21 Al 0.04	111.4	0.73
BC18	Si 0.38 S 0.27 Ca 0.18 Al 0.12 Fe 0.05	23.5	1.06
BC19	Si 0.54 S 0.38 Al 0.03	117.5	0.69
BC20	S 0.56 Si 0.36 Ca 0.03 Na 0.03	101.7	0.66
BC21	Si 0.32 S 0.31 Na 0.28 Al 0.07	27.6	0.79
BC28	S 0.38 Ca 0.31 Si 0.26 Al 0.04	32.0	0.87
BC44	Si 0.40 S 0.24 Al 0.20 Na 0.12 K 0.03	53.8	0.82
BC45	Si 0.52 S 0.25 Fe 0.14 Na 0.06	19.9	0.71
BC51	Si 0.36 S 0.22 Al 0.21 Fe 0.12 Mg 0.08	20.4	0.94
BC57	S 0.39 Si 0.33 Fe 0.14 Na 0.09 Al 0.06	11.7	0.91
E. Na + S			
BC22	S 0.46 Na 0.40 Ca 0.11 Mg 0.03	97.3	0.39
BC23	S 0.56 Na 0.38 Mg 0.06	830.4	0.27
BC24	Na 0.55 S 0.42 Mg 0.03	673.1	0.29
BC30	Na 0.86 S 0.12	72.7	0.25
BC41	Mg 0.39 Na 0.30 S 0.38	74.6	0.29
BC43	Na 0.48 S 0.27 Ca 0.16 Mg 0.08	28.3	0.42
BC46	Na 0.64 S 0.26 Mg 0.07	167.6	0.30
BC49	Na 0.39 S 0.25 Ca 0.20 Cl 0.16	18.5	0.45
F. S only			
BC25	S 0.95	2290.8	0.25
G. S + Ca			
BC26	S 0.53 Ca 0.29 Na 0.15	75.9	0.43
BC27	Ca 0.48 S 0.46 Si 0.04	129.2	0.58

Table 28 (continued)

H. Other high-S types		Conc. sum (×10^4)	Mean Av. diam.
BC63	S 0.50 K 0.43 Fe 0.03 Al 0.03	8.9	0.35
BC65	S 0.59 Mg 0.40	31.7	0.22
I. High Ca			
BC29	Ca 0.64 Mg 0.20 Si 0.10 S 0.05	29.2	0.38
J. Na-Cl types			
BC31	Na 0.63 Cl 0.35	201.6	0.27
BC32	Na 0.56 Cl 0.43	496.0	0.36
BC33	Na 0.50 Cl 0.36 S 0.09 Ca 0.04	93.8	0.49
BC34	Na 0.46 Cl 0.27 S 0.14 Ca 0.13	36.3	0.42
BC35	Cl 0.76 Na 0.19 S 0.04	107.9	0.23
BC61	Na 0.51 S 0.24 Cl 0.21 Mg 0.04	24.6	0.35
K. Cl with little Na			
BC48	Cl 0.56 S 0.38 Na 0.04	7.6	0.17
L. High Fe			
BC36	S 0.49 Fe 0.44 Si 0.04 Na 0.03	9.3	0.48
BC37	Fe 0.60 Si 0.25 Al 0.11 S 0.03	89.2	0.66
BC64	S 0.63 Fe 0.19 Na 0.10 Si 0.08	17.4	0.49
M. Other metal-bearing types			
BC39	Ti 0.68 Si 0.18 Al 0.07 Fe 0.04	22.1	0.69
BC40	Cr 0.85 S 0.08 Si 0.07	170.7	0.38
BC47	Cu 0.48 As 0.44 Mn 0.03	1.7	0.18
BC50	Cu 0.63 Ni 0.34	5.1	0.20
BC52	S 0.71 Cr 0.26	22.9	0.32
BC53	Na 0.60 S 0.30 Cu 0.09	3.1	0.26
BC54	Cu 0.78 Mn 0.12 V 0.05 S 0.03	21.5	0.21
BC55	S 0.73 Cu 0.22	6.0	0.17
BC56	Cu 0.69 Ba 0.23 Ni 0.06	1.8	0.15
BC58	Cu 0.63 Sn 0.28 Ni 0.03 Br 0.03 Mn 0.03	2.4	0.18
BC59	Cu 0.56 Pb 0.31 S 0.07 Mn 0.04	2.1	0.20
BC60	Cu 0.68 Pb 0.19 Ni 0.05 Br 0.04 Mn 0.04	1.5	0.20
BC62	S 0.57 Pb 0.39 Cu 0.03	2.8	0.35

abundant silicates, most of which are shown by air-mass trajectory analysis to be mineral dust from North Africa. Metal-bearing pollutants from Europe and North America also occur in Bermuda aerosols.

Based on air-mass trajectories, size distributions, and temporal associations with other particle types, the silica type in the Bermuda aerosol appears to be of multiple origins. In samples transported from North Africa, type BC1 particles are everywhere associated with clays and other minerals and have relatively flat size distributions (sampling periods 234, 228, and 195 in Fig. 87). Silica particles that were not associated with clays or other silicates occur in several samples that were transported from Europe. These latter silica particles

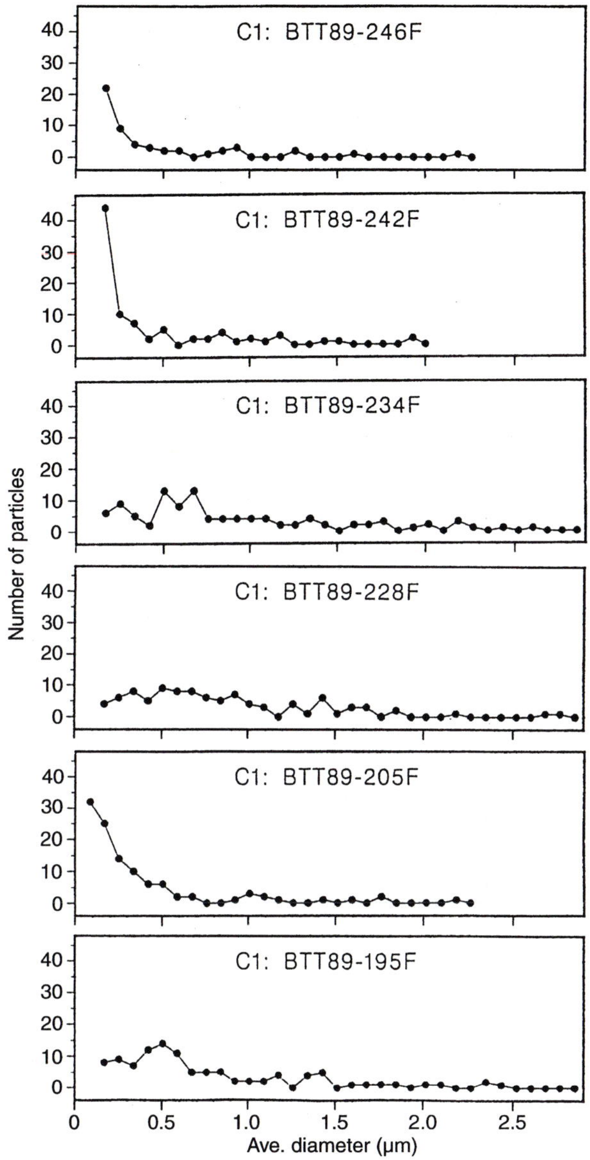

Fig. 87. Size distributions for silica particle type BC1 for six sampling periods for Bermuda 1989 aerosol samples. (Anderson et al. 1995)

greatly increase in abundance below 0.4 μm (samples 246, 242, and 205 in Fig. 87) and are interpreted to have formed as a result of SiO_2 reduction-vaporization during the combustion of high-rank coal.

We have also analyzed tropospheric aerosol samples from the equatorial Pacific that were almost entirely sea-salt particles, either relatively unaltered or almost entirely reacted with sulfate of probably biogenic origin. Samples taken in about the same area several years later contained silicate minerals transported from Asia. Regardless of how remote the site, tropospheric aerosols are highly variable both temporally and geographically.

Environmental Effects of Aerosols

The local effects of mineral dust and anthropogenic particles on visibility and human health have long been recognized. The recognition of the global influence of aerosols on climate is more recent. The effects of inorganic aerosol particles that are on a global scale occur primarily in the remote marine troposphere, the importance of which derives in part from its extent of coverage of Earth's surface. The remote marine troposphere is also an important link in the biogeochemical cycles of many trace constituents in the atmosphere, particularly sulfur.

Aerosol particles are important to the global radiative balance both because of direct scattering of light and because of their effect on global albedo through the abundance and microphysical properties of marine stratocumulus clouds. Such clouds are persistent features of marine boundary layers in the subtropics, cover about one third of Earth's oceans, and strongly contribute to the global albedo. Changes in the anthropogenic contribution to the global tropospheric aerosol can produce significant climate-forcing effects through the influence of aerosol particles on light scattering and stratocumulus cloud properties.

In clean air in the marine boundary layer, droplet size within stratocumulus clouds is a function of cloud condensation nuclei (CCN) concentration. This dependence is because the marine troposphere is generally undersaturated with respect to aerosol particles that act as CCN. Increases in CCN concentration in an undersaturated system result in more droplets with smaller sizes, thereby enhancing the cloud albedo. The smaller droplet sizes from high CCN concentrations also result in longer aerosol-particle lifetimes because drizzle is suppressed.

At the scale of individual humans the impact of aerosol particles are most significant for those particles that are inhaled. The effect an individual particle has on the pulmonary system depends upon the intrinsic toxicity of the material, the time of retention in the pulmonary system, and the area of the lung where it accumulates. Many particle types in urban aerosols are known or suspected carcinogens, including asbestos, Fe oxides, and metals such as Cd, Se, V, Cr, Ni, As, Be.

Other than some types of asbestos, most minerals are not generally believed to be carcinogenic. Many minerals, however, have shown fibrogenic effects; these include quartz, kaolinite and other clays, and talc. Quartz particles and certain types of fly ash may interfere with the immunity-inducing function of the lymph nodes in laboratory rats, suggesting a possible similar effect for other mammals. Many nominally inert mineral particles are capable of adsorbing and transporting toxic substances as surface coatings. After deposition in the pulmonary system, the toxic surface layer can react with the tissue. Inhaled particles of many common minerals are clearly not benign.

Summary

The tropospheric aerosol contains assemblages of natural and anthropogenic particle types that are highly diverse, both temporally and geographically. Many particle types are chemically active, and the physical and chemical nature of particles can change while they are in suspension. Anthropogenic activities and pollutant emissions can lead to the production of aerosol particle types that otherwise would be considered to be of natural origin. The environmental effects of aerosols are clearly significant, from their contribution to Earth's radiative balance to regional climate change and toxicity in the human lung.

References

Anderson JR, Aggett FJ, Buseck PR, Germani MS, Shattuck TW (1988) Chemistry of individual aerosol particles from Chandler, Arizona, an arid urban environment. Environ Sci Technol 22:811–818

Anderson JR, Buseck PR, Saucy DA, Pacyna J (1992) Characterization of individual fine-fraction particles from the Arctic aerosol at Spitsbergen, May–June, 1987. Atmos Environ 26A:1747–1762

Anderson JR, Buseck PR, Patterson TL, Arimoto R (1996) Characterization of the Bermuda tropospheric aerosol by combined individual-particle and bulk-aerosol analysis. Atmos Environ 30:319–338

Andreae MO, Charlson RJ, Bruynseels F, Storms H, Van Grieken R, Maenhaut W (1986) Internal mixture of sea salt, silicates, and excess sulfate in marine aerosols. Science 232:1620–1623

Bice DE, Hahn FF, Benson J, Carpenter RL, Hobbs CH (1987) Comparitive lung immunotoxicology of inhaled quartz and coal combustion fly ash. Environ Res 43:374–389

Bigg ED (1980) Comparison of aerosol at four baseline atmospheric monitoring stations. J Appl Meteor 19:521–533

Blanchard DC, Woodcock AH (1980) The production, concentration, and vertical distribution of the sea-salt aerosol. Ann NY Acad Sci 338:330–347

Charlson RJ, Lovelock JE, Andreae MO, Warren SG (1987) Oceanic phytoplankton, atmospheric sulphur, cloud albedo and climate. Nature 326:655–661

Charlson RJ, Schwartz SE, Hales JM, Cess RD, Coakley JA, Hanson JE, Hofmann DJ (1992) Climate forcing by anthropogenic aerosols. Science 255:423–430

Churg A, Wright JL, Wiggs B, Pare PD, Lazar N (1985) Small airways disease and mineral dust exposure. Am Rev Respir Dis 37:364–372

Davis BL (1984) X-ray diffraction analysis and source apportionment of Denver aerosol. Atmos Environ 18:2197–2208

Fitzgerald JW (1991) Marine aerosols: a review. Atmos Environ 25A:533–545

Gras JL, Ayers GP (1983) Marine aerosols at southern mid-latitudes. J Geophys Res 88:10,661–10,666

Leaitch, WR, Isaac GA, Strapp JW, Banic CM, Wiebe HA (1992) The relationship between cloud droplet number concentrations and anthropogenic pollution: observations and climatic implications. J Geophys. Res 97:2463–2474

Pacyna JM, Winchester JW, Li S-M (1991) Impact of carbonaceous fuel burning on Arcitic aerosols: silicon as an indicator of long-range transport of coal smoke. In: Levine JS (ed) Global Biomass Burning. MIT Press, Cambridge, Massachusetts, pp 427–430

Posfai M, Anderson JR, Buseck PR, Sievering H (1995) Compositional variations of sea-salt-mode aerosol particles from the North Atlantic. J Geophys Res 100:23,063–23,074

Post JE, Buseck PR (1985) Quantitative energy-dispersive analysis of lead halide particles from the Phoenix urban aerosol. Environ Sci Technol 19:682–685

Prospero JM; Nees RT (1986) Impact of the North African drought and El Niño on mineral dust in the Barbados trade winds. Nature 320:735–738

Pye K (1987) Aeolian dust and dust deposits. Academic Press, London 334 pp

Rojas CM, Van Grieken RE (1992) Electron microprobe characterization of individual aerosol particles collected by aircraft above the southern bight of the north sea. Atmos Environ 26A:1231–1237

Schreider JP, Culbertson MR, Raabe OG (1985) Comparative pulmonary fibrogenic potential of selected particles. Environ Res 38, 256–274

Sheridan PJ, Schnell RC, Kahl JD, Boatman JF, Garvey DM (1993) Microanalysis of the aerosol collected over south-central New Mexico during the ALIVE field experiment, May–December 1989. Atmos Environ 27A:1169–1183

Van Borm WA, Adams FC, Maenhaut W (1989) Characterization of individual partricles in the Antwerp aerosol. Atmos Environ 23:1139–1151

Winchester JW (1989) Regional anomalies in chronic obstructive pulmonary disease; comparison with acid air pollution particulate characteristics. Arch Environ Contam Toxicol 18:291–306

5.3.2 Asbestos and Other Fibrous Silicates. Health Effects

H. FÖRSTER

Asbestos

Asbestos are minerals mined for their outstanding properties (chemical stable, incombustible, strong, slender, flexible): serpentine-asbestos (*chrysotile*, white asbestos, Canadian asbestos; Fig. 88) and amphibole-asbestos [obsolete hornblende-asbestos; *crocidolite* (riebeckite-asbestos, blue asbestos, Cape asbestos), *amosite* (grunerite-asbestos, brown asbestos), *anthophyllite-asbestos* (including cummingtonite-asbestos), *tremolite-asbestos*, and *actinolite-asbestos*]. The term asbestiform should not be applied to fibrous minerals like nemalite or palygorskite. For reviews on the mineralogy and identification of asbestos, including detailed microscope documentation, see References.

Since 1860, more than 100 million tons have been mined [approx. 95% chrysotile, 3% crocidolite, 2% amosite, 0.3% anthophyllite-asbestos (mining ceased in 1966)].

Other Fibrous Silicates

Many other silicates have (or may have) fibrous morphology: zeolites, sillimanite (fibrolite), mullite, kaolinite, halloysite, garnierite, montmorillonite, vermi-

Advanced Mineralogy, Volume 3
Mineral Matter in Space, Mantle, Ocean Floor, Biosphere, Environmental Management, and Jewelry (ed. by A. S. Marfunin)

Fig. 88. Fibers of chrysotile, the most abundant asbestos mineral. Weathered asbestos cement shingle. (SEM/SE photo J. Wiegershaus). For further examples of the microscopical appearance of fibrous minerals see References

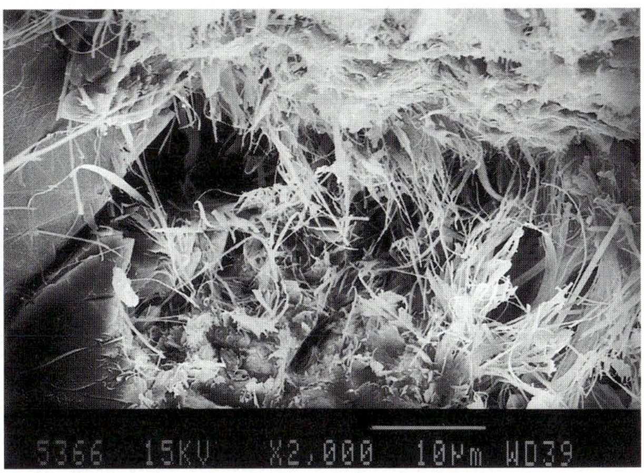

Fig. 89. Fibrous illite in a pore of a natural sandstone from southern Germany [Sander Schilfsandstein, Middle Keuper (Triassic), a common building-stone (Residence Würzburg)]. (SEM/SE photo Dr. Jens Neisel). Rocks and soils are sources for other fibrous silicates in the air. Illite may be degraded in the human lung to fibers with the EDS signal AlSi and then mistaken for man-made fibers or mullite

culite, talc, pyrophyllite, stilpnomelane, minnesotaite, illite, sericite, chlorite, sepiolite, palygorskite (attapulgite), wollastonite, and xonotlite; silica minerals [quartz, chalcedony, fibrous silica, phytoliths], and Pele's hair (natural basalt glass fibers) may be included here. In certain cases, nonasbestos fibers may be misinterpreted by the standard method described below, an example is given in Fig. 89.

Table 29. Fiber definitions

Reference	Year	L (µm)	D (µm)	L/D	Designation
Ross et al. (in Skinner et al. 1988)[a]	1984			≥20	True fiber
OSHA	1992			≥10	OSHA asbestos fiber
ILO	1990	> 5	<1.5	≥ 5	ILO fiber (proposed)
Stanton et al.	1981	> 8	<0.25		Critical fiber, Stanton fiber
Pott	1984	>10	>0.20		Pott fiber
WHO	1992	> 8	<0.20	>10	Critical WHO fiber
HSE[b]	1977	> 5	<3	≥ 3	Regulatory fiber, F-reg
		<and>5	<3	> 3	Fibers of any length, F-all

[a] Also Textbooks of Mineralogy.
[b] Also NIOSH (1981); WHO (1987); VDI 3492 (1989).

Fiber Definitions

A fiber is a small object which length is considerably greater than its width. Fibers are characterized by their length L, diameter D and aspect ratio L/D (Table 29).

An egg-shaped amphibole fragment of 9 µm length and 2.9 µm diameter or a quartz crystal of that size are regulatory fibers, but never true fibers. Doll and Peto criticized the regulatory fiber concept already in 1985. Obviously, not only the number of minerals classified as fibers but also the number of fibers counted will be influenced greatly by the fiber definition applied. For asbestos, e.g., F-reg amounts to 10–15% of F-all.

Practical Fiber Identification and Fiber Counting

Today, fibers are detected by means of an analytical electron microscope [AEM = SEM + EDS + TEM (electron diffraction)]. Standard magnifications are 2000x (VDI 3492), 2500x, 5000x, and 10000x. The number of fibers counted per sample area increases with the magnification. According to VDI, the detection limit for airborne fibers is 300 F-reg/m^3, and the sensitivity, expressed as concentration per fiber counted, ranged in one study from 130±21 to 708±49 F-reg/m^3 (Rödelsperger and Gerhard, in Bossenmayer et al. 1991). Expenses for analyzing and counting fibers increase considerably with magnification and sophistication. Fiber concentrations are reported as arithmetic mean, geometric mean, median or the 95th percentile. Risk assessment and regulations should be based on the same value.

Asbestos and Other Fibrous Silicates in the Air

Not surprisingly, reported asbestos concentrations in ambient air differ considerably. In the US, the average was 400 to 1000 F-reg/m^3 (WHO 1987). The typical asbestos concentration in urban air is 1000 to 1500 F-reg/m^3 [in the UK (HEI 1991), in The Netherlands, and in Germany]. A reliable figure for the airborne asbestos level of the unspoiled atmosphere was obtained by Kokyama (in Guthrie and Mossman 1993) on Ogasawara Islands/Pacific Ocean with 14 100 F-all/m^3. The maximum value was reported by Cooper at al. (in Guthrie and Mossman 1993) in the Clear Creek recreational area in California (adjacent to the New Idria serpentinite) for chrysotile with 5 300 000 F-reg/m^3. On the other hand, Rödelsperger and Gerhard (in Bossenmayer et al. 1991) assign the following values to ambient air in Germany: for asbestos 110 F-reg/m^3 and for other mineral fibers (OMF) 1400 F-reg/m^3 (almost identical for large cities and recreational areas). Schnittger and Förster (in VDI Berichte 1993) reported about fibers in ambient air of central Berlin: OMF gave as mean 2199 ± 1356 F-reg/m^3 [without gypsum fibers (2800 F-reg/m^3)], as median 1995 F-reg/m^3, and as range 150 to 6420 F-reg/m^3 (sensitivity 97 ± 15). Mullite + ironmullite is the most frequent component with 322 + 44 F-reg/m^3 (17% of OMF). Other important components are silica (quartz) fibers (155 F-reg/m^3 or 7%) and illite (132 F-reg/m^3 or 6%). The content of rutile is about 1%. Indoor air in asbestos-free buildings shows fiber concentrations comparable to ambient air.

The population of a group of villages in Cappadocia/Turkey suffered the highest mesothelioma mortality worldwide, which was related to airborne erionite and mordenite, fibrous zeolites with the dimensions of crocidolite, in concentrations between 1000 and 29 000 F-reg/m^3.

The indoor air of a building with spray-applied asbestos in Germany contained as average for asbestos 1600 F-reg/m^3 and for other mineral fibers (not specified) 5750 F-reg/m^3 (Rödelsperger and Gerhard in Bossenmayer et al. 1991).

According to Kane (in Guthrie and Mossman 1993), levels of asbestos exposure to the air at the workplace were high before 1960 (approx. 10 000 000 F-reg/m^3); however, current occupational levels are appr. 100 000 F-reg/m^3. Workers in an asbestos mine in Italy have been exposed to air with 600 to 8800 F-reg/m^3 (Fischer and Meyer 1983). High concentrations were observed during the production of floor tiles, textiles, and insulations, on shipyards and construction sites, and in asbestos-cement factories. As an example may serve the values measured in German asbestos-textile factories (Rödelsperger and Gerhard, in Bossenmayer et al. 1991): 1976/77 median 600 000 F-reg/m^3, 1981/82 median 930 000 F-reg/m^3. Fuehres and Heidermanns (1993) detected 10^7 F-all/m^3 of other mineral fibers (kaolinite, illite, talc, rutile) after dry blowing of cellulosis (recycled paper) for insulation purposes, exceeding twice the European limits for fine dust (5 to 6 mg/m^3).

Asbestos in Lakes, Rivers, Groundwater and Water Supplies

Natural waters, unaffected by asbestos-rich bedrock or by effluents from special mines or factories, contain per liter less than 10^6 asbestos fibers of all lengths ($< 10^6$ F-all/l). Snow (fallen ca. 1930 and 1972) and ice (> 10 000 years old) from Yamato Mts. in Antarctica have chrysotile concentrations in the range 1.1 to 6×10^5 F-all/l, indicating a quite constant global background in air and water since long before the industrial revolution (Kohyama in Guthrie and Mossman 1993). Groundwater in Germany contains $< 10^3$ F-reg/l, rivers and lakes contain between 10^5 and 0.5×10^6 F-reg/l; 20% of the German drinking water contains $> 10^6$ F-all/l (Marfels and Schörmann in Bossenmayer et al. 1991). Lakes and rivers with asbestos-rich bedrock in northern America have contents between 10^6 and 10^{13} F-all/l, mainly chrysotile, and the water supplies (municipal drinking water) contain here 10^6 to 10^8 F-all/l. A large open aqueduct near Coalinga/California contains even 10^{10} F-all/l, and the median for drinking water in California is 10^9 F-all/l. Marfels and Schörmann calculated that water with 10^7 F-all/l, if used in a household, will deliver 5×10^4 F-all/m^3 into the air corresponding to at least 1000 F-reg/m^3. Ten % of the drinking water in the US contains $> 2 \times 10^6$ F-reg/l. EPA has set the limit to 7×10^6 fibers with a length > 10 µm per liter. Water pipes made from asbestos-cement do not contribute to the asbestos burden of the water, as long as it is saturated with $CaCO_3$ (guaranteed in Germany by the regulated pH). An amphibole-asbestos concentration of 600×10^6 F-all/l has been reported in the effluents of an iron ore dressing plant in Minnesota.

Asbestos and Other Fibrous Silicates in the Human Lung

Recently, this topic has been addressed by Woitowitz (1991), Churg (1993), Rödelsperger et al. (1993) and Romer (1994), using an AEM with a magnification 10 000x and a sensitivity of 30 000 fibers per gram dry lung tissue (F/g-dry). Data by Rödelsperger et al. have been recalculated for comparability. The results are given in Table 30 (general population, N for normal) and Table 31 (occupational exposure, M for mesothelioma etc.). Churg analyzed cases of average citizens of Vancouver, BC, shipyard and insulation workers, as well as asbestos miners from Canada. Woitowitz and Rödelsperger et al. used tissue samples from Hamburg and elsewhere in Germany, Romer those from Switzerland.

Gernerally, OMF prevail in normal lungs and asbestos in occupational exposed lungs. The high OMF values obtained by Woitowitz are due to his fiber definition (even a fragment of a platy talc is a fiber). Romer looked for true fibers (mullite, fibrous rutile). He reported in addition 120×10^6/g-dry particles (except metallic particles) for normal lungs and 90×10^6/g-dry particles for mesothelioma cases (illite, quartz, feldspar, talc, chlorite, kaolinite, rutile). Romer and Rödelsperger et al. observed comparable fiber dimensions (Table 32). According to the latter authors, the ranking of OMF is mullite – rutile – quartz – Mg,Fe-silicates.

Table 30. General population, mean asbestos fiber and OMF content in the lung

N	Churg F-all/g-dry	Woitowitz F-all/g-dry	Woitowitz F-reg/g-dry	Romer F/g-dry	Rödel. et. al. F-all/g-dry	Rödel. et. al. F-reg/g-dry	Average F-reg/g-dry
Cases	>100	22	22	26	63	63	
Chrysotile	300000	342000	7000	24000	325000	121000	50000
Amphibole -asbestos	400000[a]	223000	33000	162000	108000	44000	80000
Asbestos total	700000	565000	30000	186000	433000	165000	125000
OMF		3.2×10^6	33000	485000	1.13×10^6	101400	200000
Amph/Chrys.	1.33	0.65	5	7	0.33	0.36	1.6
OMF/Asbest.		6	1.1	2.6	2.6	6	1.6

N = normal. OMF = other mineral fibers.
[a] Tremolite-asbestos.

Table 31. Workers with asbestos exposure, mean asbestos fiber and OMF content in the lung

M	Churg F-all/g-dry	Woitowitz F-all/g-dry	Woitowitz F-reg/g-dry	Romer F/g-dry	Rödel. et al. F-all/g-dry	Rödel. et al. F-reg/g-dry	Average F-reg/g-dry
Cases	168	16	16	28	47	47	
Chrysotile	200000	1.8×10^6	118000	87000	176000	124000	100000
Amphibole -asbestos	1×10^6	22×10^6	3.3×10^6	12×10^6	1.26×10^6	638000	1×10^6
Asbestos total	1.2×10^6	24×10^6	3.5×10^6	12×10^6	1.44×10^6	763000	1.1×10^6
OMF		12×10^6	180000	772000	695000	100800	350000
Amph/Chrys.	5	12	28	138	7	5	10
OMF/Asbest.		0.5	0.05	0.06	0.5	0.13	0.3

M = mesothelioma. OMF = other mineral fibers.

Table 32. Fiber dimensions in normal lungs and mesothelioma cases

Fiber type	Romer normal Mean L (D) [μm]	*Romer mesothelioma Mean L (D)* [μm]	Rödelsperger et al. normal + mesothel. Median L (D) [μm]
Chrysotile	6.9 (0.09)	*2.3 (0.09)*	2.0 (0.05)
Crocidolite	6.1 (0.17)	*4.4 (0.15)*	1.9 (0.10)
Amosite	6.9 (0.28)	*3.4 (0.16)*	1.9 (0.10)
Tremolite + Actinolite-asbestos	5.5 (0.49)	*6.4 (0.52)*	4.5 (0.50)
Anthophyllite-asbestos	5.0 (0.61)	*3.7 (0.28)*	4.6 (0.30)
Rutile	3.1 (0.44)	*3.4 (0.29)*	1.4 (0.20)
Mullite	2.9 (0.32)	*1.8 (0.19)*	0.95 (0.11)

With respect to fiber types frequency, both studies show that mullite dominates the OMF and that crocidolite + amosite outnumber chrysotile in mesothelioma cases, despite the fact that 20 times more chrysotile than amphibole-asbestos have been used in industry. Also Churg reported for occupational exposure to chrysotile, that amphibole-asbestos levels exceed chrysotile levels in the lung samples. Romer found no correlation of fiber burdens with age or duration of exposure.

Diseases Related to Asbestos and Other Fibrous Silicates

A historical review was given by Skinner et al. From the start of asbestos mining it took 30 years until the high mortality of asbestos workers was noted after the turn of the century. Doll is to be credited for first reporting in 1955 the increased occupational risk of asbestos for lung cancer and initiating epidemiological studies. Modern diagnoses are asbestosis (fibrosing alveolitis), lung cancer (cancer of the pleura) and – the least common – mesothelioma (diffuse malignant mesothelioma). These diseases are related to *inhaled* asbestos, but the expected relationship between dose and incidence of lung cancer and mesothelioma could not be obtained (Skinner et al. 1988). Mainly affected have been asbestos miners and millers, and workers in industries using asbestos (asbestos-textiles, shipyards, insulation, construction, asbestos cement). Especially risky was spray-applied asbestos. Today, the main problems are staying in sprayed rooms and removal of asbestos. Asbestos ingested with asbestos-rich drinking water causes no gastrointestinal cancer (Marfels and Schörmann).

The high incidence of mesothelioma in Cappadocia/Turkey related to air-borne erionite was mentioned above.

Ross et al. compiled relevant information about mineral dusts other than asbestos, especially with regard to pneumoconiosis. It is a matter of definition whether quartz and phyllosilicates are fibers or not. *Pyrophyllite* causes no apparent disease. Talc can cause nonmalignant respiratory disease. Rödelsperger et al. argue for exposure limits for talc fibers. Dusts of the main constituents of commercial clays (kaolinite, illite, quartz), bentonites (montmorillonite, quartz) and micas (muscovite, phlogopite, biotite, quartz) may cause kaolinosis and silicosis, but there is no increased lung cancer risk. The same holds for dusts in coal mines. Evidence for carcinogenity in humans is inadequate concerning palygorskite and sepiolite (IARC 1987). In the US mining of vermiculite, containing fibrous tremolite, lead to respiratory cancer. Dust from the vermiculite ore at Palabora/South Africa developed no deleterious effects on respiratory health. Wollastonite dust from Ihalainen/Finland (1 to 45×10^6 F-all/m^3) caused lung fibrosis, but less mortality and less lung cancer than expected. However, the advisors to the German government (DFG 1993) plan to classify wollastonite as carcinogenic.

Fibers in experimental Studies with Laboratory Animals

In the vast majority of these studies fiber suspensions were injected into the pleural or peritoneal cavities of rats (Stanton and Wrench 1972, Pott and Friedrichs 1972, among others). This non-physiological method has been heavily criticized (Nolan and Langer in Guthrie and Mossmann 1993, Schultz 1993, Bignon et al. 1995, especially p. 97: no relevance to humans, maximum tolerated dose exceeded, target tissue inappropiate, no good laboratory praxis, dose not comparable to human dose, clumping, inadequate specifity). Positive results can be obtained with ease, even with equigranular dusts. Despite these facts regulation in Germany (TRGS) is based on injection tests.

Risk of Asbestos and Other Fibrous Silicates

Risk is the probability of injury, disease or death under specific circumstances. The deadly risk is defined as one death out of a number of persons per annum, e.g., 1:5000 for traffic accidents. In the US and in The Netherlands, 1:1000000 or 10^{-6}/a is considered as virtual safe (Fischer in Bossenmayer et al. 1991), and in the UK the Royal Commission set the limit for a tolerable risk to 1:1000 or 10^{-3}/a. With respect to cofounders, Fischer evaluated smoking habits: compared to the risk for nonsmokers without asbestos exposure, the risk of a nonsmoking asbestos worker is five times, that of a smoker without asbestos exposure ten times, and for a smoking asbestos worker 50 times larger. According to Fischer, the risk for the general population in ambient air with 400 asbestos fibers per m^3 is 10^{-6}/a and therefore too small to be statistically significant. Churg (in Guthrie and Mossmann 1993) summarized the results of several clinical and epidemiologic studies of asbestos workers exposed in the range 1000000 to 5000000 F-all/m^3: “no excess of any type of asbestos related disease has been observed” and “an extraordinary burden of chrysotile and its contaminant tremolite can be tolerated without the appearance of disease”.

Regulations and Exposure Limits

Today, asbestos is one of the most heavily regulated substances. Regulators tend to propose the lowest possible exposure level. In Germany asbestos has to be removed, if the indoor air contains more than 500 F-reg/m^3. The world atmosphere has a higher level (see above). The German vocational associations [ZH 1/600.1.2 (1993) and older ZHs] apply a more realistic limit: 15000 F-reg/m^3.

Spray-applied asbestos has been banned in the US and in Germany (on shipyards since 1975, in general since 1979) and its removal required (alone 720 t in the Palast der Republik, Berlin). Asbestos in textiles and in floor tiles has been banned in Germany in 1981; the use of asbestos-cement was limited until 1991 (TRGS 519; asbestos-cement pipes until 1995). Practically no amphibole-asbestos has been used in Germany since 1990, and even the use of chrysotile in diaphragmas and gaskets (for the production of basic chemicals) will be terminated in 1999 (total phase out of asbestos). In the UK the import of crocidolite has been prohibited. Contrary to rumors, production, importation, and installation of asbestos products never stopped in the US (asbestos-cement pipes, brake pads, clutch facings, gaskets), and since 1993 asbestos-cement sheets and shingles, roofing felt, pipeline wrap, asbestos clothing and asbestos floor tiles are lawful products again. The consumption of asbestos in the US rose by 7.5% from 1992 to 1993.

In 1990 the maximum workplace concentration (TRK) for chrysotile has been lowered in Germany from 1 000 000 F-reg/m^3 to 250 000 F-reg/m^3 (TRGS 102), and any TRK was cancelled for amosite and crocidolite (zero level, during removal 50 000 F-reg/m^3 allowed). For the US OSHA lowered in 1994 the permissible exposure limit (PEL, comparable to German TRK) from 200 000 F-reg/m^3 to 100 000 F-reg/m^3 for all fiber types, not recognizing the great difference in health effects between chrysotile and amphibole-asbestos.

Erionite is banned in Canada, in Germany, and in the US (EPA 1991). Palygorskite is treated in Germany similarly to amphibole-asbestos. The German TRGS 905 (1994) classifies thin Pele's hair as carcinogenic.

Outlook

We are living in a dusty world (see also Sect. 5.3.1). Inorganic fibers from geogenic and anthropogenic sources are ubiquitous. It is impossible to remove them from the environment. The slogan "a single fiber kills" leads to confusion of the public and to frustration of nonepidemiologists and nonmineralogists (Mossman et al. in Guthrie and Mossman 1993). If we continue to put exposure limits below background levels, to consider wrong fiber definitions or to base regulations upon unsuited invivo tests, doomsaying will result. Mankind should care about fibers in the environment, but using common sense.

References

Bignon J, Brochard P, Brown R, Davis JMG, Vu V, Gibbs G, Greim M, Oberdörster G and Sébastien P (1995) Assessment of the Toxicity of Man-Made Fibres. A final report of a workshop held in Paris, France, 3–4 February 1994. Ann occup Hyg 39:89–106

Bossenmayer HJ, Schumm HP, Tepasse R (eds) (1991) Asbest-Handbuch. Erich Schmidt Verlag, Berlin

Churg A (1993) Asbestos lung burden and disease patterns in man. In: Guthrie GD Jr, Mossman BT (eds) Health effects of mineral dusts. Reviews in mineralogy 28. Mineralogical Society of America, Washington, DC, pp 409–426

Doll R (1955) Mortality from lung cancer in asbestos workers. Br J Ind Med 12:81–86

Doll R, Peto J (1985) Asbestos: effects on health of exposure to asbestos. Health and Safety Commission, Her Majesty's Stationary Office, London

Faserförmige Stäube: Vorschriften, Wirkungen, Messung, Minderung; Tagung Fulda, 7. bis 9. Sept 1993/VDI-Berichte 1075. Kommission Reinhaltung der Luft im VDI und DIN. Verein Deutscher Ingenieure, VDI-Verlag GmbH, Düsseldorf 1993

Fischer M, Meyer E (1983) The assesment of the health risk from asbestos fibres by the Federal Health Office of the FRG. VDI-Berichte 475, 325. Kommission Reinhaltung der Luft im VDI und DIN. Verein Deutscher Ingenieure, VDI-Verlag GmbH, Düsseldorf 1983

Förster H (1993) Anorganische faserförmige Partikel in der Atmosphäre. VDI Berichte No 1075, pp 211–231

Fuehres M, Heidermanns G (1993) Fasermessungen bei der Verwendung von Zellulose-Fasern. VDI Berichte No 1075, pp 345–356

Guthrie GD, Mossman BT (eds) (1993) Health effects of mineral dusts. Reviews in mineralogy 28. Mineralogical Society of America, Washington DC

Ledoux RL (ed) (1979) Mineralogical techniques of asbestos determination. Short course handbook 4, Mineral Assoc Canada, Quebec

Pott F, Friedrichs KH (1972) Tumours in rats after intraperitoneal injection of asbestos. Naturwissenschaften 59:318–332

Rödelsperger K, Woitowitz HJ, Brückel B, Arhelger R, Römer W, Pohlabeln H, Köckel KH (1993) Fall-Kontrollstudie an Mesotheliompatienten zum Gehalt anorganischer Fasern im Lungenstaub. VDI Berichte No 1075, pp 253–267

Romer M (1994) Anorganische Partikel in der menschlichen Lunge. Eclogae Geol Helv 87/2:503–512

Schnittger J (1993) Messung und Identifizierung anorganischer Fasern in der Außenluft. VDI Berichte No 1075, pp 283–295

Schultz M (1993) Die Serosatests – biologische Grundlagen – Bedeutung national und international. ErgoMed 17:170–173

Skinner HCW, Ross M, Frondel C (1988) Asbestos and other fibrous materials – mineralogy, crystal chemistry, and health effects. Oxford University Press, New York

Stanton MF, Layard M, Tegeris A, Miller E, May M, Morgan E, Smith A (1981) Relation of particle dimension to carcinogenicity in amphibole asbestos and other fibrous minerals. J Natl Cancer Inst 67:695–975

Stanton MF, Wrench C (1972) Mechanisms of mesothelioma induction with asbestos and fibrous glass. J Natl Cancer Inst 48:797–821

Woitowitz HJ (1991) Gesundheitsschäden durch Asbest. In: Bossenmayer HJ, Schumm HP, Tepasse R (eds) Asbest-Handbuch. Erich Schmidt Verlag, Berlin

Regulations (examples only)

DFG Deutsche Forschungsgemeinschaft (1993) MAK- und BAT-Werte-Liste 1993. Senatskommission zur Prüfung gesundheitsschädlicher Arbeitsstoffe, Mitteilung 29, Verlag Chemie, Weinheim, 153 + XIII pp

EPA US Environmental Protection Agency (1991) Erionite fiber; significant new use of a chemical substance. Fed Register 56, 5647056472, Washington, DC

HEI Health Effects Institute (1991) Asbestos in public and commercial buildings: a literature review and synthesis of current knowledge. HEI-Asbestos Research, Cambridge

HSE Health and Safety Executive (1977) Asbestos: guidance note from the health and safety executive. Medical series No 13, Her Majesty's Stationary Office, London

IARC International Agency for Research on Cancer (1987) IARC monographs on the evaluation of the carcinogenic risk of chemicals to humans: silica and some silicates. No 42, WHO-IARC, Lyon, 288 pp

ILO International Labour Office (1990) Safety in the use of mineral and synthetic fibres. Occupational Safety and Health Series No 64, Geneva

NIOSH National Institute of occupational safety and health (1981) Report No 81–103, Washington, DC

OSHA Occupational Health and Savety Agency (1992) 29 CFR Parts 1910 and 1926, Occupational Exposure to Asbestos, Tremolite, Anthophyllite and Actinolite. Federal Register 57, No 110, Washington, DC

TRGS 102 (1990) Technische Richtkonzentrationen (TRK) für gefährliche Stoffe. BArbBl. 2/1990, Bonn

TRGS 905 (1994) Gesundheitsgefährliche faserförmige Stäube. BArbBl. 5/1994

VDI (Verein Deutscher Ingenieure) VDI-Richtlinie 3492 Messen anorganischer faserförmiger Partikel in der Außenluft: Rasterelektronenmikroskopisches Verfahren. Blatt 1, Entwurf (1989), Blatt 2, Entwurf (1992), Beuth Verlag, Berlin

VDI (Verein Deutscher Ingenieure) (1993) Faserförmige Stäube. VDI Berichte No 1075

WHO (World Health Organization) (1987) Air quality guidelines for Europe. WHO Reg Publ, European Ser No 23, Copenhagen

WHO (World Health Organization) (1992) Validity of methods for assessing the carcinogenicity of man-made fibres. Executive summary of a WHO Consultation, Geneva

ZH 1/600.1.2 (1993) Gesundheitsgefährlicher mineralischer Staub; Teil 2: Asbesthaltiger Staub (Regelwerk der Berufsgenossenschaften). Carl Heymanns Verlag, Köln

5.3.3 Anthropogenic Heavy Metal Contaminations in Aqueous Environments

U. Förstner

Metal pollution in surface waters caused considerable public concern during the 1960s and 1970s. In Japan alone, between 1953 and 1970, more than 600 people died from poisoning by arsenic, cadmium, and mercury in air, drinking water, fish, and rice. Examples from industrialized countries indicate that aquatic metal pollution reached its peak in the 1970's, partly as a result of recent improvements in wastewater treatment. However, problems with metal pollution still exist on both a local and regional scale. Waters of large areas in southern Scandinavia and eastern North America are endangered by the combined effects of airborne pollutants and acidity. Problems arising from acid mine effluents have still not been solved satisfactorily in any part of the world. Increasing quantities of dredged materials represent long-term reservoirs for the release of heavy metals, and groundwater pollution from abandoned landfills of solid waste materials will be a major problem in the future.

A global comparison for natural and anthropogenic emissions suggests particularly strong effects of man's activities on the atmospheric metal budgets for lead (100-fold anthropogenic enrichment compared to natural emissions) and for zinc, cadmium, and copper (10- to 20-fold enrichment). The inputs into the oceans from wastewaters is about 10–20% of the respective discharges from geological weathering for these elements (Galloway 1979). On a regional scale, for industrialized areas, both wastewater and atmospheric sources occur at about the same order of magnitude. An inventory in Lake Erie (Nriagu et al. 1979) revealed that atmospheric inputs accounted for 20%, 25%, and 50% respectively of the Cu, Zn, and Pb discharges originating from the lake's catchment. The contributions from sewage effluents were 45, 30, and 20%. Additional anthropogenic inputs originate from tributaries to the lake, shoreline erosion, and deposition of dredged spoils.

A quantitative measure of metal pollution in aquatic sediments has been introduced by Müller (1979), which is called the Index of Geoaccumulation:

$I_{geo} = \log_2 C_n/1.5 \times B_n$.

C_n is the measured concentration of the element n in the pelitic sediment fraction (<2 μm) and B_n is the geochemical background value in fossil argillaceous sediment (average shale); the factor 1.5 is used because of possible variations of the background data due to lithogenic effects. The Index of Geoaccumulation consists of 7 grades, whereby the highest grade (6) reflects 100-fold enrichment above the background values ($2^6 = 64 \times 1.5$). The situation of the I_{geo} indices in longitudinal profiles of the Rhine River is shown in Fig. 90 (after

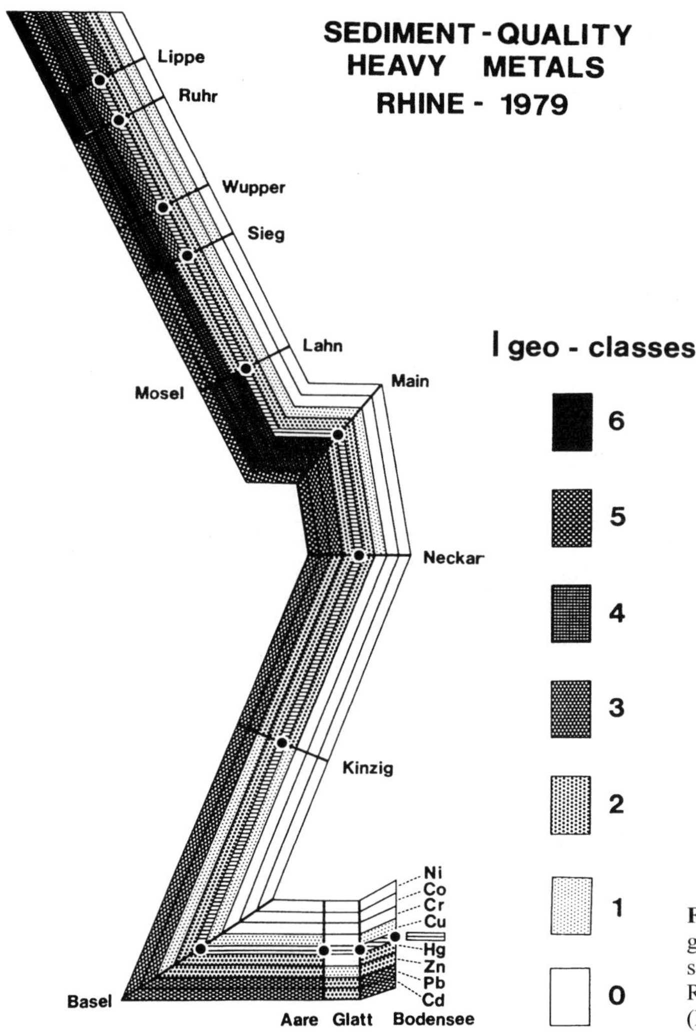

Fig. 90. Index of geoaccumulation in sediments of the Rhine River. (Müller 1979)

Müller 1979). The index values for the upper Rhine (between Basel and Mannheim, before the waters of the highly polluted Neckar and Main Rivers enter the Rhine) lie at a maximum of 3 for Cd and Pb. Cadmium reaches the highest grade (6) in the middle section of the river and the lower Rhine, which can be trace to influents from the Neckar River and from the Ruhr area. The indices for Pb and Zn attain grade 4, and those for Hg, Cu, and Cr each increase by one grade compared to the upper Rhine River section.

An example of the heavy metal budget in a densily populated region is given for the 4,488-km^2 Ruhr river catchment area, a tributary to the Rhine river, by Imhoff et al. (1980). The Ruhr river is the main source of drinking and process water for a population of about 6 million people, and industries. At the same time, the Ruhr and its tributaries receive treated wastewaters from 2.2 million people and 2 million industrial population equivalents. Among other industries, about 300 plating plants have settled in the Ruhr basin. These plants discharge a considerable amount of heavy metals, which occasionally results in metal concentrations in the river water up to critical levels (e.g., for nickel, which appears in Ruhr water in higher concentrations than in other rivers). Figure 91 shows that approximately 55% of the heavy metals (Pb, Cu, Zn, Ni, Cr, Cd) in the Ruhr river are discharged from municipal and industrial wastewater treatment plants; 45% are of geochemical origin. The municipal treatment plants receive 31% of the heavy metals from domestic wastewater and 69% from industrial wastewater. The origin of the elements and their relative fraction of the total load varies considerably. For example, 90% of the chromium in the Ruhr river is discharged by industries; 66% of nickel and cadmium derive from industries, while only 50% copper and zinc originate from industrial wastewater.

In Table 33, metal data from less contaminated and polluted water samples are presented (after Salomons and Förstner 1984). The examples for background values stem partly from large inland water masses such as Lake Michigan; however, contamination from both atmospheric and river inputs cannot be excluded. More reliable background values are those from the Amazon and Yukon Rivers, where the effect of large catchment areas meets the prerequisite of low pollution activity. The examples of the rivers Rhine and Ruhr have been selected from a long-term and dense data base. Both rivers are major sources of municipal and industrial water supply, including bank filtration and artificial groundwater recharge.

It can be seen from Table 33 that the "dissolved" metal concentrations in the Rhine River (station Cologne) were enriched by a factor 50–100 for Zn, Cd, and Pb, 10–20 for Ni, Hg, and Cu, and 4 for Cr, whereas As is near the background. Between 1970/73 and 1978, a significant decrease of dissolved concentrations was found for cadmium, copper, lead, and zinc. In the Ruhr River concentrations of Cd, Cr, Cu, Zn, and, in particular, Ni were significantly higher than in the example of the Rhine River. Even these values, as average concentrations for 1 year, are below the limits for drinking water set by the WHO Regional Office for Europe in 1970.

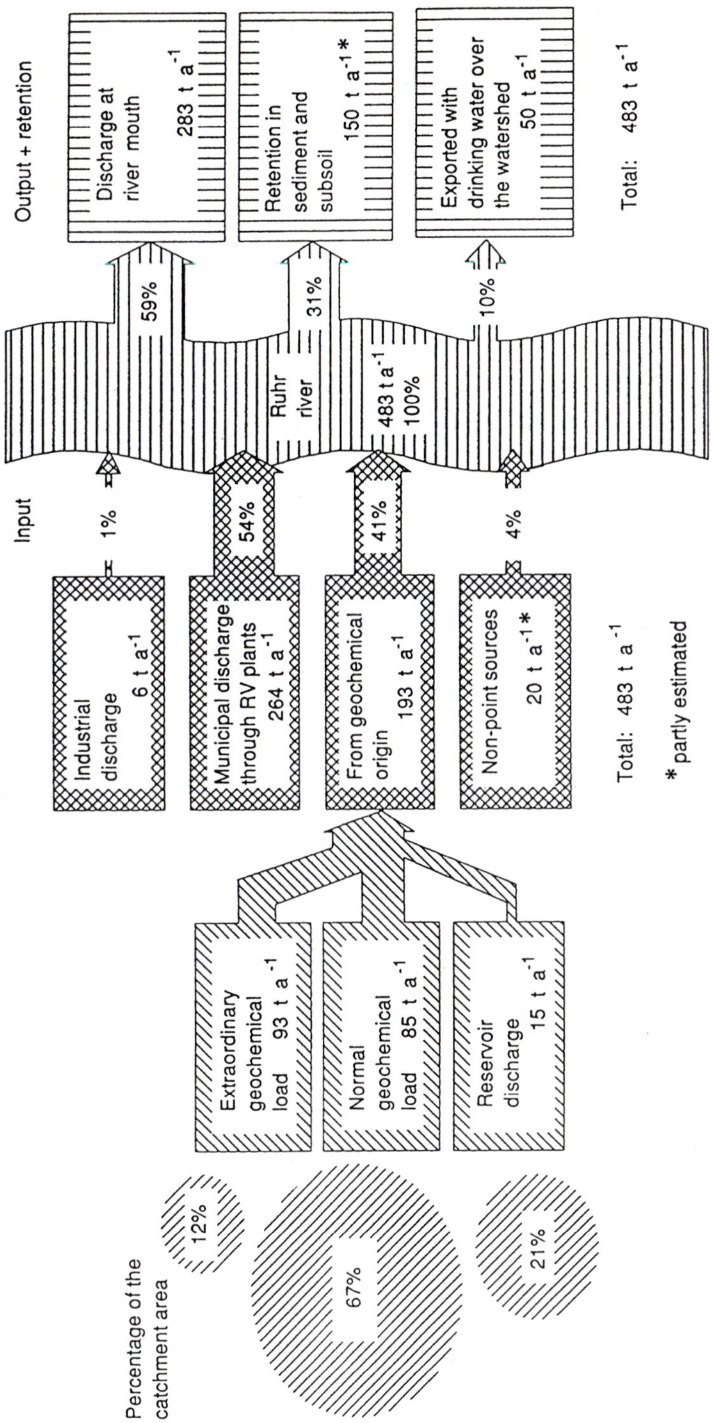

Fig. 91. The balance of heavy metal loads in the Ruhr River catchment area. (After Imhoff et al. 1980)

Table 33. Metal concentrations in river water (examples). Values in μg/l (References see Salomons and Förstner 1984)

Metal	WHO (1970)	Back-ground	Missis-sippi R.[a]	Rhine 1973	River[b] 1978	Ruhr River[c]	Witwaters-[d] rand Rivers
As	50	1[e]	3	0.7	1.2	5.9	250[f]
Cd	10	0.01[g]	0.1	1.0	0.2	2.3	50
Cr	50	0.5[a]	0.5	2	2	20	4000
Cu	50	0.5[g]	2	15	10	44	5400
Hg	1	0.01[h]	0.1	0.2	0.2	0.07	–
Ni	–	0.3[i]	1.5	8	10	62	6400
Pb	100	0.05[g]	0.2	15	3	17	300
Zn	5000	0.5–4[k]	10	65	50	226	26000

[a] Trefry and Presley (1976). [b] Dittrich and Klein in Reichert and De Haar (1982). [c] Dietz and Koppe in Reichert and De Haar (1982). [d] Wittmann and Förstner (1976). [e] Copeland and Ayers (1972). [f] Aston et al. (1975) – acid mine drainage in Cornwall/UK. [g] Sigg et al. (1977) Bodensee. [h] Förstner and Wittmann (1979) literature review. [i] Gibbs (1977) Amazon and Yukon Rivers. [k] Kennedy and Sebetich (1976) rivers and streams in California.

Much higher concentration factors have been analyzed in inland waters affected by acidic mine effluents, e.g., in the Witwatersrand Goldfields in South Africa (Table 33), where the levels of dissolved Ni, Mn, and Fe exceed the normal surface water concentrations by factor of more than 10000, values of Zn and Cr are increased 1000-fold, whereas Pb and Cd in many instances are encountered at values exceeding 100-fold enrichment.

There are many examples of deleterious effects of mine effluents on freshwater ecosystems (for details see Sect. 5.3.4 and for further references see Förstner 1981). In some streams of Wales, enrichments of Pb, Cu, and Zn leached from the outcrops of mineralized zone and spoil heaps of disused mines still cause a high mortality rate in fish and other organisms (Abdullah and Royle 1972). Oysters placed in the heavy contaminated Restronguet Creek of southwest England contained initially 250 mg/kg Cu in dry matter, rising to 1500 mg/kg after 1 month and 6000 mg/kg after 6 months (Thornton 1977). From the Upper Silesian mining areas of Poland it is suggested that the process of self-purification in some of the receiving waters is inhibited by the high concentrations of several metals. During many years no fish has appeared in the Sztola River (Pasternak 1974). Large-scale destruction of organic life has been found around Brunswick mines in Canada (Harvey 1976), from Lepanto copper mines of the Phillipines, Rum Jungle mines in northern Australia (Watson 1975) and in the South Esk River system of Tasmania (Tyler and Buckney 1973). Contamination of rice fields from mine drains is still found in Japan, e.g., in the areas of the Ichi and Maruyama river basins (Asami 1981). Maximum allowable limit for unpolished rice is 1 mg/kg Cd; however, it has been stressed by Asami (1981) that these limits may be probably too high, since the average Cd concentrations in highly endemic areas of Itai-Itai disease Cd of the Jinzu River was 0.53 mg/kg and the effects of renal dysfunction evidently occur at even lower levels; the latter disease has been found in many examples in the upper portion of the Ichi River basin.

Freshwaters, in contrast to the ocean and the atmosphere, are distinctly individual systems, in particular with respect to the behavior of heavy metals. Geochemical pecularities of rocks and water as well as point sources of contaminants are mostly dominant. In many stratified lakes, seasonal variations significantly affect concentration and biogeochemical cycling of heavy metals. In rivers, suspended matter content is important not only for the overall budget, but also for the concentration of dissolved metals. Secular changes of freshwater quality with respect to heavy metals, therefore, are mostly masked by short-term, seasonal, and local/regional effects.

During the past decade, the major objectives of research on metal-polluted waters have changed from initial surveys of sources and pathways to more detailed investigations of the mechanisms controlling the mobility and bioavailabilty of different metal species. Factors such as pH decrease, redox changes, and the effect of organic chelators affect metal behavior in both surface and groundwater to a much larger extent than was previously accepted. Of the factors affecting mobility and bioavailability of metals in aquatic systems, acid precipitation, which multiplies the influences of high-temperature emission of heavy metals, is of prime importance. The introduction of synthetic chelators, such as nitrilotriacetic acid, increases in salinity (due to salt discharges), and the change in redox conditions, all interfere with natural removal processes and may mobilize metals from aquatic particulate matter (Salomons and Förstner 1984). On the other hand, eutrophication of lakes and coastal waters can enhance the rate of removal of metals from the surface waters and the biological effects of some toxic metals can be reduced by their sequestration in complexes. It has become increasingly evident that, particularly for lakes, strong links exist between the carbon cycles, redox cycles, and metal behavior. With such linked cycles, transfer of metals into biota can be inhibited, for example, by biological reduction of proton activity and the formation of new adsorbing surfaces. However, laboratory experiments have suggested that metals mobilized from contaminated particulates by the processes mentioned above may "seek" organic matter and are significantly accumulated in this fraction. Even relatively small percentages of organic substrates, if involved in metabolic processes, may constitute a major pathway by which metals are transferred into the food chain.

Since metal cycles in aquatic systems are closely linked with the indirect influences of air and soil environments (Stumm 1986), large-scale pertubations in these systems can be expected in both surface and groundwaters. Agricultural and residential patterns are changing the rates of continental mechanical erosion. Deforestation on a large scale and the draining of major wetlands will reduce the number of sinks available for the mobilized metals. The acidification of soils enhances the rate of podzolization and consequently the release of many metals which are otherwise rather immobile. Low pH values must be expected to alter soil microbial ecology, for example, the prevalence of acidophilic microorganisms, which will, in turn, influence biogeochemical processes in soils, including the related metal cycles. The composition of groundwaters can be altered both by direct injection of pollutant metals from precipitation, waste materials and fertilizers, and by changes in soil biogeochemistry resulting from the introduction of excess nutrients. The latter processes can lead to an increase in the size of the anoxic zone in soils and consequent mobilization of elements soluble under reducing conditions. A particular threat is related to leakage waters from landfills, which are often extremely high in toxic materials.

References

Förstner U (1981) Trace metals in fresh waters (with special reference to mine effluents). In: Wolf KH (ed) Handbook of strata-bound and stratiform ore deposits, vol 9. Elsevier, Amsterdam, pp 271–303

Galloway JE (1979) Alteration of trace metal geochemical cycles due to the marine discharge of wastewater. Geochim Cosmochim Acta 43:207–218

Imhoff KR, Koppe P, Dietz F (1980) Heavy metals in the Ruhr river and their budget in the catchment area. Prog Water Technol 12:735–749

Müller G (1979) Schwermetalle in den Sedimenten des Rhein – Veränderungen seit 1971. Umschau 79:778–783

Nriagu JO, Kemp ALW, Wong HKT, Harper N (1979) Sedimentary record of heavy metal pollution in Lake Erie. Geochim Cosmochim Acta 43:247–258

Salomons W, Förstner U (1984) Metals in the hydrocycle. Springer, Berlin Heidelberg New York, 349 pp

Stumm W (1986) Water, an endangered ecosystem. Ambio 15:201–207

5.3.4 Environmental Impact of Heavy Metal Ore Deposits

G. Deissmann and G. Friedrich

The enrichment of heavy metals in the environment may produce hazards to human life and terrestrial and aquatic ecosystems. In general, heavy metals like lead, zinc, copper, or nickel are trace constituents of minerals and rocks. The production of these metals for human use requires a natural pre-enrichment of metals by magmatic, hydrothermal, sedimentary, and/or metamorphic processes to form minable ore deposits.

The accumulated metals in these ore deposits serve as a potential source for the enrichment of heavy metals in the environment. In general, two different aspects have to be considered: (1) natural dispersion of heavy metals from ore deposits and (2) dispersion of heavy metals by anthropogenic activities such as mining and smelting of heavy-metal-bearing ores. In regions with historical and recent mining activities, both geogenic and anthropogenic enrichment of

Advanced Mineralogy, Volume 3
Mineral Matter in Space, Mantle, Ocean Floor, Biosphere, Environmental Management, and Jewelry (ed. by A. S. Marfunin)

heavy metals in the environment has to be considered, although it can be difficult to distinguish between the different sources.

Natural Dispersion Halos

In many ore deposits the potentially toxic heavy metals (i.e., lead, zinc, copper, nickel, arsenic) are bound to sulfide minerals such as galena, sphalerite, chalcopyrite, pentlandite, and arsenopyrite, which are, in general, not stable under oxidizing conditions at the Earth's surface. Furthermore, these ore minerals are mostly accompanied by iron sulfides such as pyrite, marcasite, or pyrrhotite, which often form the major part of the ore. In addition, also deposits of other metals like uranium, which do not occur as sulfides, are often associated with sulfide minerals like pyrite.

Sulfide oxidation and the thus caused changes in Eh and pH are therefore an important process during the weathering of the ores and the release of metals. Furthermore, it has to be taken into account that other potentially toxic heavy metals such as cadmium, thallium, antimony, and mercury may be trace constituents of the ore minerals.

The natural dispersion of heavy metals at low temperatures and pressures in the vicinity of ore deposits at the Earth's surface is termed secondary dispersion, in contrast to the so-called primary dispersion of metals during ore formation at depth (Rose et al. 1979). The mobilization of metals from the ore due to chemical and physical processes may lead to an enrichment of heavy metals in every environmental compartment such as soils, sediments, waters, vegetation (Fig. 92), or in case of mercury even the air.

The natural enrichment of heavy metals above "normal" background values is in widespread use as tracer for ore deposits in geochemical exploration and dealt with in the relevant literature (i.e. Friedrich and Kulms 1974; Levinson 1974; Rose et al. 1979). In some cases, this enrichment of heavy metals around ore deposits is known to have toxic effects on terrestrial or aquatic ecosystems.

In general, heavy metals derived from rocks and ore deposits (geogenic sources) have a lower mobility and (bio-)availability than heavy metals from anthropogenic sources. For example, lead derived from the weathering of galena-bearing ore deposits is often bound to pyromorphite, which is the most stable lead mineral under the conditions at the earth's surface. Anyway, natural anomalies in the heavy metal contents of soils, sediments or waters on local or regional scale have to be taken into account for the determination of environmental standards and guidelines for remediation in those areas as well as for land use.

Investigations of the speciation of metals, the occurence of heavy-metal-bearing minerals, the distribution patterns of metals in soils and their parental rocks or in sediments, and the lead isotopic composition may serve as indicators to distinguish between natural heavy metal enrichment in the environment and metals derived from anthropogenic sources.

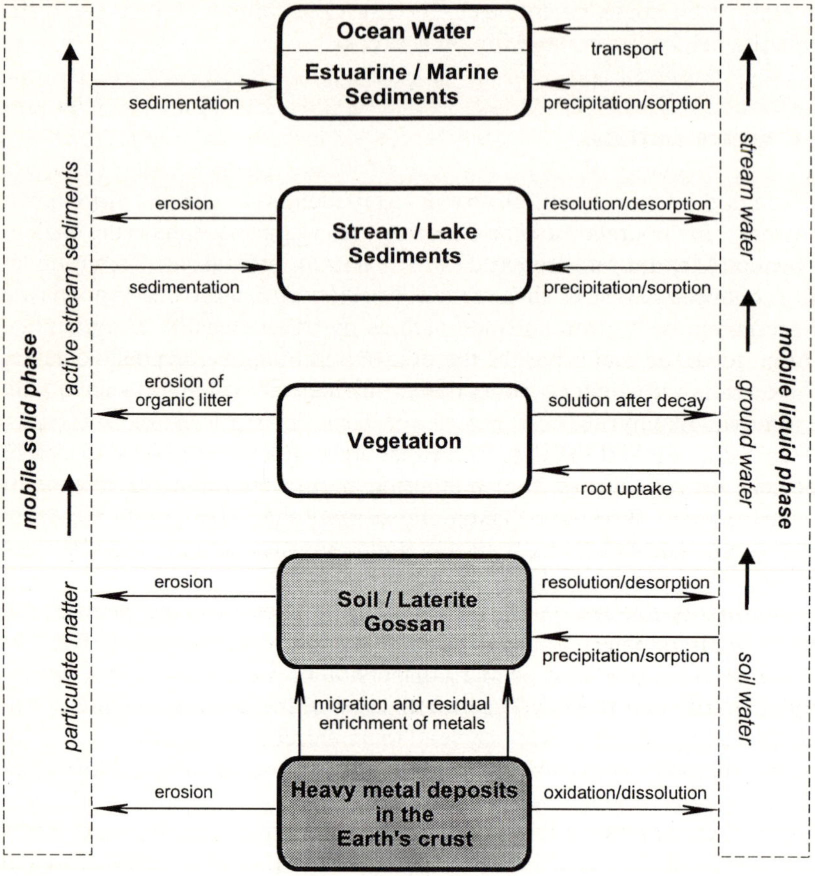

Fig. 92. Dispersion of metals from ore deposits into the natural environment. (After Rose et al. 1979)

Impact of Mining and Metallurgical Activities

Another pathway by which heavy metals enter the environment is the extraction of metals by mining activities. Mining itself affects mostly only relatively small areas, but the environmental impacts may occur on regional and larger scales. One problem is the large volumes of waste rock heaps and often sulfide-bearing tailings, which are potential sources of heavy metals. Soil pollution around mining is mainly due to (1) emissions of the metal mines, (2) disposal of waste rocks, and (3) water and wind erosion from (unvegetated) waste rock heaps and tailings. Furthermore, the transport, processing, and smelting of ores, which often (but not necessarily) takes place in the vicinity of the deposits, contribute to pollution by heavy metals. An overview on soil pollution by metals from mining and smelting activies in various countries including Japan, the USA, the UK, Australia, and others is given by Asami (1988).

The most important problem of mining is the formation of acid mine drainage (AMD) if sulfide-bearing ores are mined (Allan 1995; Salomons 1995). This AMD can contain a variety of heavy metals in toxic concentrations, depending on the nature of the ores, due to the higher solubility of most heavy metals in acidic and oxidizing waters. Each of the basic steps in the mining process may expose metal sulfides and can be a source of AMD (Fig. 93), although the main release of metals to freshwaters is from tailings and polishing ponds and emissions in the benefication stage (Ferguson and Erickson 1988).

The generation of acid mine drainage is due to chemical reactions and bacterial oxidation of sulfide minerals. In most sulfide-bearing ore deposits pyrite is the major sulfide and therefore plays a key role in the formation of AMD, although other metal sulfides are also important in determining the quality of mine drainage, especially by their release of heavy metals like lead, zinc, or copper. Other iron sulfides, i.e. pyrrhotite, marcasite, or bravoite (Ni-Co-bearing pyrite) are also amenable to oxidation and acid generation, but research and data on these minerals are sparse.

Acid mine drainage occurs when the acid generation is faster than neutralization by alkaline materials. Primary factors for the formation of AMD include pyrite and other sulfides, oxygen, water, and iron-oxidizing bacteria. The oxidation of pyrite and the formation of acid mine drainage occurs in several steps which are catalyzed by *Thiobacillus ferrooxidans* and other stalics bacteria like *Thiobacillus Thiooxidans* (Allan 1995). The formation of ferric iron, which becomes the dominating oxidizing agent, is the rate-determining step in the whole process of AMD generation.

Secondary factors control the consumption of acid either by carbonate minerals such as calcite or dolomite (short-term buffering) or by silicates (long-term buffering). Thus, in spite of the presence of pyrite and/or other sulfide minerals. AMD may not be present at a specific mine site. The buffering of AMD by carbonates is relatively fast and controlled by solution equilibria, although the total amount of carbonates is not always available for buffering due to the formation of iron hydroxide or calcium sulfate coatings around the carbonate particles. Other secondary factors include reactions like gypsum precipitation, acid-induced dissolution of minerals, and ion exchange processes. These reactions can change the character of the drainage water by adding heavy metals as well as aluminum and alkaline earth cations (Ferguson and Erickson 1988).

Physical characteristics of waste rocks and tailings and the hydrological regime are the tertiary factors that influence acid production, migration, and consumption. Although fine-grained materials have a higher surface area, which promotes reactions, the limited availability of oxygen and water may reduce acid production. On the other hand, oxygen advection and diffusion is promoted by coarse-grained materials, which can result in higher acid generation.

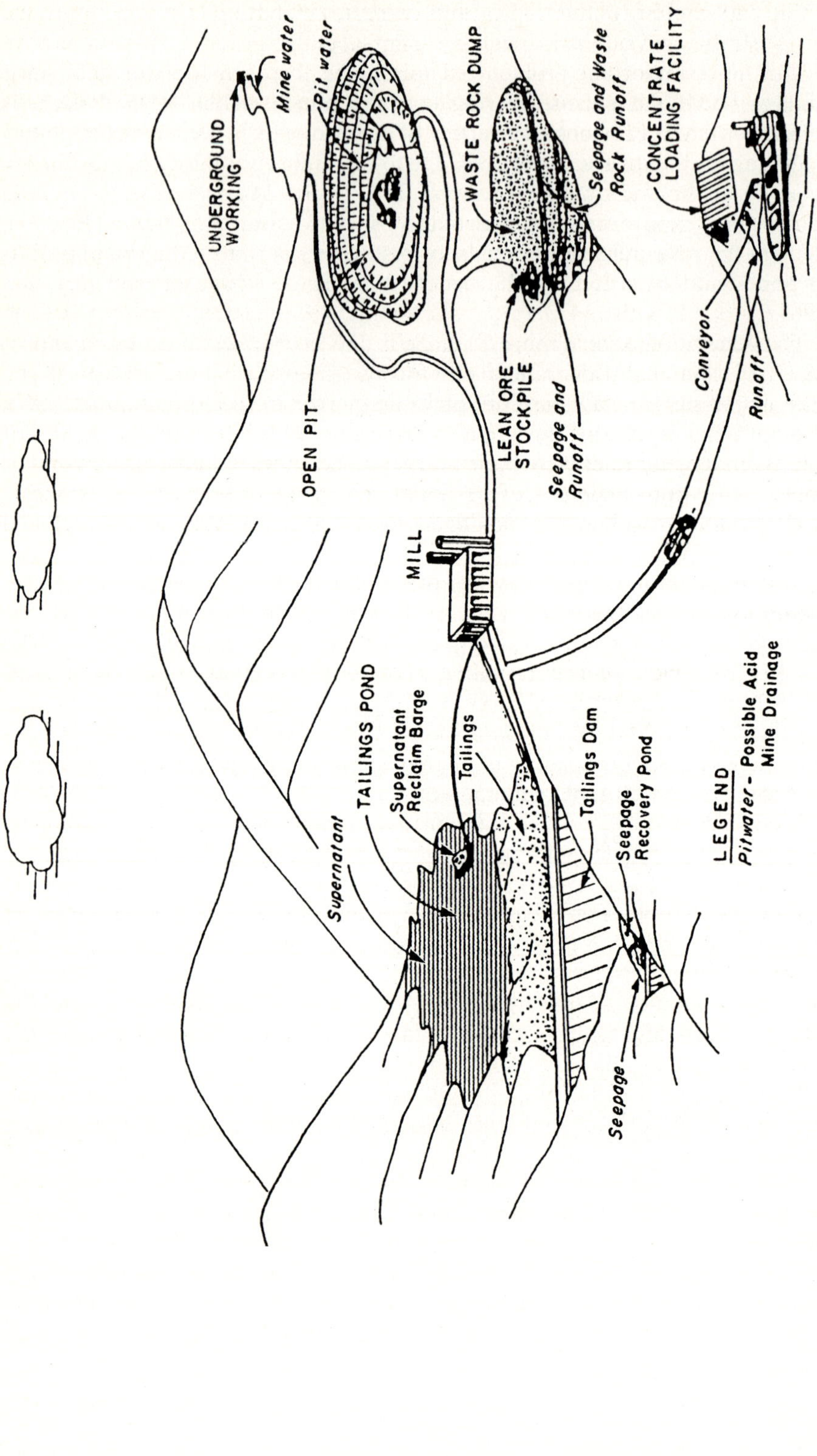

Fig. 93. Major components of a metal mine and sources of acid mine drainage AMD. (Furguson and Erickson 1988)

The acid and heavy-metal-bearing waters derived from mining complexes, tailing ponds, or waste rock heaps might especially influence aquatic ecosystems. The behavior and fate of metals derived from mining in streams and rivers, as well as the response of aquatic ecosystems to AMD, is described, e.g., by Salomons and Förstner (1984) and Salomons (1995), and is the subject of another chapter in this volume.

One of the key factors for the prediction of metal release of mining is the prediction of acid mine drainage before, during and after mining activities (Fig. 93). For this purpose, different test methods have been developed, including kinetic titration tests, lysimeter tests, soxhlet extraction procedures, and shake flask tests with and without microbiological inoculation, amongst others (Ferguson and Erickson 1988; Allan 1995). Although the basic geochemical and biogeochemical processes which control the production of AMD and the behavior of heavy metals are known, the application of this know-how to field conditions and the modeling and prediction of AMD is difficult. The prediction of AMD and metal release during and after mining based on a combination of hydrological, (bio-)geochemical and physical processes requires still much basic research, as well as verification studies of already developed models.

References

Allan RJ (1995) Impact of mining activities on the terrestrial and aquatic environment with emphasis on mitigation and remedial measures. In: Salomons W, Förstner U, Mader P (eds) heavy metals – problems and solutions. Springer, Berlin Heidelberg New York, pp 119–140

Asami T (1988) Soil pollution by metals from mining and smelting activities. In: Salomons W, Förstner U (eds) Chemistry and biology of solid waste – dredged material and mine tailings. Springer, Berlin Heidelberg New York, pp 143–169

Ferguson KD, Erickson PM (1988) Pre-mine prediction of acid-mine drainage. In: Salomons W, Förstner U (eds) Environmental management of solid waste – dredged material and mine tailings. Springer, Berlin Heidelberg New York, pp 24–43

Friedrich G, Kulms M (1974) Geochemische Untersuchungen im Bereich der Blei-Zinkerz-Lagerstätte Maubach/Nordeifel, Forschungsbericht Land Nordrhein-Westfalen 2395. Westdeutscher Verlag, Düsseldorf, pp 1–52

Levinson AA (1974) Introduction of exploration geochemistry. Applied Publishing Calgary, 612 pp

Rose AW, Hawkes HE, Webb JS (1979) Geochemistry in mineral exploration. Academic Press, London, 657 pp

Salomons W (1995) Environmental impact of metals derived from mining activities: processes, predictions, prevention. J Geochem Explor 52:5–23

Salomons W, Förstner U (1984) Metals in the hydrocycle. Springer, Berlin Heidelberg New York, 349 pp

5.3.5 Treatment of Contaminated Dredged Sludge – Disposal Strategies

W. CALMANO, U. FÖRSTNER and R. KHORASANI

Many problems exist with contaminated sediments in rivers, estuaries, and coastal areas. In order to maintain ports and channels accessible to shipping, sediments have to be removed regularly by dredging. In the harbors around the North Sea, approximately 100 million m^3 have to be dredged annually – about ten times the average annual sediment discharge of the Rhine River. Typical problems with these sediments are increasing volumes of dredged materials, and high concentrations of toxic substances, and these problems are concentrated mainly at the mouth of large rivers and in harbors. Traditional disposal alternatives of dredged materials are subaqueous (open water) disposal, spoiling on intertidal sites and deposition on land. In the past years, coastal marine disposal in capped mound deposits above the prevailing sea floor, disposal in subaqueous depressions, and capping deposits in depressions have been applied for contaminated sediments (Kester et al. 1983). These categories differ primarily in the exposition paths of toxicants to biological populations, plants, or groundwater, in physicochemical conditions and transport processes potentially capable of removing contaminants from dredged material at the disposal site.

In the discussion of environmental problems associated with contaminated sediments, it is now widely accepted that the ecological significance of toxicants is determined by the specific form and reactivity of the compound more than by its accumulation rate. This effect typically refers to heavy metals which may partly be mobilized by changes in pH and redox conditions, changes in ionic strength, formation of soluble complexing organic compounds, and microbial interactions. While the first two factors typically relate to land disposal of dredged materials, the effect of salinity is more important for the deposition of freshwater sediments in intertidal and coastal marine environments because metals (e.g., Cd) sorbed on particles may be displaced by Na^+, Mg^{2+}, or Ca^{2+} and complexed by Cl^--ions.

For most sediments from maintenance dredging the question is disposal or treatment subsequent to removal. An interesting answer to this question is given in Hamburg. The soft fine-grained sediments in the harbor basins are excavated by a bucket chain dredger. Extracted sediments are transferred into a barge, mixed with water, and the sludge suspension is pumped to the first treatment step. Formerly, this was a classification running and drying basin for the separation of coarse-grained and more fine material, the latter being stored in confined disposal ponds. Since space is no longer available, this technique has been changed. Now the ponds are used only as stockpiling basins, from which the sediments are transferred to a mechanical classification plant. Separation of sand and silt takes place in two stages. The first step is a hydrocyclone classification. In the second step the underflow is retreated in an upcurrent sorter. The separated sand is dewatered on screens, and the silt suspension formed from the overflows of the hydrocyclone and upcurrent sorter is thickened in two high-performance clarifiers. Deposition is aided by flocculants, and dewatering takes place in screen belt presses, where a residual moisture content of approximately 45 % is obtained. This material is stored on land in the form of a heap, and most of the sand is used for stabilization and draining of the harbor sludge deposit. In total, approximately half a million tonnes per year of dredged material are treated in this way.

A more general conceptual scheme related to excavated sediments has been proposed by the TNO, The Netherlands Scientific Technological organization (Van Gemert et al. 1988). A- and B-techniques are distinguished: A is for larger scale concentration techniques like mechanical separation as performed in Hamburg. A-techniques are characterized by relatively low costs per unit of residue, low sensitivity to variations, and they may be applied in mobile plants. B-techniques are decontamination procedures which are especially designed for relatively small-scale operations. They involve higher operating costs per unit of residue, are more complicated, and are usually constructed as stationary plants. B-techniques include biological treatment, acid leaching, solvent extraction, etc.

Such remediation techniques on contaminated sediments generally are much more limited than for most other solid waste materials, since the widely diverse contamination sources in larger catchment areas usually produce a mixture of pollutants which is more difficult to treat than an industrial waste. Here, geochemical engineering, applying geochemical principles such as concentration, stabilization, solidification, and other forms of long-term, self-containing barriers, emphasizes the efforts to use natural resources and processes for reducing negative environmental effects of pollutants, e. g., by immobilizing toxic metals. Long-term immobilization of critical pollutants can be achieved by promoting less soluble chemical phases, i. e., by thermal and chemical treatment, or by providing respective milieu conditions. Selection of appropriate environmental conditions predominantly influences the geochemical gradients, whereas chemical additives are aimed to enhance capacity controlling properties (e. g., buffer capacity) in order to bind micropollutants. In general, formation of mineral precipitates in the pore space of a dredged sediment body will be employed rather than using large-scale enclosure systems such as clay covers or wall constructions. A common feature of geochemically designed deposits, therefore, is their tendency to increase overall stability in time, due to the formation of more stable minerals and closure of pores, thereby reducing water permeation.

Most of the stabilization techniques aimed for the immobilization of metal-containing wastes are based on addition of cement, water glass, coal fly ash, lime, or gypsum (Malone et al. 1982; Wiedemann 1982; Goumans et al. 1991). Laboratory studies on the evaluation and efficiency of stabilization processes were performed by Calmano (1988) and Khorasani et al. (1988).

They studied the effects in dredged material after addition of cement, lime, calcium hydroxide, calcium carbonate, gypsum, trass, and waste kiln dust. Analysis of the stabilized products by polarization microscopy, scanning electron microscopy, X-ray diffraction, electron microprobe, proton-induced X-ray emission (PIXE), microprobe and chemical extraction methods showed that new minerals had been formed, which solidified the material and changed the metal-binding forms by incorporation and fixation in new-formed mineral phases like silicates, carbonates, sulfates, and hydroxides (see Sect. 5.2.2). Good results were attained by the use of calcium carbonate as additive since the pH conditions are not changed significantly. Generally, maintainance of neutral pH values favors adsorption or precipitation of soluble metal compounds. On the other hand, it can be expected that both low and high pH values will have unfavorable effects on the mobility of heavy metals and organic contaminants.

Regarding the various containment strategies, it has been argued that land deposition could provide a more controlled management than disposal in the marine environment. However, contaminants released either gradually from an imperfect impermeable barrier or catastrophically from failure of the barrier could produce substantial danger for the groundwater. In a review of various marine disposal options, Kester et al. (1983) suggested that the best strategy for disposing contaminated sediments is to isolate them in a permanently reduced environment. This type of waste deposition under stable anoxic and sulfidic conditions, where large masses of polluted dredged materials are covered with clean, sandy sediments is known as subsediment-deposit. Under subsediment conditions there is a particularly low solubility of metal sulfides compared to the respective carbonate, phosphate, and oxide compounds. One major prerequisite is the microbial reduction of sulfate. Thus, this process is particularly important in the marine environment, whereas in anoxic freshwater sediments there is a tendency for enhancing metal mobility due to the formation of stable complexes with ligands from decomposing organic matter. Marine sulfidic conditions, in addition, seem to repress the formation of monomethyl mercury, one of the most toxic substances in the aquatic environment (Craig and Moreton 1984). There are also indications that the degradation of halogenated hydrocarbons is enhanced in the sulfidic environment (Sahm et al. 1986).

Generally, remediation techniques in the closer sense are economically unacceptable because of the large volumes of contaminated sediments. There is a long retention time for sediments in larger catchment areas. Improvement at the source may need decades to become effective in the sediments at the lower reaches and harbors close to the river mouths. For most sediments from maintainance dredging there are more arguments in favor of disposal than treatment. Final storage conditions would imply that these materials should be deposited in a favorable chemical and long-term stabile environment. In the actual state of knowledge, this could only mean permanent anoxic sulfidic and well-buffered conditions. Such conditions can be made artificially or be selected from natural environments.

References

Calmano W (1988) Stabilization of dredged mud. In: Salomons W, Förstner U (eds) Environmental management of solid waste – dredged material and mine tailings. Springer, Berlin Heidelberg New York, pp 80–91

Craig PJ, Moreton PA (1984) The role of sulphide in the formation of dimethyl mercury in river and estuary sediments. Mar Poll Bull 15:406–408

Goumans JJJM, Van der Sloot HA, Aalbers ThG (eds) (1991) Waste materials in construction. Studies in environmental science 48. Elsevier, Amsterdam

Kester DR, Ketchum BH, Duedall IW, Park PK (1983) Wastes in the ocean, vol 2: Dredged material disposal in the ocean. John Wiley, New York

Khorasani R, Calmano W, Förstner U (1988) Stabilization of dredged sludge by chemically and mineralogically different additives. In: Wolf K, Van den Brink WJ, Colon FJ (eds) Contaminated soil '88. Kluwer, Dordrecht, pp 1465–1470

Malone PG, Jones LW, Larson RJ (1982) Guide to the disposal of chemically stabilized and solidified waste. Report SW 872. Office of Water and Waste Management, EPA, Washington, DC

Sahm H, Brunner M, Schobert SM (1986) Anaerobic degradation of halogenated aromatic compounds. Microb Ecol 12:147–153

Van Gemert WJT, Quakernaat J, Van Veen HJ (1988) Methods for the treatment of contaminated dredged sediments. In: Salomons W, Förstner U (eds) Environmental management of solid waste – dredged material and mine tailings. Springer, Berlin Heidelberg New York

Wiedemann HU (1982) Verfahren zur Verfestigung von Sonderabfällen und Stabilisierung von verunreinigten Böden. Berichte Umweltbundesamt 1/82. Erich Schmidt Verlag, Berlin

5.3.6 Flue Gas Purification Products of Coal-Fired Power Plants and Municipal Waste Incinerators. Characteristics and Waste Management

H. U. BAMBAUER

Residues of Coal-Fired Power Plants

Coal-fired power plants produce huge amounts mainly of bottom and fly ashes, and increasing amounts of gypsum from flue gas desulfurization (FGD). For example, rough estimations of solid residues for German power plants in the year 2000 are (in million tons/year): hard coal bottom ash 4, fly ash 3, and FGD gypsum 2.5; lignite bottom ash 3.5, fly ash 10.5, and FGD gypsum 6.5.

The powdery fly ashes produced by power plants may show a broad chemical and mineralogical variation (Tables 34 and 35). This variation depends mainly on the structure and composition of the coal type, and on the combustion process. Therefore, even fly ash from a single power plant may vary to some extent over both short- and long-term periods.

For instance, among hard coal-fired power plants high-temperature slag tap furnaces give fly ashes consisting mainly of glassy (<90%) and opaque

Table 34. Variation in chemical composition of the main types of fly ashes, mainly on the basis of German power plants and waste incineratiors

	Kind of coal		Municipal waste	
	Hard coal	Lignite	Fly ash	Fly ash + reaction products
Main elements in wt.%				
SiO_2	40–55	10–70	10–80	10–26
Al_2O_3	23–35	4–28	1–17	3–15
Fe_2O_3	4–20	1.5–30	2–10	<1–2
MnO		0.1–0.2	0.1	<0.08
MgO	<1–4.8	0.5–8	1–3	1–2
CaO	1–8	2–55	12–21	22–60
Na_2O	<1–1.2	0.1–2	1–4	1.5–3.5
K_2O	1.5–5	0.1–1.2	2–4	1.3–2.8
P_2O_5	0.4–3.5	1–3	0.3–4	
SO_3	0.1–3.5	0.1–15	2–10	2–7
Cl			1–7	9–15
CO_2			0.6–6	3–11
CaOfree	<1–20	<1–14	0.5–2	1–31
A selection of heavy metals in ppm				
As	500	3–190	1–30	1–10
Sr	1800	400–2100		
Ba	2000	180–2300	2000–2600	400–1700
Cd	2–35	<1–6	30–1700	80–190
Co	150	5–13	800–7000	8800–13800
Cr	120–650	5–65	20–600	10–100
Cu	160–300	2–70	300–1300	250–800
Mo	250	<3–8	1–70	1–13
Ni	170–650	5–50	50–400	10–80
Pb	200–2400	3–100	<1–10500	900–3800
Sb	100	<1–6	80–900	170–470
V	800	35–90	30–200	5–40
Zn	300–2700	10–80	3600–36800	9800–19000
H_2O-soluble fract. (wt.%)		2–3	3.5–16	16.5–27.5

(<10%) spheres, whereas low-temperature fluidized bed furnaces may give fly ashes consisting mainly of thermally altered detritic minerals, in particular quartz (<10%) and clays, illite (<50%), as reaction products anhydrite (<20%) and free lime (<20%). On the other hand, medium-temperature dry bottom furnaces give several of the possible constituents shown in Table 35 which may be completely realized by lignite fly ashes.

Fineness of, for instance, German coal fly ashes may vary as a function of the kind of coal, combustion process, kind and number of dust precipitators. Grain size of hard coal fly ashes may often range between D(50) ~5 and 30 μm, and specific surface between <1 to 3 m^2/m^3. Grain size of German lignite fly ash was found to range mainly between D(50) ~30 and 120 μm, and specific

Table 35. Ash particle structures and more frequent mineral phases in hard coal and lignite fly ashes from German power plants

Particle type	Mineral phases		Agglomeration/shape
Quartz particles	Quartz (i)	SiO_2	Clastic fragments
Feldspar particles	K-Feldspar (i)/Na-Feldspar (i)		Clastic fragments
Opaque particles	Magnetite (i)	$FeFe_2O_4$	Usually as finely intergrown microspheres, partly intergrown with glass and other components, partly as isolated tiny crystals
	Hematite (i)	α-Fe_2O_3	
"Fe-oxides"	Maghemite (i)	γ-Fe_2O_3	
	Wüstite (i)	$Fe_{1-x}O$	
	Mg(Al,Fe)-Spinel (i)		
	Rutile (i)	$(Ti,Fe)O_2$	
Glassy particles	Variable silicate glasses (lh-p-i); possibly mullite "$3Al_2O_3.2SiO_2$" (i) content.		Usually as microspheres, also irregular, and fused with other components
Clay particles	Thermally altered clay (illite) minerals („metakaoline") (p).		Usually irregular
Diverse particles	Anhydrite (h)	$CaSO_4$	Tiny isometric crystals, or microspheres, or irregular intergrowths with other components
	Lime (h)	CaO	
	Ca-ferrites (h)	$Ca_2Fe_2O_5$ $CaFe_2O_4$	
	Brownmillerite (h)	Ca-aluminate ferrite	
	Gehlenite *(i)*	$Ca_2Al_2SiO_7$	
	Ye'elimite (h)	$Ca_3Al_6O_{12} \cdot CaSO_4$	Finely intergrown
	Oldhamite (h)	CaS	
	Alite (?) (h)	Tricalcium silicate	
	Belite (?) (h)	Dicalcium silicate	
	Tricalciumaluminate (h)	$Ca_3Al_2O_6$	
	Periclase (h)	MgO	Microspheres, finely intergrown
	Barium oxide (h)	BaO	Finely intergrown
	Na-K-chloride (h)	(Na, K) Cl	Tiny cubic crystals
"Coal particles"	Thermally altered coal (i)		Irregular, usually very porous

Underlined names: frequently main constituents. (i) = mainly inert, (lh) = latent hydraulic active, (p) = puzzolanic active, (h) = highly hydraulic active, or highly reactive or water soluble. A large number of accessory mineral phases, for instance, ye'elimite, may play an important rôle (see also Bambauer et al. 1988). The full spectrum of composition is not present in each of the known ashes.

surface between 0.1 and 0.6 m^2/cm^3. Shape and kind of agglomeration/intergrowth of coal fly ash particles may vary to a great extent, as shown in Table 35 and Figs. 94 and 95.

Normally, FGD gives a rather constant bulk composition of the resulting gypsum masses, and, in general, a load of trace elements similar to natural gypsum. Moisture content is kept at <10 wt.%. But, depending on coal type and FGD process, there may be some variation in morphology and grain size [frequent range D(50) $\sim$20–60 µm] of the gypsum crystals; an example is given in Fig. 96.

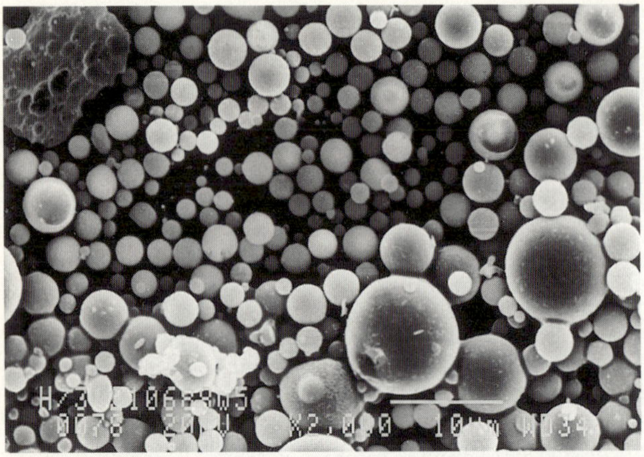

Fig. 94. Fly ash sample form German hard coal-fired power plant (slag tap furnace, Ruhrgebiet), mainly composed of (glassy) microspheres. SEM/SE

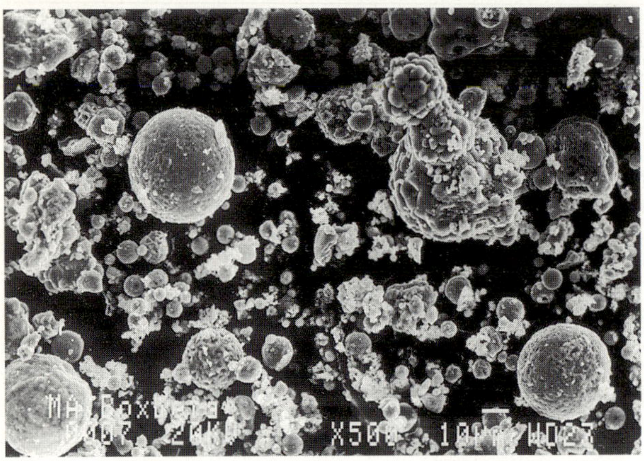

Fig. 95. Fly ash sample from German lignite-fired power plant (Niederlausitz), composed of microspheres and various agglomerates. SEM/SE

Fig. 96. Sample of FGD gypsum from a lignite-fired power plant (Middle German district). SEM/SE

FGD gypsum is increasingly used for various construction purposes, more and more replacing natural gypsum.

Also, a considerable amount of hard coal bottom ash is used for several road construction purposes. Hard coal fly ash is increasingly being used as admixture to Portland cement, and may strongly improve the properties of concrete. In addition, it is being attempted to separate technically valuable constituents (glassy sphere fillers, iron oxides, etc.). There are also attempts for a similar utilization of lignite fly ash, but to date German ashes are chiefly used for landfill purposes.

According to the low-temperature environment reaction scheme in Sect. 5.2.2, a more or less modified raw mix of mainly lignite fly ash is mixed (liquid/solid ratio ~0.2–0.7) with pure water or FGD circulation water (max. 1.3 Mt/a) to give a stabilized solid material by hydraulic reactions. As a result of neomineralization and condensation of microstructure, the solid material may display the following technological properties after a duration of 180–360 days: density of 1.6–2.0 g/cm^{-3}, compressive strength from <5 to ~40 N/mm^2, and a water permeability Kf ~ 10^{-7} to ~10^{-9}. Together with the formation of reservoir minerals, mainly phases of the types ettringite, calcium monosulphate, and calcium silicate hydrate (CSH), brucite, calcite etc., remarkable lowering of leaching of hazardous elements below the limiting values of the German regulations may be effected by this mineralogical barrier system (Sects. 5.1.2, 5.2.2, 5.3.7). In addition, stabilized solids of lignite fly ash, as well as modified slag cement mortar compositions used for the immobilization of low radioactive compounds may "... act both as source and sink for a wide range of ions present in the groundwater" (Glasser 1986). Long-term groundwater resistance of stabilized solids is frequently supported by a strong alcaline buffer capacity and may range down to pH >6. Stabilized solids of certain composition may be used as material for outer or inner barrier systems (see Sect. 5.2.2).

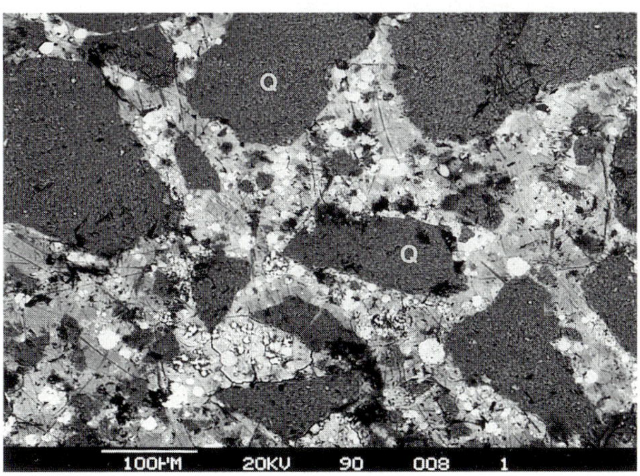

Fig. 97. Example of the dense microstructure of a stabilized solid on the basis of fly ash of a lignite-fired power plant (Rhineland, Germany). Q = Quartz. Polished section SEM/BE

Municipal Waste Incinerator Ashes

Waste incineration reduces municipal waste (see Table 17, Sect. 5.1) by about 85%. Resulting residues are slags and products of physical and chemical flue gas purification, i.e., fly ashes and chemical reaction products ("reaction salts"). According to waste composition, incineration process and flue gas purification installation used, a broad range of chemical and mineralogical compositions of the residues may result (Tables 34 and 36).

For most of the pure powdery fly ashes, particles of <0.5 mm contribute to >92% of the ash volume. Grain size of this main fraction was found to range between D(50) ~75 and 150 μm, and specific surface to be 0.08–0.35 m^2/cm^3, whereas mixed samples fly ash + reaction products were found to range between D(50) ~7 and 73 μm, and 0.25–1.4 m^2/cm^3.

Shape and kind of agglomeration/intergrowth of corresponding ash particles may vary to a great extent, as shown in Table 36.

Whereas the glassy bottom ashes of waste incinerators are often used as material for road construction, the flue gas purification products are classified (according to German regulations) as hazardous waste. Therefore, they have to be safely dumped, for instance in underground mines or open casts, unless effective procedures are available for at least partial extraction of usable constituents. In order to avoid groundwater contamination, conventional outer barriers are in use (e.g., plastic bags), but there are also efforts for the development of additional inner barrier systems for the immobilization of hazardous components (see Sect. 5.2.2). In certain cases, the residues were totally fused to a glass; this energy-consuming procedure is effective in saving dumping space and probably has an immobilization effect. Another example is a strategy similar to that used for lignite fly ash. A raw mix of waste incinerator fly ash (≤50%), with hard coal fluidized bed ash and blast fournace slag cement, both

Fig. 98. Example of fly ash from a municipal waste incinerator (Germany); various ash particles overgrown by various salts. SEM/SE

Table 36. Ash particle structures and more frequent mineral phases in flue gas purification products from German municipal waste incinerators. (Steffes-Tun et al. 1995)

Relicts from untreated waste and mineral phases neoformed during the combustion process		Agglomeration/shape
<u>Aluminum</u>	Al	Irregular-rounded
<u>Quartz</u>	SiO_2	Clastic fragments
Feldspar	Alkali feldspar/plagioclase	
Glass	Frequently Ca-silicatic, variable composition	Microspherers, hollow spheres, irregular shapes
Gehlenite	$Ca_2Al_2SiO_7$	?
Cordierite	$Mg_2Al_4Si_5O_{18}$	?
"Ca-Al-silicate"		?
Hematite	α-Fe_2O_3	Mostly as microspheres, frequently finely intergrown with glass
Magnetite	$FeFe_2O_4$	
<u>"Coal particles"</u>	Themally altered coal, graphite-like	Irregular, porous
<u>Anhydrite</u>	$CaSO_4$	Frequently as isometric crystals
Mineral phases neoformed in flue gas and physical flue gas purification		**Agglomeration/shape**
<u>Calcite</u>	$CaCO_3$	Frequently isometric
<u>Anhydrite</u>	$CaSO_4$	Frequently isometric crystals, crusts
<u>Sylvite</u>	KCl	Salt crusts and isometric crystals
<u>Halite</u>	NaCl	
Bassanite	$CaSO_4 \cdot 0.5H_2O$	?
Mineral phases neoformed during chemical flue gas purification		**Agglomeration/shape**
<u>Basic Ca-chloridehydrate</u>	$CaCl_2 \cdot Ca(OH)_2 \cdot H_2O$	Fine powder-like aggregates
Basic Ca-chloride	$CaCl_2 \cdot Ca(OH)_2$	?
Basic Zn-chloride	$ZnCl_2 \cdot Zn(OH)_4$	Tiny isometric crystals
Ettringite	$Ca_6Al_2(SO_4)_3(OH)_{12} \cdot 26H_2O$	Fine hex.-acicular crystals
CaAl-monochloride-hydrate	$Ca_4Al_2O_6Cl_2 \cdot 10H_2O$	Hex.-tabular crystals
Defernite	$Ca_3(CO_3)(OH,Cl)_4 \cdot H_2O$	?
NaCa-sulphate-hydrate	$Na_2Ca_5(SO_4)_6 \cdot 3H_2O$	?
Relicts of the absorber substances		**Agglomeration/shape**
<u>Portlandite</u>	$Ca(OH)_2$	Fine powder-like aggregates, hex.-tabular crystals
<u>Calcite</u>	$CaCO_3$	Mostly isometric cleavage fragments

Underlined names: frequently main constituents. In addition, a number of undetected accessory mineral phases (beyond X-ray detection limit) is to be expected.

as binder material, is mixed with water (liquid/solid ratio ~0.3–0.5) to give a stabilized solid by hydraulic reactions. Also, lignite fly ash may be used as binder. At the expense of the reactive and easily soluble compounds of Table 36 and the binder materials, in general, the following neoformed main mineral phases develop: ettringite, thaumasite, Ca-aluminate monochloride hydrate, Ca-silicate hydrates (CSH), which mainly act as reservoir minerals. According to bulk composition and amount of neomineralization and condensation of microstructure, compressive strength of the solid material may range from ~5 to up to ~30 N/mm^2 after a duration of 180–360 days; standard leaching tests mainly in the alcaline pH range (>6) give leachate concentrations below the limiting values of German regulations (Schmitt-Riegraf et al 1997). The procedure presented above cannot be applied to the pure reaction salts!

Waste management concepts are often believed to be an engeneering problem only; but it has to be pointed out that a successful management concept for fly ashes has to take their chemical and mineralogical composition into consideration (besides technological properties).

References

Baier H, Steffes-Tun W (1993) Aufbereitung von Steinkohlenflugaschen aus Trockenfeuerungen für Bauzuschlagsstoffe. Brennstoff-Wärme-Kraft 45/5

Bambauer HU, Gebhard G, Holzapfel Th, Krause Chr, Willner G (1988) Schadstoffimmobilisierung in Stabilisaten aus Braunkohlenaschen und REA-Produkten, I. & II. Fortschr Mineral 66, 2:253–290

Glasser FP (1986) Chemical and Geochemical basis for the immobilisation of nuclear waste materials in cements. Fortschr Mineral 64, 1:19–35

Helmuth R (1987) Fly ash in cement and concrete. Portland cement Assoc Res & Development Lab 5420 Old Orchard Road, Skokie, Illinois

Holzapfel Th, Bambauer HU (1987) Flugasche-Aufbereitung: Neue Rohstoffe und Recycling. Tonindustrie-Z, 111 (2):78–83

RWE AG. Zentralbereich Forschung u Entwklg. (Hrsg) (1995) Handbuch d. Verwertung v. Braunkohlenflugaschen in Deutschland. Essen

Schmitt-Riegraf C, Bambauer HU, Baumgarte C, Dülmer M, Heetjans H, Kermer R, Steffes-Tun W (1997) Verfestigung von Rückständen aus der Rauchgasreinigung von Müllverbrennungsanlagen (MVA). II. Experimentelle Versuche zur Schadstoffimmobilisierung. BMFB Statusber Deponiekörper, Wuppertal; UBA/BMFB, Band 2

Steffes-Tun W, Bambauer HU, Baumgarte C, Heetjans H, Rieping D, Schmitt-Riegraf C (1995) Verfestigung von Rückständen aus der Rauchgasreinigung von Müllverbrennungsanlagen. I. Mineralogisch/chemische Zusammensetzung und Eigenschaften der Rückstände. BMBF Statusber Deponiekörper, Wuppertal; UBA/BMFT, Band 1

Thomé-Kozmienski KJ (Hrsg) (1987) Müllerverbrennung und Umwelt 2, EF-Verlag für Energie- und Umwelttechnik, Berlin

VGB Technische Vereinigung der Großkraftwerkbetreiber (ed) (1992) Verwertungskonzept für Reststoffe aus Kohlekraftwerken in der Bundesrepublik Deutschland – VGB-TW 702 –. VGB Kraftwerkstechnik GmbH, Verlag technisch-wissenschaftliche Schriften, Klinkestraße 27–31, Essen

Wirsching F (1991) FGD-Gypsum in research, development, production and application. Proc 2nd Int Conf on FGD and chemical gypsum. Toronto.

For further literature see Section 5.2.2.

5.3.7 Solid-Solution Aqueous-Solution Equilibria in Cementitious Waste Stabilization Systems

M. KERSTEN

Solidification/stabilization (S/S) technology is the process whereby hazardous wastes are treated in a way to reduce their toxicity and to prevent dissolution of the toxic components and their release into the environment. Solidification generally implies the physical encapsulation of toxic residues with cementitious or pozzolanic materials. Solidification may also result in altering the chemical nature of the residues by providing a highly buffered solid matrix which further limits the solubility of toxic components (e.g., neutral to sightly alkaline, oxic conditions with respect to trace metals). On the other hand, stabilization is generally defined as the process to chemically transform soluble contaminants into less soluble form. As a result, many of these processes are referred to as S/S techniques; however, the distinction between the two should be recognized. With respect to the long-term fate of S/S systems, the latter should be preferred, since most of the artificially formed physical barriers or buffer capacities have a limited residence time under ambient environmental conditions.

The principle and most commonly followed way of stabilization of toxic components like trace metals is incorporation into the bulks of less soluble host minerals such as hydrous silicates or phosphates ("reservoir minerals" Sects. 5.2.1 and 5.2.2). The formation of a solid solution is the basic mechanism whereby aqueous solubility of trace components is decreased. Natural minerals always contain substitutional impurities which influence the chemical evolution of natural waters to varying extents, depending on the amount of substitution and its effect on solid-phase free energies. A variety of concepts have been the subject of much debate during the past two decades to define the state of a solid-solution aqueous-solution (SSAS) system at thermodynamic equilibrium, after a dissolution, precipitation, or recrystallization process. It has been shown that the thermodynamic properties of equilibrium and stoichiometric saturation states can be derived in a unifying way, since they correspond to global or constrained minima of the total Gibbs function of the system. Binary SSAS systems can be represented on equilibrium phase diagrams developed on mixed carbonate systems by Lippmann in the late 1970s. The introduction of Lippmann diagrams was an important advance in aquatic geochemistry, enabling prediction of SSAS reaction paths and endpoints. Binary SSAS equilibria in the low-temperature system $CaCO_3$-$MgCO_3$-$FeCO_3$-$MnCO_3$-$SrCO_3$-CO_2-H_2O are among the best-investigated examples, but more recently there have also been increasing studies of toxic metal systems, such as $CdCO_3$-$CaCO_3$-CO_2-H_2O or $PbSO_4$-$BaSO_4$-H_2O. These systems are important also in applied geochemistry, because metal contamination in soils and aquifers often resides as impurities in carbonate, hydroxide, or sulfate phases. In the relatively low-temperature natural water systems, however, prediction of SSAS system behavior is limited by (1) a lack of solid-solution free-energy data, (2) the dependence of SSAS reaction paths and equilibrium endpoints on the solid/solution ratio, and (3) kinetic conditions that cause multicomponent solid solutions to react as one-component solids of fixed composition.

Phase diagrams (for binary gas/binary liquid systems or for solid/melt systems) are typically constructed in terms of mole fractions (and partial pressures in the case of gas/liquid equilibria). Lippmann diagrams (Fig. 99) differ from traditional phase diagram representations in that they use aqueous activity products and activity fractions (rather than elemental concentrations and mole fractions) to characterize the aqueous phase composition, especially with respect to the extent of ion pairing and/or complexation with other ions in the aqueous solution. The effect of speciation is important in the alkaline solutions of S/S systems, where ions that even form complete or nearly complete solid-solution series may not have similar complexing affinities. As a result, an aqueous thermodynamic model will be required in the construction of

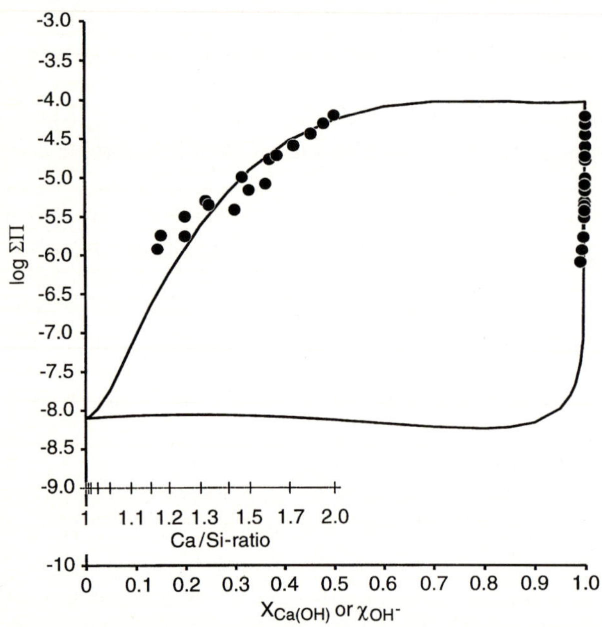

Fig. 99. Lippmann equilibrium diagram of the aqueous solubility in the SSAS system C-S-H $-H_2O$ with the end-member pKs of $CaSiO_3 \cdot 2H_2O$ and $Ca(OH)_2$ and a mixing gap at $Ca/Si > 2.5$. The model curves represent the solidus line of the total solubility product: $\Sigma\Pi_{Solidus} = (\{H_3SiO_4^-\} + \{OH^-\}) \cdot \{CaOH^+\} = (X_{Ca} \cdot \gamma_{CH} \cdot K_{CH}) + (X_{Si} \cdot \gamma_{CSH} \cdot K_{CSH})$ for the solid solution composition, and the solutus line: $\Sigma\Pi_{Solutus} = [\chi_{OH^-}/K_{CH}\gamma_{CH}) + (\chi_{H3SiO4^-}/K_{CSH}\gamma_{CSH})]^{-1}$ for the aqueous solution composition. The black dots represent some of the most cited experimental literature data on aqueous suspensions of C-S-H phases with varying Ca/Si ratios for verification of this model

Lippmann diagrams from experimental aqueous concentration data. Furthermore, two different scales are used for the abscissa on a Lippmann diagram: (1) an activity-fraction scale χ_{aq} relating to the aqueous-phase composition, and (2) a mole-fraction scale X_{SS} defining the solid-phase composition. In contrast, traditional phase diagrams use a single mole fraction scale to define the composition of all phases in the system. User-friendly computer codes such as the MBSSAS have been developed recently to calculate thermodynamic equilibrium, pure-phase saturation, and stoichiometric saturation states in binary (or at least pseudobinary) SSAS systems using two-parameter Margules-type excess free energy of mixing equations. Lippmann phase diagrams can be displayed from the output data files, but also accessory information, such as the location of miscibility gaps, spinodal gaps, critical-mixing points, alyotropic extrema, Henry's law solid-phase activity coefficients, and limiting distribution coefficients. Alternatively, this program can use such mineralogical information to calculate the appropriate excess free energy of mixing equation for any given binary SSAS system.

When Portland cement (OPC) reacts with water, the main product is a rigid gel, of which the principal solid component (about 60%) is a nearly amorphous calcium silicate hydrate (C-S-H). A generally accepted model for the structure and composition of the C-S-H phase in mature hardened OPC pastes is based on a highly disordered layer structure comprising finite silicate dreierketten (chain repeating at intervals of three tetrahedra, with wollastonite, β-$CaSiO_3$, as the simplest of these from the compositional standpoint), in solid solution with a variable amount of portlandite, $Ca(OH)_2$. With the known solubilities of these two end-members, and the fact that an asymmetrical mixing gap is known to occur at Ca/Si ratios >2.5 (i.e., X_{Ca} >0.6), it is relatively simple to draw a Lippmann diagram as a good approximation to describe the SSAS equilibrium behavior in a C-S-H system (Fig. 99). The SSAS equilibrium approach enables prediction of the dependence of the incongruent solubility of the C-S-H phase on the Ca/Si ratio, which is a basic physiochemical property of hardened OPC pastes in contact with aqueous solutions. Some minor components in cementitious S/S systems like ettringite (Aft) and monosulfate (AFm) also form extended multi-component solid solution series (see Sect. 5.2.2). Lippmann diagrams of the SSAS equilibria in these systems may, in principle, be drawn, because most of the end-member phases are congruently soluble and the mixing gaps are relatively well known, but these systems are too complex to be represented by just a few binary SSAS diagrams.

Most trace elements which substitute in C-S-H phases at a crystal-chemical level are octahedrally coordinated and hence substitute for Ca in the $Ca(OH)_2$ layers (or closely admixed AFm phase layers in the case of three-valent ions). The aqueous solubility of these metal impurities can therefore be best represented by a pseudobinary SSAS Lippmann diagram of the metal-calcium hydroxide system. This finding was important for the assessment of the long-term fate of cementitious S/S systems, because the effect of carbonation of these systems may then be represented as just a change from the hydroxide to the carbonate SSAS system. A comparison of the $Ca_XZn_{1-X}(OH)_2$ and $Ca_XZn_{1-X}CO_3$ systems, for example, revealed that no significant change in aqueous Zn solubility is to be expected upon this change in the host phase composition.

References

Glynn PD (1991) MBSSAS: a code for the computation of Margules parameters and equilibrium relations in binary solid-solution aqueous-solution systems. Comput Geosci 17:907–966

Glynn PD, Reardon EJ (1990) Solid-solution aqueous-solution equilibria: thermodynamic theory and representation. Am J Sci 290:164–201 (Comment by Königsberger E, Gamsjäger H (1992) in Am J Sci 292:199–214, Reply by Glynn PD, Reardon EJ (1992) in Am J Sci 292:215–225

Kersten M (1996) Aqueous solubility diagrams for cementitious waste stabilization systems. 1. The CSH solid-solution system. Environ Sci Technol 30:2286–2293

Lippmann F (1980) Phase diagrams depicting the aqueous solubility of binary mineral systems. N Jb Mineral Abh 139:1–25

Spence RD (1993) Chemistry and microstructure of solidified waste forms. Lewis Publ, Boca Raton, 241 pp

Taylor HFW (1992) Tobermorite, jennite, and cement gel. Z Kristallogr 202:41–50

5.3.8 Clays and Zeolites: Sorption and Exchange Properties and Their Implication in Environmental Problems

E. Kohler †

Clays were initially defined as consisting of particles less than 2 µm in diameter, beyond the limit of microscopic resolution. The English word clay is used for a variety of very different geological materials. The most important constituents of clays are clay minerals with a distinguished crystallographic habit which is sheetlike. Other minerals present in the clay fraction less than 2 µm in diameter include quartz and oxide minerals as well as minor portions of all other weathering products found in the geological environment. Zeolites are often found together with clay minerals in clayey sediments. Both groups – clay minerals and zeolites – are members of different mineral families which should be strictly distinguished. Zeolites are aluminosilicates with a framework structure enclosing cavities occupied by large ions and water molecules, both of which have considerable freedom and movement, thereby permitting ion exchange and reversible dehydration. Some of the clay minerals have similar properties: the sepiolite-palygorskites contain water in channels in their structure which is easily released by heating. We call it zeolite water. The devitrification of volcanic ash can produce other types of materials, one highly zeolitic with varying amounts of clay minerals (tuffs) and another dominated by swelling clay minerals, notably dioctahedral smectites (bentonites). One of the

Table 37. Table of clay minerals

	Dominant elements
Swelling types	
Smectites	
Beidellite	Al
Montmorillonite	Al (Mg, Fe^{2+} minor)
Nontronite	Fe^{3+}
Saponite	Mg, Al
Vermiculite	Mg, Fe^{2+}, Al (Fe^{3+} minor)
Mixed layer minerals[a]	
Nonswelling types	
Illite	K, Al (Fe, Mg minor)
Glauconite	K, Fe^{2+}, Fe^{3+}
Celadonite	K, Fe^{2+}, Mg, Fe^{3+}, Al^{3+}
Chlorite	Mg, Fe, Al
Berthiérine	Fe^{2+}, Al, (minor Mg)
Kaolinite	Al
Halloysite	Al
Sepiolite	Mg, Al
Palygorskite	Mg, Al
Talc	Mg, Fe^{2+}

[a] Two or more types of basic layer interstratified in the same crystal.

striking characteristics of the bentonites is the monomineral assemblage of almost pure aluminous smectites. Smectitic bentonites are extensively used in industry for their swelling and gel-forming properties. The relatively low purity of most zeolite deposits and the absence of appropriate natural zeolitic species prevent sedimentary zeolites from being used to replace the more expensive synthetic zeolite minerals in industrial use such as catalysis and laundry detergents. Clays like bentonites and sedimentary zeolites, however have achieved great importance in any kind of pollution control, e.g., radioactive-waste disposal, conventional waste deposit construction, sewage-effluent and agricultural waste-water treatments, stack-gas and oil-spill clean-up, and in other kinds of chemical engeneering. The effects of clays and clay minerals as well as of zeolites are based on the reversible dehydration, ion exchange, and adsorption properties and on the ability of the minerals to exchange selectively large cations from aqueous solutions.

According to the major characteristics of clay chemistry and structural relations, one can construct a rough table of clay minerals (Table 37).

The zeolites are not as yet adequately classified. The following general classification is useful:

1. Natrolite group, which includes natrolite, mesolite, scolecite, thomsonite, gonnardite, and edingtonite. A chainlike unit is fundamental to the structures within this group, of which all members have a fibrous morphology.

2. Harmotome group, which includes harmotome, phillipsite, gismondine, and garronite. The silica tetrahedra of this group form chains similar to those in the feldspars, except that the linkage of the chains is different. The network contains many fourfold and eightfold rings, the latter constituting the channel openings.
3. Chabazite group, which includes the zeolites chabazite, gmelinite, levyne, and erionite. Single and double sixfold rings of tetrahedra perpendicular to a triad or hexad axis are found in this group. Eight- and 12-fold rings also occur, forming wide channel systems.
4. Faujasite group, which includes faujasite and some synthetic zeolites, e.g., Type A. Here, the tetrahedra are linked to form cubo-octahedral cage-like units which, in turn, are joined so as to give the structure a cubic symmetry.
5. Mordenite group, which includes the minerals mordenite and dachiardite. The characteristic feature of these structures is a chain containing five-fold rings of tetrahedra. These chains can be linked in a variety of ways. Wide channels are formed by 12-fold rings in mordenite and by 10-fold rings in dachiardite.
6. Heulandite group, which includes heulandite, clinoptilolite, stilbite, epistilbite, ferrierite, and brewsterite.

Clay minerals and zeolites are dominantly cation exchangers. The anion exchange of these materials involves the structural hydroxyl ions of the sheet-silicates and the hydroxyl ions from the silanol (Si-OH) groups on the zeolite surface.

Certain cations can be fixed so firmly on clay minerals that they are virtually fixed. This phenomenon is known to occur with K^+, NH_4^+, Rb^+ and Cs^+. Caesium is generally very strongly held by major clay minerals, which is of advantage for contamination of water containing radioactive caesium. The nature of the exchangeable cations may often be determined approximately by measurement of the pH:

pH 9 : Na^+
pH 7.5 : Ca^{2+}
pH 7 : H^+.

At neutral conditions the rate of cation exchange varies with the type of clay mineral, cation concentration and type and concentration of the anions:

Kaolinite:	very rapid
Montmorillonite and vermiculite:	slow
Illite and chlorite:	very slow.

Order of exchange capacity of some cations:

$$H > Al > Ba > Sr > Ca > Mg > NH_4 > K > Na > Si.$$

Cation exchange capacity	Mineral	Anion exchange capacity
3–5 mval/100 g	Kaolinite	6–20 mval/100 g
5–10 mval/100 g	Halloysite	
80–150 mval/100 g	Montmorillonite	20–30 mval/100 g
10–40 mval/100 g	Illite	
100–150 mval/100 g	Vermiculite	4 mval/100 g

The cation exchange capacity of zeolites ranges from about 230 mval/100 g (mordenite) to about 620 mval/100 g (thompsonite). There are enough metal cations, such as Na^+, K^+, Ca^{2+}, Mg^{2+} or Sr^{2+} present in the interstices of the cross-linked, polymeric macromolecules of the zeolites to make the crystal electrically neutral. The anionic lattice of zeolites is manifold, complex and usually not very acid-resistant. This prevents their application in acid solution below pH 5. However, the acid resistance rises with increasing Si/Al ration in the zeolite structure. Therefore erionite (Si/Al = 3) is more acid-resistant than chabazite (Si/Al = 2). For application as radioactivity scavengers, it is of the greatest importance that zeolites are resistant towards ionizing radiations and to neutral waters at temperatures up to 100 °C.

Clays and clay minerals are of great importance in the construction of waste disposals. Nonswelling kaolinitic clays and high-swelling bentonites represent excellent materials for technical and/or natural barrier systems due to the small grain size of the clay particles, the large specific area and their chemical reactivity. The long-term stability of clay materials is much more favorable than that of zeolitic sediments. One of the most important parameters concerning the safety and long-term working of any landfill construction is the capacity of adsorption and permeability.

References

Gottardi G, Galli E (1985) Natural zeolites. Springer, Berlin Heidelberg New York, 409 pp

Müller-Vonmoos M, Kohler EE (1993) Geotechnik und Entsorgung. In: Jasmund K, Lagaly G (eds) Tonminerale and Tone – Struktur, Eigenschaften, Anwendung und Einsatz in Industrie und Umwelt. Steinkopff Verlag, Darmstadt, pp 312–357

Saha AK (1985) Natural zeolites and clays – the promising media for selective radioisotope partition/immobilisation. Doc Symp Waste Mange vol 3, pp 197–201

5.3.9 Clay Liners for Waste Dumps

D. HELING

The purpose of the base seal of a waste deposit is to prevent the contaminated leachates from entering the groundwater circuits. This is achieved by a high imperviousness of the seal for aqueous solutions as well as by retention of the contaminants by adsorption. The sealing liner must resist any reduction of its barrier functions for 100 years or more.

Reduction of the imperviousness of clay mineral liners can be caused by alterations of the mineralogical composition or the microstructure. Diagenesis of clay liners includes all alterations of both the mineral contents and the fabrics of the liner soils after construction. Diagenesis is effected by the reactions between the soil components and the aqueous pore solutions, i.e., the percolating leachates. Initially present mineral components of the liner soil (primary minerals) will dissolve if they become instable due to the change in the chemical composition of the pore solution. If the pore solution becomes oversaturated with respect to a certain compound, the components of this compound will recombine forming a new (secondary) mineral by precipitation.

The rate of diagenetic reactions is controlled mainly by the flow rate of the solutions through the pore channels, implying that both the hydrostatic head and the permeability coefficient of the liner are the determining factors. The rate of flow is controlling both the supply of dissolved matter and – in the case of a dissolution – the removal of the primary matter. Due to the permeability, a certain diagenetic reaction can proceed at widely varying rates, e.g., instantaneously or very delayed.

The most common diagenetic reactions in waste deposit liners are cation exchanges on the surfaces of clay minerals (predominantly with smectitic clays) and/or of humic particles. The exchanging cations would diffuse to the exchange sites along a concentration gradient, whereas the exchanged ions would take the opposite direction. Most commonly, Ca-smectite would be altered to Na-smectite, when the pore solutions contain sufficient NaCl, the most frequent component of waste deposit leachates. Na-smectites are more finely dispersed – and therefore create less permeable fabrics – than the Ca-clays (Jasmund and Lagaly 1993).

The exchange of Na^+ by Ca^{2+} with a smectitic clay increases its shear strength, and consequently decreases the plasticity, diminishing the self-healing-potential of the clay.

Swelling can also occur in Ca-sulfate-bearing soils. When anhydrite is transformed into gypsum there is a volume increase of up to 61 %. By this process the compacted fabric is disintegrated. The transformation normally takes a long period of time. It is accelerated if thin streaks or laminae of clay are incorporated in the sulfatic soil. To minimize the detrimental effects of swelling gypsum on the clay microfabric, the use of anhydrite-bearing soils for the construction of liner beds should be avoided as far as possible.

The most important diagenetic alteration of smectite is the illitization of smectite. Echle et al. (1988) report a striking example of diagenetic illitization from a waste deposit having been in operation for 10 years. The leachates were highly concentrated with K^+. The illitization depends mainly on the supply of potassium from circulating pore solutions. The process includes the exchange of Ca^{2+} – as the most frequent interlayer cation of smectites in natural environments – for K^+, and the substitution of some of the tetrahedral Si^{4+} by Al^{3+}. In this way, the smectite is transformed to illite passing through an intermediate phase of mixed-layered smectite/illite (I/S). Fig. 100, taken from Echle et al. (1988), shows the increase in K^+ and decrease in the SiO_2/Al_2O_3 ratio.

By the uptake of potassium, the smectites gain in layer charge but lose their expandability, and part of their plasticity. In the long run, small smectite particles can aggregate to larger stacks of illite.

Since all the cation exchanges are diffusional processes, the illitization of smectite is greatly stimulated by elevated temperatures, a factor which is, however, rather unimportant in near-surface waste deposits. The rate of illitization is the greater the smaller the grain size of the smectite, because with small grain sizes the K^+-ions will reach the exchange sites over much shorter distances than larger crystallites. Illitization of smectite has been reported even from superficial soils where the necessary K^+ was supplied by fertilizers.

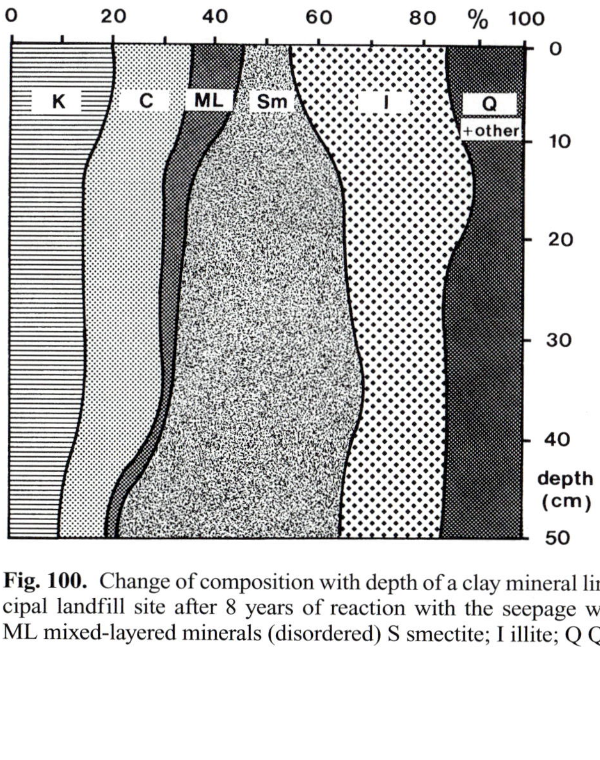

Fig. 100. Change of composition with depth of a clay mineral liner (fract. < 6 μm) from a municipal landfill site after 8 years of reaction with the seepage waters. K Kaolinite; C chlorite; ML mixed-layered minerals (disordered) S smectite; I illite; Q Quartz (After Echle et al. 1988)

Frequently, the leachates are acid. In such cases, particulate carbonates, which are often present in liner soils, will be readily dissolved, producing additional (secondary) pore spaces by leaving molds of the dissolved carbonate particles. Moreover, the escaping CO_2 can enlarge existing or even break up new flow channels by a process called fracing. Consequently, the permeability is unfavorably increased further. However, before such fabric damage can commence, the acid solutions have to penetrate through the overlying thickness of the liner. With a low permeability liner in combination with a low hydrostatic head of the leachates, extensive periods of time will elapse before the leachates can penetrate through a liner of normal thickness. By this consideration it becomes evident that a well-operated drainage system, keeping the hydrostatic head of the leachates continuously low, is essential for the long-term effectiveness of a waste deposit liner.

The precipitation of Fe-hydroxides deserves special attention. The solubilities of Fe-hydroxides are pH/Eh-sensitive. According to the stability relations of the system Fe-CO_2-H_2O, siderite ($FeCO_3$) is the stable phase at both pH >7 and reducing conditions (Garrels and Christ 1965). $FeCO_3$ is transformed to $Fe(OH)_3$, from which, in turn, $FeCO_3$ can be reconstituted by addition of CO_2 (for instance from the decay of organic matter or dissolution by acids), provided a negative Eh is prevailing. The dissolution of Mn- and Fe-oxides can also be increased by lowering the Eh.

Fe- and Mn-compounds can adsorb heavy metal ions from the leachates. Thus, processes of either coprecipitation or – on the contrary – remobilization of heavy metal ions along with the Fe/Mn-hydroxides are rather common and can become important reactions within waste liners. The precipitation of Fe-hydroxides would decrease the permeability favorably, but would diminish the plasticity properties of the liner unfavorably, thus reducing the self-healing potential of plastic clays.

With leachates bearing inorganic electrolyte, the natural interlayer cations, in particular those of the three-layer silicates, will be exchanged for the prevailing cations of the solution. From sulfatic permeates, low soluble sulfates would precipitate. With acid solutions the octahedral Al is subject to weak dissolution. Furthermore, Fe-hydroxides are dissolved completely by acids. With alkaline solutions the tetrahedral layers preferably are subject to decomposition. In spite of such corrosion processes, all the laboratory tests quite unexpectedly show that the permeabilities of clay beds permeated by either acid or alkaline solutions (pH ranging from 3.5 to 9) appear to remain unaffected. The reasons for such behavior are not fully understood as yet.

The most hazardous compounds of waste deposit leachates to spoil groundwater reservoirs are of organic nature (e.g., PCBs, CKWs, PAKs). The effects of organic compounds on the liner clays are influenced mainly by the water solubility of the organics. As proved by many laboratory tests, pure organic liquids will increase the hydraulic conductivity by shrinkage and cracking. However, dilute organic solutions have essentially no effect (Madsen and Mitchell 1989).

In order to increase the adsorption of nonaqueous organic compounds, layers of organophilic bentonites have been incorporated into the base seal, forming multilayered composite liner systems. Organophilic bentonites hold high adsorption capacities for organic compounds. There are great differences between the adsorption effects with regard to the combination of adsorbents and adsorbates. For known pollutants a best-suited adsorbent can be determined. The sealing effect of a liner system can be increased by plugging narrow flow channels when organophilic clays combine with high-molecular organic compounds forming expanding organobentonite complexes. The shear strength of soils was found to be unaffected when blended with organo-bentonite as compared to soil samples lacking organo-bentonite (Stockmeyer 1993).

Dealing with various types of clays in connection with very different leachates, it should be emphasized that none of the diagenetic alteration processes involving supply or removal of ions by the pore solutions can commence before the ions to be supplied arrive at the reaction sites. The ions can migrate by either viscous flow or diffusion. The importance of diffusional mass transport is often overestimated, in particular, when considered to explain mass transport over distances of more than 1 mm (NYE 1979).

Both modes of mass transfer in a porous fabric, advective flow and diffusion, depend principally on permeability and tortuosity. Thus, as long as the front of the penetrating leachates has not passed through the total liner thickness, the seal will stay tight. In other words: the moment of the breakthrough point will be delayed the more, the lower the primary permeability. The dominant requirement for the construction of a liner should be minimizing its permeability as greatly as possible. To accomplish this, an intensive compaction of the liner soil, as well as the selection of a finely dispersed Na-bentonite for the liner, are the most important requirements.

Today's common state of the art in constructing waste deposit base seals is to make use of a combination seal, i.e., a clay liner is combined with a tight sheet of plastic (HDPE) foils to be installed above the clay (Fig. 101). At least when being applied to hazardous waste deposits, the clay liner itself should be composed of a (fine-grained) kaolinitic clay layer, which is overlain by a layer of strongly swelling Na-bentonite. The smectite of the bentonite is supposed to retain heavy metals by adsorption and precipitation. Thereby, the bentonite layer might lose some of its imperviousness by shrinkage fissures or cracks, but then the underlying kaolinite, being less susceptible to chemical corrosion, will protect the deeper bed rock strata from being entered by the leachates. To extend the long-term persistency of extremely hazardous waste deposits, Weiss (1989) has suggested including an additional layer of organophilic smectite in the sealing system in order to retain heavy organic molecules. The organophilic smectite will keep organic compounds from proceeding through the pore spaces of the clay fabric. Another advantage of organophilic smectites is the sorption of metabolic products of microbes, which can populate densely within clay liners under anaerobic conditions. Such compounds can affect normal smectites unfavorably in certain cases.

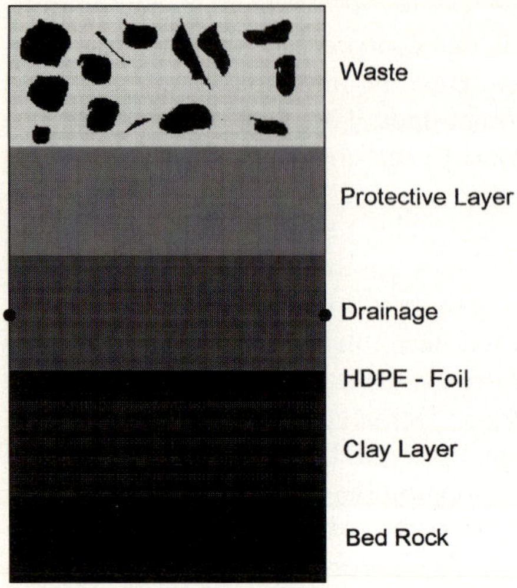

Fig. 101. Schematic section through a combination base seal of a waste deposit

Opposing this model of a multilayered clay seal system, it has been argued that a well-compacted layer of smectite alone, but of the same thickness as the combined kaolinite and smectite layers together, would retard a breakthrough of the leachates for a longer time than the corresponding double clay layer seal. Thus, it is considered a more effective practice to increase the thickness of the smectitic bed by the thickness of the kaolinite layer and to delete the kaolinite. However, smectite or bentonite of liner soil quality (for instance Deponit or Montigel) are more expensive than kaolinite.

In order to calculate the thickness of the clay layers required to resist a breakthrough of the leachates for a maximum of time, it must be taken into consideration that the total adsorption capacity of the clay included in the seal should meet with the supply of contaminants released by the leachates entering the seal. The concentration of the contaminating compounds in the leachates would decrease exponentially during the period of operation. Since neither the concentration of the contaminants carried by the wastes to be deposited, nor their solubility, is known at the time of construction, only an average value for the amount of contaminants can be taken for a rough estimation.

For industrial waste deposits in the Federal Republic of Germany the instructions of the TA Abfall provide a minimum thickness of 150 cm with a permeability coefficient of 5×10^{-10} m s^{-1}. For municipal waste deposits, a minimum of 50 to 75 cm thickness of the same permeability as above is laid down as a rule. Both specifications hold for dual-layered composite seals (TA Abfall; TA Siedlungsabfall, vgl. GDA 1993). The clay should be spread in sheets of not more than 0.3 m in thickness, and then be compacted intensely before the next sheet is to be superimposed.

In selecting the most appropriate sealing system for a given deposit, the geologic barrier (i.e., the underlying bed rock) must be included in the consideration. Any clayey basement seal has to be the thicker, the higher the gross permeability of the bed strata, (possibly due to fractures, coarse grain size, subrosion, or the like). Resuming the above statements, it is stressed again that it is most essential for a reliable seal to select clays of high quality and to compact the clay thoroughly and carefully when constructing the site.

References

Echle W, Cevrim M, Düllmann H (1988) Clay mineralogical, chemical, and physical changes in a basal clay liner – test area of Geldern-Pont waste deposit. In: Czurda KA, Wagner JF (eds) Tone in der Umwelttechnik. Schr Angew Geol Karlsruhe 4:99–121

Garrels RM, Christ CL (1965) Solutions, minerals, and equilibria. Harper & Row, New York

GDA (1993) Empfehlungen des Arbeitskreises "Geotechnik der Deponien und Altlasten" – GDA. – 2. Aufl, Ernst & Sohn, Berlin

Jasmund K, Lagaly G (eds) (1993) Tonminerale und Tone. Steinkopff, Darmstadt

Madsen FT, Mitchell JK (1989) Chemical effects on clay hydraulic conductivity and their determination. Mitt Inst Grundbau u Bodenmech, ETH Zürich, Nr 133, Zürich

Nye PH (1979) Diffusion of ions and uncharged solutes in soils and soil clays. Adv Agron 31:225–272

Stockmeyer MR (1993) Organophile Bentonite als Komponente in Deponiebarriere Systemen. Mitt Inst Grundbau u Bodenmech, ETH Zürich, Nr 202, Zürich

Wagner JF (1988) Mineral transformations during the migration of heavy metal solutions through clay rocks. In: Czurda KA, Wagner JF (eds) Tone in der Umwelttechnik. Schr Angew Geol Karlsruhe 4:47–62

Weiss A (1989) About sealing of waste disposals by clays with special consideration of organic compounds in percolating water. Appl Clay Sci 4:193–209

5.3.10 Conservation of Building Stones – Keystones of Their Deterioration and Conservation

R. Snethlage

Since early antiquity, the deterioration of building stones has been a well-known problem in the preservation of the cultural heritage. The Great Sphinx, which had been covered by sand for centuries was excavated and restored for the first time by Thutmosis IV (1400 BC). Vestiges of this intervention can still be seen today. The cathedral stonemasons workshops – some of them existing since the erection of the cathedral – demonstrate the necessity of steady maintenance and repair. According to the opinion of most scientists, the damage rate to historical objects has exponentially increased since the beginning of industrialization in the last century. In fact, many photographs taken around the turn of the century seem to prove the enormous progress of deterioration in monuments during the past 100 years. However, many monuments can also be men-

Fig. 102. Sandstone cross at Oberpleichfeld near Würzburg Germany. (Photo Lantz 1978)

tioned whose state was already alarming in the last century, e.g., the sculptures of the Fürstenportal at Bamberg Cathedral. Because there is not sufficient documentation on the state of monuments, our knowledge today is too small to decide whether an increased deterioration has taken place during the past 100 years or not (see Fig. 102).

Historical monuments reflect the development of air pollution during their life history. The damages that we observe today are the consequences of a long-term accumulation of damaging substances and growing damage factors. Exposure programs started during the past 15 years show a correlation between the measured material loss and air pollution. Several authors estimated damage functions for SO_2 and H^+, especially for calcareous stones. A synergistic effect of SO_2 and NO_x on the damage risk was found. The susceptibility of different stones to various gaseous air pollutants becomes evident by the different deposition rates, e.g., for SO_2, NH_4^+ and NO_x, which depend on the physical and chemical properties and on the moisture content of the stone surface. Both parameters are very much influenced by the formation of biofilms, which enhance the deposition of particulate matter and gaseous pollutants. Further information can be found in Livingston (1997) and Mirwald and Brüggerhoff (1997).

Scientific investigations in the deterioration and conservation of stones began in the last century and depend on the progress in chemical research. In particular, conservation was mainly concerned with the preservation of the works of art in museums. Bibliographies on the use, deterioration, and conservation of natural stone were presented by Lewin (1966) and Riederer et al. (1993).

Deterioration of Natural Stones

Deterioration in the open air is characterized by strong temperature and moisture changes that cause displacements of the components in the grain structure and weaken the intergranular forces. To understand the deterioration of natural stones, it is therefore important to know the driving forces of the different weathering mechanisms and how intrinsic and external parameters interact with each other.

Intrinsic Parameters

The durability of a stone is primarily determined by its grain structure. Stone with direct line or area contacts from grain to grain are more durable than those with point contacts. Weatherable minerals like carbonates, feldspars, and clays enhance their susceptibility for environmental influences due to a release of alkaline ionic species and subsequent formation of salts. Among the sandstones, those with a quartzitic cement are more resistant than those with carbonatic or clay matrix. The carbonatic matrix can easily be dissolved by acidic pollutants. Clay minerals swell and shrink in the course of the drying and wetting cycles. The displacements caused by this process loosen the grain-to-grain cohesion and lead to a reduction in strength.

The behavior of limestones is different from that of sandstones in the respect that for many types the carbonate crystals are connected by real calcitic bridges. Therefore, the grain size of the calcitic cement is of major importance. The smaller the crystals, the more they are prone to react quickly to acidic air pollutants. Because of the rapid reaction of calcite with acidic air pollutants, the deterioration of limestones is mostly limited to the surface.

The marble structure is characterized by plane crystal surfaces, platelike pore shapes, and a very low porosity. A marble structure normally shows a high degree of indentation of the individual grains. To destroy a marble, the crystals must be widely separated from each other so that they can be removed from their place. Probably the anisotropic thermal expansion of calcite is a deciding parameter.

Pore-size distribution influences the durability in a very complex manner. A high porosity does not necessarily mean lower durability of the stone. Coarse porous stones with a high porosity are very resistant to salt weathering and frost, whereas those with a bimodal pore size distribution are especially endangered. High contents of micropores cause frequent capillary condensation of moisture. This is always accompanied by high swelling during wetting and contraction during the drying. With time, these displacements lead to a weakening of the whole grain structure.

External Parameters

Natural stones on buildings are exposed to the surrounding climate. Moisture changes in the hygric region (air humidity changes) as well as in the super-hygric region (liquid water) cause a hygric dilatation of the stone. For sandstones, the hygric dilatation is in the order of 500 µm/m but can also reach several thousands µm/m for clay-rich varieties (see Fig. 103). The displacements due to moisture changes are regarded as one of the most important damage parameters.

The impact of air pollutants like SO_2, NO_x, and dust particles is very harmful to the stone. In water, SO_2 and NO_x react to sulfuric and nitric acid, which dissolve the stone substance. A very well-known example is the formation of gypsum from calcite. However, more resistant minerals like feldspars are also attacked. The formation of mixed-layer chlorite-smectite in the acidic environment of building stones is clearly proved.

The reaction of limestone surfaces with air pollutants is different from that of sandstones, due to the high buffering capacity of the carbonatic surface. Additionally, many limestones show a low capillarity so that acidic pollutants are already neutralized on the very surface. The formation of sulfated black crusts on limestones has been very often described. A special galvanic cell

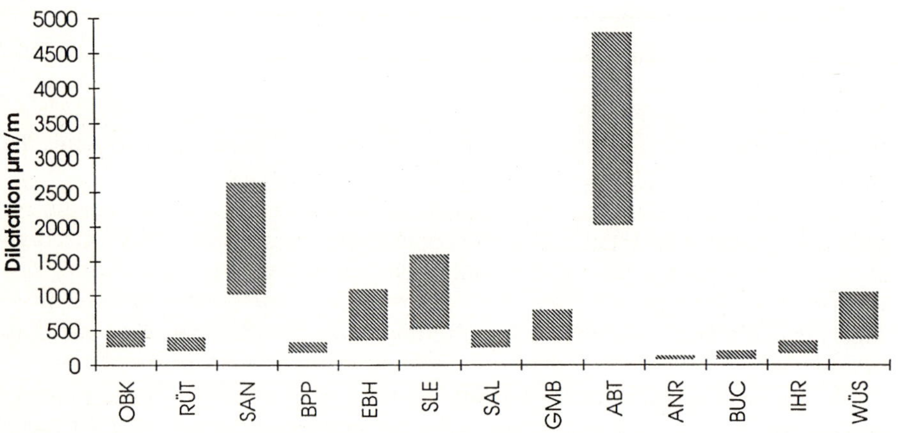

Fig. 103. Range of hygric dilatation of selected sandstones

model for the sulfatation of calcite has been developed by Skoulikides and Papakonstantinou (1981). In agreement with this model, the pseudomorphosis of gypsum to calcite on marble surfaces was described. A major problem in the conservation of limestone and marble in the mediterranean countries is the formation of oxalate crusts, whose origin is not yet fully understood.

The absorption of dust particles changes the water transport ability of the surface so that the drying process is delayed. Additionally, the absorbed dust layer is a high absorber for gaseous pollutants and humidity due to its large internal surface. Mass balance calculations of outdoor experiments show that the dry deposition of pollutants exceeds the wet deposition by rain.

The salts formed by the acidic air pollutants dissolve, migrate, and crystallize in the pore space, depending on the moisture content. By observation, it is evident that stone damage on buildings is always correlated with increased salt concentrations. Many authors ascribe salt damage to the linear crystal growth pressure and the hydration pressure of salts. Additionally, a correlation between pore size distribution and the resistance of sandstones to salt damage has been found. A comprehensive compilation of the different opinions and the present state of knowledge is given by Duttlinger and Knöfel (1991).

Recently, Snethlage et al. (1996) and Snethlage and Wendler (1997) derived a new model for salt weathering on the basis of colloid chemistry. They found that stones contaminated with NaCl contract during wetting and expand during drying, whereas salt-free stones behave in the opposite way (see Fig. 104). This behavior can be explained by the contraction of hydration shells due to the salinity of the pore solution. The displacements in the grain structure are not reversible, but increase from cycle to cycle. Therefore, it can be concluded that displacements in the grain structure due to highly saline solutions alone are the fundamental reason for every kind of deterioration.

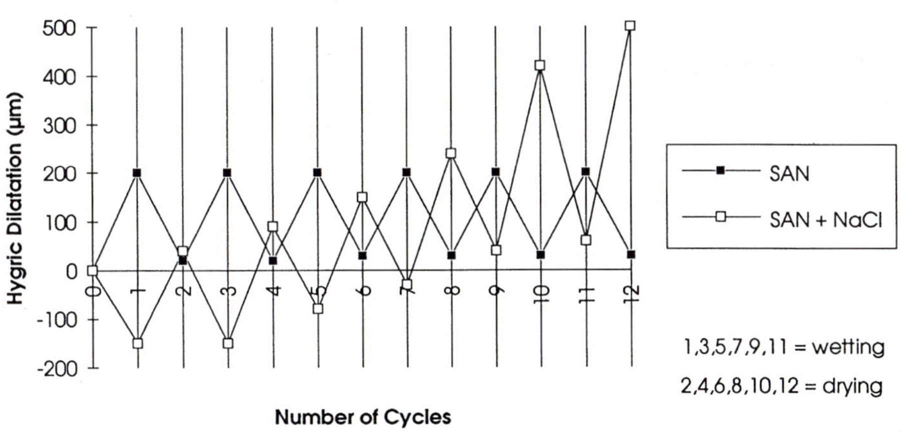

Fig. 104. Hygric dilatation of Sander Schilfsandstein in cycles between 35 and 90% relative humidity

Biological colonization is especially important at the beginning of the deterioration process. Wet biofilms act as absorbers for particles and gaseous components and enhance the formation of black deposits. Biofilms can change the appearance of buildings completely. Organisms are also harmful by the formation of acids. The formation of nitric acid by nitrifying bacteria, and of organic acids, which can cause biopitting is of particular interest.

Biological matter is a major constituent of black crusts. Two types of black crusts occur on buildings; the one bound on sheltered areas very frequently occurs on limestones, marbles and carbonatic sandstones (so-called gypsum crust); the other is found on exposed and wet building parts like pillars and buttresses and occurs only on stones with very resistant surfaces like quartzitic sandstones. Contrary to this, limestone is clean at exposed parts because dust and gypsum are washed away by rain. The biological matter, which seasonally dries out and revives, is responsible for the tough consistency of the black crusts.

During the past 10 years, a comprehensive model for the mapping of decay phenomena on buildings has been developed. The phenomena are divided into five major groups: relief, sanding, exfoliation, scaling, and crust. Transition forms and damage intensities further specify the possibilities of description. Each damage type is defined by a characteristic color so that the kind and degree of damage can easily be read.

In addition to the general observation that moisture in combination with salts is the fundamental cause of every kind of deterioration, the place where the damage will happen can be predicted. Computer simulation calculations, as well as measurements on test walls, show that during wetting-drying cycles the mean moisture distribution shows a maximum not at the surface but at a certain depth (see Snethlage and Wendler 1997). The position of this maximum depends on the moisture transport properties of the stone itself and on the surrounding climatic conditions (see Fig. 105). The maximum also coincides with the zone where last pore solutions evaporate and salts crystallize. If the maximum of the mean moisture distribution curve in Fig. 105 is situated below the surface, a scale will form. If the maximum is situated in or very near the surface, the stone surface will sand off. In the case of black crusts, the position of the maximum lies under the black crust and is responsible for the destruction of the stone surface and the detachment of the crust. The position of the moisture maximum has also important consequences for the treatment of stones with conservation products because it determines the required penetration depth.

Methods of Conservation

Conservation follows the main aim of preserving the original substance for future generations. The monument should show its aged state. A historic monument is not only a witness of former periods of art but also a document

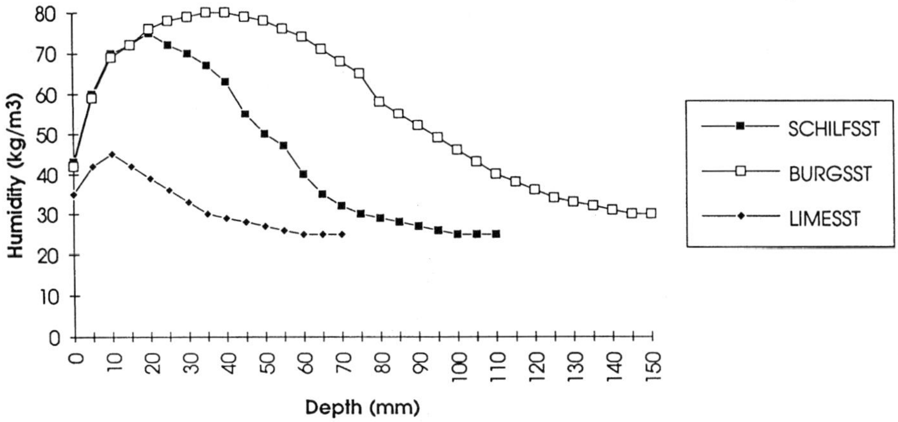

<u>Burgsandstone porous</u>	<u>Sander Schilfsandstone</u>	<u>Lime-Sandstone dense</u>
maximum water content 150 kg/m^3	maximum water content 130 kg/m^3	maximum water content 70 kg/m^3
equilibrium water content 25 kg/m^3	equilibrium water content 25 kg/m^3	equilibrium water content 25 kg/m^3
temperature air 20 °C	temperature air 20 °C	temperature air 20 °C
rel. humidity air 25 %	rel. humidity air 25 %	rel. humidity air 25 %
1 cycle: rain 2 h, drying 22 h	1 cycle: rain 2 h, drying 22 h	1 cycle: rain 2 h, drying 22 h
finite element thickness 1 mm	finite element thickness 1 mm	finite element thickness 1 mm

Fig. 105. Curve of the mean moisture distribution in selected sandstones after three cycles of wetting and drying

of technical and technological history. Methodologies of quality control must be introduced into the conservation of historic monuments. VDI guideline 3798 (1989), Snethlage and Wendler (1996) Sasse and Snethlage (1997), and Snethlage (1997) describe the laboratory investigations which are indispensible to secure the quality of conservation treatments. Practical advices for conservators-restorers are given and illustrated by Ashurst and Dimes (1990). The main aspects of conservation measures and their quality requirements are discussed in the following sections.

Stone Cleaning

The cleaning of stone has to follow two aims

- removal of deleterious substances, especially salts
- opening of pores to reestablish the water vapor transport and the capillarity for absorption of liquid stone-protecting agents.

The visible effect is no appropriate measure to assess the quality of a cleaning.

Cleaning with pure water spraying can be sufficient in the case of gypsum-rich crusts, which are softened by water. High-pressure water vapor (140–180 °C, 20–40 bar) increases the effect of dirt removal because oily substances are more easily dispersed. High-pressurized cold water does not improve the cleaning. All water-cleaning methods cause a risk for loose parts, scales, and disintegrated surfaces. Water can also deeply penetrate into the masonry at defect joints.

Dry cleaning methods show several advantages. However, sand blasting with coarse and split quartz and pressures up to more than 10 bar are not allowed for historic monuments. Modern dry-particle blasting techniques work with pressures up to 2 bar, fine sand with a grain size smaller than 100 μm. Small nozzles allow the cleaning of hidden parts of ornaments. Other recommendable particles are corundum, limestone powder, glass spheres, crushed furnace slag, plastic granulates, etc. The grain size of these materials is in the order of 50 μm. The blown dust is directly soaked up by vacuum cleaners. In some cases, sharp-edged particles clean more carefully, as they remove tough dirt and soot with less pressure than rounded particles. On light stones like marble and limestones, good results are obtained with Nd-YAG laser beam cleaning.

Joint and Repair Mortar

In general, there is no difference between joint and stone repair mortars. Both the mechanical and physical properties of the mortar have to be adapted to the properties of the stone. The dynamic E-modulus, the compressive strength, the thermal dilatation coefficient, the water uptake coefficient (capillarity), the value of water vapor resistance and the pull-off strength should be considered. In special cases it is necessary to consolidate the friable stone surface to obtain a sufficient pull-off strength. Because the mortar should always be softer than the stone, the E-modulus should be as low as possible.

Joint and stone repair mortars mostly consist of lime + cement as binding materials. To improve the adhesive properties all commercial stone repair mortars contain dispersion powder, mostly acrylic or copolymer dispersions. The use of puzzolanic additives must be carefully considered. Puzzolanic additives react very slowly so that the mortar has to be kept wet for a long time.

Recently developed mortars based on silica sol as binder can be tailored to the properties of very different stones. They have the advantage that the contact zone between mortar and stone can be consolidated with ethyl silicate so that a smooth transition zone is formed where the properties change gradually. A comprehensive compilation of all necessary tests to tailor appropriate joint and repair mortars is given by Knöfel and Schubert (1993).

Stone Consolidation with Ethylsilicate

Weathering processes cause a loss of strength either close to the surface (grains are sanding off from the surface) or at a ceratin depth (formation of a scale). Very often, a strength profile with increasing strength towards the interior can be observed. Consolidation is necessary to strengthen the grain-to-grain cohesion, and should return the strength of a stone to its original state. An over-strengthening ist neither necessary nor desired.

Silicic acid esters (ethylsilicate) are by far the most frequently used chemicals. They hydrolyze in the pore space, forming an amorphous SiO_2 gel (Opal A). The SEM photograph shows that new grain-to-grain bridges are formed (see Fig. 106). Monomer or diluted or prehydrolyzed products can be used to tailor the conservare according to the actual stone properties. Figure 107 shows an example of a consolidated stone before and after treatment. A comprehensive compilation of the results of sandstone consolidation with ethylsilicate is given by Sattler (1992).

The consolidation with ethylsilicate changes the mechanical properties of sandstones from a plastic to an elastic behavior (see Fig. 108). If the E-moduls of the consolidated surface zone is much higher than that of the untreated stone, the formation of a sacale cannot be excluded (see also Fig. 107). Therefore abrupt material changes must not occur. The quality requirements are described in detail by Sasse and Snethlage (1997) and Snethlage (1997).

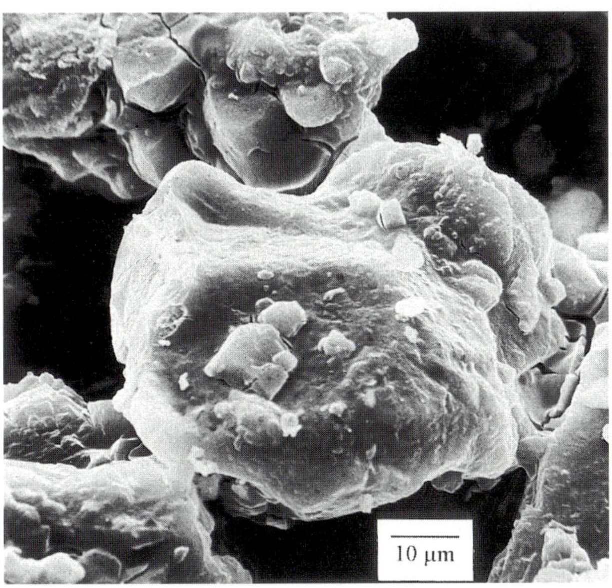

Fig. 106. SiO_2 gel, formed from ethylsilicate in a sandstone pore space. SEM/SE photograph, total length of photo 90 µm

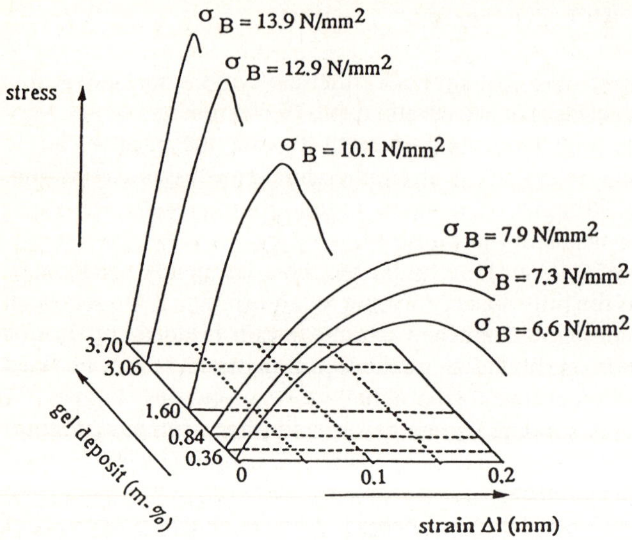

Fig. 107. Change from plastic to elastic deformation of a sandstone in biaxial flexural strength test as a consequence of consolidation with ethylsilicate

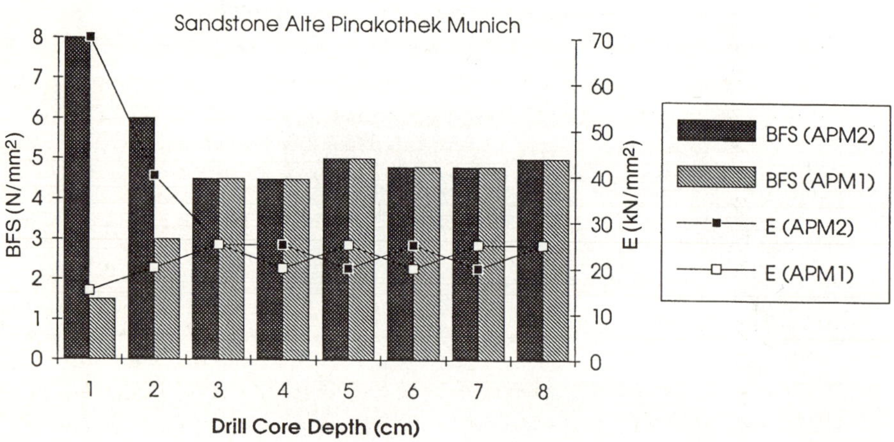

Fig. 108. Flexural strength (ring load on drill core slabs) and E-modulas before (APM1) and after (APM2) treatment with ethylsilicate

To secure detached layers and exfoliations on limestones, the use of acrylic resins or mixtures of acrylic resins with alky-trialkoxysiloxanes is recommended.

A special method to preserve extremely decayed objects from further destruction is the acrylic total impregnation method (Ibach method). Monomer MMA is used to soak the whole pore space of the stone in a vacuum chamber in which the monomer is then polymerized to solid PMMA under increasing temperature and pressure.

Hydrophobation

Hydrophobation is used to prevent the capillary absorption of liquid water, whereas the water vapor transport is almost not reduced. The hydrophobic agents belong to the group of alkyltrialkoxysilanes, Monomer, oligomer, or polymer products are available. A deciding parameter which determines the durability of the hydrophobation is the alkyl group. Modern products mainly contain butyl- or isobutyl alkyl groups because they are stable in alkaline solutions so that they can be also applied to concrete. Methyl products were first on the market but are less stable against acid and alkaline attack.

The effectiveness of a hydrophobic agent must be tested in every single case because the effectiveness of the agents varies from stone type to stone type. In particular, the hydrophobation of limestones is sometimes difficult. Although the hydrophobation prevents the access of liquid water, the activity of salts is not stopped because salts also dissolve and crystallyze in the hygric region, i.e., under different moisture conditions of the ambient air. Also the hygric dilatation of stones is not inhibited, but in some cases even enhanced.
The following requirements can be formulated to assess a hydrophobation:

- Hydrophobation is useful if a bulding part is exposed to driving rain and if the w-value of stone is greater than 1 $kg/m^2 \sqrt{h}$.
- Hydrophobation must be applied to a whole facade or sculpture, not only to certain exposed parts.
- Hydrophobation must lower the w-value below 0.1 $kg/m^2 \sqrt{h}$.
- An aged hydrophobation whose w-value has increased to more than 0.5 $kg/m^2 \sqrt{h}$ is insufficient and must be renewed.

Recently, efforts were made to develop impregnation agents on the basis of polyurethanes and polyurethane-disiloxane copolymers which combine strengthening and hydrophobic properties. Mixtures of etylsilicate and alkyltrialkoxysilanes which have strengthening and hydrophobic properties can be also commercially obtained.

Paints on Stones

In former times, stone monuments were mostly painted to portray a real person or imitate a more precious stone material. Also buildings were painted. Paints can protect a stone against rain and dirt deposition. However, they must not block the pores and prevent the transport of water vapor. Sealing paints like oil paints crack when they age, so that moisture can penetrate into the interior of the stone. As the trapped moisture cannot evaporate, the stone will be damaged and the paint layer will peel off.

The main parameter to assess a paints is the water vapor resistance number μ and the diffusion equivalent air layer thickness $s_d = \mu \cdot d$. The μ and the s_d value must remain the same before and after the application of the paint. Because paints behave differently on different stones, their behavior must be determined by laboratory investigations. Historic paints were based on lime with additives like casein or albumin or on linseed oil with lead white as siccative. Modern paints are based on water glass (silicate paints), on silicon resins, or on dispersions like acrylic or acrylic-polyurethan copolymers.

References

Ashurst J, Dimes FG (1990) Conservation of building and decorative stone, vol 1 + 2. Butterworth – Heinemann, London

Duttlinger W, Knöfel D (1991) Salzkristallisation und Salzschadensmechanismen. Jahresberichte Steinzerfall-Steinkonservierung, vol 3. Verlag Ernst & Sohn, Berlin, pp 197–214

Knöfel D, Schubert P (eds) (1993) Handbuch der Mörtel und Steinergänzungsstoffe in der Denkmalpflege. Sonderheft der Publikationsreihe BMFT-Verbundforschung zur Denkmalpflege. Verlag Ernst & Sohn, Berlin, 225 pp

Lewin SZ (1966) The preservation of natural stone, 1839–1965. Art Archaeol Techn Abstr 6 (1): 185–277

Livingston RA (1997) Development of air pollution damage functions. In: Baer N, Snethlage R (eds) Saving our cultural heritage, the conservation of historic stone structures, Dahlem Workshop Reports ES 20. John Wiley, Chichester, pp 37–62

Mirwald PM, Brüggerhoff S (1997) Requirements for and interpretation of accelerated and field testing. In: Baer N, Snethlage R (eds) Saving our cultural heritage, the conservation of historic stone structures, Dahlem Workshop Reports ES 20. John Wiley, Chichester, pp 251–264

Riederer J, Beeger D, Kühne E (1993) Verwendung, Zerfall und Konservierung von Naturstein in Architektur und Plastik. Bibliographie 1986–1990. Berliner Beiträge zur Archäometrie, vol 12. Staatliche Museen zu Berlin, Rathgen Forschungslabor, Berlin, pp 115–202

Sasse HR, Snethlage R (1997) The methods for the evaulation of stone conservation treatments. In: Baer N, Snethlage R (eds) Saving our cultural heritage, the conservation of historic stone structures, Dahlem Workshop Reports ES20. John Wiley, Chichester, pp 223–243

Sattler L (1992) Sandsteinfestigungen mit Kieselsäureester. Dissertation Universität München, publ in: Bayerisches Landesamt für Denkmalpflege, Zentrallabor, Forschungsbericht 9/1992, 156 S mit Anhang

Skoulikides T, Papakonstantinou-Ziotis P (1981) The mechanisms of sulfatation by atmospheric SO_2 of limestones and marbles of the ancient monuments and statues. I.: Observations in situ and measurements in the laboratory, activation energy. J Br Corros 16: 63–69. II.: Hypothesis and proofs and the rate-determining step; glavanic cell model. ibidem, pp 70–77

Snethlage R (1997) Leitfaden Steinkonservierung. IRB Verlag, Stuttgart

Snethlage R, Wendler E (1996) Methoden der Steinkonservierung, Anforderungen und Bewertungskriterien. In: Snethlage R (ed) Denkmalpflege und Naturwissenschaft, Natursteinkonservierung I. Verlag Ernst & Sohn, Berlin, pp 3–40

Snethlage R, Wendler E (1997) Moisture cycles and sandstone degradation. In: Baer N, Snethlage R (eds) Saving our cultural heritage, the conservation of historic stone structures, Dahlem Workshop Reports ES20. John Wiley, Chichester, pp 7–24

Snethlage R, Wendler E, Klemm DD (1996) Tenside im Gesteinsschutz – bisherige Resultate mit einem neuen Konzept zum Schutz von Denkmälern aus Naturstein. In: Snethlage R (ed) Denkmalpflege und Naturwissenschaft, Natursteinkonservierung I. Verlag Ernst & Sohn, Berlin, pp 127–146

VDI Richtlinie 3798 (1989) Untersuchung und Behandlung von immissionsgeschädigten Werkstoffen, insbesondere bei kulturhistorischen Objekten. VDI Kommission Reinhaltung der Luft, Ausschuß Wirkung von Luftverunreinigungen auf Werkstoffe, Düsseldorf, 55 pp

5.4 Radiation Mineralogy

5.4.1 Natural Radioactive Materials and Radiometry

H. von Philipsborn

Uranium, thorium, and their radioactive decay products, as well as potassium, are the three most important naturally occurring radioactive elements. They are found as constituent or accessory elements in many minerals and mineral products and hence in their aqueous and atmospheric environment. The emission of ionizing radiation is a characteristic property of this large group of materials. This property, in turn, is characterized by several quantities, such as nuclide-specific activity and half-life, matrix- and geometry-dependent energy dose rate and time- and person-dependent equivalent dose.

Radiometry, the science and technology of measuring these quantities, is in a very advanced state. One of the reasons is the general, justified and unjustified, apprehension of personal exposure to elevated or even low doses of man-made or natural ionizing radiation. The availability of both simple and sophisticated radiometric instruments and methods and their manifold possibilities for original research, teaching, and services in applied mineralogy are not yet fully realized and used by mineralogists.

Mineralogy and the geosciences in general have as much to contribute to radiometry as they may profit from it. This discipline has largely been shaped by physics, chemistry, biophysics, and radioecology, and by a mixture of these, called radiation protection or health physics. The occurence and spreading of radionuclides is closely linked to geochemistry, crystal chemistry, economic geology, mineral processing, and other disciplines. Radiometric instruments often make use of mineral or crystalline materials for their radiation detector effect, such as fluorescence, scintillation, thermoluminescence, and radiation tracks. Studying any property of materials requires close knowledge both of

Advanced Mineralogy, Volume 3
Mineral Matter in Space, Mantle, Ocean Floor, Biosphere, Environmental Management, and Jewelry (ed. by A. S. Marfunin)

measuring methods and of the materials measured. To evaluate the meaning of the results requires knowledge in additional fields.

This contribution sets out to outline and to illustrate four aspects:

1. Natural radioactive materials, their occurence and distribution.
2. Radiometric quantities and units, their use and misuse.
3. Radiometric instruments, their purpose and qualification.
4. Problems and solutions in research and teaching.

Natural Radioactive Materials, Their Occurrence and Distribution

The natural abundance of uranium and thorium with a crustal average of 2.7 ppm, respectively 8.5 ppm, varies by orders of magnitude in different types of rocks and mineral assemblages. In the following, we exclude uranium and thorium minerals, deposits, and ores, their processing and use, as well as their dumps, and everything belonging to the nuclear fuel cycle.

Rather, we concentrate on three types of materials listed in Tables 38, 39 and 40. Such materials may be considered in their primary location, during their processing as raw materials, as part of industrial products, by-products, and building materials, as part of dumps and, in all locations, as sources for contamination of the aqueous and atmospheric environment.

Table 38 lists minerals of economic importance with appreciable amounts (ppm to percent) of uranium or thorium as accessory, nonconstituent elements serving as isomorphous replacement of a constituent element. U^{4+} has a large ionic radius of 97 pm, Th^{4+} of 95 pm, and can replace similar large ions such as Zr^{4+} (80 pm), Y^{3+} (93 pm), Ce^{4+} (101 pm), and the lanthanides.

Table 39 listes mineral raw materials containing minerals of Table 38 as constituent or as accessory minerals or uranium/thorium minerals as accessory minerals. Other sources of unwanted and often unsuspected uranium/thorium contents are paragenetic admixtures such as hydrothermal fluorite and barite

Table 38. Minerals of economic importance with accessory, nonconstituent uranium or thorium and their decay products

Apatite $Ca_5\,[F\mid(PO_4)_3] \pm U$	Columbite, niobite, tantalite $(Fe, Mn)\,(Nb, Ta)_2O_6 \pm U, Th$
Zircon $Zr\,[SiO_4] \pm U, Th$	Pyrochlore $(Ca, Na)_2\,(Nb, Ta_2)\,O_6\,(O, OH, F) \pm U, Th$
Baddeleyite $ZrO_2 \pm U$	Euxenite $(Y, Ca, Ce)\,(Nb, Ta, Ti)_2O_6 \pm U, Th$
Monazite $Ce[PO_4] \pm Th$	Fergusonite $(Y, Ce, La)\,(Nb, Ta)O_4 \pm U, Th$
Xenotime $Y\,[PO_4] \pm Th, U$	Samarskite $(Y, Ca, Fe)_3\,(Nb, Ta, Ti)_2O_{16} \pm U, Th$
Cerianite $CeO_2 \pm Th$	Yttrotantalite $(Y, Fe)\,(Ta, Nb)O_4 \pm U, Th$
Bastnäsite $Ce[F\mid CO_3] \pm Th$	

Table 39. Mineral raw materials, containing minerals of Table 38 as constituent or accessory minerals or uranium/thorium minerals as accessory minerals

- Heavy mineral sands mined for rutile or ilmenite for production of TiO_2 white pigment, or for the rare earth elements
- quartz/feldspar/china clay sands; by „refinement" heavy mineral sands are separated
- Zircon sands used for zirconium production, refractories, foundry sands
- Sedimentary phosphates used for fertilizer
- Bauxite for alumina or aluminum production
- Hydrothermal fluorite and barite, paragenetic with uraninite
- Sandstones, granites, phonolites as building stones
- Black shales, formerly used in Sweden for porous building stones
- Black shales considered as oil reserves
- exploration and production of offshore oil and gas
- Coal from certain deposits

Table 40. Technical products, by-products, or wastes produced from materials of Table 39

- Chemical ZrO_2, not properly refined, for refractories and fuel cells
- Refractory zircon stones
- Chemical gypsum from processing of phosphorites, if not properly refined
- Red mud from alumina production, used for brick-stones
- Slag and slag-stones from copper smelters using black shale ores
- Slag from V, Nb, Ta, Sn smelters, used for road building
- Ash and ash-stones from certain coal deposits
- scales with radium bearing barite in oil and gas pipes
- Building materials containing any of the materials of Tables 38, 39, 40
- Radon and radon decay products in water and air

with uraninite or pegmatitic feldspar with primary or secondary uranium minerals or ordinary quartz sands with heavy minerals. Uranium exists in two valency states, U^{6+}, water-soluble, depending on available ligands, and U^{4+} insoluble. Uranium can thus be transported in oxidizing milieu or precipitated in reducing milieu, often found in sediments with organic carbon. Thorium may be found absorbed in bauxites.

Table 40 lists technical products, by-products, or wastes produced from materials of Table 38. Special attention should be given to the fact that materials of Tables 38 to 40 may contaminate the aqueous or atmospheric environment by weathering or simple emanation of radon-222 and radon-220. Tables 38 to 40 are nonexhaustive.

Uranium, thorium, and their decay products are nonphysiological and potentially harmful radioactive elements and are therefore monitored in some cases for practical reasons of radiation protection. Potassium is a physiologically essential element – adults contain ca. 150 g K with 4500 Bq K-40 – and is not considered further. Radiometric studies of K may nevertheless be of interest for academic reasons because they are simple and informative.

The general abundance of potassium with a crustal average of 2.6 wt.% is the major source of natural radioactivity in most rocks, soils and mineral assemblages. A representative soil sample is quoted to contain per Kilogram dry mass 360 Bq K-40 (1.2% K-nat), 24 Bq Pb-214 and 20 Bq Bi-214 (1.9 ppm U-nat), 25 Bq Ac-228 (6.3 ppm Th-nat.). Even most granites owe most of their radioactivity to potassium feldspars and not to legendary uranium. Simple radiometric instruments, such as GM-counters and most scintillation counters and other field instruments, do not distinguish radionuclides.

Radiometric Quantities and Units, Their Use and Misuse

Radioactivity is one of three sources of ionizing radiation – X-rays and cosmic radiation are the two others. Radiometric quantities and units serve to characterize ionizing radiation by its source and by its physical and biological effects.

The primary radiometric quantity is pulse rate, i.e., the number of events in the detector per time. By calibration, pulse rates may be made to indicate the activity, i.e., the number of events (nuclear decays) in the probe per time, or to indicate dose rates, as defined later. The response, sensitivity, or efficiency of most radiation detectors strongly depends on the type and energy of radiation. For example, a given activity of K-40 with the gamma line at 1.40 MeV generates a smaller pulse rate in a detector than the same activity of uranium with its decay products Pb-214 and Bi-214 with gammalines between 250 and 600 keV.

A radioactive source is characterized by its radionuclides and their activity and its matrix. The activity gives the number of nuclear disintegrations per time. The unit is 1 Becquerel = 1 decay per second in time average. Each one of the many radionuclides has a specific scheme of decay, emitting alpha or beta particles or gamma radiation or a mixture, with a specific half-life and specific energies of the radiation. Quoting activities without radionuclides is as meaningless as quoting traffic accidents without saying if automobiles, trains, or airplanes collided. The matrix of a radioactive source needs to be defined, because it makes a big difference for measurements and for physical and biological effects according to whether the source is a solid, liquid, powder, or gas, diluted or concentrated, large or small, flat or compact, far or close, outside or inside a human body.

Physical effects of ionizing radiation are characterized by the quantities energy dose = radiation energy absorbed per mass with the unit of 1 Gray = 1 Joule per 1 kg and energy dose rate = energy dose per time (s, min, h, a). Sievert is the unit of the quantity dose equivalent and is not a measurable physical quantity such as activity and energy dose. In addition, several types of dose equivalent are used in radiation protection, each calculated from activity or energy dose using biological and medical assumptions, still under discussion, and various parameters for exposure.

For all purposes except for radiation protection (though this is often the ultimate purpose of a study and its measurements), one can stick to pulse rates or nuclide-specific activity or to energy dose rate. These are then compared to levels of natural concentration or exposure or used with standard conversion factors to arrive at a judgment as to whether and under what conditions the results indicate a potentially harmful situation or not.

Radiometric Instruments, Their Purpose and Qualification

Any measuring instrument must fit the conditions of the probe, the purpose of the study and the qualifications of the examiner. It is not surprising that many results are questionable. Some conditions of the probe are covered by the matrix of the radioactive source, mentioned above. In addition, there are different instruments for laboratory and for field work, for large-scale or small-scale studies, for large and for small budgets, though big money often destroys ingenuity. Desk, lab, and field work should integrate. The purpose of a study may be analytical, comparative, screening, or dynamical. For analysis, low-level, high-sensitivity, nuclide-specific alpha- or gamma-spectrometry is highly desirable. For comparisons, simple pulse rates (count rates) in selected instruments are often sufficient. For screening, simple low-cost methods are required. For dynamic studies the time resolution and hence the efficiency of the instrument is a key factor. Finally, different radiometric samples and aims are as different as northern and tropical flowers, and require qualified examiners with practical experience.

Instruments for measuring radioactivity may be grouped in at least ten categories according to their purpose:

1. Portable instruments for gamma dose rate in the field.
2. Pulse counters with energy discrimination of U, Th, K, for exploration.
3. Contamination monitors.
4. Gamma-spectrometers for laboratory, with lead shielding.
5. Alpha-spectrometers for laboratory, with vacuum chamber.
6. Low-level monitors with alpha-beta discrimination.
7. Instruments for radon and radon decay products.
8. Dosimeters.
9. Photographic films for autoradiographs.
10. Ultraviolet lamp for fluorescent uranium minerals.

In most categories there are five to ten different types of instruments, differing in type of probe or detector and other specifications. Hardly any instrument is qualified for more than one category. Gross errors result if pulse-counting detectors are used for activity measurements without proper calibration, or contamination monitors without knowing the nuclides present. The detector of the instrument and its entrance window determine which type of radiation (alpha, beta, gamma) can be detected. The detector and its mode of operation and the electronic processing of the primary signals determine if nuclides can be distinguished. The geometry of probe and detector and their shielding determine the limit of detection and the necessary counting time.

Each category of instruments would require a separate paper, even for non-exhaustive treatment. The reader must be referred to the literature. One convenient, multipurpose instrument, both for solid and air- and waterborne samples, for research, teaching and services, will be described in the next section.

Problems and Solution in Research and Teaching

Natural radioactivity, expect for commercial mining and processing of uranium and thorium ores, is, or was up to very recently, exempt from legal regulations in most countries. This situation is changing rapidly. Physical and biological effects of elevated doses of ionizing radiation are the same for natural and artificial radionuclides, if the proper nuclide-specific activity-to-dose conversion factors are used. The limits of exposure legally set for artificial, man-made radionuclides are so low that they are easily exceeded by natural radionuclides present under certain conditions in many locations. Regulations which limit exposure to natural radionuclides at workplaces are in preparation. Recommendations for lower limits for housing already exist in many countries.

It is now realized that exposure to radon and radon decay products – normally contributing more than half to the average natural radiation exposure of the general public – may be much higher under certain conditions than previously thought. Possible harmful effects of so-called low doses of radiation are still under lively discussion. "Low" means doses which are of the same order of magnitude as natural dose levels. The origin of elevated doses of natural radiation may be one or several of the materials of Tables 38 to 40. Considering their mineral origin, it is clear that interesting work in research and teaching of advanced mineralogy does exist.

Figures 109 and 110 show the decay schemes of uranium-238 and thorium-232 necessary to understand widespread natural radioactivity, originating from the materials listed in Tables 38 to 40, and necessary to make the best radiometric choice for different materials. The two decay schemes contain 25 radioisotopes of ten elements: U, Pa, Th, Ac, Ra, Rn, Po, Bi, Pb, and Tl. The different radionuclides are distinguishable by three methods: gamma energy-spectrometry, alpha energy-spectrometry, and beta time-spectrometry. Which method should be chosen depends on the matrix of the probe. Alpha radiation, because of its strong self-absorption in the matrix, requires for its measurement special sampling or preparation of the probe. Gamma radiation, because of its penetrating nature, at least at high energies, is best suited for large solid or liquid probes. Beta measurements, if properly designed, are best suited for air- and waterborne probes, because of highly efficient sampling on filters and of measurement with proportional detectors.

To meet the demands of research and teaching, a multipurpose low-level monitor was recently developed by the author. The instrument, called Living

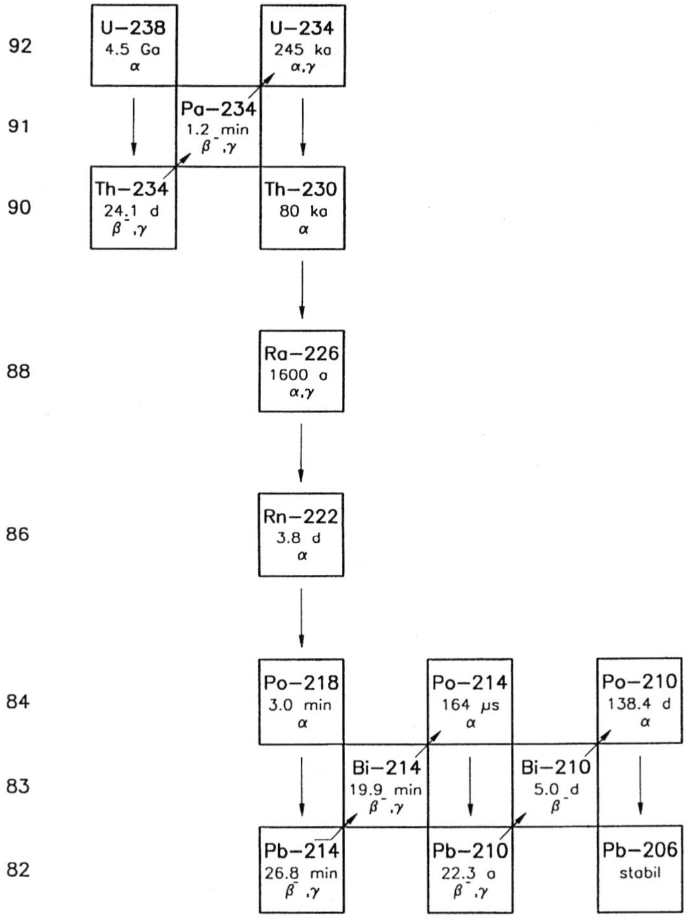

Fig. 109. Radioactive decay scheme of uranium-238. On the left side the atomic number, in the squares the element symbol and mass number, the half-life and the type of radiation: alpha, beta, or gamma

Level Monitor mab LLM 500, has been successfully produced by Münchener Apparatebau. It is the only instrument capable of measuring powder samples, such as mineral sands, as well as air- and waterborne radionuclides. Its simple, versatile operation and low price make it the ideal instrument for teaching and laboratory courses for students of advanced mineralogy. Its high sensitivity enables us to measure within minutes even a few Bq/m^3 of radon anywhere present. For this purpose, a novel type of large area tribo-electrostatic collector, the Philion Plate, was developed (von Philipsborn and Hoffmann 1995).

The Living Level Monitor mab LLM 500 is portable, height 17 cm, width 42 cm, depth 40 cm, weight 7 kg. It contains a 120-cm^2, beta-sensitive, sealed proportional detector under an opening for the diskette-type sample holder for

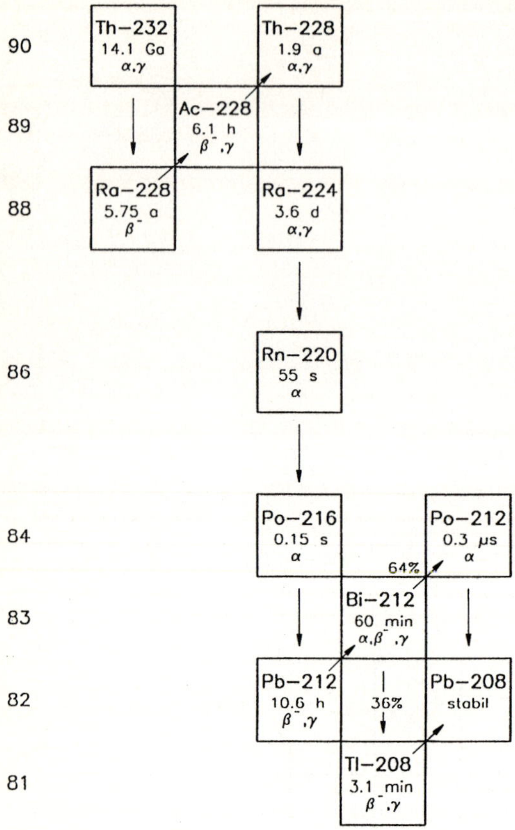

Fig. 110. Radioactive decay scheme of thorium-232

filters, powders, or Philion Plates. The hardware and software for counting and evaluating the impulses allow for selecting different values for the gate time, for volume or weight of the sample, and for the calibration factor for converting impulses to activities. A unique feature of the instrument is the repetition mode in which a measurement is automatically repeated up to 100 times, the data being stored and released to a printer for plotting a time spectrum of beta activity. By simple beta time-spectrometry, radon and thoron decay products are distinguished from artificial radionuclides such as Cs-137 and I-131, well known from the Chernobyl accident, or natural radionuclides with long half-life.

Building materials are often the source of elevated indoor radon concentrations. Sand makes up a good part of prefabricated walls. As mineralogists know, even ordinary quartz sand contains variable amounts of heavy minerals, among these monazite with accessory thorium. For finding radon-220 (thoron) decay products in the air, originating from such walls, a room with the largest ratio of surface to volume was chosen. In my university area it happens to be

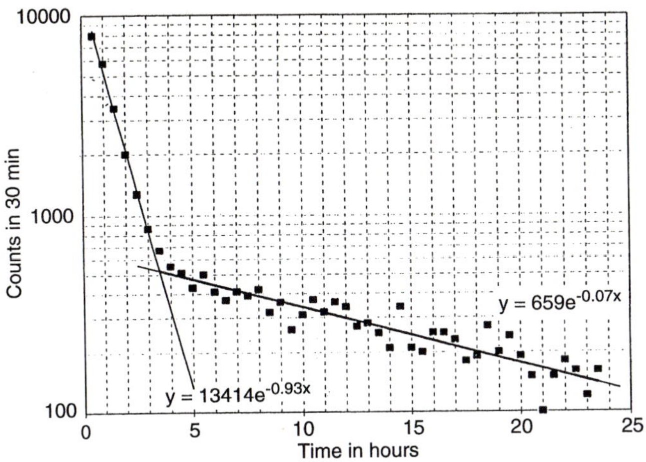

Fig. 111. Indoor airborne radioactivity originating from walls. Sample: air from a photographic dark room, collected with a Philion Plane, exposed for 30 min. Instrument Living Level Monitor (shielded) in repetition mode, giving the time-spectrum of net total beta counts. Background 2178 counts in 30 min gate time. Branch 1: half-life T = 0.69/0.92 = 45 min from Pb-214 T = 27 min + Bi-214T = 20 min, beta-emitting decay products form Radon-222. Branch 2: half-life T = 0.69/0.073 = 9.5 h from Pb-212 T = 10.6 h + bi-212 T = 61 min, beta-emitting decay products form radon-220

the photographic dark room without ventilation. The gamma dose rate in the dark room, measured with a calibrated scintillation instrument mab 604, is only 70 nSv/h and not noticeably increased.

Our patented Philion ("lover of ions") Plate was exposed for 30 min in this dark room. The triboelectrically charged 120 cm^2 surface attracts radon-222 and thoron decay products, charged electrically positive by having lost electrons by virtue of the alpha emission recoil energy. Small ions, not attached to aerosol particles, have a high mobility, defined as velocity per field strength. Such radioactive decay products are collected with high efficiency by the Philion Plate. Figure 111 shows the time spectrum of net counts of the said Philion Plate measured with the Living Level Monitor, clearly showing the presence of thoron decay products.

The gamma energy spectrum of a 100-g probe from the wall of the dark room, taken with a lead-shielded high-purity germanium detector, is shown in Fig. 112. It reveals the presence of thorium decay products. X-ray diffraction or other simple mineralogical phase analysis methods are not sensitive enough to reveal the presence of mother monazite in the wall.

To complete the triplet of basic radiometric methods with still the same example, an alpha-energy spectrum is shown in Fig. 113. It was taken from the Philion Plate with a 30-cm^2 silicon detector in a vacuum chamber. Note the sharp lines, thanks to an absence of self-absorption on the Philion Plate.

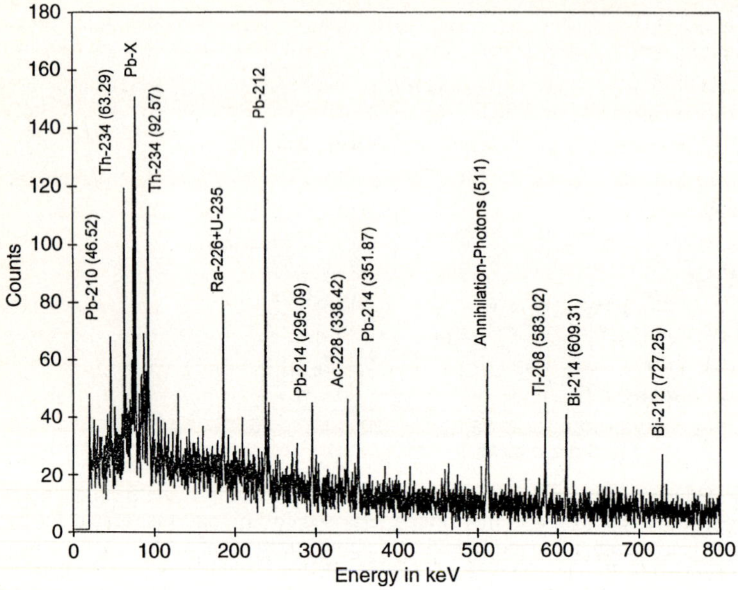

Fig. 112. Gamma energy-spectrum of a 122-g probe of wall material, taken in a high-purity germanium detector, counting time 4 h. Note the lines of the decay products of the thorium decay scheme (et. Fig. 110).

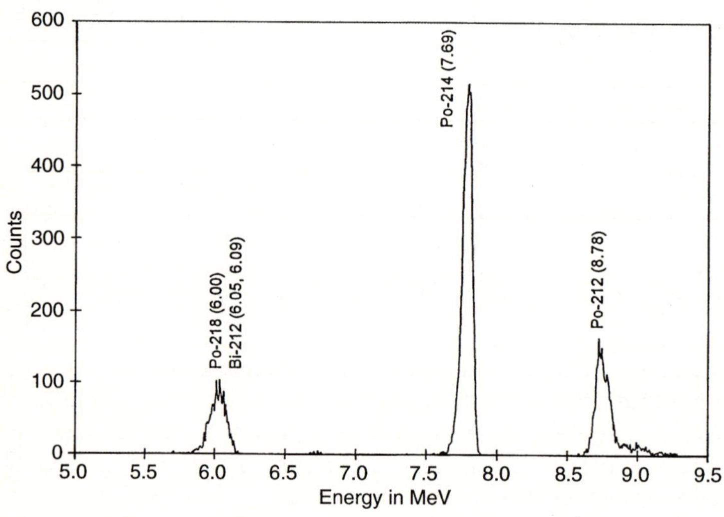

Fig. 113. Alpha energy-spectrum of a Philion Plate, exposed for 30 min in the same room, taken with a 30-cm^2 silicon detector in vacuum, counting time 20 h. Note the sharp lines, thanks to absence of self-absorption on the collector. From left to right Po-214 at 6.00 MeV, Po-214 at 7.69 MeV, and Po-212 at 8.78 MeV

The release of thoron, despite its short half-life of 55 s, from nonporous building materials, and the indoor presence of thoron decay products come to health physicists as a surprise and as an alarm. Thoron decay products are considered to be more harmful than radon decay products and were not looked for in previous indoor survey. Geochemists have long known thorium to be three times more abundant than uranium and to be found in unsuspected materials. Sensitive and convenient sampling and measurement of radon and thoron decay products even at low levels is now possible with the Philion Plate and the Living Level Monitor.

In most recent experiments, glass fibre filters were found to be very efficient (retention >95%) for rapid filtration of radon decay products in water (von Philipsborn 1997). This comes unexpectedly, and may be the reason why it was not tested before. Wet filters are quickly dried on a hot stage and evaluated in the Living Level Monitor. Limit of detection is 1 Bq/l. Compared to studies of radon in air of dwellings, studies of radon in groundwater and wells are of much greater geological interest.

A novel, simple and efficient method to measure radon exhalation from building materials in existing houses and in the laboratory has been developed (Just and von Philipsborn 1997).

References

Living Level Monitor mab 500. Datasheet. Münchener Apparatebau GmbH, Mehlbeerenstr. 2, D-82024 Taufkirchen

von Philipsborn H, Hoffmann Chr (1993) Living Level Monitor LLM 500. 25. Jahrestagung Binz 1993, Fachverband Strahlenschutz, Band I. Verlag TÜV Rheinland, Köln, pp 409–416

von Philipsborn H, Hoffmann Chr (1995) Messung von Radonfolgeprodukten – ein einfaches Verfahren. Strahlenschutzpraxis 4:56–58

Just G, von Philipsborn H (1996) Radonexhalation aus Bau- und Rohstoffen: In-situ-Einschätzung und Analytik im Labor. Strahlenschutzpraxis 4:78–82

von Philipsborn H (1997) Efficient adsorption of water-borne short-lived radon decay products by glass fiber filters. Health Phys 72:277–281

Further Reading

Debertin K, Helmer RG (1988) Gamma- and X-ray spectrometry, with semiconductor detectors. Elsevier, Amsterdam

Eisenbud M (1981) Environmental radioactivity. Academic Press, London

Ivanovich M, Harmon RS (1992) Uranium-series disequilibrium: applications to earth, marine an environmental sciences. Clarendon Press

Knoll GF (1989) Radiation detection and measurement. John Wiley, New York

Siehl A (ed) (1996) Umweltradioaktivität. Ernst & Sohn, Berlin

Vael PCJ (editor) (1997) Proceedings int. symposium on radiological problems with natural radioactivity in the non-nuclear industry. Amsterdam Sep 8–10, 1997. N V Kema 6800 ET Arnhem NL

5.4.2 Management of High-Level Nuclear Waste (HLW)

H. PENTINGHAUS

Nuclear waste management provides knowledge and means that enable safe handling, temporal storage, and final disposal of radionuclides (RN) in order to exclude them from the biosphere for their life-times. Radioactivity can easily be traced by its individual radiation almost always on the level of the nuclear event, a beneficial advantage over chemical analyses. During the first decade when nuclear technologies were introduced, the feasibility of the processes under development was of prime importance, so especially the high-level wastes (HLW) have been stored and the problems of their disposal left to the future. Mineralogy, with its disciplines and materials science, became involved much later performing research and development (R & D) work on waste forms and their longterm behavior. Especially the interactions in aqueous contact in different geological settings the characteristics of which are revealed by geology, geophysics, and hydrology have been studied all together in continuous effort worldwide. In this chapter, emphasis is put on the management of HLW from energy production, and examples are taken from the vitrification of HLW by means of the liquid-fed ceramic melter technology (LFCM) the development of which has been mineralogically supported over the past 14 years. The subject is treated in three parts: emergence of RN and time scales; waste forms, and their technologies; final disposal of HLW, chemical properties of waste forms, and future activities.

Emergence of RNs and Time Scales

RN mainly emerge from the production of energy, of nuclear weapons (up to 1950), from medical applications, and, of course, from mining and milling of uranium, where hundreds of millions of tons of tailings await remediation. Today about 300 t of ^{239}Pu alone are the estimated inventory of nuclear weapons worldwide. The huge amount of accompanying defense waste accumulated over nearly 50 years is widely spread over many sites and heavily blended with a variety of chemicals added during reprocessing of spent fuel and nuclide extraction. National programs for the remediation of contaminated sites exist which cause huge efforts; however, there is no general recipe at hand that can be followed. From energy production about 135 000 t spent fuel is presently in temporary storage throughout the world awaiting reprocessing or direct disposal, respectively. As radioactivity decays with time and the heat generated thereby as well, interim storage of HLW is beneficial because it diminishes the thermal loading of the final repositories. As regards this aspect, interim storage times up to 100 years or even more would be very advantageous.

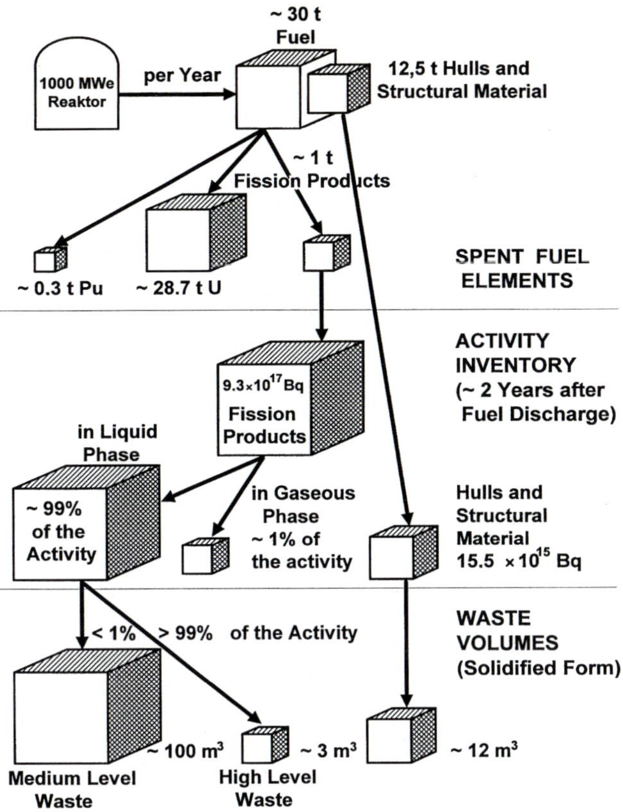

Fig. 114. Annual amount of spent fuel elements, inventories of activity and waste volumes arising from a nuclear power station (1000 MW E)

The annual production of spent fuel, inventories of activity and waste volumes from a nuclear power station (1000 MWe) is shown in Fig. 114. Dependent on the burnup, beside 0.3 t Pu, about 1 t of fission products is fixed in the fuel. After reprocessing and solidification, 99% of the activity is fixed in only about 3 m^3 of a vitreous waste form. Future techniques, which are already developed, could avoid the stream of medium level waste by chemical separation procedures. Solidified HLW by no means poses any volume problems, even after running energy production for 50 years and more with conventional reactors.

The time scales for the management of RN can be taken from Fig. 115, in which the radioactivity decay of 1 t heavy metal is shown. Three main groups of nuclides determine the total activity as a function of time: fission products, actinides and decay products. A future separation of ^{137}Cs and ^{90}Sr and also the actinides from the HLW could open the chance of different waste forms in different scenarios of final disposal.

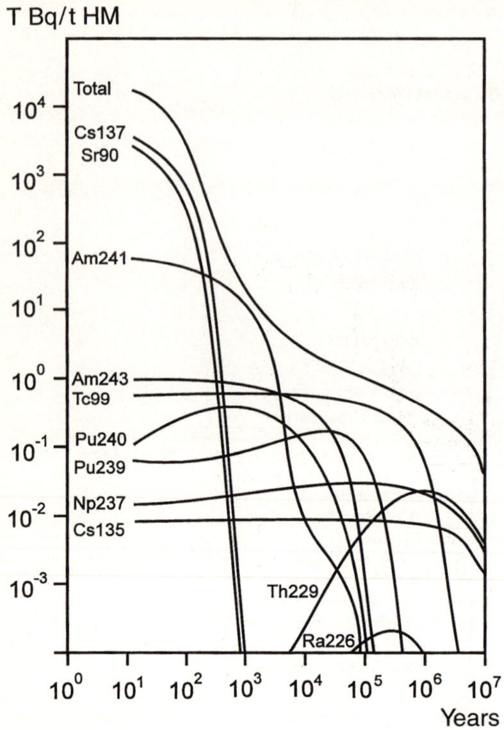

Fig. 115. Radioactivity of 1 t HM as a function of time after discharge from the reactor. The function is determined by three groups of main nuclides, fission products, actinides, and decay products

HLW arising from the reprocessing of spent fuel elements is a 3–5 molal acidic solution of HNO_3 containing the fission products as well as chemicals added during the separation processes (organics and mainly sodium) and corrosion products from the stainless steel containers and tubings of the plant. In Fig. 116, as an example, the composition of a real HLW from the experimental reprocessing plant at Karlsruhe (WAK) is presented expressed in wt.% of the oxides which are formed when the solution is calcined, the organic and the sodium contents are omitted. The main elements are shown groupwise to envisage the formation of possible main phase assemblages involving the rare earth oxides and the anion forming MoO_3 when these elements are able to react during aqueous corrosion of a waste form.

Waste Forms and Their Technologies

HLW has to be protected safely against dispersion in the biosphere. In the case of reprocessing of spent fuel elements, the nuclide inventory of liquid waste is transferred into solid matrices by thermal processes. The addition of oxidic matrix material becomes necessary because the oxide mixture of HLW alone

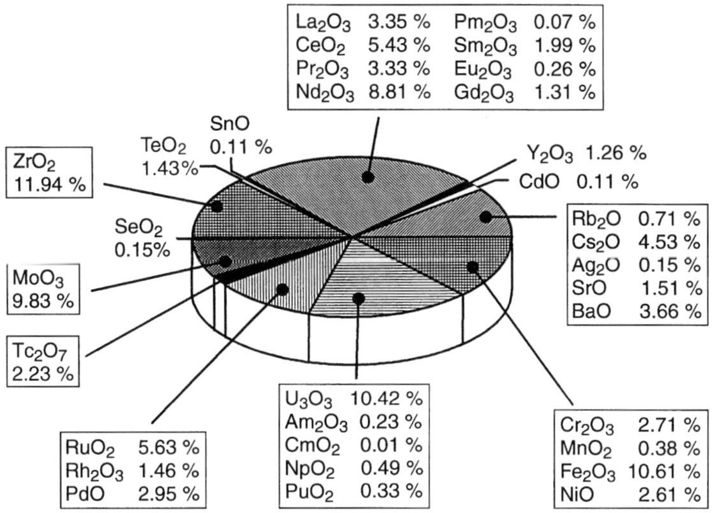

Fig. 116. Composition of HLW from reprocessing of spent fuel elements calculated as oxides of the main waste elements which are formed during calcination of the high active waste solutions (WAK). The oxides from the corrosion of stainless steel containers and tubes are included

does not form chemically resistant solids. Silica-based additives lead to glassy waste forms, whereas those based on titania result in crystalline ceramics.

In both cases, RN become part of complex amorphous or crystalline structures, respectively. These waste forms possess reproducible properties which can be optimized and studied in all detail and they determine the chemical and physical long-time behavior in different repositories. R&D work on waste forms is legion, and the results are mainly published in a report literature which is difficult to trace, so the monograph on waste forms by Lutze and Ewing (1988) is very helpful and still topical.

The glass-making process is called vitrification and, simply stated, it is the way in which RN are dissolved in an alkaliborosilicate melt. Two technologies are available, the two-step French AVH process applied in France and the UK, which consists of a calciner and a metallic melter, and the one-step LFMC process, in which evaporation, calcination, and vitrification are performed by feeding the HLW solution and the glass frit on the melt surface. The melt is directly heated by means of submerged electrodes. The LFCM allows a higher throughput than the AVH system. After a certain residence time in the melter, the melt is poured discontinuously into thin-walled canisters made from stainless steel. These are cooled down in a controlled manner to produce the glassy waste form. Alkaliborosilicate systems have been introduced mainly because of the high solubility for a wide variety of oxides, their beneficial chemical properties, and their low melting temperatures. Recipes of some basic glass frits which have been applied for technical vitrification are given in Table 41. Their

Table 41. Composition (wt.%) of some basic glass frits, all of which have been used for the vitrification of HLW simulates and/or HLW on technical scale by means of the LFCM or by the AVH ([a]) technology, respectively

Glass frit	SiO_2	TiO_2	B_2O_3	Al_2O_3	Li_2O	Na_2O	MgO	CaO	ZnO
VG98/12.2	53.2	4.3	14.8	2.4	–	18.6	2.2	4.8	–
SM58FR	64	5.0	13.8	1.3	4.2	5.2	2.3	4.3	–
WAK1D5	60	1.2	17.5	3.1	3.5	7.1	2.2	5.3	–
SON68FR[a]	54.9	–	16.9	5.9	2.4	11.9	–	4.9	3.0

compositions are derived from the basic system Na_2O-B_2O_3-SiO_2 by introducing minor chemical substitutions in order to modify especially those properties which are constrained by the vitrification technology (LFCM): processing temperatures, viscosity, and electrical conductivity, including their thermal dependencies. In addition, these substitutions also keep or even improve the chemical properties of the waste form. In Fig. 117 the liquidus surface is shown which determines the location of the mean composition of a melt, when certain minimum temperatures for the stable liquid state are necessary. Furthermore, the basic system Na_2O-B_2O_3-SiO_2 shows metastable unmixing, which is not a tolerable property of any waste form. The solvus of the basic system is enlarged when Li_2O is introduced. In Fig. 118 the solvus isotherms are shown for (NaLi)O-B_2O_3-SiO_2 as well as the data points for three basic frits containing LiO_2. Indeed, these frits unmix fast well above (SM58FR) or about the transformation temperatures (Tg) in experimental times. Today, it is not

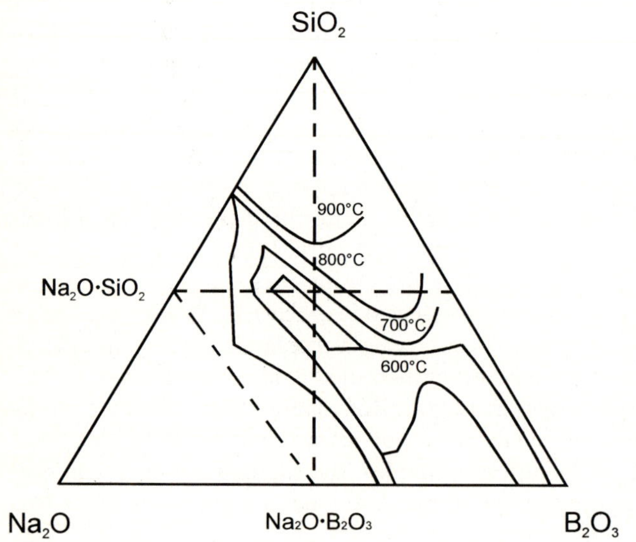

Fig. 117. Isotherms on the liquidus surface of the lower temperature portion of the system Na_2O-B_2O_3-SiO_2

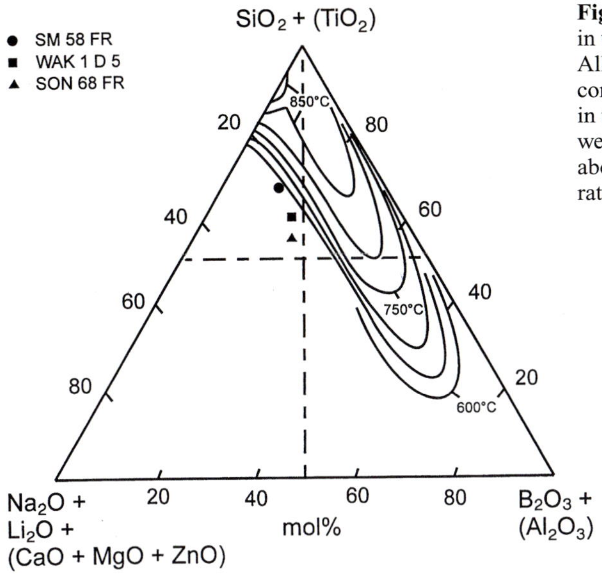

Fig. 118. Metastable unmixing in the system LiNaO-B_2O_3-SiO_2. All three basic glass frits the composition of which are plotted in the diagram show unmixing well above (SM 58 FR) and about the transformation temperature (Tg)

known whether this phase transformation of the basic glasses (WAK1D5, SON68FR) will macroscopically influence the properties of the corresponding waste glass during the first 100 years when central temperatures in the waste canisters reach their highest values (cf. Figs. 120 and 125).

Addition of LiO_2 to the glass recipes, on the one hand, helps to adjust empirically the electrical properties of the melt to processing demands; on the other hand, it improves the chemical durability of the waste form, as evidenced from measured corrosion rates in aqueous solutions.

The waste loading is of prime economical importance. It is limited by those elements having the lowest solubility in the melt. These are Mo and Cr, expressed as MoO_3 and Cr_2O_3 for short, not considering the existing yields of redox pairs in the melt. As MoO_3 predominates, it is the limiting factor allowing loads up to 20 wt.%. As the maximum surface temperature of the canistered waste form is limited to 200 °C in Germany by legal regulations the actual waste loadings are lower, up to about 16 wt.%, and are dictated by the thermal power of the HLW under consideration. In this way, chemical saturation effects of RN are avoided. During idling periods of the melter, however, Cr can be dissolved and accumulated from the ceramic lining of the melter, which consists mainly of $(Al,Cr)_2O_3$ and zirconia ZrO_2. In areas having high surface area to volume ratios (overflow system), oversaturation of Cr may be reached and, instead of Cr_2O_3 (eskolaite), complex chevkinites $(REE,Ca)_4(Cr,Zr,Ti,Al)_5O_8(Si_2O_7)_2$ will crystallize. Their settling gives rise to the formation of suspensions of much higher viscosity, influencing the overflow operation after idling periods. The impact on the increase of viscosity is so strong because the volume of chevkinites is ten times larger than that of Cr_2O_3, which forms

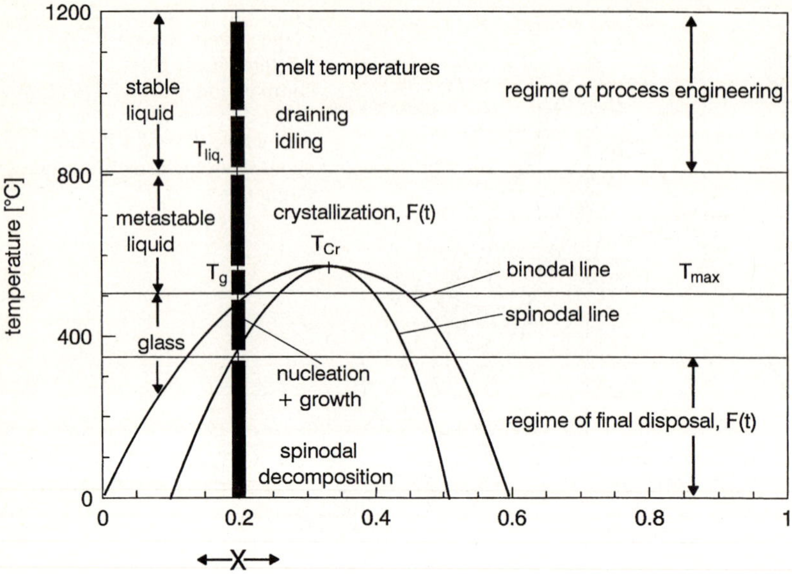

Fig. 119. Temperature regimes of process engineering, phase transformations and final disposal of Li_2O-bearing HLW-glass products. The position of the glass composition relative to the two-phase field has been chosen arbitrarily. The binodal temperature is below Tg

above about 1050 °C. Chevkinite formation can be avoided if TiO_2 is omitted from the recipes.

The regimes of process engineering, phase transformations, and final disposal of Li_2O-bearing HLW-glass products are visualized in Fig. 119. The vitrification processes (LFCM, AVH) are operated between 1200 and 1000 °C. Taking into account the idling periods of the LFCM, the process melt as the solvent for HAW must be stable down to about 800 °C. This demand drastically limits the mean silica (SiO_2) content of the melts and the glassy waste forms, respectively, and thereby reduces their chemical durability. The process chemistry of the vitrification of HLW is well understood today (Pentinghaus 1989) and the kinetics involved lead to very homogeneous melts and glasses. Problems were mainly caused by the low solubility of platinum metal elements which form solid phases (Ru, Rh)O_2 and complex $Pd_xRH_yTe_z$. Both phases are metallic conductors and can interfere with the electric field in the melter if accumulated by sedimentation. This causes changes in the dissipation of heat and overheating in the sediment. Steep walls of the melter and a special flow-accelerating design of the bottom outlet overcome these problems. The platinum metal phases develop as a function of temperature and change their compositions time dependently (Pentinghaus 1994). The solubility of Pd is of great importance. It is strongly dependent on the effective oxygen fugacity, a fact which is also considered with implications for core formation in the Earth (Borisov et al. 1994).

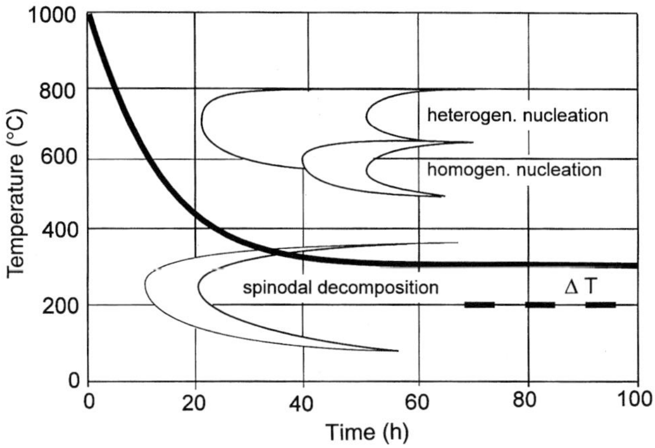

Fig. 120. Development of the central temperature in a thin-walled HLW-melt/glass canister during free cooling (∅ 42 cm, thermal power 4 W/l). Equilibrium temperature is reached after some 80 h, then the surface temperature is 200 °C, Possible phase transformations are plotted in this time-temperature field

Crystallization of the melt during cooling after filling the canisters at about 1000 °C does not occur, as is schematically indicated in the T,T,T-diagram of Fig. 120, where the evolution of the central temperature within a waste canister is plotted (Kienzler 1989). After about 80 h, a stable gradient between the core and the surface has been developed, the value of which depends on the thermal power of the glass product and the extent of cracking, which normally is quite moderate. Cracking cannot be avoided completely. The absolute temperature depends on the heat flow at the canister's surface.

Crystallization occurs only during annealing periods. The amount of crystals formed is very limited and does not exceed a few wt.%. Mainly powellite, $CaMoO_4$, with some substitution is formed. If fluorine is dissolved in the melt, then even fluorotremolite may crystallize. The waste products' chemical properties are not influenced by partial crystallization of these phases.

Spent fuel itself can provide a waste form and the chemical state of fission products in oxide fuels is well established (Kleykamp, 1988). Four groups can be distinguished:

- Oxides dissolved in the matrix: Sr, Zr, Nb, Y, La, Ce, Pr, Nd, Pm, Sm.
- Metallic precipitates: Mo, Tc, Ru, Rh, Pd, Ag, Cd, In, Sb, Te.
- Oxide precipitates: Ba, Zr, Nb, Mo, (Rb, Cs, Te).
- Gases and other volatile elements, e.g.: Kr, Xe, Br, J, (Rb, Cs, Te).

Materials characterization and studies of the chemical behavior of spent fuel, however, are still in progress (e.g., Johnson and Werme 1994). High burnup UO_2-based nuclear fuel can be simulated (Simfuel) with respect to its phase assemblage and microstructural features by sintering UO_2 and up to 11 stable oxides (between 3 and 6 at % at 1650 °C for 2 h under flowing H_2). The

UO_2 matrix, comparable to spent fuel, contains metallic alloys ε-(Mo, Ru, Pd), α-(Mo, Ru, Pd), intermetallic σ-$MO_{0.7}Ru_{0.3}$, and an oxide precipitate (Ba, Sr)ZrO_2-perovskite (Simfuel: Lucata et al. 1991; ε-phase properties: Pentinghaus 1990).

As spent fuel has a comparatively low chemical durability, it has to be kept in heavy casks for final disposal. This alternative waste form is now applicable after the cask technology has been developed (e.g., Bechthold et al. 1987). Even if reprocessing of spent fuel is performed, its direct disposal becomes mandatory when the yield of recovered uranium and fissible plutonium fades.

Synroc is an acronym coined by A.E. Ringwood (1978) for his concept applying synthetic rocks as hosts for RN. The crystalline phases the rocks are composed of can take up all the RN in solid solution and thereby behaving as reservoir minerals (see Sects. 5.2.1 and 5.2.2). The distribution of RN on the crystalline phases is governed by the crystal chemical potentials of their sites. Titania-based synroc recipes comprise, simply stated, members of the following structure families:

hollandites (Ba $(Al, Ti)_4O_8$),
zirconolites ($CaZrTi_2O_7$), and
perovskites ($CaTiO_3$).

Not all RN can be incorporated crystal chemically when higher waste loadings are needed. The preferred fabrication under sufficient reducing conditions helps to overcome this problem. Hot-pressing is performed in stainless steel bellows with an excess of TiO_2 and Ti, and this leads to the formation of metallic alloys based on the elements Mo, Ru, Rh, Pd, Ag, Te, Ni, and some Fe. Mo-rich, Ru-rich and Pd, Te-rich phases can be distinguished. The composition and mineralogy of synroc C, a main formulation, is displayed in Fig. 121 whereby the HLW loading is 20 wt.%. The microstructure is determined by mainly submicro-sized crystals and metallic inclusions and the hot pressed disks are fully densified. Titania-based synrocs can also be continuously cast from the melt (Pentinghaus 1981); the resulting textures, however, are quite

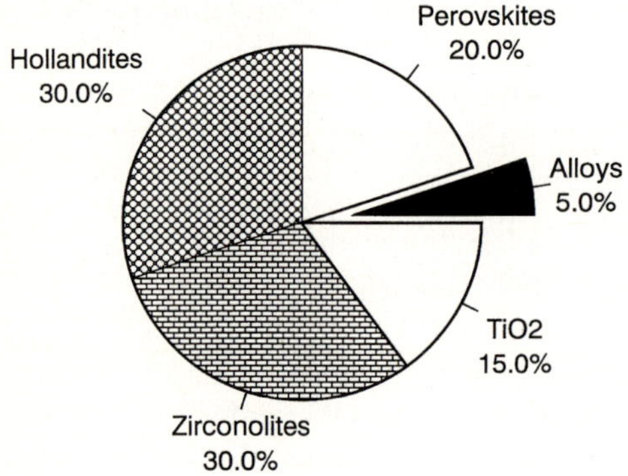

Fig. 121. Composition and mineralogy of synroc C, waste loading 20 wt.%

Fig. 122. Synroc B: coarse texture developing during continuous cooling from the melt (length 2 mm). Pol. microscope

coarse (see Fig. 122) and some amorphous material is always formed. The fuse-cast synroc contains a lot of shrink holes and shows enrichment of some elements in the amorphous residues. This simple route of synroc fabrication may become interesting for the solidification of certain individual RN. Recent studies on actinide solidification in synroc alone evaluate it as a suitable waste form (Vance 1994).

The synroc concept has brought the most comprehensive research activities on the main phases involved, which is reflected by the number of annual entries in *Chemical Abstracts*.

Synrocs can well tolerate the consequences of transmutations in due course of the decay (e.g., $^{137}Cs \rightarrow Ba$, $^{90}Sr \rightarrow Zr$) and of radiation damages which cause metamictization without significant changes in their chemical properties. The chemical durability of synroc formulations is superior to glasses as waste forms; however, there are no scales available telling how good the durability must be. Geology gives examples of prime importance of complex glasses as well as of titanates (e.g., zirconolites) which survived long geologic times.

Nevertheless, one has learned that the barrier function of the waste form itself is a considerable fact and must become part of safety assessments.

The technology of synroc fabrication is much more complex than the vitrification processes and poses many problems which have not yet been overcome, because so far no pilot plant operation with real waste could be performed.

Final Disposal of HLW, Chemical Properties of Waste Forms and Future Activities

Final disposal of HLW relies on the multibarrier concept (see Sect. 5.2.2), where the waste form is placed about 1000 m from the surface within a rock. The solid waste forms are encapsulated in stainless steel canisters or heavy casks (spent fuel) and specific overpacks and/or buffering surroundings are applied in order to keep these assemblages dry over the time period of the nuclides' activity.

Figure 123 schematically shows the different possible scenarios. In rocksalt as the host rock one can distinguish two cases. Under the action of heat transferred from the canisters and casks, the flow of the surrounding rock salt is accelerated. Figure 124 illustrates the dimensions of a heavy cask which will become the standard container for different waste forms.

Within months, the full contact between container surfaces and the rock is then obtained and the pressure is built up. Heat flow now will keep the surface temperatures at 200 °C at the most, and the whole rock will expand homogeneously, depending on the thermal power installed. Final disposal in hard, fractured basement rocks like granite or in thick sequences of extrusive volcanic rocks makes individual buffer surroundings necessary, bentonite being

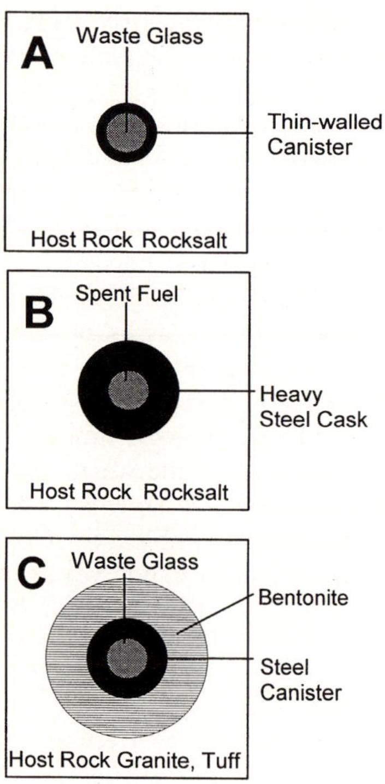

Fig. 123. Multibarrier concept of final disposal of HLW. Three basic scenarios

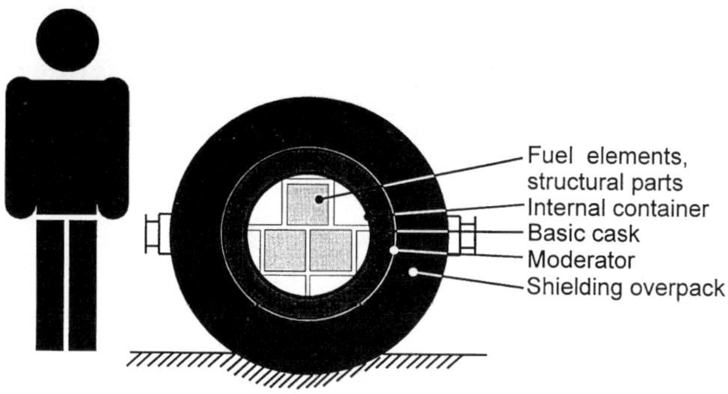

Fig. 124. Sketch of a heavy cask for different waste forms illustrating the dimensions, outer diameter 1.4 m, length about 6 m

the best solution (e.g., Grauer 1994). Even low permeability sediments can provide the host rock, depending on the national availability. In order to keep the thermal loading of the geologic body as low as possible for a given capacity, interim storage of the conditioned HLW is mandatory. This affords minimum times between 25 to 50 years. Longer times up to 100 years would be very advantageous; however, several reasons make it advisable not to leave the decision on unretrievable disposal to the next generation. The severest argument comes from the unknown political stability in the future, which should guarantee the technical maintenance of the storage site and the safeguard of the nuclide inventory.

So far, there are no final repositories for HLW waste forms under operation worldwide. Their availability is directly dependent on the age of the HLW, and for commercial waste (from energy production) they will be needed after the next 20 years.

If a failure in engineered barriers has to be considered and aqueous corrosion of the waste forms becomes possible, then knowledge of their individual properties will be necessary in order to predict any release of RN into the aqueous phase.

With advancing age of a repository, the temperature will decrease, as exemplified in Fig. 125. Down to about 100 °C, innumerable experiments with simulated and, more recently, with active glasses either isothermally or over gradients have been performed under stagnant or flowing conditions. The yield of these research efforts is quite a sound understanding of the dissolution reactions and the kinetics involved. Today, we know what we have to know (Grambow 1994). Aqueous interactions with waste glasses develop specific layered textures. This holds as well for the aqueous corrosion of the basic glass frits. They have been studied separately in order to understand which secondary phases form progressively from the glass constituents. In Fig. 126, such a sequence of pristine glass, amorphous silica, smectite, and lime-silicate hydrates is visible, which was formed after 6 days at 200 °C from the basic frit

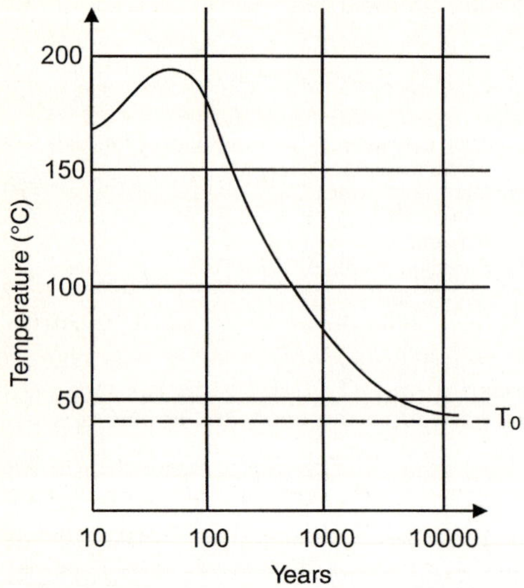

Fig. 125. Change of surface temperature with time on a HLW-glass canister in a rock salt repository (Calculation is based on 10 wt.% waste load, ∅42 cm and thermal lower of 4 W/l, after interim storage)

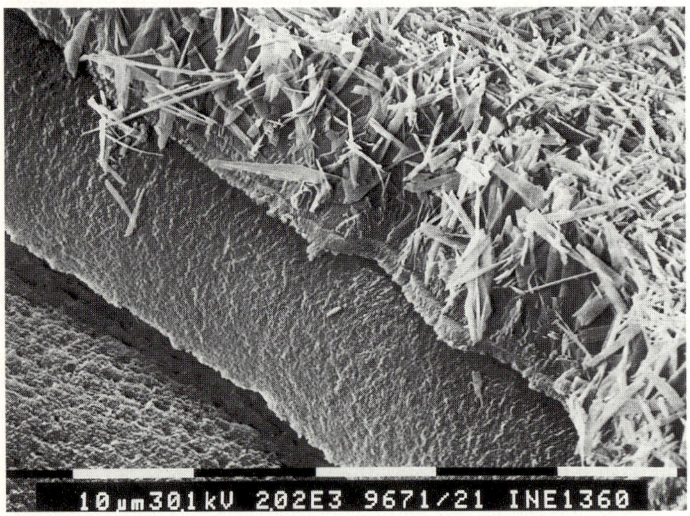

Fig. 126. Sequence of phases formed in due course of corrosion of basic glass frit VG 98/12.2 in water at 200 °C under equilibrium pressure after 6 days. SEM/SE

VG98/12.2. With increasing time, the crystalline phase assemblage changes into the final state consisting of silica-lime hydrates, zeolites, due to the Al-content, and quartz. Basic frits are corroded faster than the waste glasses. In any case, the potential of fixing RN in these phases is very limited. The effective activities of the glass constituents will influence the formation of phases from the dissolved RN. Dominating are pseudobinary or -ternary REE-molybdates for the largest group of RN. The phase assemblages that determine the second barrier of RN after a corrosion of the waste form are of paramount importance for the validation of the long-term behavior of the waste form. This is a demanding field of future research.

This direct experimental access covers about 1000 years when major RN have decayed. Low-temperature corrosion and the emerging phase assemblages cannot be studied in this way. Instrumented diffusion reactors operated under a broad variety of boundary conditions (chemical gradients, redox states, etc.), however, indirectly open a route to the syntheses and crystal growth of those phases that might be formed in the far future. This experimental strategy already been successfully applied to REE and molybdate solutions.

Further kind of information on long-term behavior of natural materials comes from the study of so-called natural analogs (e.g., Miller et al. 1994). The main advantage of these studies lies in the known age of the materials so that the observable changes can be scaled. This kind of research will generally help to decipher low-temperature reactions and their kinetics, and thereby indirectly support experimental work on the chemically different waste forms. Today, we are already on our way from description to prediction.

References

Bechthold W, Closs KD, Knapp U, Papp R (1987) Direct disposal of spent nuclear fuel. Radioactive Waste Management Series, Publ. by Graham & Trotman Ltd. London of the Commissions of the European Communities, EUR 11268 EN

Borisov A, Palme H, Spettel B (1994) Solubility of palladium in silicate melts: Implications for core formation in the earth. Geochim Cosmochim Acta 58:705–716

Grambow B (1994) Borosilicate glass: future research requirements or "what we don't know" Mater Res Soc Bull XX, 12:20–23

Grauer R (1994) Bentonite as a backfill material in a high-level waste repository. Mater Res Soc Bull XX, 12:43–46

Johnson LH, Werme LO (1994) Materials characteristics and dissolution behaviour of spent nuclear fuel. Mater Res Soc Bull XX, 12:24–27

Kienzler B (1989) Cooling and cracking of the technical HLW glass products: Experimental and numeral studies. Mater Res Soc Symp Proc 127:191–198

Kleykamp H (1988) The chemical state of fission products in oxide fuels at different stages of the nuclear fuel cycle. Nuclear Technol 80:412–422

Lucuta PG, Verrall RA, Matzke HJ, Palmer BJ (1991) Microstructural features of SIMFUEL – simulated high-burnup UO_2-based nuclear fuel. J Nuclear Mater 178:48–60

Lutze W, Ewing RC (eds) (1988) Radioactive waste forms for the future. North-Holland, Amsterdam

Miller W, Russel A, Chapman N, McKinley I, Smellie I (1994) Natural analogues studies in the geological disposal of radioactive wastes. Studies in environmental science 57. Elsevier, Amsterdam.

Pentinghaus H (1981) To Synroc through melting: thermal analysis, thermogravimetry and crystal chemical characterisation of phases. Proc Jülich Conf 42:713–732

Pentinghaus H (1989) On the process chemistry of HLLW-vitrification in a liquid-fed ceramic melter (LFCM) In: Ebert K, v. Ammon R (eds) Safety of the nuclear fuel cycle. VCH, Weinheim

Pentinghaus H (1990) Dissolver residues of fuel elements from nuclear reactors: predicting metallic ε-phase (Mo-Tc-Ru-Rh-Pd) behaviour during solidification processes. Am Nuclear Soc Proc SPECTRUM '90:240–242

Pentinghaus H (1994) Behaviour of platinum metal elements during the vitrification of HLLW by means of the LFCM-technology. Am Nuclear Soc Proc SPECTRUM '94:2237–2239

Ringwood AE (1978) Safe disposal of high level nuclear wastes. A new strategy. Australian National University Press, Canberra

Vance ER (1994) Synroc: a suitable waste form for actinides. Mater Res Soc Bull XX, 12:28–32

CHAPTER 6

Gemology and Jewelry: Scientific and Technological Bases. Artistic Value of Minerals

6.1 New Concept of Gemology and Jewelry; Crucial Changes and Contemporary Situation

A. S. Marfunin and J. Shigley

A multisided branch of knowledge has arisen during the past decades as a result of the crucial historical changes in industry, art, and mineralogy of diamonds, gold, and colored stones. There is no a single word defining this branch, but its general essence is a special aspect of mineral matter: its quality as an artistic value.

Three kinds of significance of minerals exist: (1) as a source of metals, energy, materials; (2) as geological materials; (3) as mineral type of matter in the Universe. The fourth kind of significance of minerals is represented by their artistic (esthetic) value. The qualities of minerals include their physical parameters, origins, ore component contents, properties in technological and material science, chemical bonding and spectroscopic characteristics, and finally, their esthetic properties corresponding to special aspects in the human perception of inorganic nature.

The conventional definition of gemology as a field of science about gemstones must be extended to include the connection between (1) its objects and (2) the approaches to the objects.

The objects of gemology can include:

- mineral species which occur as gemstones or even gem-quality crystals;
- diamonds and faceted brilliants;
- transparent colored stones, faceted: emerald, ruby, sapphire, other colored transparent stones;
- colored stones not faceted;
- collection minerals;
- rare gem-quality crystal of minerals;
- nonconventional precious stones;
- organic minerals: pearl, amber, coral;
- synthetic gemstones and imitations;
- gem materials treated to improve their color or appearance;
- stones not faceted, stones cut, and stones mounted in jewels which give them their final artistic significance;
- glyptics (cameo, intaglio);
- gold and platinum as native minerals and jewelry materials whose significance cannot be separated from the significance of gemstones in jewelry;
- jewelry;
- ornamental stones.

The approaches to this large field include its significance to industry, market, art, jewelry, and aspects of mineralogy.

The collection of all these objects on the basis of the artistic value of mineral matter is divided between these different parts of gemology, each representing a special kind of art and jewelry, their industry and market, and their special professional methods and technologies. Diamonds and colored stones form quite different fields of gemology (different volumes of production, methods of identification and grading). Rough diamonds and brilliants compose separate parts of the industry and market. Gemology and jewelry, gold and stones, glyptics and pearls, synthetic crystals and collection minerals each represent quite different parts of the main branch, which can be called gemology in its broadest sense.

The combination of a series of crucial changes has led to a new situation in the industry, market, art, and mineralogy of diamonds, gold, and colored stones. These changes as a whole are characterized by the greatly varying artistic value of gemstones.

Discoveries of the principal diamond provinces and greatest deposits: in Africa: South Africa (Kimberly, Premier, Venetia), Zaire and Botswana (Orapa, Yetlhakane, Zwaneng), then Namibia, Guana and Angola; in Russian: Yakutian pipes (Mir, Udachnaya, Yubileynaya) and, in perspecive, Arkhangelsk; in Australia (Argyle). Several thousand kimberlite (and lamproite) pipes have been found, but only about 60 are economically diamondiferous, and only 10–15 are major producers. Only five countries are responsible for 92% of the world's annual production of natural diamonds.

As a result of these discoveries, the peak of diamond production (up to 108 million carats) was reached in 1995, which is comparable with the first half of this century (millions carats). Diamonds have progressed to an other scale of production, and have reached still greater prevalence over colored stones (Fig. 127).

- Important changes in diamond production have been supported by still greater changes in the role and production of gold, the other component of the jewelry industry:
- increasing gold production from a maximum of 1280 t in 1965–1966 to 1890 t in 1995 owing to developments in extraction of the metal from ore (nevertheless gold production in South Africa has decreased);
- most important is a historical change of the role of gold: after 2700 years of its monetary role, when gold entered state finance and private treasuries, after its final demonetization in 1971–1978, gold became directed to jewelry use. In 1995 jewelry consumed 2537 tons of gold! (Electronics consumed 185 t, gold reserves have taken 281 t, instrument 44 t, other 285 t. In all, 3355 t were consumed in 1995: 1890 t production, 102 t sales, 232 t official sales, 583 t gold scrap, forward sales 461 t, hedge 87 t).

The extraordinary and probably maximum growth in the production and marketing of colored stones. Only a few occurrences of all the gemstones have been intensively explored. Import of the "big three" – emerald, ruby, sapphire – alone in the USA and Japan in 1994 were valued at more than a billion dollars

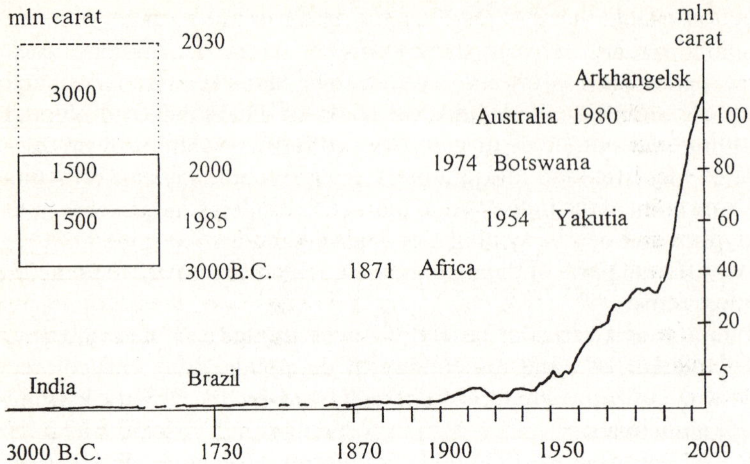

Fig. 127. Total world diamond production from ancient time to the second millinary; crucial increase in production during the past decades

(half for emerald, about one quarter for ruby, and one quarter for sapphire) while all other colored stones were valued at 200–250 million dollars. Topaz, amethyst, cytrine, aquamarine, tourmaline, chrysoberyl, garnets, spinel, kunzite, opal, pearl and new gem minerals (tanzanite, tsavorite, and others) became widely available.

- Along with the extraordinary increase in diamond, gold, and colored stone production, the new phenomenon which arose with widespread synthesis of gem materials and imitations, created problems in identification and certification.
- Production of industrial synthetic diamonds reached 300 million carats, three times more than all natural diamond annual output, and about ten times more than natural industrial diamond output.
- Gem-quality synthetic diamonds up to 14 carats have been obtained; “superdiamonds” isotopically enriched for purposes of electronics, and special synthetic diamonds with a single selected H3 type color center for laser materials.
- Production of CVD (chemically vapor-deposited) diamond thin films will reach 3.5–5 billion dollars in the year 2000.
- Outstanding developments in several laboratories (Chatham, Gilson, Linde, Novosibirsk, Alexandrov, and many others) led to the commercial production of all major colored stones (“created minerals”) and the unlimited synthesis of imitations.
- Gemstone enhancement (heat treatment, irradiation, chemical treatment and impregnation, surface modifications, coating by thin films) has enormously increased during the past decades: a quarter – a third in 1984 in two

thirds-three quarters of treated gemstones on the market in 1994 (green diamonds, blue topaz, and others).

- Close competition between the attempt to synthesize gemstones identical to natural stones, and the sophisticated developments of methods of distinction of natural and synthetic crystals. Further deliberate efforts to obviate these distinctions that are not connected with quality improvement can raise the question: will precious stones continue to be precious? These efforts can be equated with making spurious coins.
- Developments in specific methods of identification and evaluation of gemstones, nondestructive and nondamaging: visual, crystalloptical, infrared, Raman, luminescent, express detectors for diamond and gold, arbitrage methods (EPR, microprobes).
- Problems in the certification of faceted gemstones: distinction from synthetic and imitations, irradiated and heat-treated, quality grading. Hallmark for gold, certificate for diamond. High standard of certificate service determines the level of confidence and represents a necessary condition for well-being in jewelry.
- Developments in faceting technologies: cutting of 20–30 million carats every year (more than 500 million crystals), severe restrictions in the "ideal brilliant" parameters, their conservation for fancy cut crystals, capability of full cut for extremely small crystals. There is also the increasing use of automated faceting methods.
- Entry of diamonds and gold in a new contemporary financial culture; INTERNET and the electronics market.
- Establishment of the system of the diamond market:
  - "ideal brilliant" as a standard for the system,
  - a system of the 4Cs in diamond grading as a basis for the system,
  - a system of professional training ensuring identical criteria in diamond sorting and grading,
  - a system of the world diamond-brilliant jewelry market with three separate group of countries,
  - a system of market regulation by De Beers,
  - an entry in the system of the Russian diamond industry.
- The third aspect of the significance of minerals as an artistic value.
- Two transformations of jewelry: into a special kind of figurative art and into one of the largest world enterprises (47 billion dollars). Democratization of jewelry.
- Development of the culture of stone and education of perception: distinguishing 6000 color graduations, clarity, perceiving the highest level of perfection in gem crystals. Jewelry as collections of rare precious stones.

6.2 Diamond in Gemology

A. S. Marfunin, M. Gaft, G. Panczer and J. Shigley

6.2.1 General

In spite of its simple structure and simple nominal composition, diamond takes a special place by its scientific and applied significance among all minerals and materials. This is reflected in the vast literature, which can be divided into several groups.

1. Diamond and other gem stones in gemology and jewelry:
   - monographs on the precious stones, their mineralogical and gemological characteristics; general texts on gemology; gemological dictionaries; art of jewelry, its styles, periods, specializations; world jewelry market;
   - professional manuals for sorting, cutting, grading, and price evaluation;
   - physicomineralogical gemological investigations of composition, structure, properties, and origin of gem minerals as a basis for understanding their gem qualities.
2. Physicotheoretical surveys on diamond, summarizing thousands of highly specialized studies of structure, spectroscopy, properties, applications, and synthesis of diamond as one of the most studied systems of solids and material of modern electronics and technologies (Davies 1994; Fields 1992; Proceedings of several international conferences on the New Diamond Science and Technology, including diamond films).
3. Petrological and mineralogical studies of different types of diamond origin and diamond deposits (in particular in Proceedings of several international kimberlite conferences).

   Diamond as a multifunctional material of electronis has been described in the first volume of *Advanced Mineralogy*. In this chapter the condensed summary of the results of physicomineralogical studies of diamond is presented to the extent of their significance for understanding the properties of diamond in jewelry. On the other hand, the specialized, highly professional vision of diamond in jewelry reveals mineral characteristics not revealed by other approaches.

The new situation in the significance of diamond results from a series of events and factors; but the first bases for investigating this new significance and the whole sense of its existence were developments in diamond synthesis.

The predestination of diamond is different for synthetic and natural crystals. Synthetic diamonds represent materials applicable in electronics and technology: only synthetic crystals can be obtained with different stable parameters; this is impossible for natural crystal with specimen properties that vary from one to another.

Three types of synthetic diamonds can be distinguished.

Industrial: their production is more than 300 million carats per year, ten times more than the output of natural diamonds which are not of gem quality or near gem quality.

Synthetic diamond crystals of gem quality: nitrogen-free; semiconductors (nitrogren-free with boron); implanted; isotopically enriched; with intentionally selected center (one from the 100 possible and coexisting centers) which may be used as laser material; precisely these crystals are the basis for diamond use as a multifunctional material (Vol. 1, Chap. 7.3).

Diamond films: constantly growing production and unique properties as materials.

Natural diamonds, which cannot compete with synthetic materials for industry and electronics, justify their existence as jewels. Synthetic diamonds do not compete in this field because the first property of a jewel is to be of natural origin. Thus, the purpose of natural diamonds is to have gem quality, to exist in the form of ideal brilliants, the only form that reveals diamond qualities, and to express the artistic value of the stone.

Both types of diamonds (synthetic and natural) in both fields of application allow the utmost results to be reached: the best parameters of the materials in industry and electronics, and the highest artistic quality in jewelry; this is unique among the other stones.

The main costs of diamond production are connected with the natural high quality diamonds (Fipke et al. 1995):

	Volume (%)	Value (%)	$ million
Gem quality	15	70	4290
Nearly gem quality	40	20	1100
Industrial	45	2	110

The only form of diamond as a gem stone is as cut crystal (ideal brilliant and related), since only its optical-geometrical conditions allow its artistic quality to be emphasized.

All diversity in fancy forms must be taken into account in these geometrical conditions. The artistic properties of diamond are not apparent in its natural crystallographic forms (octahedron, cube, rombododecahedron) but in the forms of brilliants, which are given to diamonds by cutting.

6.2.2 Mineralogical Types of Diamonds and Professional Classification of Real Crystals

The annual diamond production of 108 million carats represents more than half a billion individual crystals. The first step in the technological cycle of diamond from mining to jewel selling is the grading of each of these crystals. This grading is carried out at least three to four times for every crystal for jewelry or industrial application.

Parallel to this professional grading, the mineralogical description of the typical groups of diamond crystals is carried out.

Thus, the natural forms of diamond (before cutting into brilliant forms) are most important from two points of view: diamond sorting for different applications and mineralogical-genetic characterization.

The laborious and repeated process of diamond sorting is based on a unique classification system of individual diamonds, taking into account all their real features. The systematization is based not on generalized crystal types and crystallographic forms, but on the description of actually existing individual crystals according to their size, geometry, distortions, defects, facets, and colors. The main aim of such a classification is to forecast the brilliant yield after cutting each diamond of gem quality, or to predetermine their optimal use in different technical applications.

First the crystals are sorted into five groups by weight: two groups less than 0.65 ct, then 0.66–0.89 ct, 0.90–1.79 ct, and more than 1.80 ct. Lesser crystals are measured in grainers (0.25 carat).

Groups 1 and 2 are further subdivided into 10 mesh classes. Also subdivided into 10 mesh classes are crystals greater than 1.80 carats. In all, there are 24 groups by weight dimensions. Within each of these groups the crystals are sorted according to geometry (17 groups: crystals, fragments, twins, intergrowths), distortions (6 groups: along the third order axes or along the maximum distortion directions), facets (5 groups: smooth surface and different grades of rough surface), defects (5 groups), and colors (more than 21 groups).

For groups with different sizes and weights, the manner of classification is slightly different. Crystals are also selected for special technical applications.

In technological applications, the crystals are sorted according to their geometry into stones, shapes, cleavages, flats, macles, drilling, dressers, etc. The classification encompasses 4000 to 5000 types. The sorting is done by the wholesale trade and major customers.

Mineralogical varieties of diamonds are first described according to morphological features which are correlated with inner structural features (in particular with X-ray topography).

Morphological characteristics of diamonds include, on the one hand, the presence of only three or four crystallographic forms (mostly octahedra, then cubes and rhombododecahedra), but on the other hand, the exceptional variety of morphological types is connected with different growth and resorption forms, complicated concentric structures, curved octahedroids, cuboids, dodecahedroids, etch figures and microlaminations, striations, pyramidal hillocks, sculptured faces, block ornaments, growth and intergrowth macles, surface cover, plastic deformation, fissures, and inclusions.

Already in 1912, Fersman and Goldschmidt composed an atlas of 125 diamond crystals. Now two detailed classification schemes are accepted: those by Orlov (1977) and Harris (1987). The scheme by Orlov includes five types of

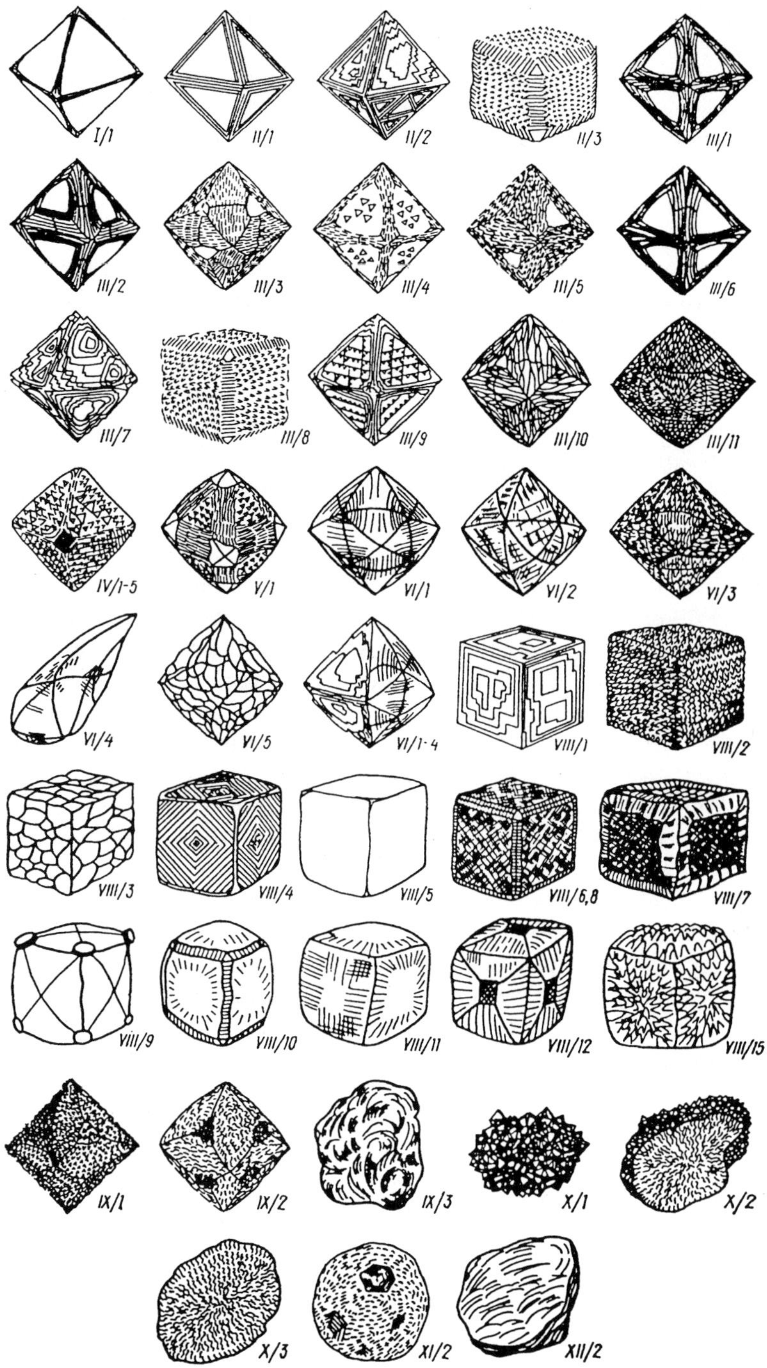

Fig. 128. Mineralogical (morphological) types of diamonds (Bartoshinsky 1983)

crystals and five types of aggregates. Type I: transparent flat-faced octahedra more or less curved. Most of the diamond crystals belong to this type (90–99% in the majority of deposits). For this reason, subdivision for the type I has to be based on real composition, structure, impurities, and centers which determine color and luminescence, IR and EPR spectra. Type II: yellow and green cubes. Type III: combinations of octahedron, rhombododecaedron, and cube. Type IV: coated diamonds (independent of crystal form). Type V: black diamonds with syngenetic graphite.

The polycrystalline types: VI: ballas (spherolites); VII: includes intergrowth of semitransparent yellowish crystals; VIII: bort, aggregates of small crystals; IX: bort, irregular pieces of grained aggregates of diamond crystals; X: carbonado, cryptocrystalline fragments of specific varieties of diamond.

The more detailed morphologic classifications have been presented by Bartoshinsky (including 12 groups and 55 types), J. Harris (based on diamond studies in South Africa), and I. Sunagawa (correlating the morphology with the growth conditions). Beside morphology, all these classifications take into consideration color, luminescence, surface features, and inclusions (Fig. 128).

An experienced specialist can distinguish between diamonds from different provinces and different kimberlite pipes.

Morphological forms of synthetic diamonds are a function of pressure and temperature: cube-cubo-octahedron-octahedron (with rising T).

6.2.3 Ideal Brilliant: Revealing Diamond Artistic Value. Division of World Diamond Trade and World Market: Diamonds and Brilliants

The esthetic perception of diamonds differs from that of all other precious stones: all are colored stones, while most diamonds are colorless. Their property as precious stones is determined by light inside the crystal, not color.

Thus, faceting is an indispensable condition to reveal the precious character. Thus, the purpose of diamond cutting is different from that for color stones, and consequently the geometry of cutting is different.

It follows that diamond quality is not a property of its natural crystallographic form, but has to be created by the optical-geometrical proportions of cut forms with narrow restrictions of angles and facet sizes.

The geometry of the classical form of 57 facet "ideal brilliants", calculated by Marcel Tolkovsky in 1919 is shown in Figs. 129 and 130. There are many types of "fancy" cuts of diamond, but in all cases the geometrical restrictions have to be taken into account.

The yield of brillants fashioned from a rough diamond crystal can be estimated theoretically as 55.5%; "practical" cutting corresponds to a yield of 52.7–57.9%. This means that about 44.5% of diamond weight is lost in sawing

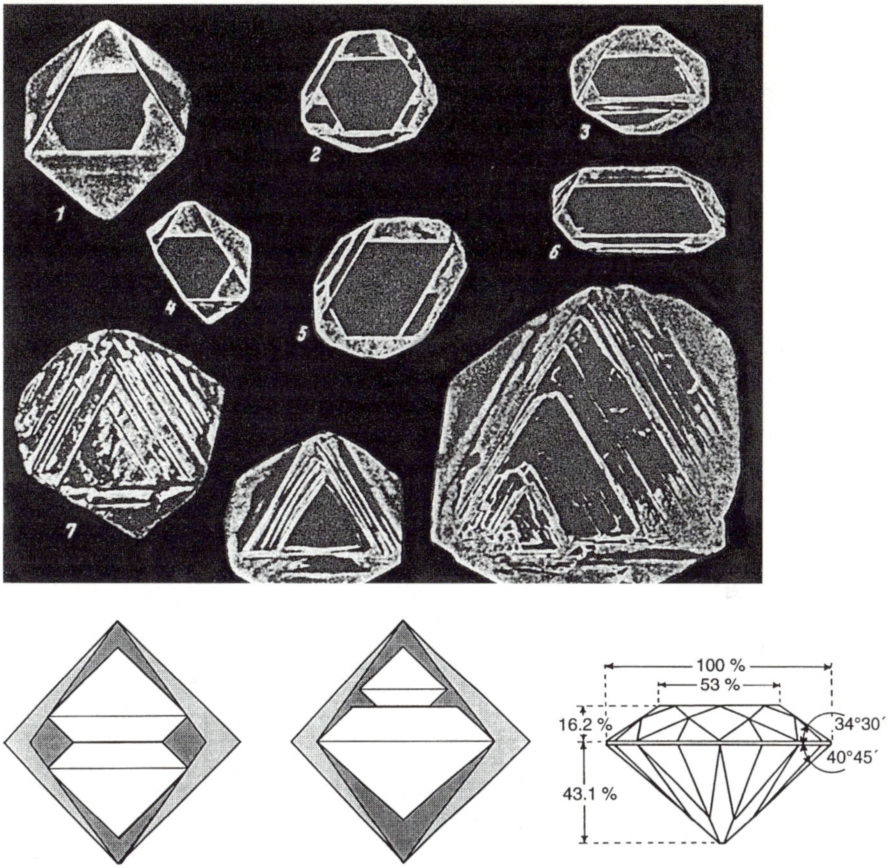

Fig. 129. From natural crystals to ideal brilliant. *Above* Natural irregular octahedral diamonds; below theoretical yield of two ideal brilliants sawn from one regular octahedron: 55.5 %; total yield estimation = world production of gem diamonds/world brilliant production = 40–41 %; *right* theoretical proportions of Tolkovsky ideal brilliant

and cutting. These losses have to be taken into account when prices of rough diamond and cut brilliant are compared: a carat of a brilliant costs at least twice as much as a carat of rough diamonds (Fig. 129).

The difference between rough diamonds and brilliants is of basic significance for the division of world trade and the world diamond market: one group of countries (Africa, Australia, Russia) are the principal rough diamond producers, and other group (Belgium, Israel, India, Thailand, USA) are the main brilliant producers.

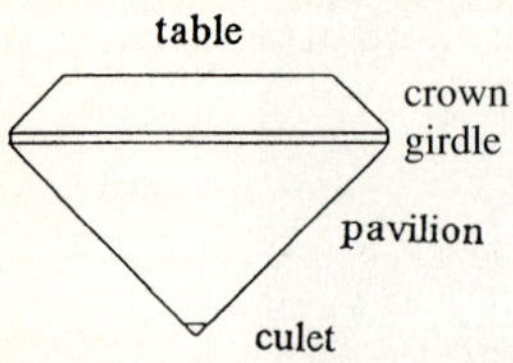

table	1
crown:	
table facets	8
uper main facets	8
uper girdle facets	16
pavillion	
lower main facets	8
lower girdle facets	16
culet	1

	58

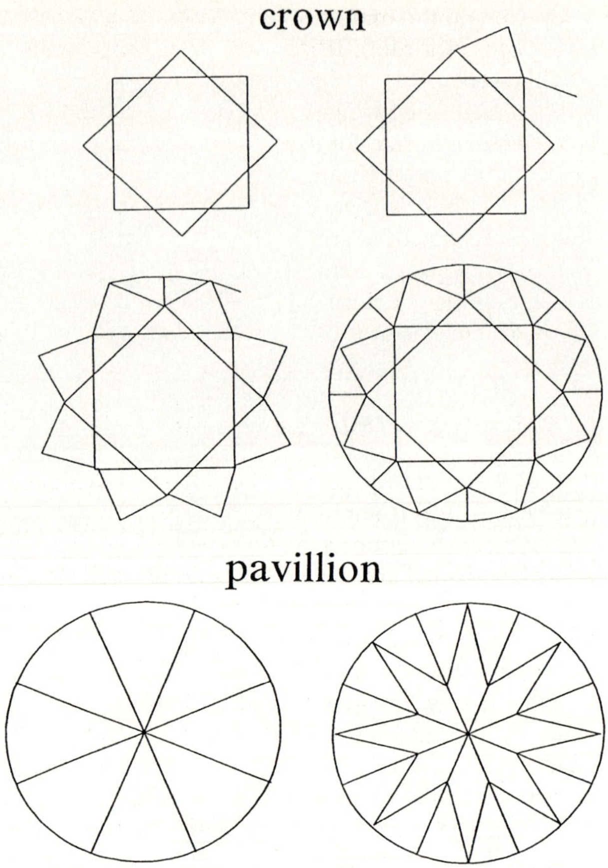

Fig. 130. For estimating and understanding brilliant you have to be able to see all its facets and for to see all the facets you have to know how to draw them

6.2.4 System of Cut Diamond Grading: Physicomineralogical Aspect

Cut diamond (brilliant) grading represents a unique procedure: it deals with the most subtle gradations of those principal characteristics of faceted diamonds which can be defined as approaching the ultimate perfection of matter.

Perception and evaluation of brilliant quality comprises four components, which give the 4C system: carat (the weight of the crystal in carats; i.e., 0.25 g), clarity, color, and cut.

Brilliants are evaluated visually by their size, but estimated by their weights measured in carats. The ratio between dimensions and carats is determined by the empirical formula: $M = 0.006122\, d^2 h$, where M is mass in carats, d diameter, and h height (for round brilliants). This may be used together with tables which contain the mass in carats for the different diameters and heights (with thick and thin girdle). It is worth noting the quadratic relation between mass and diameter.

With increasing weight, the visible size enlarges more slowly. For example, the difference in size for diamonds of 2 and 3 carats (with a huge difference in price) is only about 1 mm.

Since carbon is a very light element, small crystals of 0.03 carat already have dimensions of nearly 2 mm, 0.25 carat – 4 mm, 0.5 carat – 5 mm, 1 carat – 6 mm, 2 carats – 7.5 mm, 4 carats – 1 cm and 6 carats – 1.1 1.2 cm.

Cut grading is determined by deviations from the ideal proportions and the symmetry and features of the faces.

6.2.5 Clarity Grading and Microinclusions in Diamonds

The degree of clarity depends on the presence of microscopic inclusions and traces of fissures (their number, size, position, and visibility).

There are two aspects of microinclusions in diamond. In gemology, microscopic inclusions are defects diminishing the crystal value. In petrology, the inclusions are priceless sources of information about the Earth mantle and conditions of diamond formation. Petrologists study the inclusions in diamonds rather than diamond crystals. This has become an important field in the petrology, mineralogy, and geochemistry of the upper mantle. The microinclusions in diamond are much more informative in understanding their genesis than diamond itself.

Gemological assessment of the microinclusions in diamond is carried out visually: by the naked eye (in the same way as the artistic value of diamond is evaluated) or with professional (×10) loupe or gemologial microscope.

In petrological and geochemical assessment, the most sophisticated equipment is used: X-ray microanalyzer, ion-microprobe, proton microprobe, scanning electron microscope, and advanced mass-specrometry methods.

Petrologists complain that they found only 500–600 crystals with inclusions of 1–2 mm among 10000 studied diamonds. Gemologists are dissatisfied that the majority of diamonds contain one-two-three microinclusions of 1 mm and less (up to 1 nm). Analysts suspect that submicroinclusions of extremely small size can also occur.

The clarity characteristics are graded depending on size, position, and number of inclusions: loupe-clean, Fl (flawless) VVS (minute inclusions), VS, SI (small inclusions), P (Pique), or I (inclusions visible to the naked eye).

Diamonds in Precambrian magma of the mantle form a closed geochemical space detached by their faces. With increasing number of microinclusions, these "microminerals in diamond" became a specific mineralogical phenomenon. The list of microinclusions in diamonds now includes 59 minerals:

– Ultrabasic paragenesis (olivine-pyroxene rocks): olivine (forsterite), enstatite, Cr-diopside, Cr-pyrope, phlogopite, Cr-spinelide, Mg-ilmenite, magne-

siowustite, ferropericlase, zircon, native iron, chromium, nickel, and early generation of diamond (inclusions of "diamond-in-diamond" type).
- Eclogitic paragenesis (garnet-pyroxen rocks): omphacitic clinopyroxene, pyrope-almandine, kyanite, coesite, sanidine, rutile, ruby, native iron, moissanite, wustite, and diamond).
- Unclear paragenesis: biotite, muscovite, amphibole, magnetite, apatite, mineral of chevkinite type, and mineral of the loparite type.
- Sulfides, which are dominant inclusions in diamonds: pentlandite, troilite, pyrrhotite, chalcopyrite, Zn-chalcopyrite, cubanite, millerite, violarite, polydimite, heazlewoodite, pyrite, djerfisherite, solid solutions based on pyrite, solid solutions based on cubanite, and Ni-sulfides: $(Ni, Fe)_9S_8$, $(Ni, Fe)(NiFe)_{3+x}S_2$.
- Epigenetic: serpentine, calcite, graphite, hematite, kaolinite, acmite, richterite, perovskite, spinel, xenotime, hematite, and sellaite.
- Additionally, microinclusions of the mantle melts; fluid inclusions (with H_2, CO, CO_2, CH_4, H_2O, N_2), occluded noble gases He, Ne, Ar, Xe, Kr, and cloud-type inclusions (possibly with apatite).

Only in the past 10–12 years have microinclusions in diamonds been investigated with all the resources of contemporary methods (X-ray, proton, ion microprobes, mass spectrometry).

There are two main directions of the studies.

Isotopic analysis: (1) Age determination of syngenetic microinclusions using trace impurities of U in sulfides and K, Rb, Sr, Sm, and Nd in pyroxenes and amphiboles (incompatible with the composition and structure of the minerals). It was shown by the Sm-Nd method that the age of diamonds syngenetic with inclusions may be 3200 Ma, while the time of kimberlite intrusion is 85 Ma (Kimberley), or 1580 and 1126 Ma (Argyle). These may be interpreted as mantle xenocrysts of diamonds which later were taken and carried out by kimberlite magma. (2) Isotopic composition of occluded noble gases: $He^3 - He^4$, $Ne^{20} - Ne^{22}$, $Ar^{40} - Ar^{36}$, $Xe^{129} - Xe^{132}$, $Xe^{136} - Xe^{132}$. (3) Isotopy of diamonds themselves: $\delta^{13}C$, $\delta^{15}N$.

Geothermometry and geobarothermometry by Mg-Fe distribution between coexisting garnet and clinopyroxene, garnet and olivine, orto- and clinopyroxene: temperature ranges 950–1370 °C, pressure 5–7.6 Gpa. Inclusions with dimensions of 1–2 mm (chromite, olivine, garnet, pyroxene), which are very appreciated by petrologists and geochemists, are present in only 2–5% of the crystals.

Micron-sized microinclusions observed by gemologists are present as: (1) black sulfides accompanied by disklike and rosettelike cracks; (2) black (without cracks) chromite (often 1–2 mm), or ilmenite; (3) purple and rose Cr-pyropes, red-orange pyrop-almandines; (4) colorless and greenish pyroxenes and olivines. The microinclusions occur in all zones of diamond crystals: small in the center part (seeds) and larger in the external areas.

The range of the strains introduced by inclusions is sulfides-olivine-enstatite-garnet-chromite. The strongest strains around sulfide inclusions may relax by cracks and do not influence the cutting of diamonds. On the contrary, garnet, and especially olivine and pyroxene inclusions, are accompanied by strains which lead to destruction of the crystal during cutting. Gemologists have to consider that microinclusions provide evidence of the diamond's source: they are evidence of mantle genesis, age, and natural (not synthetic) origin.

6.2.6 Four Types of Diamond Color. Real Composition and Real Structure. Color Centers

Color is the central and most subtle part of brilliant grading. Different crucial aspects of its artistic impression are connected with its color.

However color is only the visible manifestation of the most complicated inner structure of diamond. Its most subtle grading reflects the whole complex of physicotheoretical, crystallochemical, and analytical studies of the color centers in diamond.

Four types of diamond color can be generally distinguished:

- Colorless to yellowish: this is a "regular" color of jeweler diamonds connected with nitrogen impurities and certain types of nitrogen centers.
- Brown to yellow Australian (Argyle) called Cognac and Champagne, related to dislocation centers.
- Fancy colors:
  - Rare blue diamonds: the color is connected with boron impurity.
  - Green diamonds: color is radiation-induced (resulting from either natural or artificial irradiation) and is connected with vacancies (formed not only after irradiation, but during growth and as result of transformation of color centers in the solid state).
  - Red, pink, or purple, rarest crystals of Australian diamonds among prevalent brown and yellow; the color in both types is connected with dislocation centers.
  - Yellow, single nitrogen atoms substituting for carbon.
- Black and dark violet diamonds with epigenetic finely dispersed graphite.

Different specimens of diamond crystals acquire or change color by radiation. Further annealing changes the acquired color in a different way, depending on the preceding history of the crystal.

All spectroscopic centers which appear in absorption, luminescence, IR, Raman, EPR, and double resonance spectra belong to color centers.

All diversity in natural gem quality diamonds, all diamond properties in contemporary technology and electronics, and the understanding of real diamond as the most-studied physical system, are based on studies of the color centers. These studies include:

- Real composition, including all impurities and their aggregated forms.
- Real structure, types and models of the centers and their positions in the structure.

- Energy level schemes and elelctronic structures (chemical bonding characteristics) for each type of center.
- Interpretation of luminescence, IR, Raman, and EPR spectra.
- Mechanisms of center formation and transformation during growth, natural and artificial irradiation and heating.
- Influence on diamond properties which are important in electronics, technology, and gemology.
- Identification of centers in natural, artificial, and treated diamonds.

Gemologists and owners have to understand what is behind diamond clarity and color.

Real Composition of Diamond

Characteristic of diamond composition is determined by the combination of the nearly ultimate purity and the exceptional sensitivity of all its properties to the smallest contents of impurities. The diversity of natural diamonds and the possibility of using synthetic and natural diamonds in many different fields are due to these quite small impurities. This entails two problems: the necessity of using the most sensitive analytical methods to distinguish the mode by which detected impurities enter the diamond structure (and influence properties) or the form of submicroinclusions of other phases.

Unique investigations on the limits to the analytical capabilities of even the most sophisticated methods have been carried out (Sellschop, Bibby, Fesg); using gem-quality diamond crystals with no visible inclusions, even with 10× and even 50× magnification. This means that the problem of impurities in diamond has been transferred to the submicrolevel.

Analytical investigations have been carried out by neutron activation analysis (for elements with atomic numbers more than 20), gamma-activation analysis, charged particles activation, proton- and electron microbes, spectrochemical analysis with ion beams, gas chromotography, mass spectroscopy, and Rutherford backscattering.

In total, 64 Elements have been detected in diamonds. However, consideration of the list of elements which have not been found shows that those absent are the elements with the smallest geochemical abundance or which are "unsuitable" for existing element-specific analytical methods. (Among the rare-earth elements only Pr, Pm, and Tm are absent, among actinides only Ac and Pa, among alkaline and alkaline-earth elements only Fr and Ra, among noble gases only Ru, among halogens only I and At, among platinoids Re, Os, Ru, Rh, Pd, etc.).

Detected elements can be divided: ppb elements (from As to U, i.e., heavier and less abundant) and ppm elements (lighter, including usual rock-forming elements). Among the ppm elements there are the groups of inter-correlated elements whose compositions corresponds to the most usual micro-inclusions: olivine, pyroxene, garnet, ilmenite, Cr-spinelide, and sulfides.

It was most important that Sellschop and Bibby proved that the distribution of these elements is uniform and similar in all crystals. This may be explained by the uniform distribution of disperse submicroscopic "magma droplets" (of the parental magma) with dimensions from 1–3 microns up to nanoparticles.

In diamonds with cloudy type inclusions, voids were found by the scanning electron microscope. It is presumed that "voidites" probably also belong to this category.

Among all the elements, nitrogen is the most common impurity in diamond. Its content usually amounts to 20 to 200 ppm ($\sim 10^{19}-10^{20}$ at cm^{-3}), and approaches 5500 ppm ($\sim 10^{21}$ at cm^{-3} or 0.55%). The boron content is lower by 2 or 3 orders (0.02–0.26 ppm).

The decisive importance of nitrogen (and boron) for the structure and properties of diamond is determined by its presence in the diamond structure but not in microinclusions. It is unambiguously confirmed by EPR, and is manifest in optical, luminescence, and IR spectra.

Can it be that other elements of the 64 detected in diamonds enter the crystal structure? Most of them occur in the form of microinclusions of microminerals, melt drops, or fluid inclusions; but some are actually incorporated into the crystal structure. This was shown for substitutional Ni in natural and synthetic diamonds by EPR, and by implantation of such incompatible elements as Li, P, He, Al, Si, Ne, Ti, Cr, Ni, As, Ag, Sb, Tl, and Xe.

The only reliable tool showing the nature of incorporation is EPR. Supersensitive EPR techniques (indirect detection) and transformation of certain elements in paramagnetic states enable detection of $10^{8}-10^{10}$, or even only 10–100 centers in a sample.

EPR, luminscence, IR, and optical spectroscopy have shown that the total analytically detected content of nitrogen (and other substitutional impurities) is divided into several different forms. Each form (color center) has its own spectrum, or it does not manifest itself in a certain type of spectroscopy. Nonparamagnetic centers do not appear in EPR, other forms are not detected in optical and luminescence spectra. The main part of nitrogen in diamonds (usually ~0.02%) is present in a form which does not influence the color. Only the smallest part of nitrogen forms the centers which determine the diamond's color. These extremely small contents of several color centers (0.001–0.0001%) are visible to the naked eye. If all nitrogen had been present in diamond only in an optically active form, diamond would not have existed as a completely transparent colorless precious crystal, but as a brown-yellow mineral.

Thus, it is possible to suppose that in the closed space of diamond crystal all the elements of the Periodic Table are present, in amounts up to 1 ppm for the lighter elements (up to As, N 33) and 1 ppb for all heavier elements, and in diverse forms.

The contents of such heavier elements (from As to Au) are usually 0.01–0.001 ppb, i.e., 0.000000001% and 0.0000000001%.

For nitrogen only it is proved that is present in the diamond's structure from a maximum of 0.55% (5500 ppm), average 0.02% (20 ppm), and minimum

0.00000006% (0.0006 ppm in "nitrogen-less" diamonds). Boron contents in diamond structure are usually less than 0.06 ppm.

Besides N, B, and Ni, all elements, at least to a great extent, are dispersed in the space of diamond crystals as submicro- and nanoinclusions of "magma drops", fluid and gaseous inclusions. If these dispersed elements do not enter into the diamond structure, then each of these captured drops of the parent magma must contain all the elements in ppm and ppb quantities.

The composition of the most clean "nitrogen-less" diamond can be expressed as $C_{99.9999}\ N_{0.0001}\ [le]_{ppm}\ [he]_{ppb}$, while for the usual diamonds the composition is $C_{99.98}\ N_{0.02}\ B_{0.001}\ Ni_{0.001} \cdot [le]_{ppb}$, where le is lighter elements, and he heavier elements (in the form of submicroinclusions).

Real Structure of Diamonds. Types of Color Centers

A complete presentation of diamonds' real structure must consist of systematic placement in the structure of all compatible substitutional impurities, all their aggregate forms, vacancies, interstitials, dislocations, aggregates of impurities with vacancies, interstitial atoms and dislocations, physically nonequivalent varieties of the centers, and geometrically equivalent repetitions of each center.

The models must be consistent with the EPR ideology and have to include: substitional position in a specific polyhedron of the structure, local symmetry of the center, orientation of the axes of the center (of the g_x, g_y, g_z factors) relative to the polyhedron, orientation relative to crystal axes, parameters of fine structure depending on interaction with the nearest neighbors, superfine structure parameters, depending on interaction with the nucleus of the center, superfine structure parameters, depending on interaction with nucleii of the atoms in the first and second coordination spheres, and number of physically and geometrically non-equivalent positions.

The parameters of the spectra must be compared with quantum chemistry calculations. Each model must be correlated with IR, optical, and luminescence spectra.

At present, the number of centers detected by different spectra is about ~100 types. For more than 50 types the parameters have been measured by EPR. More than 350 lines have been established in electronic, vibrational, and EPR spectra from the UV and visible up to IR regions. These centers are only partly correlated.

The particularity of the study of these centers is that, in spite of having obtained the most detailed spectroscopic measurements, there are models for only few centers. From more than 100 settled centers, models based on EPR interpretations have been determined for only seven.

In principle, the description of the real structure presents all possible ways of placement of defects in the structure type for all possible structure centers.

For this purpose, diamond structure can be presented in the form of coordination polyhedra (Fig. 131). The coordination tetrahedra must be shown not

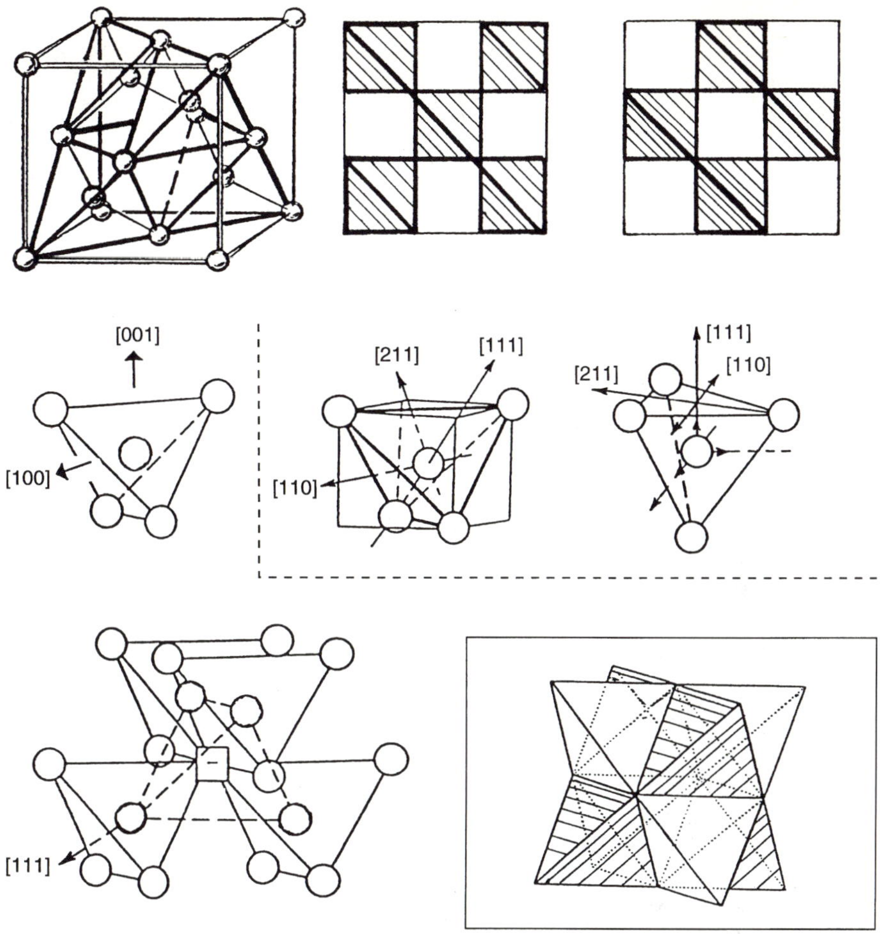

Fig. 131. Tetrahedral carbon groups in diamond structure. Alternation of occupied (by carbon) and empty tetrahedral at two levels. Crystallographic orientations of carbon tetrahedra. Position of a vacancy at the intersection of four occupied and four empty tetrahedra. Four occupied and four empty tetrahedra around an empty octahedron

only for those occupied by carbon atoms with four carbons in vertices, but also empty carbon tetrahedra which can be occupied by interstitial carbon or impurity atoms.

The occupied tetrahedron is connected by each of its four vertices with four occupied tetrahedra in the [110] direction. The occupied tetrahedron is connected by each of its six edges with six empty tetrahedra in the [100] direction. The faces of the four occupied tetrahedra and the four empty tetrahedra form empty octahedral positions. The empty tetrahedra and empty octahedra comprise neighboring empty volumes suitable for placement of complex aggregates of interstitial atoms.

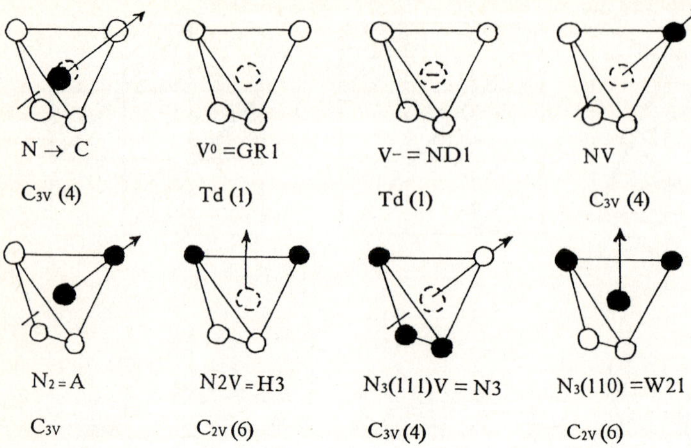

Fig. 132. Models of the centers in diamond, their spectroscopic designations, local symmetry and number of symmetry-related positions

By means of this presentation of diamond structure, the description of local symmetry of the centers, orientation of the centers in the crystal, and the number of physically and geometrically nonequivalent varieties of center can be shown. These structurally nonequivalent varieties of the same center (e.g., N_2V) explain the existence of many centers in the presence of only nitrogen as usual impurity (Fig. 132).

The origin of diamonds imposes certain restrictions on the distribution of the types of centers. It reduces the vast number of possible centers to several sets in diamonds of different origin.

The conventional physical classification of diamonds is based on classification by nitrogen content and form:

Ia contain nitrogen in aggregates form;
Ib with single substitutional nitrogen atoms;
IIa nitrogenless;
IIb containing boron.

More detailed presentation of the diamond types can be as follows.

1. All natural crystals of diamonds contain nitrogen impurity; in "nitrogenless" diamonds, nitrogen content decreases up to the sensitivity limits of optical and IR absorption (less than 10^{17} at cm^{-3}, or 0.6 ppm), but it may be detected by EPR and luminescence. Nitrogen is the only impurity which is always present in all types of diamonds. Other impurities (B, Ni) are present in special cases.

In al natural diamonds, nitrogen is present in aggregated forms together with small amounts of single nitrogen atoms. Thus, sets of three to four and up to seven to eight and even more nitrogen centers are always present. To detect their existence, a combination of several methods is needed: IR spectroscopy for nitrogen pairs and more complicated aggregates; optical, luminescence, and EPR spectra for other types of centers.

Table 41. Types of centers in diamonds

Nitrogen		Aggregation	Irradiation		Dislocation	Impurities	
	N	N+V	Vacancy aggregation	C_iN_i	D NDVCi	B	Ni
Ib	"C"(N)	(NV)	$GR_{1-8}(V)$	$01(C_iV)$	ND	B	Ni^+
	N.. N..	$W_{15}(NV)^-$	$ND1(V^-)$	NC_i	N_2D	BN	Ni_S^-
			V^+				
EPR		$640(NV)^+$	$3H(V_2)$	$575(N_iV)$	D_1		(NiE-1)
		$440(NV_2)$	$R(V_3)$	TR_{12-17}	D_2		(NiE-2)
Lum		S1	$R(V_4)$		D_3		(NiE-3)
		$S2(N_2V)$	$R(V_5)$		R_F		NIRIM-1
O_{Pt}		$S3(N_2V)$	[A1, A2, A3, N6, N7,		R_D		NIRIM-2
		$H3(N_2V)$	O2, R_1–R_{18}, W_4W_5,		R_A		
		$H4(N_2V)$	W_{11-20}, W_{29}, $W_{33}W_{34}$]				
		$H2(N_2V)^-$					
	W21(N3)	P2 = $N3(N_3V)$					
IR	$A(N_2)$	Platelets					
IR	$B(N_4)$	Voidites					

Energy transfer between excited states of the coexisting centers may enhance the luminescence and EPR of certain centers at the expense of the others. Laser-induced time-resolved luminescence must be used to study short-lived excited states.

2. Classification of natural diamonds may be reduced to the following types: with dominant specific type of nitrogen center; "nitrogenless" with low nitrogen content; "nitrogenless" with boron: semiconductor, blue color; with nickel, yellow color; with dislocations: (a) brown and yellow color; (b) red, pink, and purple color.

3. All synthetic diamonds contain nitrogen only as isolated atoms. The implanted and isotopically enriched (^{13}C, ^{15}N) diamonds can be related to this group. Thus, the difference between natural and synthetic diamonds is the nitrogen form: aggregates in natural, and isolated atoms in synthetic. This classification is shown in Table 41.

Comments to Table 41:

A) Every center has a double definition which includes (1) spectroscopic (different methods) nomenclature (working, historically established designations of bands, lines, or systems of lines); (2) center model (for many of them the improved interpretation is needed).

B) Comments to spectroscopic designations:
 - By IR bands the following centers have been detected: A nitrogen pairs, B_1 more complex aggregates, C isolated nitrogen atoms.
 - The systems of luminescence lines in the irradiated crystals: GR (general radiation) for all irradiated crystals, from GR_1 (1.67 eV) to GR_8 (3.00

eV), R (radiation-induced) from R_9 (3.04 eV) to R_{11} (3.99 eV), TR (radiation type 2) from TR_{12} (2.64 eV) to TR_{17} (2.83 eV), H (heated after irradiation) from H_1 (0.18 eV) to H_{18} (3.56 eV), N (natural) from N_1 (1.56 eV) to N_9 (5.26 eV).

- From EPR spectra: W_1–W_{34} centers detected in Witwatersrand University; R_1–R_2 centers detected in Reading (England), N_1–N_4 centers detected in Novosibirsk (Russia), NIRIM centers detected in the National Institute for Research of Inorganic Materials (Japan).

C) The first step in systematization is group determination: with nitrogen (N), with vacancies (V), with dislocations (D), with interstitial atoms (C_i), with boron (B), with nitrogen and vacancies (NV), with nitrogen and dislocations (ND), and with nitrogen and boron (NB). (Diamonds with implanted atoms can be considered special cases).

D) These groups also correspond to a subdivision of centers according to their formation:
- Nitrogen- and boron-containing form during the diamond's growth; different types of such centers may be formed in the course of crystallization or as result of high-temperature transformation in the solid state (aggregation and dissociation of the centers).
- Centers with vacancies and interstitial atoms form as result of irradiation of crystals by electrons, neutrons, or high-energy ions; all such centers are detected after artificial irradiation of "nitrogenless" diamonds (type IIa), but some of them are also known in natural diamonds.
- Centers with dislocations are connected with natural or artificial plastic deformations of diamonds crystals.

E) Within the groups, the types of centers can be distinguished:
- Among the nitrogen-containing: according to the complexity of nitrogen aggregation of three, four, and more nitrogen atoms in the (111) plane or nitrogen and carbon aggregates in the (100) plane forming "platelets" visible by electron microscopy, and giving additional reflections in X-ray diffractions ("spikes").
- Vacancy-containing: according to the extent of aggregation: isolated vacancy (V^0, V^+, V^-, pairs of vacancies, chains containing three, four, and five vacancies along [100] (it is possible that the utmost manifestation of such aggregation is voidities – the voids formed by the (111) planes.
- Dislocation-containing: according to the number of interatomic distances between the atom pairs with broken bonds and unpaired electrons (nearly all with spin S = 1 and usually situated on dislocations along [110]).

The centers are further subdivided according to their content, and consequently to the method of determination:
- "Main" centers with prevalent content (up to 10^{20}–10^{21} at cm^{-3} or 0.55–0.06%) which all present nitrogen-containing diamonds: $A(N_2)$, $B_1(N_n)(111)$, $B2(N_n)(100)$ detectable by IR spectroscopy (their content is determined by the absorption coefficient of the IR band at 7.8 μm). All other centers are present at lower concentrations (with the exception of

several dislocation-containing centers), and they are not detectable by IR spectroscopy. The "main" form determined by the IR method is the most important for different material applications, but it is not detectable by EPR spectra, while only the luminescence spectra appear in the IR region. This form does not influence diamond color, and many colorless diamonds of gem quality can have a high nitrogen content in the form of A, B_1, and B_2 defects.

All other types of centers never occur with a content of more than $10^{18}–10^{19}$ at cm^{-3}, and always compose a small part of the total nitrogen; but these centers cause diamond luminescence, and their relations determine the different types of luminescence spectra. These centers can be observed only because of the higher sensitivity of EPR and luminescence, but it is worth emphasizing that this is only the smallest part of the overall nitrogen. This group includes mostly nitrogen-vacancy centers, but also nitrogen and sometimes vacancy types (the latter connected with natural irradiation). The presence of such centers causes the appearance of yellow tints, and diminishes the diamonds grading, since they have to be absolutely colorless.

	Total nitrogen				$N_3V + N$	
	at cm^{-3}	ppm	%		at cm^{-3}	
I	10^{20}	600	0.06	IR-	$10^{16}–10^{19}$	colorless, yellow
	10^{18}	6	0.0006	UV	$10^{14}–10^{16}$	colorless, yellowish
II	10^{16}	0.06	0.000006	"nitrogen-		
	10^{14}	0.0006	0.00000006	less"		

(The number of carbon atoms in diamond is 1.764 10^{23} at cm^{-3}).

Optical Absorption and Luminescence Spectra

The nature of color in diamonds is different from in all other minerals: chromophores in diamonds are various forms of nitrogen impurities, dislocation centers, and rarely boron. This is the only case when N and B are the origin of coloration. Another difference is in the quality of the color: it is not, strictly speaking, the color, but a faint hue, which slightly tints the light passing through the inner space of the cut diamond.

While diamond colors are so unusual, their absorption spectra are monotonous: a slight slope from IR to UV in the yellowish coloration by isolated nitrogen, of from UV to IR in the bluish tint connected with boron.

As was explained above, nitrogen in diamond exists in several forms. Most nitrogen forms $A(N_2)$ and $B(N_4)$ centers, with the bands in IR and UV, but not in the visible. The important consequence is that even diamonds with considerable nitrogen content, but in the form $A(N_2)$ and $B(N_4)$, remain colorless.

The principal yellowish chromophore is a center N_3V, sometimes combined with isolated nitrogen and a N_2V center. Absorption spectra can be discussed together with luminescence spectra related to the nitrogen centers in diamonds, since they are described by the same energy level schemes.

In the spectra of color centers in diamond there is a narrow line corresponding to pure electronic transitions (zero-phonon line: ZPL) and a broad band composed of overlapping phonon repititions of pure electronic transitions. The centers are defined by the position of the zero-phonon transitions, but the color of luminescence and the color of crystal are determined by this broad band (Figs. 133, 134, 135).

Another peculiarity is a mirror symmetry relative to the zero-phonon line of the absorption band (in the UV direction) and luminescence (IR direction).

Fancy Colors and Fancy Cuts of Brilliants

Colorlessness and clarity are the essence of the precious character of a diamond. Any hint of faint color is considered as a deviation from ideal perfection. Yellow and brown diamonds are excluded from gem diamond color grading. On the other hand, clear blue, green, pink or purple and yellow brilliants are as rare and precious among diamonds as the diamonds themselves are among other gem minerals.

Among more than 100 million carats of world production, only 2–2.5 millions become brilliants larger than 1 carat, with D-H coloration and F-VS

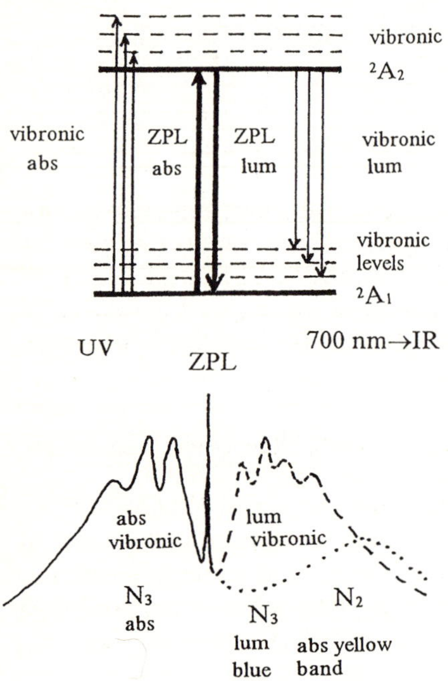

Fig. 133. Schemes of energy levels (see also Fig. 135) and absorption-luminescence spectra for N3 center in diamond. *ZPL* (Zero phonon line), electronic transition between $^2A_1$ and $^2A_2$ levels in absorption and luminescence; *vibronic* (vibrational + electronic) *levels* (phonon-assisted): visible luminescence and UV-absorption bands are determined not by narrow ZPL but by broad phonon repetitions of the ZPL transition. Vibronic luminescence band (*vibronic lum*) is shifted from ZPL towards lower energy in visible; it is responsible for blue luminescence of diamonds; vibronic absorption band (*vibronic abs*) is shifted form ZPL toward higher energy in UV; absorption and luminescence spectra are (almost) mirror images; *yellow* color of N_3 center depends on N_2 band which has no symmetrical band in luminescence

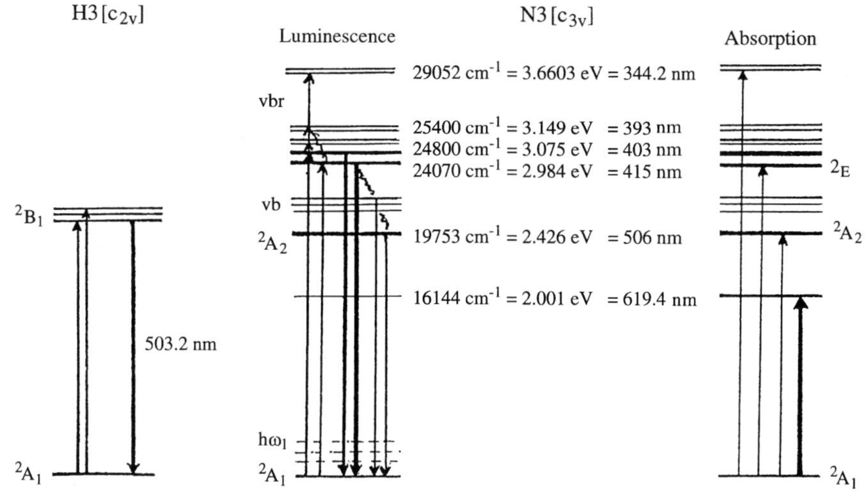

Fig. 134. Energy levels in absorption and luminescence of N3 and H3 centers in diamond

clarity. Of this quantity, only about 600 crystals with D color and F clarity are obtained in a year.

At the same time, only a few dozen fancy color diamonds can be found. Their grading oversteps the usual scales of value and is beyond the usual price lists. Record prices per carat have been paid for fancy brilliants.

A green color results from vacancy-containing centers formed due to natural irradiation, but also during the crystal's growth in the process of aggregative transformations of the centers (N_3V and N_2V type centers with vacancies).

A pink and purple color in diamonds is generally associated with dislocation-containing centers.

Four groups of diamond cutting can be distinguished: (1) old simple cut (those having no historical value are generally cut anew), (2) classic round brilliant with 57 facets (or 58 facets with culet), (3) brilliant cut modifications (oval, marquise, pear, heart, rose) or square cut (emerald, baguette, square), (4) fancy, highest grade with 86 facets (King), 102 facets ("magma"), 144 facets (princess), which are derivatives of round brilliant cut, and barion, troide, and other cuts derived from the square cut.

Stones greater than 20 carats are often cut to individual fancy shapes.

Scale of Artistic Value of Diamonds

Each scale of a brilliant's grading in the 4C system has its own gradations and characteristics. In gemology the descriptive characteristics are based on visual evaluation and comparison with standard crystals.

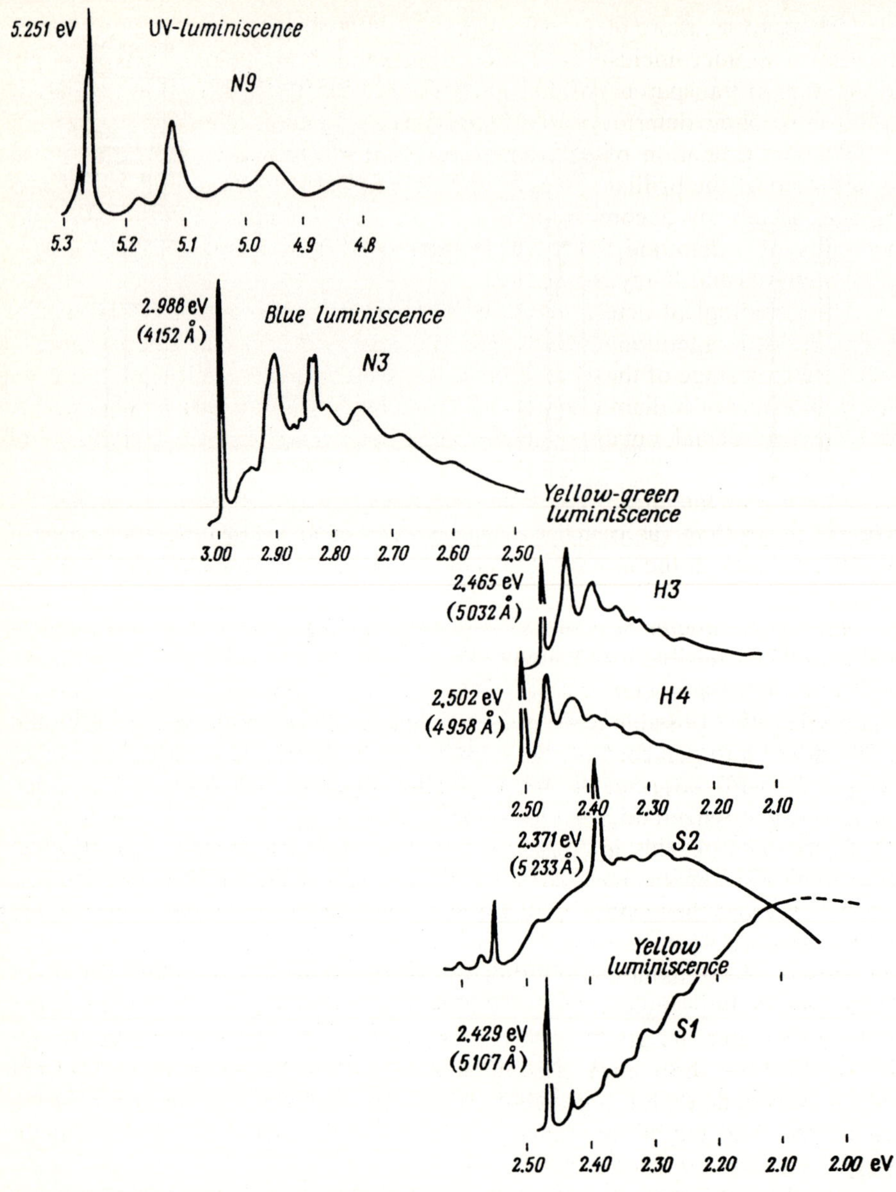

Fig. 135. Luminescence spectra of diamonds

Nevertheless, more objective estimations may be attempted: clarity – connection with microinclusions, dislocations, and cracks; color – spectroscopic estimation of transparency independent on the reason of absorption or analytical spectroscopic determination of the types and content of color centers.

The determination of all four characteristics represents the gemological description of the brilliant's quality. The result of such a description is a certificate, which now accompanies all brilliants of more than 0.5 carats. The individuality of a diamond is sometimes identified by laser inscription or laser reflectograms and X-ray topograms.

The gemological determination of a diamond's quality is finished at this stage, and it is intentionally separated from the price determination, which is the following stage of the evaluation.

In the case of a diamond-brilliant the determination of price acquires not only its own special, but also a general meaning, exceeding merely commercial significance.

First, it is an integrating parameter summarizing the results of the whole 4C-system of grading (as one point on the colorimetric diagram integrates all bands and lines in the absorption spectrum which determines the color of the crystal).

It is the parameter of the stone's quality, the characterization of its artistic value as an indivisible impression.

This "price parameter" has a quantitative expression and maximum "resolving power". It is possible to determine the price of each color tint and each ten millionth of nitrogen content in a diamond, the price of a fraction of carat of its weight, the different prices for carat in different weight intervals, and the price of each micron-sized microinclusion (Fig. 136).

In the huge worlds jewelry trade, with billions of diamond crystals, the single price systems is indispensable to prevent chaos in the diamond market. For this reason, the system of diamond price determination is an inseparable part of the whole system of the diamond trade. The basis of this system is composed of the natural property of specific artistic value of diamond; its reduction to the single form of "ideal brilliant" (and its derivatives); the strictly regulated 4C grading system (without which the world market would be impossible); the common assertion of the 4C system single professional education standards for gemologists; the price determination system which is connected with the 4C grading system. In weekly publications, the brilliant prices per carat are given in a form corresponding to the 4C system. Tables are presented for different weight intervals (in carat) with the coordinates: color grading (D, E, F, G, H, I, K, L, M); clarity grading (loupe-clean $VVS_{1,2}$, $VS_{1,2}$, $S_{1,2}$, $I_{1,2}$).

The different cuts (pear, marquise, etc.) and cut grading are taken into account by special coefficients. It is a world price which serves as a starting point for the actual trade in all countries. The information on diamond prices is included in INTERNET, which leads to the formation of an electronic diamond market. There is a dual aspect to the price of brilliants: it is its objective price

Fig. 136. Color in the scale of diamond grading

	Color	$/ct	$N_3 + N_2V$ at cm^{-3} (at%)
colorless	D	15500	10^{16}(0.000006%)
	E	10800	
	F	9400	
near colorless	G	7600	10^{17}(0.00006%)
	H	6600	
	I	5800	
	J	5300	
faint yellow	K	4800	10^{18}(0.0006%)
	L	4000	
	M	3300	
very light yellow	N	2500	
	O		
	P	1600	
	Q		
	R		10^{19}(0.006%)
light yellow	S- -Z		

reflecting its artistic value, but it is simultaneously its market price expressed in different monetary systems and connected with the demand-supply ratio, rates of currency exchange, and conditions of world economics. The ratio between the prices of diamond and brilliant plays the key role in the mechanisms of the world diamond-brilliant complex. It is necessary to take into consideration the three components which determine the added value of the brilliant per carat compared with the value of rough diamond carat: irretrievable losses of weight in diamond (52–55%); the cost of cutting; and the rate of profit. For these reasons, the cost per carat of brilliants is more than twice as high as of the original rough diamond. However, the cost of brilliants differs from the cost of original diamonds only by the cost of cutting and the profit. The usual profit is 6–10%. The cutting industry exists on account of this rate of profit.

The basis of the brilliant's price is the price of diamond, but it is also the cost of brilliants paid by the purchaser, which includes also the price of diamond. The price of diamond from which the brilliants have been cut may be reproduced retrospectively. However appraisal of the diamond has to take place before cutting. The best way to determine the price of diamond is to prognosticate the price of the brilliant which can be obtained from the original crystal, which is then inscribed in this crystal. The procedure of diamond appraisal following sorting into 4000–5000 groups differs strongly from the appraisal of brilliants based on the 4C grading system and using published price lists. The greatest part of world rough diamond supply (up 85%) is set by the De Beers Central Selling Organization (CSO) in London. It is a closed broker organization which connects main diamond producers and regular wholesale buyers (sight-holders, about 160 worldwide).

De Beers' company is not involved in brilliants but only in rough diamonds. CSO purchases the majority of rough diamonds and sells then after resorting into lots and parties. The fixing of the diamond price is the affair of De Beers, based on their own price list. The price is not discussed, but fixed (depending on the demand-supply ratio for diamonds and brilliants), and sight-holders have to accept it. Thus, De Beers controls the whole world diamond-brilliant system, distributing diamonds and fixing the prices for diamonds and brilliants by using diamond stocks, which take diamonds at times of market surplus, and supply them in periods of deficiency. It cannot be said that this is merely monopolistic, since here the natural quality of diamonds is regulated which materializes the value contained in the crystal. As a result, all participants in the industry and market are gaining, including even retail buyer and owners who are interested that diamonds keep their value.

6.2.7 World Diamond-Brilliant Market

It has unique features among other large enterprises. Its peculiarities are composed of the three groups of factors:

1. It is subdivided into:
   a) Three parts: market and industry of diamond, market and industry of brilliants, jewelry trading.
   b) Three groups of the main countries participating in the market:
      - Countries and regions of principal diamond producers, but having no cutting industry and not included as main jewel consumers: Africa (Botswana, South African Republic and also Zaire, Namibia, Angola, Tanzania), Russia, Australia.
      - Countries with leading cutting centers, but without diamond production: Israel, Belgium, India, USA (New York), and also Thailand, (Indonesia), Sri-Lanka and China.
      - Countries and regions with main diamond consumers but without their own diamonds: Japan, USA, Europe.

2. The factors which amalgamate it into a single world enterprise because of the interdependence between the production of diamonds, which receive their value as brilliants, and the cutting industry, which depends upon diamond supply, and the interdependence between the groups of the countries above. Thus, the world diamond-brilliant trade and market appear.
3. These various components of the market and their unification into a single world enterprise can effectively exist only if the system of the world diamond-brilliant complex is created. The mechanism of its activity contains, on one hand, De Beer's system, which controls the prices of the rough diamonds and diamond supply and, on the other hand, three systems which control the natural value of diamonds-brilliants: the geometrically restricted shape of the ideal diamond, brilliant grading based on the 4C system, price fixing also based on the 4C system.

De Beers Company carries out a versatile activity, promoting stability in the world diamond-brilliant complex. The key role in the control and regulation of this complex belongs to the Central Selling Organization in London, which determines the prices of the rough diamonds and establishes the connection between the principal diamond producers (Africa, Russia, Australia) and the principal producers of brilliants (Israel, Belgium, India).

6.3 Problems of Gemology of Precious Stones

A. S. Marfunin, V. S. Balitsky and J. Shigley

The property of artistic value can be endowed, in principle, to all minerals. It is a specific side of their geological nature. Manifestations of this property are extremely diverse, and not arranged in any single scheme. They can exist as natural transparent crystals, faceted stones, in settings of precious metals, precious materials of jewels, in the form of objects and decorations, or in the self-contained value of minerals, druses, giant crystals and microminerals, transparent and opaque, colored, black, and colorless.

However, among more than 4500 minerals, only slightly more than 200 are selected in gemology. From this number, only about a dozen minerals form transparent pieces, faceted and jewel stones which are most highly valued. In this group only four – diamond, emerald, ruby, and sapphire – constitute the main part of the world market. They are precious stones with a historically developed "conception". Diamond has the first position and the "big three": emerald, ruby, and sapphire follow.

Gemology considers not the mineral species but their precious varieties: not beryl, but emerald, aquamarine, morganite, vorobyevite, heliodore, rosterite; not corundum, but ruby and sapphire. Moreover, blue, orange, violet, and green sapphires are considered as separate precious stones. On the other side, vanadium containing green beryl and maxixe-emerald colored by bleaching the color centers differs from true Cr-containing emerald.

6.3.1 Methods of Precious Stone Identification

The first stage is the problem of identification of precious stones. The difficulties are stipulated by the variety of stones with the same colors and similar cut, the impossibility of using traditional mineralogical features (color, scratch hardness, cleavage, etc.) and methods (immersion, thin section, X-ray diffraction), and the necessity of employing nondestructive and nondamaging methods to large faceted crystals, including stones in setting.

As a result, a specific system of gemological methods has been set up. It is based mainly on crystallooptic measurements by specially constructed gemological modifications of instruments: polariscope to differ between isotropic and birefringent crystals, one-axial, and biaxial crystals; refractometer or reflectometer (for crystals with refraction more than 1.7 – 1.8); dichroscope (or polarizing microscope) for pleochroism observation.

Different types of spectrometers are used in the next stage: from the spectral ocular for usual binocular microscope and simplest gemological spectroscopes to the most sophisticated contemporary spectrophotometers. The additional characteristic is luminescence, which may be detected visually under UV excitation or studied by contemporary spectrometers suitable to study large crystals.

Density is one of the most important parameters in gemology, and is measured by the heavy liquids method or hydrostatic weighing. This standard complex of gemological methods enables the effective identification of minerals and their distinction from imitations. The diversity of colored stones and the creation of many imitations no longer present unsolved problems for this systems. In gemological manuals, the precious stones are usually classified in color groups (green, blue, violet, red, yellow), then by parameters of hardness, optical character, refraction, and density. Diamond testers based on heat conductivity measurements are effective in distinguishing diamond from all imitations.

Quality grading is carried out visually by (10× loupe) or gemological microscope. In the same way, the traces of certain treatments may be identified (tracks cure, removing of inclusions by lasers). In the second stage, the problem of distinguishing between natural and synthetic minerals appears, as well as the detection of traces of enhancement by irradiation, heating, or coating by diamond films.

6.3.2 Competition of Synthesis and Methods of Distinguishing Between Natural and Synthetic Stones. Status of Synthetic Stones

Several new stages in the synthesis of minerals have led to new situations in this field and introduced a new aspect into the approach to traditional stones. The new history of synthesis, beginning with ruby synthesis from the melt by the Verneil method (1905), has preserved its role for certain crystals up to now. Contemporary achievements started from 1950–1960, while the most effective discoveries were obtained in the 1970s. During the past 10 years, synthesis technology has been greatly improved, perfect crystals synthesized, and the output substantially enlarged. As a result, this field has assumed new contemporary significance.

Two branches of crystal synthesis exist: (1) as analogs of precious stones and (2) crystals with special properties: hardness (ruby supports in apparatus), stabilization of radio-frequency (piezoquartz), with special parameters for lasers, optoelectronics, acoustooptics, and other fields of contemporary technology. The demands of technology determine the growing outpout and highest development of the synthesis. Clear crystals with high hardness and different colors which may be used as precious stone imitations are by-products of crystal synthesis for technological aims. However, the synthesis of analogs of natural precious minerals is a new specialized branch.

The crystals which have no natural analogs and are used as imitations (mainly for diamond, because for other stones their synthesized analogs are used) are:

- strontium titanate ("fabulite") $SrTiO_3$;
- gallium-gadolinium garnet (GGG) $Ga_3\ [Gd_2[GdO_4]_3]$;
- lithium niobate $LiNbO_3$;
- "cubic zirconia" ("phianite"*[1], "djevalite") $ZrO_2{:}Hf$ or $ZrO_2{:}Ca$;
- rare-earth borates, lithium tantalite, yttrium oxide, and many others.

The synthesis of each analog of precious minerals has its specific history with sensational achievements which strongly influenced the market and problems of identification: amethyst, citrine, emerald, alexandrite, and also opaque polycrystalline turquoise, opal, malachite, lazurite, nephrite, jadeite, and coral.

Several stages can be distinguished in the synthesis of diamond: from the first sensational synthesis of small diamonds of industrial quality (1954) to their commercial production. Then followed: clear crystals of gem quality but with yellow and brown colors; then colorless crystals of gem quality up to 14 mm; "super diamond" enriched isotopically; and diamonds with H3 centers used as laser materials.

At the present time, analogs of many minerals with desirable properties can be synthesized, and their production is limited only by demand on the market.

[1] "Phianite" from the Physical Institute of Academy of Science [Nauk] of Russia.

Synthesis of crystals for industry and electronics, and as by-product, as analogs and imitations of gem stones is estimated now at hundreds of millions of carats (phianite $-60 \div 300$ million carats, YAG – more than 40 million carats). The synthesis of colored minerals reaches hundreds of millions carats (emerald more than 200000 carats).

Thus, alongside natural colored stones, the mass of synthetic minerals and imitations exceeds that of natural minerals. In consequence, two problems have arisen: (1) to distinguish between natural and artificial gem stones; (2) status of synthetic minerals.

Concerning imitations, the problems of status and identification do not exist: the fact that they are not natural minerals determines their status similarly to that of glass and plastic imitations; standard gemological instrumentation enables easy identification.

In spite of any possible quality, imitations have no "image", they are only imitations of the quality of gem stones created by geological processes and the assembled perceptions accumulated during the long history of the culture of stones in different epochs and civilizations.

The situation with artificial analogs of precious minerals is different. Outstanding achievements in the synthesis of clear, flawless, large crystals of ruby, sapphire, emerald, alexandrite, diamond, and other minerals, and the ability to give them any color, change their composition, control their inclusions, all this generated in many laboratories the intention to put such "created" stones in the same category with natural stones. Do they not, in fact have the same composition and structure as natural stones? Do they not, in fact, have the same origin of color? Do they not have the same absorption and luminescence spectra and optical parameters?

Two principal qualities differentiate the natural precious minerals: natural origin and rarity. The natural origin introduces into the artistic value of the stone not only color and transparency, but also geological time and geological events belonging to a certain country and mineral deposit; and rarity is the indispensable condition for the higher price of a stone.

The relationship natural stone-artificial analog-imitation can be compared with the relationship between an original masterpiece, its copy, and a reproduction in painting.

The decisive significance in the possibility of distinguishing these relationships is a competition between (1) the contrivances of synthesis in trying to reproduce all specific features of the natural stone, and to circumvent all obstacles in distinguishing natural and synthetic stones and (2) the most sophisticated methods and careful observations which can indicate this difference.

The results of this competition will determine if precious stones will continue to be precious. The situation is different for different stones. The cut synthetic amethyst cannot be distinguished from the natural one even by EPR. The relatively low price of the natural amethyst is practically equal to the price of the synthetic one, and they have a common market.

In contrast, the synthesis of ruby during nearly 100 years has not affected the status of the natural stone, which belongs to the group of the four most expensive minerals: synthetic ruby synthesized from melt is readily distinguished from natural by the microinclusions.

A complicated problems present in the distinction between hydrothermal synthetic and natural emerald (which is the second after diamond by its price per carat and total volume of sale): the crystals created in the laboratory are rather difficult to distinguish from the natural.

However, all this is not comparable with the consequences which the appearance of millions of carats of synthetic diamonds of gem quality would have. These crystals are logically prospective materials for new branches of electronics and technology. It is already impossible to retard the growth of this production and unlimited improvement of the crystal quality. This is already a real threat to the whole huge diamond-brilliant trade. (Albeit the growth of a 1-carat diamond takes more than 5 days and at present the number of synthetic diamonds of gem quality is relatively small.)

The future of the diamond trade depends on the ability to distinguish between natural and synthetic diamonds. Comparative studies have shown that synthetic diamonds differ from natural ones by several features: inclusions of metals (Fe, Ni, Mn) used as catalysts and solvents; specific parameters of luminescence, zone distribution of cathodoluminescence colors; growth and strain structures.

The variability of nitrogen and other centers, growth structures, and plastic deformation traces is characteristic for natural diamonds and, to a lesser degree, for synthetic ones. Thus, empirical comparison is not enough, and determinations of center types and analysis of growth and deformation structures are needed in order to find the logical connections with the formation conditions.

For example, forms of nitrogen in diamond structure (isolated nitrogen atoms in synthetic and nitrogen aggregates in natural diamonds) can be used, although nitrogen aggregates have been found in synthetic diamonds also. High-temperature treatment of synthetic diamonds can produce the "aging" of nitrogen centers with resulting aggregation (like artificial aging – patina formation – in painting).

6.3.3 Enhancement of Gem Stones

Along with the increasing production of natural gem stones and the vast amount of synthetic crystals, another important event which contributes strongly in the situation in this field is the development of art and technology, and increasing progress in the enhancement of gem stones.

The influence of the latter on the destiny of different stones is quite different. For ruby, sapphire, and topaz it has become as necessary as cutting, while for other minerals (turquoise, malachite, nephrite, jadeite, opal, chalcedony) it is used for the enhancement of low-quality samples. In the case of diamond, these are experiments which raise to doubts concerning the authenticity of the treated crystals. Altogether, it represents a new aspect in the approach to gem stones.

In many cases, enhancement so improves and even transforms the artistic and jewel qualities of different stones that doubts must give way to acceptance. The term "rough" diamond (before cutting to brilliant) and "rough" stones can be opposed not only to cut and polished, but also to enhanced stones.

Two ways of enhancement can be distinguished: textural (the structure and composition of mineral are not changed) and structural. The following treatments belong to the textural group:

- Filling of cavities in opaque stones (turquoise, malachite, lazurite, charoite, coral, nephrite) by cement prepared from the powder of the same stone mixed with resin.
- Crack filling in transparent stones: in emerald (by cedar oil – "Canadian balsam", plastic, rubber), ruby and sapphire (glassy materials, colorless or colored plastics).
- Impregnation of porous and cracked stones (in particular weathered turquoise) by organic resins, plastic, wax, and siliconorganic compounds.
- Surface coloring (beryl, jadeite, nephrite, lazurite, turquoise) by organic colorant.
- Dyeing of fine-grained- and cryptocrystalline agate, chalcedony, and opal by water solutions of chromium-, iron-, nickel- and cobalt salts; they fill the structural pores and channels with subsequent heating or treatment by different acids, alkalis, and salts; as a result of the chemical reactions in the porous space, the stones are colored green, red, orange, yellow, and black.
- Color overgrowth of slightly colored emerald, beryl, ruby, and sapphire under hydrothermal conditions.
- Overgrowth by thin-colored diamond films of colorless brilliants to give them a blue fancy color; covering pale sapphires and topazes with blue films; unusual quartz colors (green, blue, yellow, orange) as a result of covering with a thin layer of gold or boron nitride and iron oxide.
- Laser burning out of microinclusions, for example, in diamond.
- Preparing composite or assembled stones (doublets and triplets).

The structural methods of enhancement include complicated mechanisms of manipulation with structural color centers: chromophore diffusion into the structure; formation of color centers by neutrons and electron irradiation; formation of color centers by gamma- and X-ray irradiation at existing "precursor" centers (impurities, vacancies, interstitial atoms); destruction, intransformation, diffusion, aggregation (pairs and complicated complexes) and dissociation of color centers by heat treatment; change of valence states of ion-chromophore by heat treatment.

The following examples are most important:

- Irradiation by neutrons, high-energy electrons, and gamma rays transforms colorless and yellow topazes into blue and deep blue.

- Gamma irradiation transforms colorless quartz to black morion, "rauch-topaz" or yellow-green citrine; the violet-purple color of amethyst bleaching in the light is totally restored by gamma irradiation.
- Slightly colored corundum becomes yellow, orange, or green after gamma-irradiation.
- Practically all colored transparent stones change color as a result of irradiation; in many cases radiation-induced colors are light- and thermostable (deep blue maxis-beryl, spodumen-giddenite, yellow and brown-red topaz).
- By changing the temperature and duration of heating, it is possible to destroy certain color centers while retaining others.
- A combination of heating with preliminary irradiation enables the creation of new color centers: after gamma irradiation, colorless or wine-yellow topaz becomes red-brown or brown-red as a result of the combination of two color centers; yellow and wine colors are thermostable and disappear after heat treatment at 200–300 °C for half an hour, giving a clear sky-blue color; yellow and green-yellow topazes obtained from colorless, after heat treatment in nuclear reactors, have a deep blue color which is not known in natural stones.
- Heat treatment of natural and artificial ruby and sapphire to improve their colors has been widely used during the past 15 years to improve the color.
- Changes in diamond colors as a result of different irradiation, heating, and irradiation followed by heating under different temperature regimes are extremely variable: variations in diverse original centers (vacancies nitrogen in different associations with or without vacancies, interstitials, nickel, boron) undergo different transformations as a result of the treatments.
- Estimating the influence of the enhancement of gem stomes on the state of the industry and market, it is possible to distinguish the following.
- Approximately 95 % of all sapphires and rubies which have appeared on the world market during the past fifteen years have been subjected to treatment.
- Nearly all blue topazes are obtained as a result of treatment by irradiation and heating; amethysts and citrines are synthesized or subjected to irradiation and heating.
- More than 75 % of all colored stones have been treated.

6.3.4 Hallmark for Gold and Certificate for Diamonds

For brilliants of more than 0.5 carat, gemological laboratories usually give certificates which confirm the diamond identification and its grading by the 4C system. This grading forms the basis for brilliant price determination. Taking into account the extremely complicated situation with gem stones, the necessity to determine the subtle differences between natural and synthetic stones and to detect the traces of different enhancement and traces of radioactivity, the certificates have to be obligatory for diamonds and colored stones as the hallmark for gold. It is necessary to progress from the visual observation of color changes to systematization of color centers in gem stones, especially in diamonds, and a systematic representation of mutual transformations due to different kinds of irradiation and heating.

References

Anderson BW, Jobbins AE (1990) Gem testing. 10th edn. Butterworths, Boston

Bartoshinsky ZV (1983) Mineralogical classification of natural diamonds. Mineral J 5:94–103 (in Russian)

Bockstael van M (1992) Diamond grading and identification. Institute of Gemology, Antwerp, Belgium

Bruton E (1978) Diamonds. Chilton, Radnor, Pennsylvania

Clark CD, Collins AT, Woods GS (1992) Absorption and luminescence spectroscopy. In: Field JE (ed) The properties of diamond. Academic Press, New York, pp 36–79

Collins AT (1982) Colour centers in diamond. J Gemm 18:37

Davies G (ed) (1994) Properties and growth of diamond. INSPEC, London

Diamond grading manual (1992) The Gemological Institute of America, Santa Monica, California

Facing the future (1992) Proc Int Gemological Symp, Los Angeles, June 20–24, 1991. Gemological Institute of America, Santa Monica, California

Field JE (ed) (1992) The properties of natural and synthetic diamond. Academic Press, New York

Fipke CE, Gurney JJ, Moore RO (1995) Diamond exploration techniques emphasizing indicator mineral geochemistry and Canadian samples. Geol Surv Can Bull 423

Fritsch E, Phelps AW (1993) Type IIb diamond thin films deposited onto near-colorless natural gem diamonds. Diamond Related Materials 2:70–74

Fritsch E, Shigley JE (1993) The separation of natural from synthetic gem-quality diamonds on the basis of crystal growth criteria. J Crystal Growth 128:425–428

Harris JW (1987) Recent physical, chemical, and isotopic research of diamond. In: Nixon PH (ed) Mantle xenoliths. Wiley, New York

Hurlbut CS, Kammerling RC (1991) Gemology. Wiley, New York

Liddicoat RT (1989) Handbook of gem identification. GIA, Santa Monica

Marfunin AS (1979) Spectroscopy, luminescence and radiation centers in minerals. Springer, Berlin Heidelberg New York, 352 pp

Marfunin AS (1988) L'Or. Edition Gaston Lachurie, Paris

Messier R, Glass JT, Butler JE, Roy R (eds) (1990) Proc 2nd Int Conf New diamond science and technology, Washington

Nassau K (1980) Gems made by man. GIA, Santa Monica

Nasau K (1983) The physics and chemistry of color. Wiley, New York

Nassau K (1994) Gemstone enhancement. History, science and state of the art. Butterworth, Oxford, 252 pp

Orlov Yu L (1977) The mineralogy of the diamond. Wiley, New York

Proc 3rd Int Conf on the New diamond science and technology (1992), Heidelberg

Shigley JE, Fritsch E, Koivula JI, Sobolev NV, Malinovsky JY, Palyanov YN (1993) The gemological properties of Russian gem-quality synthetic yellow diamonds. Gems Gemol 29:228–248

Sinkankas J (1983) Gemology: an annotated bibliography, 2 vols. Scarcrow Press, Metuchen, New Jersey

Sixth Int Kimberlite Conf (1995) Extended Abstracts, Novosibirsk

Vleeschdrager E (1993) Durete 10: Diamant Histoire-Taille-Commerce. Gaston Lachurie, Paris

Webster R (1983) Gems: their sources, descriptions and identification, 4th edn, revised by BW Anderson, Butterworths, London

Wilks J, Wilks E (1991) Properties and applications of diamond. Butterworth-Heinemann, Boston

Periodicals:

Gems and Gemology (GIA, Santa Monica)

Journal for Gemmology (London)
Zeitschrift der deutschen Gemmologischen Gesellschaft (Idar Oberstein, Germany)
The Australian Gemmologist (South Yarra, Australia)
Lapidary Journal (Devon, Pensylvania)

Rapaport Diamond Report (New York)
A Gemstone Pricing Index. S.S. Michelsen (Bayonne, New Jersey)
A Gemstone Pricing Guide. R.B. Drucker (Northbrook, Illinois)

Diamond World Review (The World Federation of Diamond Bourses)
Diamond Research (UK)
Diamond and Related Materials

Subject Index

Contents of Volume 1: Composition, Structure, and Properties of Mineral Matter

Introduction

Chapter 1 The Chemical Nature of Minerals

Chemical Composition of Minerals. Crystallochemical Constraints and the Nature of Impurities (Mitchell, Novgorodova, Semenov); Treatment of Chemical Composition Data for Minerals. Structural Formilae (Bulach, Zussman), Multivariate Statistical Analysis for Processing Chemical Data of Minerals (John Mann, Ryakhovsky); The Mineral Composition of the Earth's Crust, Mantle, Meteorites, Moon and Planets (Yaroshevsky, Bulakh)

Chapter 2 Crystal Structures of Minerals

General Results of Crystal Structure Analysis of Minerals. Stages of Scientific and Technical Development (Drits, Liebau, Prewitt), Symmetry Bases. The Contemporary Symmetry Theory in Solids (Koptsik); Empirical Relationships in Structural Geometry (Baur); Systems Emprical Radii of Atoms and Ions; Orbital and Pseudopotential Radii (Baur); Representation of Crystal Structures as Packings of Spheres, Coordination Polyhedra, or Nets of Bonds (Baur, Kassner); Computer Simulation of Crystal Structures (Price, Urusov); Precision Electron Density Calculations: Relation to Chemical Bonding and Localization of Impurities in Crystal Structures (Tsirelson, Frank-Kamenetskaya); High-Temperature and High-Pressure Crystal Chemistry (Filatov, Hazen); Types of Structural Relations in Minerals Modulated and Incommensurate Structures Spinodal Unmixing (Organova); Hybrid Layer Structures, Commensurate and Incommensurate (Allmann); Tunel Structure Oxide Minerals (Turner, Gorchkov, Buseck); Polytipism in Minerals (Christy, Zvyagin); Mixed-Layer Minerals (Drits, Reynolds); Structural and Magnetic Phase Transitions (Ghose); Poorly Crystallized Minerals (Besson, Drits); The Metamict State (Ewing, Akimoto); Systematics of Crystal Structures and Crystallochemical Classifications (Bokiy); Two Approaches Toward Crystallochemical Classification of Minerals and Inorganic Crystals (Bokiy); Crystallochemical Classifications of Minerals (Bokiy); The Structural Classification of Minerals (Lima-De-Faria); Counting Theory and Classification of Crystal Structures (Moore)

Chapter 3 Real Structures of Minerals

Point Defects in Minerals (Duba, Schock); Direct Imaging of Point Defects by HRTEM (Veblen, Cowley); Point Defects and Diffusion in Minerals (Ryerson, Condit); Point Defects as Precursors for Electron-Hole Centers: Systematics and Theories of Radiation Centers in Minerals (Dusausoy, Weil); Structural Types of Nonstoichiometry in Minerals (Veblen); X-Ray Topographic Study of the Real Structure of Minerals (Authier, Zarka); Twinning Due to Phase Transformations and Plastic Deformation (McLaren)

Chapter 4 Natural Classes

Structure and Properties of Silicate Glasses and Melts; Theories and Experiment (Mysen, Virgo); X-Ray Studies of Glass Structure (McKeown, Zotov); Terrestrial and Lunar, Volcanic and Impact Glasses, Tektites, and Fulgurites (Bouska, Feldman); Nuclear Waste Glasses; Recent Advances in the Spectroscopic Investigation of Their Structure (Petit-Maire, Petiau, Calas)

Chapter 5 Chemical Bonding in Minerals

Survey of Quantum Chemistry and Methods Used to Calculate the Electronic Structures of Minerals (Dolin, Sherman); Chemical Bonding in Silicates (Dubrovinsky, Sherman); Chemical Bonding in Sulfide Minerals (Bullett); Bonding in Oxides, Oxyanions, and Halides (Urch); Electronic Structures of Iron Oxides and Silicates (Scherman); Hydrogen Bonding in Minerals (Hawthorne, Baur)

Chapter 6 Properties of Minerals

Elastic Properties of Minerals (Agochkov, Pankov); Hardness of Minerals. Hardness, Compressibility, Cohesive Energy (Titkov); Problems of Extra Hard Materials (Brookes); Deformations in Minerals. Mechanisms of Plastic Deformation of Minerals; Role Dislocations (Willaime, Gandais); Fracture Mechanics and Fracture mechanisms in Minerals (Lloyd, Ferguson); High Resolution Electron Microscopic Techniques in the Study of Defects (Amelinckx); Hydrolytic Weakening of Quartz (Doukhan and Cordier); High Temperature Flow in Minerals and Ceramics and Its Earth's Silicate Mantle (Drury); Electrical Properties of Minerals. Electrical Conductivity in Iron-Bearing Minerals and Materials (Mason); Superionic Conductors (Maximov); Electrical Properties of Ore Minerals (Gorbatov); Dielectric Properties of Minerals and Rocks; Applications to Microwave Remote Sensing (Cervelle; Xiao Jin-Kai); Lattice Dynamics, Vibrational Spectra, Thermodynamic and Heat Properties of Min-

Chapter 7 Mineralogical Material Science

Contents of Volume 2: Methods and Instrumentation

Chapter 8 Isotopic Mineralogy

Radioactive Isotopes in Mineralogy and Geochemistry (Shukolyukov, Wetzel); Isotopic System in Geochronology (Shukolyukov, Lippolt); ^{40}AR/^{39}AR and its Laser Variant (Karpenko, Sutter); The Rb-Sr Method of Isotopic Dating (DePaolo, Anderson, Vinogradov); The Sm-Nd Methods of Isotopic Dating (Vinogradov, DePaolo, Anderson); The U-Pb System and Zircon as Minerals Geochronometer (Bibikova, Aleinikoff); Noble Gas Isotopes in Planetary and Earth Minerals (Shukolyukov, Ozima); Radiogenic Isotopes as Indicators of Soruces of Mineral Matter. Pb Isotopy; The Lea Sources (Chernyschev, Gulson); Light Stable Isotope Ratios Indicators for Conditions of Mineral Formation. Theoretical Aspects of Isotopic Fractionation (O'Neil, Galimov); Natural Variation in Stable Isotopes (Hoefs; Vinogradov); Oxygen and Hydrogen Isotopes in Mineralogy (Pokrovsky, Anderson); Carbon Isotopes in Mineralogy and Geochemistry (Galimov, Rumble); Sulfor Isotopes in Mineralogy (Vinogradov, Anderson); Nitrogen Isotopes in Mineralogy (Hanedel, Pokrovsky), Geochemical Significance of ^{87}Sr/^{86}Sr Isotopic Ratios (Anderson, DePaolo, Vinogradov); Geochemical Significance of ^{143}Nd/^{144}/Nd Isotopic Ratios (DePaolo, Anderson, Vinogradov)

Chapter 9 Computer Databases in Mineralogy (Smith)